HEAT TRANSFER IN PROCESS ENGINEERING

ABOUT THE AUTHOR

DR. EDUARDO CAO is a professor of heat transfer unit operation in the chemical engineering departments of Buenos Aires University and ITBA (Instituto Tecnologico de Buenos Aires). He also works as a consultant in the process department of Techint, one of the largest engineering groups in the world.

HEAT TRANSFER IN PROCESS ENGINEERING

Eduardo Cao

New York Chicago San Francisco Lisbon London Madrid
Mexico City Milan New Delhi San Juan Seoul
Singapore Sydney Toronto

Library of Congress Cataloging-in-Publication Data

Cao, Eduardo.
    Heat transfer in process engineering / Eduardo Cao.
         p.   cm.
    Includes index.
    ISBN 978-0-07-162408-4 (alk. paper)
    1. Heat—Transmission.   2. Chemical processes.   3. Chemical
engineering.   4. Heat exchangers.   I. Title.
    TP363.C27   2009
    660'.28427—dc22   2009015096

1 2 3 4 5 6 7 8 9 0   DOC/DOC   0 1 5 4 3 2 1 0 9

ISBN 978-0-07-162408-4
MHID 0-07-162408-2

Sponsoring Editor
  Taisuke Soda

Editing Supervisor
  Stephen M. Smith

Production Supervisor
  Pamela A. Pelton

Project Manager
  Preeti Longia Sinha, International
  Typesetting and Composition

Copy Editor
  James K. Madru

Proofreader
  Ragini Pandey, International
  Typesetting and Composition

Indexer
  WordCo Indexing Services, Inc.

Art Director, Cover
  Jeff Weeks

Composition
  International Typesetting and Composition

Printed and bound by RR Donnelley.

CONTENTS

Chapter 6. Shell-and-Tube Heat Exchangers   115

Chapter 7. Thermal Design of Shell-and-Tube Heat Exchangers   147

Chapter 8. Finned Tubes   217

Chapter 9. Plate Heat Exchangers   255

Chapter 10. Condensation of Vapors   275

Chapter 11. Boiling

Chapter 12. Thermal Radiation

Chapter 13. Process Fired Heaters

Appendix A. Distillation

Appendix B. LMTD Correction Factors for E-Shell Heat Exchangers

Appendix C. LMTD Correction Factors for Air Coolers

Appendix D. Tube Count Tables

CHAPTER 1
INTRODUCTION

This book has been written with the double purpose of being used as a textbook for university courses in unit operations and as a consulting book in the professional field of process engineering. Thus every attempt has been made to maintain a difficult equilibrium in the weights assigned to theory and practice.

It is assumed that the reader has taken courses in thermodynamics and is familiar with energy balances and the calculation of physical properties in simple systems with and without change of phase. Thus only a few basic concepts considered convenient for the fluidity of the redaction were included. It is also assumed that the reader have some basic knowledge of fluid mechanics and can write mechanical energy balances and calculate pressure drops in piping systems.

A difficult decision was to choose the approach to Chap. 4 on convection. In the chemical engineering curricula in many universities, unit operations courses are preceded by a course in transport phenomena. In other universities, they are not. Some knowledge of the theory of transport phenomena helps in an understanding of the principles of convective heat transfer and allows a more elegant treatment of this subject, but a treatment based on transport equations also increases the mathematical complexity and is not particularly useful in daily engineering practice. Therefore, after rewriting Chap. 4 a few times, it was finally decided to avoid presenting a theoretical approach based on transport equations so as not to drive away readers who are not friends of mathematics.

Other chapters present subjects associated with the thermal design of different kinds of heat transfer equipment. Nowadays, in the professional field, this task is performed almost exclusively by commercial software. There are many programs on the market, some of them developed by important companies that have well-known heat transfer researchers working for them. Additionally, since these programs are used worldwide, many users supply important feedback that allows error corrections and fine-tuning of the correlations.

Thus it is difficult to conceive of heat transfer equipment design without the help of these programs. However, most of these programs are presented as a "black box," and the suppliers offer very little information about their content and the correlations used. This makes their use difficult. It is necessary for users to know how the program will use its input data to evaluate the importance of each input data field.

For example, to calculate the boiling heat transfer coefficient, some programs use correlations based on critical properties. Other programs do not use the critical properties and calculate boiling heat transfer coefficients based on properties that are difficult to predict, such as surface tension of the boiling liquid. This means that some variables that are important in some programs are not important in others.

Boiling heat transfer is probably the most complicated subject. There are limits to the maximum heat-flux density that sometimes define equipment design. Prediction of these limits is ambiguous and varies according to the approach of different authors. Frequently, when comparing designs created with different software packages, one discovers important differences in the calculated heat transfer areas due to different criteria in the adoption of these limits. In some cases, the limits are incorporated as simple "rules of thumb" without any theoretical background and can be changed by the user if desired, which obviously changes the program results. Thus it is important that users have the necessary knowledge to be capable of investigating the calculation path on which the design is performed

In heat transfer equipment design, many independent variables must be adopted. To simplify the use of this software by nonexperienced users, the programs usually have default values for many of the variables. There is a natural tendency to accept the first results offered by the program without investigating the effect of modifications in some of these variables. Sometimes, however, the modification of the default values allows a significant improvement in the design.

For example, a change in the number of tube rows or the air face velocity in an air cooler can be translated in a substantial decrease in the required heat transfer area. This is why it is very important that users know the theoretical background behind the programs to make an efficient use of them. On the other hand, the complexity and diversity of situations that may exist in the design of heat transfer equipment make it impossible to cover all the possible situations with general and simple correlations such as those presented in this book.

This book includes design methods that, based on the author's experience, allow readers to obtain reasonable results in most cases. However, it is not possible to guarantee that in certain specific situations the use of these general methods do not result in appreciable deviations in the heat transfer coefficients.

Therefore, the recommendation is to use the commercial software, but doing so with enough knowledge of the subject to be able to evaluate and analyze the results. To that end, having a grasp of the simple tools presented in this book will allow process engineers to perform their own calculations and to detect the critical aspects of the design.

CHAPTER 2
HEAT TRANSFER FUNDAMENTALS

2-1   CONCEPT OF UNIT OPERATIONS

A complex chemical process can be divided in elemental stages involving physical or chemical transformations. Stages where materials suffer physical changes are called *unit operations*. If transformations of a chemical nature are involved, they are usually called *unit processes*. Unit operations can modify the momentum, energy, or composition of the different phases of a system. Let us consider the system in Fig. 2-1.

Tank A contains a two-component liquid mixture. Pump B circulates the liquid through heat exchanger C, where it is heated with steam. The heated liquid goes into flash vessel D, where the pressure is reduced, and part of the liquid vaporizes. The vapor, richer in the more volatile component than the liquid, is condensed in water-cooled condenser E, whereas the liquid fraction exits the separator as a bottom-product stream

In this simple sketch, several elemental units can be identified. They are

1. The pumping of the fluid by means of pump B
2. The heat exchange in heat exchanger C
3. The phase separation in separator D
4. The condensation of vapor in condenser E

Each of these units is a unit operation. Unit operations normally are classified in three groups, which generally are studied separately. They are

- Momentum-transfer operations
- Heat transfer operations
- Mass-transfer operations

The first group includes all operations in which mechanical transformations are involved. They are associated with variations in system pressure, momentum, and kinetic energy due to the action of pumps or compressors or mechanical separations such as filtration, centrifugation, etc.

The second group, which is the object of this book, corresponds to systems in which thermal energy is transfer to fluids, with or without changes of phase, such as heat exchange between two fluids, condensation, vaporization, etc.

The third group refers to systems in which one or more components of a multicomponent system are transferred between two phases, thus achieving a change in the composition of both phases. Examples of this type of operation include absorption, distillation, solvent extraction, etc.

Many times, classification into one of these three groups is not very clear, and some operations must be studied using the principles of both heat and mass transfer simultaneously, such as the condensation of multicomponent mixtures.

2-2   HEAT AND THE FIRST LAW OF THERMODYNAMICS

Heat can be defined as energy that is transferred because of the existence of a temperature difference between two systems or between two parts of a system. When we speak about heat, we always refer to energy in transit. It cannot be said that a system accumulates heat. Heat is transferred to a system, and once

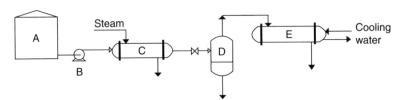

FIGURE 2-1   Different unit operations in a process.

it enters the system, this energy is transformed in other types of energy, such as kinetic or internal energy or mechanical work.

The relationship among these types of energies is known as the *first law of thermodynamics* or the *energy conservation law*. The mathematical expression for a closed system is as follows:

$$Q = W + \Delta U + \Delta Ec \qquad (2\text{-}2\text{-}1)$$

where Q = heat added to the system
   ΔU = change in internal energy
   ΔEc = change in kinetic energy
   W = work done by the system

We shall clarify a few aspects of this expression:

1. The sign convention adopted is the usual in thermodynamics, which is to consider positive the heat added to the system and the work performed by the system.

2. The action of electric or magnetic fields is not considered. However, in such cases it is possible to consider their effect by introducing the mechanical or thermal equivalent of their energy.

3. It must be noted that we didn't include the potential energy, since we consider the weight of the object as an external force, so its action is treated as mechanical work performed by an external agent. For example, let's consider an object in free fall. Since in this case there is no internal energy change or heat added to the system, Eq. (2-2-1) reduces to

$$W + \Delta Ec = 0 \qquad (2\text{-}2\text{-}2)$$

W is the work performed by the force $m \cdot g$ (mass × acceleration due to gravity); this means that

$$W = m \cdot g \cdot \Delta h$$

The expression of the first law of thermodynamics for this system therefore is

$$m \cdot g \cdot \Delta h + \Delta Ec = 0 \qquad (2\text{-}2\text{-}3)$$

The first term is what sometimes is called *potential energy*.

4. Usually, in process engineering, when heat terms are present, the kinetic energy terms are of lower order of magnitude and can be disregarded. The expression of the first law in this case is

$$Q = W + \Delta U \qquad (2\text{-}2\text{-}4)$$

5. It must be noted that Eq. (2-2-1) may be considered a definition of the internal energy if it is written as

$$\Delta U = Q - W - \Delta Ec \qquad (2\text{-}2\text{-}5)$$

We can see that a duality of concepts exists because we cannot define the internal energy without accepting the validity of the first law, but at the same time, to express the first law, we must postulate the existence of this type of energy. Thus the internal energy and the first law of thermodynamics constitute a single concept because they cannot be made independent of each other.

2-3 TEMPERATURE AND INTERNAL ENERGY

The internal energy, since it is a state function, is related to the thermodynamic parameters of the system—pressure, temperature, and composition—and can be calculated from them. When the composition of the system remains unchanged, the internal energy is a function of pressure and temperature.

Usually, the changes in internal energy with pressure are not significant as long as these pressure changes do not produce changes of phase in the system, so the internal energy sometimes may be considered a function of temperature exclusively. This is why the term *thermal energy* is used sometimes. This expression, although not having a precise thermodynamic meaning, is both graphic and intuitive.

Internal energy is stored in a system at the molecular or atomic level. This means that atoms or molecules can exist in different energy levels, which correspond, for example, to different velocities or frequencies of their movements or oscillations.

Temperature is the macroscopic observable that is more directly related to these energy levels. Thus molecules or atoms with more energy are observed as more "hot." For example, in the case of monatomic gases, the kinetic energy of the molecule is related to temperature by the expression

$$\frac{1}{2} m \cdot u^2 = \frac{3}{2} k \cdot T \tag{2-3-1}$$

where m = molecular mass
u = molecular velocity
k = Boltzman constant
T = absolute temperature

Equation (2-3-1) can be explained by saying that the internal energy of the gas is stored as kinetic energy and that it can be measured through its temperature, both magnitudes being related by that expression.

The energy units employed in different units systems are

International system: joule (J)

Technical metric system: kilocalorie (kcal)

English system: British thermal unit (Btu)

And their equivalence is

$$1 \text{ kcal} = 3.968 \text{ Btu} = 4183 \text{ J}$$

2-4 CONTINUOUS AND STEADY-STATE SYSTEMS

Most process plants nowadays operate in continuous mode. This means that material circulates through the system, and a steady state is achieved, in which all the operating parameters (i.e., pressure, temperature, composition, and flow rate) at any point in the system remain constant over time. For example, in the system in Fig. 2-1, from the piping inlet section up to the condenser outlet, a steady state exists, which means that at any point in the system the operating parameters do not change with time.

In continuous systems, mass and energy balances are expressed per unit time, so we speak about kilograms per hour of flow rate or British thermal units per hour or joules per second (watts) of energy added to the system. The equivalences then become

$$1 \text{ kcal/h} = 3.968 \text{ Btu/h} = 1.162 \text{ W}$$

From now on, and unless specifically stated, we shall use the symbol Q to designate heat flux, this is, the heat delivered to a system per unit time.

2-5 FORMULATION OF THE FIRST LAW FOR CONTINUOUS SYSTEMS

To understand how the first law of thermodynamics can be applied to a continuous system, let us consider the system presented in Fig. 2-2. This represents a portion of a system in steady state in which a fluid circulates at a flow rate W(kg/s). Within the boundaries we are considering, a continuous heat flux Q (J/s) is added, and a mechanical work per unit time Ws (J/s) is performed.

In section 1 of the system, the fluid parameters—pressure, temperature, density, velocity, internal energy, and geometric height—are p_1, t_1, ρ_1, v_1, u_1, and z_1. In section 2 of the system, the fluid parameters will be different, and the corresponding increments are indicated in the figure.

Since the fluid is in motion, if we consider the mass that in a certain instant is enclosed within the boundaries of the control volume (sections 1 and 2) after a certain time increment Δt, this mass will occupy the volume contained between sections 1' and 2'.

Let Δx_1 and Δx_2 be the displacements of the system boundaries. Since in all the intermediate points the thermodynamic parameters remained constant, we can imagine this evolution as if m kg of the fluid, originally contained between sections 1 and 1', were moved to the volume contained between sections 2 and 2', changing its pressure, temperature, internal energy, density, kinetic energy, and height.

We shall apply the first law to this evolution. The change in the internal energy of the displaced mass is

$$m \cdot \Delta u \qquad (2\text{-}5\text{-}1)$$

The kinetic energy change is

$$m \cdot \Delta \left(\frac{v^2}{2} \right) \qquad (2\text{-}5\text{-}2)$$

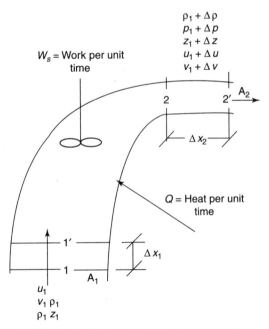

FIGURE 2-2 Application of the first law to a continuous system.

To calculate the work received or performed by the system, it must be considered that in addition to the external work, there is work performed by the pressure forces acting on the boundaries of the system (inlet and outlet sections) and work performed by the gravitational force.

We shall first consider the pressure forces. In the inlet section (section 1), this force equals the product of the pressure times the area over which the force is exerted, which is the area of the cross section A_1. Since the boundary of the system moved from section 1 to section 1', it can be considered that this force performed an amount of work given by

$$p_1 \cdot A_1 \cdot \Delta x_1 \qquad (2\text{-}5\text{-}3)$$

The product $A_1 \cdot \Delta x_1$ is the volume that the mass m occupied at the thermodynamic conditions of the inlet section. The work then can be expressed as

$$m \frac{p_1}{\rho_1} \qquad (2\text{-}5\text{-}4)$$

This work is done on the system.

In the outlet section (section 2), the system performs work against the force exerted by the pressure $p_1 + \Delta p$. This work can be expressed as

$$-m \frac{(p_1 + \Delta p)}{(\rho + \Delta \rho)} \qquad (2\text{-}5\text{-}5)$$

The difference between both terms can be written as

$$m \cdot \Delta \left(\frac{p}{\rho} \right) \qquad (2\text{-}5\text{-}6)$$

Regarding the work performed by the gravitational forces, again, it can be considered that the evolution suffered by the system is the translation of a mass m from position z_1 to position $z_1 + \Delta z$. The work is

$$m \cdot g \cdot \Delta z \qquad (2\text{-}5\text{-}7)$$

If Q is the amount of heat received or given per unit time, the heat per unit mass will be the quotient of Q and the circulating mass flow rate. And the heat received or given by the mass m will be

$$\frac{Q}{W} m \qquad (2\text{-}5\text{-}8)$$

Similarly, the circulation work performed or received by mass m will be

$$\frac{Ws}{W} m \qquad (2\text{-}5\text{-}9)$$

Then the expression of the first law can be written as

$$m \frac{Q}{W} = m \left[\frac{Ws}{W} + g\Delta z + \Delta u + \Delta(v^2/2) + \Delta(p/\rho) \right] \qquad (2\text{-}5\text{-}10)$$

or

$$Q = Ws + W \left[g\Delta z + \Delta u + \Delta(v^2/2) + \Delta(p/\rho) \right] \qquad (2\text{-}5\text{-}11)$$

The function $u + p/\rho$ is the specific enthalpy i, so we get the final expression of the first law for a circulating system as

$$Q = Ws + W \left[g\Delta z + \Delta i + \Delta(v^2/2) \right] \qquad (2\text{-}5\text{-}12)$$

In systems where heat transfer operations are performed, the terms Q and $W \cdot \Delta i$ are usually of a higher order of magnitude than the rest of the terms. For example, in the system shown in Fig. 2-1, the power delivered by the pump or the changes in hydrostatic height or fluid velocity have little influence on the temperature change experienced by the fluid. Then the energy balance can be expressed as

$$Q = W \cdot \Delta i \qquad (2\text{-}5\text{-}13)$$

This is the equation we shall use in this course for the energy balance in heat transfer equipment.

2-6 THERMODYNAMIC VARIABLES

The thermodynamic state of a system is defined when its state variables—pressure, temperature, and composition—are known. If these variables are defined, it is possible to calculate all the other intensive properties, such as

u = specific internal energy (J/kg)

i = specific enthalpy (J/kg)

$\bar{v} = 1/\rho$ = specific volume (m³/kg)

as well as all other properties that thermodynamics define for mathematical simplicity (e.g., entropy, free energy, etc.).

All these thermodynamic properties are related by a set of relationships in partial derivatives called *Maxwell equations*. If the relationship between any of these intensive thermodynamic variables and the state variables (pressure, temperature, and composition) is known, then it is possible to calculate the rest of them. But this first relationship must be obtained experimentally. The thermodynamic property that is the easiest to measure is volume. If we have a function of the type

$$\bar{v} = \bar{v}(p, T, \text{composition}) \qquad (2\text{-}6\text{-}1)$$

it will be possible to calculate entropies, enthalpies, and any other property. Functions of the type of Eq. (2-6-1) are called *equations of state*.

There are some systems for which it is possible to find an equation of state that adequately represents the system volume in both the liquid and vapor phases. From this equation, a thermodynamic-mathematical package can be developed that allows the calculation of all other properties. These systems normally consist of nonpolar hydrocarbon mixtures. In other cases, when it is not possible to find a suitable equation of state, it is necessary to use other types of experimental information.

Modern process-simulation programs calculate all thermodynamic properties from an equation of state or other experimental information using these mathematical relationships.

However, it is usual to define some other auxiliary variables that allow manual calculations to be simplified. For example, for a system that performs an evolution without a change of state, the specific heat at constant pressure is defined as

$$c_P = \left(\frac{\partial i}{\partial T} \right)_P \qquad (2\text{-}6\text{-}2)$$

or the specific heat at constant volume is defined as

$$c_V = \left(\frac{\partial u}{\partial T} \right)_V \qquad (2\text{-}6\text{-}3)$$

In many constant-composition systems without change of phase, enthalpy can be considered a function of temperature exclusively (the variation with pressure is weak). It is then possible to calculate the enthalpy change between two states at different temperatures as

$$\Delta i = cp \cdot \Delta t \qquad (2\text{-}6\text{-}4)$$

(In this book we will only use the specific heat at constant pressure, so we will drop the subscript and will designate it as c.)

The specific heats may be considered approximately constant within certain temperature ranges. It is then possible to put them into tables or graphs that are useful tools for hand calculations. Specific heat values for several substances can be found in the tables of App. I.

Another useful auxiliary variable is the latent heat of vaporization of a pure substance λ. This is defined as the enthalpy difference between the saturated vapor and the liquid at its boiling temperature and same pressure. Latent heats of vaporization for different substances are also included in App. I.

2-7 *HEAT TRANSFER MECHANISMS*

It is normally accepted that there are three basic heat transfer mechanisms—conduction, convection, and radiation. In what follows we shall make a brief presentation of them, and they will be examined in more detail in following chapters

2-7-1 Conduction

Description of the Mechanism.   When there is a temperature gradient within a substance, heat is transferred through it by molecular contact. In the case of solids, it is normally accepted that the molecular energy is associated with some form of vibration. Each molecule vibrates in a fixed position and can interchange energy with its neighbors. If heat is supplied to one part of the solid, the molecules vibrate faster. As they vibrate more, the bonds between molecules are shaken more. This passes vibration on to the next molecule and so on. Thus heat spreads through the solid in the direction of decreasing temperature.

It has been observed that the heat flux is, within certain limits, proportional to the temperature variation per unit length and to the area through which the heat flows. That is,

$$Q = -kA \frac{dT}{dx} \tag{2-7-1}$$

where x is the axis corresponding to the direction of the temperature variation, Q is the heat flux, A is the area perpendicular to the heat flux, and k is a proportionality constant. In this expression, the negative sign means that the direction of the heat flow is the direction in which temperature decreases.

In the rest of the book we will work frequently with heat-flux densities, meaning heat transferred per unit time and per unit area. The heat flux density will be designated as q. Thus it will be

$$q = -k \frac{dT}{dx} \tag{2-7-2}$$

In the case of gases, the model that explains conduction heat transmission is quite different from that for solids. Gas molecules move with a velocity related to their temperature [see Eq. (2-3-1)].

Let's assume that we have a stagnant gas enclosed within two walls, A and B, at different temperatures, with $T_A > T_B$ (Fig. 2-3). Let C be an imaginary planar surface located between both walls. This planar surface is crossed in both directions by gas molecules. If the system pressure is constant, the same number of molecules crosses the surface in one direction as the other. Then, at this surface, the average velocity of all the molecules is zero. This means that the gas is stagnant even though the molecules move individually.

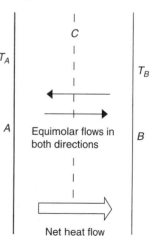

FIGURE 2-3 Thermal conductivity in gases can be explained by molecular transport of energy.

However, since the molecules crossing the surface from left to right come from a higher-temperature region, they have more energy than those crossing the surface from right to left. The result is a net energy flux through the surface. The higher the thermal gradient, the higher will be the heat flux. Therefore, again,

$$q = -k\,\frac{dT}{dx}$$

By means of the kinetic theory, it is possible to calculate a value for the proportionality constant, which is a function of the temperature and the nature of the gas.

Note   Actually, it is not possible to have a completely stagnant fluid in a heat transfer situation. The temperature differences always create density differences that provoke fluid motion. The movement of a mass of fluid between regions of different temperature provides an additional mechanism for energy transport. The "pure conduction" model we have just described for the gas is an ideal situation for a case in which the distance between the walls is very small.

In the case of liquids, the model is more complex, and we shall not try to explain it. But it can be seen that although the phenomenologic descriptions of these mechanisms look very different, the final expression that allows the calculation of the heat flux is the same.

Thus, even though the mechanisms are different for solids, gases, and liquids, the fact that proportionality between the heat flux and the thermal gradient can be established makes possible that all these mechanisms can be grouped together under the common designation of *conduction*. The mathematical expression of Eq. (2-7-2) then is the only possible definition of this mechanism.

Thus we can say that *conduction* is a heat transfer mechanism whose characteristic is a proportionality between the heat flux and the thermal gradient and which, at the microscopic level, can be explained with different models depending on whether the conduction takes place in a gas, a liquid, or a solid. The proportionality coefficient in Eq. (2-7-2) is called *thermal conductivity*.

As it will be explained later, in the case of fibrous or porous materials, the thermal conductivity must include many effects, such as conduction through the solid fibers, convective effects in the air separating the fibers, and radiation heat transfer between the fibers. What we really do is transfer to the proportionality constant our inability to explain these phenomena with more rigor. Maybe someday we will have a better knowledge of the nature of matter, and what today is generically called "conduction heat transfer" will be studied with more specific models in each case.

The thermal conductivity units in the different systems are

$$1\ \text{kcal/(h}\cdot\text{m}\cdot{}^\circ\text{C)} = 0.671\ \text{Btu/(h}\cdot\text{ft}\cdot{}^\circ\text{F)} = 1.162\ \text{W/(m}\cdot\text{K)}$$

Some Characteristics of the Mechanism of Conduction

1. For conduction to occur, the presence of matter is necessary. We shall see that there is another mechanism, radiation that allows the propagation of heat through a vacuum.

2. This mechanism is restricted to heat propagation through stagnant matter. In the case of fluids in motion, the heat transfer mechanism is convection.

3. Conduction is the only possible heat transfer mechanism in opaque solids.

2-7-2   Convection

Description of the Mechanism.   *Convection* is a heat transfer mechanism present in fluids in motion. Let's consider a solid wall at temperature T_w in contact with a moving fluid at temperature T_f (Fig. 2-4).

We can see that even though we say that the fluid temperature is T_f, there is a small thickness layer of fluid in which the temperature varies continuously from T_w to T_f. This is known as thermal boundary layer. Owing to the existence of this temperature gradient, heat will flow from the solid wall into the fluid.

Let's assume that the fluid is in motion, with velocity in the direction parallel to the solid wall. It is possible to observe experimentally that a velocity profile, as shown in the right side of the figure, appears

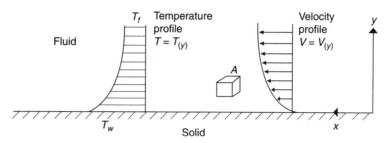

FIGURE 2-4   Boundary layer and velocity distribution in a fluid flowing over a flat plate.

in the system. This graph shows the fluid velocity at any point as a function of the distance to the wall. The fluid velocity is zero at the solid-fluid interface and reaches a constant value beyond a certain distance from the solid. This region of variable fluid velocity is called the *velocity boundary layer.*

In this model, we are assuming that the fluid velocity has only a component in the direction parallel to the solid surface and that the streamlines are straight lines. This is known as *laminar movement,* and it takes place only at low fluid velocities. If the fluid velocity increases, eddies or turbulences appear in the fluid, provoking erratic fluid movements in other directions.

Since the velocity in a region immediately adjacent to the solid wall is nil, the fluid in this region behaves like a solid, and the heat flows into it by conduction. Then,

$$q = -k \left(\frac{dT}{dy} \right)_{y=0} \tag{2-7-3}$$

Let's consider an imaginary volume element located at a certain distance from the wall, at position A in Fig. 2-4. This situation is shown enlarged in Fig. 2-5.

Heat enters the volume element through the lower side by conduction and exits through the upper side. If conduction were the only heat transfer mechanism, both heat flows would be equal. However, the fluid that is entering into this volume absorbs part of the energy and leaves at a higher temperature. Thus the heat flux through the lower and upper sides of the volume must be different.

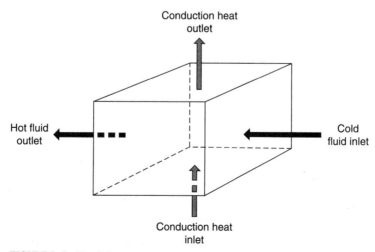

FIGURE 2- 5   Heat balance on a volume element within the boundary layer.

Therefore, the dT/dy values also will be different. This means that the fluid movement distorts the temperature profile. Then, even though Eq. (2-7-3) is absolutely general, the value of the derivative is affected by the hydrodynamics of the system.

For simple geometries and laminar flow, it is possible to obtain an expression for the temperature distribution function analytically, and then the heat flux can be calculated by means of Eq. (2-7-3). However, laminar flow is seldom found in process engineering, and the most common situation is that dT/dy cannot be calculated.

It can be observed experimentally that within a certain range of system parameters, the heat flow is proportional to the temperature difference between the fluid and the wall. Thus

$$q = h(T_w - T_f) \tag{2-7-4}$$

where h is a constant of proportionality called the *convection film coefficient*. This coefficient is a function of system geometry, the properties of the fluid, and fluid velocity. The relationship between h and the system parameters is usually found experimentally.

As a general rule, the higher the fluid velocity, the higher is the convection film coefficient. This is due to the fact that the fluid streams are an additional mechanism for heat removal, the magnitude of which increases with the increase in the velocity of fluid renewal over the surface. The units of h in the several units systems are

$$1 \text{ kcal/(h} \cdot \text{m}^2 \cdot {}^\circ\text{C}) = 0.205 \text{ Btu/(h} \cdot \text{ft}^2 \cdot {}^\circ\text{F}) = 1.162 \text{ w/(m}^2 \cdot \text{K})$$

It is important to emphasize that the heat delivered by a solid to a fluid enters the fluid initially by conduction, and then the fluid streams absorb this energy and transport it into and through the fluid. It does not make sense trying to distinguish between heat transferred by conduction and heat transferred by convection. It is the same energy that is transported by the combined action of fluid movement and thermal conductivity of the fluid. The global phenomenon is expressed by Eq. (2-7-4), which allows calculation of the total heat flux through the interface.

Natural and Forced Convection.   Fluid movement may be caused by the application of external pressure gradients, for example, by the action of pumps or compressors. It is said that in this case the convection is *forced*, meaning that the velocity profile is imposed externally on the system independent of the heat transfer.

However, in the absence of fluid movers, if heat is applied to a fluid, the temperature gradients provoke density differences that cause the fluid to move. This movement is the result of heat flux and is called *natural convection*. This means that every time heat is transferred to or from a fluid, the mechanism will be convection, either natural or forced.

2-7-3 Thermal Radiation

Explanation of the Mechanism.   Radiation is probably the primary heat transfer mechanism between molecules. In contrast to the other mechanisms, the heat source does not need a physical medium to transfer its energy. Energy can be transmitted from an emitting to a receiving body through a vacuum. This is why it is possible to receive solar energy on the earth.

Radiation is propagated by electromagnetic waves. The emission of electromagnetic energy is not continuous. It is emitted as a multiple of a minimum amount of energy called a *quantum* or *photon*. This means that the body emits a certain number of quantums. The energy of the quantum is related to the wavelength of the emitted radiation.

To emit energy, a body must have been excited previously from the exterior. Owing to the natural tendency of things to return to equilibrium, the body emits energy so as to return to its original state. This decrease in the energy of the body corresponds to the radiation emitted. The emission takes place by the same mechanisms by which the energy was absorbed. The change in the energy level of the molecules may be related to an increase in the rotation or vibrational movements of its atoms or electrons as they move to higher-energy orbits or even to changes in the energy of the atomic nucleus. These different mechanisms produce radiation of very different wavelengths. The electromagnetic spectrum ranges from wavelengths of 10^{-11} cm (cosmic rays) to 10^5 cm (low frequency radio waves).

There is a small portion of this spectrum in which the emission and absorption of energy produce changes in temperature of the related bodies. This is the region between 0.5- and 50-μm wavelengths. This is called *thermal radiation.*

Any body at an absolute temperature T emits an amount of energy within this wavelength range that is proportional to the fourth power of its absolute temperature. This is

$$q = \sigma \cdot \varepsilon \cdot T^4 \tag{2-7-5}$$

where q = energy emitted by the body per unit time and per unit area

σ = Boltzman's constant = 4.88×10^{-8} kcal/($m^2 \cdot h \cdot K^4$) = 5.672×10^{-8} W/($m^2 \cdot K^4$)

ε = emissivity of the body (This is a parameter that is always less than 1 and depends on the type and conditions of the surface.)

Later in this text we shall see that when a body at temperature T_A is in an infinite ambient at temperature T_B, it exchanges heat with the surroundings at a rate that can be expressed as

$$Q = \sigma \cdot \varepsilon \cdot A(T_A^4 - T_B^4) \tag{2-7-6}$$

Some Characteristics of the Radiation Mechanism

1. The mechanism of heat transfer by radiation is superimposed on that of heat transfer by convection. Both mechanisms can be considered to be in parallel. Thus the energy exchanged between a body and the ambient that surrounds it can be expressed as

$$Q = \sigma \cdot \varepsilon \cdot A(T_A^4 - T_B^4) + h \cdot A(T_A - T_B) \tag{2-7-7}$$

where T_A is the temperature of the body and T_B is the temperature of the surroundings.

2. We see that as a difference between conduction and convection, the heat exchanged does not depend on the temperature difference but on the fourth power of the temperatures.

This mechanism thus is particularly important in systems operating at high temperatures, such as process furnaces or steam generators. On the other hand, in low-temperature systems, sometimes it can be neglected, unless the convection film coefficient is also particularly low, as in natural convection systems in the air.

2-7-4   Other Heat Transfer Mechanisms

Even though it is commonly accepted that conduction, convection, and radiation are the *primary* heat transfer mechanisms, there are some more complex situations that must be studied separately and that could be considered in the same category as these three. These are cases where a fluid experiences a change of state, such as boiling and condensation.

In these cases, the heat transfer rate is associated with the velocity by which the change of state takes place, which frequently is related to the kinetics of a mass-transfer process or requires more complex models to be predicted. These mechanisms will be studied in specific later chapters.

GLOSSARY

Ec = kinetic energy (J)

g = acceleration of gravity (m/s^2)

h = convection film coefficient [W/($m^2 \cdot K$)]

h = height (m)

i = specific enthalpy (J/kg)

k = thermal conductivity [W/(m · K)]

k = Boltzman's constant

m = mass (kg)

Q = heat (J) or heat flow (W)

q = heat flux density (W/m^2)

T = temperature (K)

u = molecular velocity (m/s)

u = specific internal energy (J/kg)

U = internal energy (J)

v = velocity (m/s)

$\bar{v}$ = specific volume (m^3/kg)

W = mass flow (kg/s)

Ws = work per unit mass (J/kg)

x, y, z = coordinates—distance (m)

σ = Boltzman's constant

ε = emissivity

Subscripts

w = wall

f = fluid

CHAPTER 3
HEAT CONDUCTION IN SOLIDS

Conduction is a heat transfer mechanism resulting from the energy exchange between neighbor molecules as a consequence of molecular contact. In conduction heat transfer, atoms or molecules interact by elastic and inelastic collisions to propagate the energy from regions of higher temperature to regions of lower temperature.

Conduction is the only possible heat transfer mechanism in opaque solids. In liquids and gases, pure conduction cannot exist. The fluid motion always offers a vehicle for energy transport, and the transfer of heat is a combination of both effects generically designated as *convection*. This is why, in the present chapter, we shall study only heat conduction within solids. Only steady-state systems will be studied. For conduction of heat in nonstationary systems, see ref. 2.

3-1 GENERAL EQUATION OF CONDUCTION IN SOLIDS

In Chap. 2, we presented the equation of heat conduction in one direction as

$$q_x = -k\frac{dT}{dx} \tag{3-1-1}$$

This is a simplified expression. In a general sense, the conduction heat-flux density can be represented by a vector $\vec{q}$, which direction is the principal direction of the propagation of heat in the solid medium. This direction coincides with the direction of the temperature-gradient vector $\overrightarrow{\nabla T}$.

The temperature gradient, in turn, is a vector whose direction is that of the greatest temperature variation per unit length and whose module is the value of this derivative. In a Cartesian coordinates system, the components of the gradient of a scalar field T in the coordinate axes are the partial derivatives of T with respect to each coordinate. This is

$$\overrightarrow{\nabla T}_x = \frac{\partial T}{\partial x} \qquad \overrightarrow{\nabla T}_y = \frac{\partial T}{\partial y} \qquad \overrightarrow{\nabla T}_x = \frac{\partial T}{\partial z} \tag{3-1-2}$$

And the most general conduction equation is

$$\vec{q} = -k\overrightarrow{\nabla T} \tag{3-1-3}$$

This vector equation can be decomposed in the following three scalar equations:

$$q_x = -k\frac{\partial T}{\partial x} \qquad q_y = -k\frac{\partial T}{\partial y} \qquad q_z = -k\frac{\partial T}{\partial z} \tag{3-1-4}$$

However, in most cases it is possible to choose the coordinates axes so that the direction of one of them coincides with the direction of the gradient vector. For example, in the case of an infinite planar wall whose faces are kept at different temperatures (as in Fig. 3-2), it is evident that the gradient direction is perpendicular to the wall. Then, if we choose the z-axis coincident with that direction, it will be

$$\frac{\partial T}{\partial x} = \frac{\partial T}{\partial y} = 0 \qquad \text{and} \qquad q_z = -k\frac{dT}{dz} \tag{3-1-5}$$

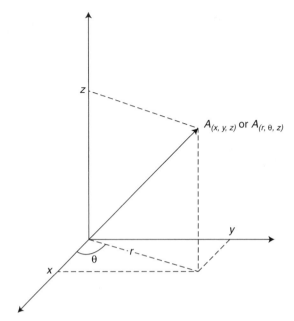

FIGURE 3-1   Rectangular and cylindrical coordinates.

In a cylindrical system, the coordinates of any point are r, θ, and z. In Fig. 3-1, the coordinates of a point A are shown in both systems. The components of the gradient vector in cylindrical coordinates are

$$\nabla T_r = \frac{\partial T}{\partial r} \qquad \nabla T_\theta = \frac{1}{r}\frac{\partial T}{\partial \theta} \qquad \nabla T_z = \frac{\partial T}{\partial z} \tag{3-1-6}$$

The most usual case in process heat transfer involving cylindrical geometry is heat conduction through the wall of a cylindrical pipe. In this case, the direction of the vector $\vec{q}$ is radial. Then the partial derivatives with respect to θ and z are zero, and it is

$$q_r = -k\frac{dT}{dr} \tag{3-1-7}$$

3-2   HEAT CONDUCTION THROUGH A WALL

3-2-1   Case of Planar Geometry

Let's consider a planar wall of thickness L and infinite dimensions in the other two directions (Fig. 3-2). On each face of the wall, a constant temperature is maintained. Let T_A be the temperature of the left face and T_B be the temperature of the right face.

If we make the z axis coincide with the direction normal to the wall, the temperatures will be

$$T = T_A \qquad \text{at } z = 0$$
$$T = T_B \qquad \text{at } z = L$$

The direction of the vector $\vec{q}$ will be perpendicular to the wall, so it will only have a component in the z axis.

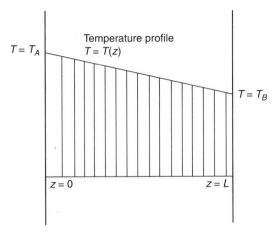

FIGURE 3-2 Temperature profile in a homogeneous planar wall in steady state.

Since we are dealing with a steady state, the amount of heat per unit time crossing any plane normal to z is the same. Then q_z is constant with changes in z. Thus we have

$$\frac{dq_z}{dz} = 0 \tag{3-2-1}$$

Then
$$\frac{d^2T}{dz^2} = 0 \quad \text{and} \quad \frac{dT}{dz} = C_1$$

Therefore, finally,

$$T = C_1 z + C_2 \tag{3-2-2}$$

The integration constants are evaluated with the boundary conditions, and this results in

$$C_1 = (T_B - T_A)/L$$
$$C_2 = T_A$$

The temperature profile is a straight line, as shown in Fig. 3-2. Then

$$q = k(T_B - T_A)/L \tag{3-2-3}$$

Composite Walls. In the case of a solid wall composed of two layers of different materials I and II, as shown in Fig. 3-3, the equation deduced for the simple wall can be applied to each material. Then, if T_i is the temperature of the plane separating both layers,

$$q_I = k_I(T_1 - T_i) / \Delta z_I = \text{constant} = C_I \tag{3-2-4}$$

$$q_{II} = k_{II}(T_i - T_2)/\Delta z_{II} = \text{constant} = C_{II} \tag{3-2-5}$$

Since in the plane of union q is the same for both materials, $C_I = C_{II} = q$, and then

$$q\frac{\Delta z_I}{k_I} = T_1 - T_i$$

$$q\frac{\Delta z_{II}}{k_{II}} = T_i - T_2 \tag{3-2-6}$$

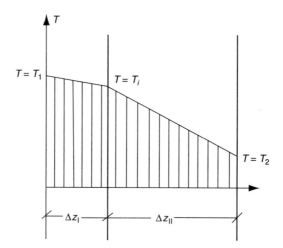

FIGURE 3-3 Temperature profile in a composite planar wall in steady-state heat transfer.

Adding both equations

$$q\left(\frac{\Delta z_{\mathrm{I}}}{k_{\mathrm{I}}} + \frac{\Delta z_{\mathrm{II}}}{k_{\mathrm{II}}}\right) = T_1 - T_2 \tag{3-2-7}$$

this can be written

$$q = U(T_1 - T_2) \tag{3-2-8}$$

where

$$\frac{1}{U} = \frac{\Delta z_{\mathrm{I}}}{k_{\mathrm{I}}} + \frac{\Delta z_{\mathrm{II}}}{k_{\mathrm{II}}} \tag{3-2-9}$$

If the wall is composed of more than two materials, the general expression is

$$\frac{1}{U} = \sum \frac{\Delta z_i}{k_i} \tag{3-2-10}$$

U is called the *overall heat transfer coefficient,* and its utility lies on the fact that it allows a direct calculation of the heat flux without calculating the temperatures at the intermediate planes.

Heat Transfer Through a Wall Separating Two Fluids. In the preceding case we assumed that the temperatures of both faces of the wall were known. However, the usual case is that these temperatures are unknown. Let's consider the case of a composite solid wall separating two fluids at different temperatures.

For example, it could be the wall of a vessel containing a warm liquid. The wall may be composed of a steel plate with a mineral wool thermal insulation layer. The temperature of the contained liquid is T_i, whereas the temperature of the exterior air is T_o. Heat will flow from the liquid to the exterior through the composite solid wall (Fig. 3-4). The heat-flux density q_z will be constant through the wall.

Heat is transferred from the liquid to the internal wall by convection. For heat transfer to exist, a temperature difference between the bulk of the liquid and the solid wall is necessary. We will call T_1 the interior wall temperature, which, as we can see in the figure, is lower than T_i. Then

$$q = h_i(T_i - T_1) \tag{3-2-11}$$

where h_i is the convection heat transfer coefficient.

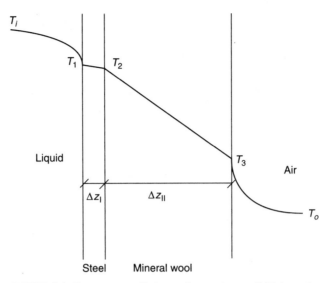

FIGURE 3-4   Temperature profile in a wall separating two fluids in steady-state heat transfer.

Since heat must flow by conduction through the steel and the mineral wool, a temperature gradient must exist in both materials. Let T_2 be the temperature of the separation plane between the steel and the insulation, and let T_3 be the temperature of the insulation external face. We then can write

$$q = \frac{k_I}{\Delta z_1}(T_1 - T_2) \tag{3-2-12}$$

$$q = \frac{k_{II}}{\Delta z_2}(T_2 - T_3) \tag{3-2-13}$$

From the external face of the insulation to the ambient air, heat is transferred by a combined convection-radiation mechanism, so

$$q = h_o(T_3 - T_o) \tag{3-2-14}$$

where h_o is the combined convection-radiation coefficient.

Since the heat-flux density is constant through the wall, we can isolate the temperature differences in Eqs. (3-2-11) through (3-2-14), and this results in

$$T_i - T_1 = \frac{q}{h_i} \tag{3-2-15a}$$

$$T_1 - T_2 = q\frac{\Delta z_I}{k_I} \tag{3-2-15b}$$

$$T_2 - T_3 = q\frac{\Delta z_{II}}{k_{II}} \tag{3-2-15c}$$

$$T_3 - T_o = \frac{q}{h_o} \tag{3-2-15d}$$

Adding member to member these equations, the intermediate temperatures cancel, and the result is

$$T_i - T_o = q\left(\frac{1}{h_i} + \frac{1}{h_o} + \frac{\Delta z_1}{k_I} + \frac{\Delta z_2}{k_{II}}\right) \tag{3-2-16}$$

The term within parentheses is the reciprocal of the overall heat transfer coefficient, so generalizing,

$$q = U(T_i - T_o) \tag{3-2-17}$$

and

$$U = \left(\sum \frac{\Delta z_1}{k_I} + \frac{1}{h_i} + \frac{1}{h_o}\right)^{-1} \tag{3-2-18}$$

Controlling Resistance. It is useful to analyze Eqs. (3-2-17) and (3-2-18) making an analogy with an electric circuit in which the heat flux is analogous to the electric current, the temperature difference acts as potential difference, and U is the reciprocal of the resistance. Then each term within the parentheses can be considered as a partial resistance representing any of the solid-fluid interfaces or any material layer.

All these resistance are in series, and the total resistance is the sum of the partial resistances. Frequently, one of these resistances is much smaller than the others and can be neglected without causing an appreciable error in the calculations. For example, the resistance of the metallic wall of a vessel is usually very small because the thickness is small and the thermal conductivity is very high (metals are good heat conductors).

When one of the partial resistances is much larger than the others, the total resistance is similar to it, which is then called the *controlling resistance*. For example, when a liquid container is insulated to avoid heat losses, low-thermal-conductivity materials are employed, and most of the total resistance and most of the temperature drop take place in the insulation layer.

Example 3-1 A vessel contains a liquid at 160°C and is exposed to air at 20°C. The thickness of the steel wall is 10 mm. Calculate the heat loss and the temperature profile through the system if the vessel is insulated with mineral wool [thermal conductivity 0.072 W/(m·K)] with the following thicknesses:

a. 25 mm
b. 38 mm

Assume that the convection heat transfer coefficient between the liquid and the metallic wall is 500 W/(m²·K) and that the combined convection-radiation coefficient between the insulation and the ambient air is 5 W/(m²·K). The thermal conductivity of steel is 43 W/(m·K).

Solution First of all, we must clarify that the convection heat transfer coefficient between the insulated wall and the ambient air is a function of the external temperature of the wall (as we shall see in Chap. 4). To assume this value as constant is a simplification that in this case is acceptable because between both cases this temperature will not change significantly. Therefore, referring to Fig. 3-4,

a. For 25-mm thickness,

$$U = \frac{1}{\sum \dfrac{\Delta z_i}{k_i} + \dfrac{1}{h_i} + \dfrac{1}{h_o}} = \frac{1}{\dfrac{0.01}{43} + \dfrac{0.025}{0.072} + \dfrac{1}{500} + \dfrac{1}{5}} = 1.82 \ \text{W/(m}^2 \cdot \text{K)}$$

$$q = U\Delta T = 1.82(160 - 20) = 254.8 \ \text{W/m}^2$$

The temperature drop between the liquid and the internal wall (see figure) is

$$T_i - T = q/h_i = 254.8/500 = 0.51°C$$

Then
$$T_1 = 160 - 0.51 = 159.5°C$$

The temperature drop through the metal is

$$T_1 - T_2 = q/(k_1/\Delta z_1) = 254.8/(43/0.01) = 0.059°C$$

Then
$$T_2 = T_1 - 0.059 = 159.4°C$$

The temperature drop in the insulation is

$$T_2 - T_3 = q/(k_2/\Delta z_2) = 254.8/(0.072/0.025) = 88.47°C$$

Then
$$T_3 = T_2 - 88.47 = 70.92°C$$

Finally, the temperature drop between the insulation and the external air is

$$70.92 - 20 = 50.92°C$$

(which must coincide with $q/h_o = 254.8/5$ within rounding errors). It must be noted that practically there is no temperature drop between the liquid and the external face of the steel wall because the two involved resistances are very small.

The highest temperature drop, as may be expected, is in the insulation layer; even so, an important temperature drop also takes place between the external face of the insulation and the air. The temperature of the external face of the insulation is 70.92°C. This is too high and may be objectionable from a safety point of view because it may provoke burns if a person touches the external wall. (Normally, temperatures higher than 60°C are not accepted in surfaces that may be in contact with plant operators.) One way to reduce this temperature is to increase the insulation thickness, as we shall see.

b. For 38-mm thickness,

$$U = \frac{1}{\sum \dfrac{\Delta z_i}{k_i} + \dfrac{1}{h_i} + \dfrac{1}{h_o}} = \frac{1}{\dfrac{0.01}{43} + \dfrac{0.038}{0.072} + \dfrac{1}{500} + \dfrac{1}{5}} = 1.37 \text{ W/(m}^2 \cdot \text{ K)}$$

$$q = U\Delta T = 1.37(160 - 20) = 191.8 \text{ W/m}^2$$

The temperature drop between the liquid and the internal wall is

$$T_i - T_1 = q/h_i = 191.8/500 = 0.38°C$$

Then
$$T_1 = 160 - 0.38 = 159.6°C$$

The temperature drop in the metal is

$$T_1 - T_2 = q/(k_1/\Delta z_1) = 191.8/(43/0.01) = 0.044°C$$

Then
$$T_2 = T_1 - 0.044 = 159.55°C$$

The temperature drop in the insulation layer is

$$T_2 - T_3 = q/(k_2/\Delta z_2) = 191.8/(0.072/0.038) = 101.2°C$$

Then $$T_3 = T_2 - 101.2 = 58.3°C$$

It can be noted that there is now a larger temperature drop in the insulation owing to the higher thickness.

3-2-2 Cylindrical Geometries

The other important case in process engineering is that of cylindrical geometry. For example, let's consider the case of a cylindrical tube with a warm internal fluid flowing inside (Fig. 3-5). The tube is exposed to air at a lower temperature, so there is a loss of heat through the tube wall. The internal wall of the tube is at a temperature T_i, and the external wall is at a temperature T_o. Let R_i and R_o be the internal and external radii.

In steady state, the amount of heat flowing through any concentric surface must be the same. This means that

$$Q = 2\pi r q_r = \text{constant} \tag{3-2-19}$$

Then

$$r q_r = C_1 \tag{3-2-20}$$

or

$$q_r = \frac{C_1}{r} \tag{3-2-21}$$

Since

$$q_r = -k\frac{dT}{dr} \tag{3-2-22}$$

we get

$$-k\frac{dT}{dr} = \frac{C_1}{r} \tag{3-2-23}$$

Integrating this differential equation with the following boundary conditions,
For $r = R_i \rightarrow T = T_i$ and for $r = R_o \rightarrow T = T_o$, the result is

$$-k(T_i - T_o) = C_1 \ln\frac{R_i}{R_o} \tag{3-2-24}$$

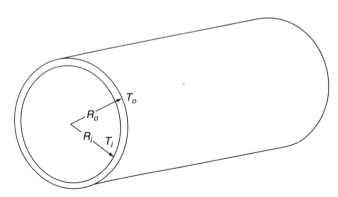

FIGURE 3-5 Heat conduction through a cylindrical wall.

Thus

$$C_1 = \frac{k(T_i - T_o)}{\ln \dfrac{R_o}{R_i}}$$

(3-2-25)

Then

$$q_r = \frac{k}{r} \frac{(T_i - T_o)}{\ln \dfrac{R_o}{R_i}}$$

(3-2-26)

where q_r is the heat-flux density through a concentric surface radius r.

The product of this heat-flux density times the flow area is the total heat flow through that surface. This is

$$Q = 2\pi L r q_r$$

(3-2-27)

And from Eq. (3-2-26), we get

$$Q = 2\pi L k \frac{T_i - T_o}{\ln \dfrac{R_o}{R_i}}$$

(3-2-28)

This equation allows calculating the heat flow through the tube wall.

Composite Cylindrical Walls.   As in the case of the composite planar wall, we can assume that we have a cylindrical wall composed of two different materials, for example, a metallic tube with an insulating layer (Fig. 3-6). Here again, we can consider that instead of the internal and external wall temperatures, we know the bulk temperatures of the interior fluid T_f and the external air T_o.

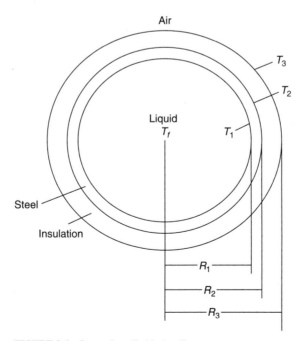

FIGURE 3-6   Composite cylindrical wall.

Let's call T_1 the internal wall temperature, T_2 the temperature of the metal-insulation separation surface, and T_3 the temperature of the external insulation face. The convection heat transfer from the internal fluid to the tube can be expressed as

$$Q = 2\pi L R_1 h_i (T_f - T_1) \tag{3-2-29}$$

where h_i is the convection coefficient. For both solid media, Eq. (3-2-28) is valid, so we can write

$$Q = 2\pi L k_1 \frac{T_1 - T_2}{\ln \dfrac{R_2}{R_1}} \tag{3-2-30}$$

$$Q = 2\pi L k_2 \frac{T_2 - T_3}{\ln \dfrac{R_3}{R_2}} \tag{3-2-31}$$

The same heat flow transfers from the external insulation face to the air by a combined convection-radiation mechanism. This can be expressed as

$$Q = 2\pi R_3 L h_o (T_3 - T_o) \tag{3-2-32}$$

In Eqs. (3-2-29) through (3-2-32), we can isolate the temperature differences and add the equations so that the intermediate temperatures are canceled and we get

$$\frac{Q}{2\pi L}\left[\frac{1}{h_i R_1} + \frac{1}{h_o R_3} + \frac{\ln(R_2/R_1)}{k_1} + \frac{\ln(R_3/R_2)}{k_2} \right] = (T_f - T_o) \tag{3-2-33}$$

If we multiply and divide by R_3, we get

$$\frac{Q}{2\pi L R_3}\frac{1}{U_o} = (T_f - T_o) \tag{3-2-34}$$

where

$$\frac{1}{U_o} = \left[\frac{1}{h_i (R_1/R_3)} + \frac{1}{h_o} + \frac{R_3 \ln(R_2/R_1)}{k_1} + \frac{R_3 \ln(R_3/R_2)}{k_2} \right] \tag{3-2-35}$$

where U_o is the overall heat transfer coefficient based on the external surface.

Example 3-2   A 3-in Schedule 40 pipe ($D_i = 77.9$ mm, $D_o = 88.9$ mm) conducting gas oil at 190°C is insulated with 25 mm of mineral wool [conductivity 0.05 W/(m · K)]. The combined radiation-convection heat transfer coefficient for the external wall is estimated as 10 W/(m² · K). Calculate the heat loss per meter of pipe and the temperatures of the external face of the insulation and of the metal wall, assuming that the film convection coefficient for the internal side is 200 W/(m² · K) and the ambient temperature is 20°C. The metal conductivity is 80 W/(m · K).

Solution   With reference to Fig. 3-6,

$R_1 = 38.95$ mm $h_o = 10$ W/(m² · K)
$R_2 = 44.45$ mm $k_1 = 80$ W/(m · K)
$R_3 = 69.45$ mm $k_2 = 0.05$ W/(m · K)
$h_i = 200$ W/(m² · K)

We shall calculate the individual resistances

$$\frac{1}{h_i(R_1/R_3)} = \frac{1}{200x(38.95/69.45)} = 0.0089 \text{ m}^2 \cdot \text{K/W}$$

$$\frac{1}{h_o} = \frac{1}{10} = 0.1 \text{ m}^2 \cdot \text{K/W}$$

$$\frac{R_3 \ln(R_2/R_1)}{k_1} = \frac{0.06945 \ln(0.04445/0.03895)}{80} = 1.14 \times 10^{-4} \text{ m}^2 \cdot \text{K/W}$$

$$\frac{R_3 \ln(R_3/R_2)}{k_2} = \frac{0.06945 \ln(0.06945/0.04445)}{0.05} = 0.61 \text{ m}^2 \cdot \text{K/W}$$

From an analysis of the results, we can see that

1. The resistance corresponding to the metal wall is clearly negligible in comparison with the other terms, so it is not necessary to include it in the calculations. This means that there will be no temperature drop through the metal, and we can consider $T_1 = T_2$.
2. The term corresponding to the internal liquid-solid resistance is also two orders of magnitude lower than the insulation resistance. It must be considered that a value of $h_i = 200$ W/(m$^2 \cdot$ K) is quite low for liquids flowing inside pipes, and it could correspond to a very viscous fluid or a fluid circulating at a very low velocity. Then, in the majority of actual cases, the relative importance of this term will be even lower. This explains the usual practice of also neglecting this term when dealing with insulated pipes and to consider that the metal temperature coincides with the inner fluid temperature.
3. The terms corresponding to the insulation resistance and to the exterior convection-radiation resistance are of the same order of magnitude, and both should be considered.

The overall heat transfer resistance then is

$$\frac{1}{U} = 0.0089 + 0.1 + 0.61 = 0.7189 \text{ m}^2 \cdot \text{K/W}$$

And the heat loss per meter of tube is

$$Q = 2\pi R_3 U(T_f - T_a) = \frac{2\pi \cdot 0.06945(190 - 20)}{0.7189} = 103 \text{ W/m}$$

The temperature drops can be obtained by multiplying the heat flow times the corresponding resistances, that is,

$$\Delta T_{i,\,i+1} = \frac{Q}{2\pi L R_3} R_{i,\,i+1} = \frac{103}{2\pi \cdot 0.06945} R_{i,\,i+1} = 236 R_{i,\,i+1}$$

$$T_f - T_1 = 236 \times 0.0089 = 2.1 \text{ K}$$

$$T_1 - T_3 = 236 \times 0.61 = 143.9 \text{ K}$$

$$T_3 - T_a = 236 \times 0.1 = 23.6 \text{ K}$$

Then $T_3 = 43.6°C$ and $T_1 = 187.9°C$.

3-3 STEADY-STATE CONDUCTION HEAT TRANSFER IN MULTIDIMENSIONAL SYSTEMS

Up to now, we have considered cases in which the direction of the vector representing the heat flux is known beforehand, and it is possible to make this direction coincide with one of the coordinate axes. In these cases, the vector equation Eq. (3-1-3) reduces to a simple differential equation that can be integrated as in the analyzed cases. However, there are situations in which this procedure is not possible. For example, let's consider a buried pipeline with a hot fluid circulating into it (Fig. 3-7). We want to calculate its heat loss.

If we write the conduction equation in cylindrical coordinates for the medium in which the pipeline is buried, we get

$$\frac{1}{r}\frac{\partial(rq_r)}{\partial r} + \frac{1}{r}\frac{\partial q_\theta}{\partial \theta} + \frac{\partial q_z}{\partial z} = 0 \tag{3-3-1}$$

Assuming that there is no heat flow in the axial direction, we get

$$\frac{1}{r}\frac{\partial(rq_r)}{\partial r} + \frac{1}{r}\frac{\partial q_\theta}{\partial \theta} = 0 \tag{3-3-2}$$

The boundary conditions for the integration of this differential equation can be

$T = T_o$ on the pipeline surface

$T = T_A$ on the surface of the ground

It can be appreciated in Fig. 3-7 that it is not possible to make the direction of heat flow coincide with the direction of any of the coordinate axes, and Eq. (3-3-2) must be integrated as a differential equation in partial derivatives. The analytical treatment is quite complex, so this type of problem normally is handled with numerical or graphic methods.

However, in bidimensional systems, where only two limit temperatures are involved, it is possible to define a conduction shape factor S as

$$Q = S(T_1 - T_2) \tag{3-3-3}$$

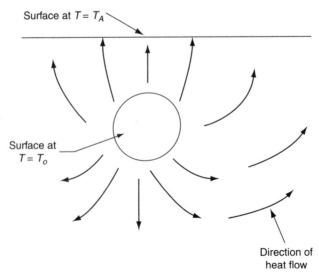

Surface at $T = T_A$

Surface at $T = T_o$

Direction of heat flow

FIGURE 3-7 Heat flow from a buried pipeline.

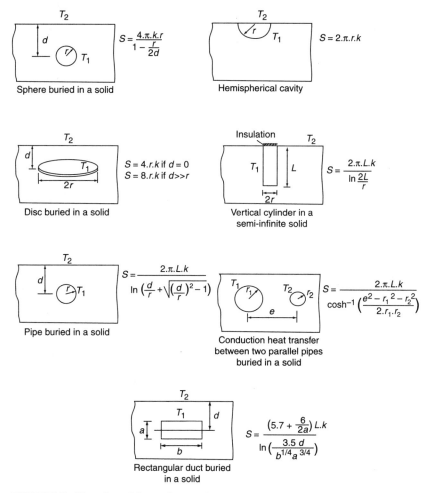

FIGURE 3-8 Shape factor S for usual geometries.

S values for the usual geometries are included in Fig. 3-8 with the corresponding values of T_1 and T_2. For other geometries, ref. 1 can be consulted.

Example 3-3 A 6-in underground pipeline ($D_{ext} = 0.168$ m) transports crude oil at 45°C. The centerline of the pipe is 1.5 m below grade. The conductivity of the soil is 0.9 W/(m·K). Assuming that the ground surface is at 0°C, calculate the heat loss per kilometer. If the mass flow is 120000 kg/h (33.33 kg/s) and its specific heat is 2500 J/(kg·K), what will be the temperature drop of the oil?

Solution The thermal conductivity of a soil depends on many factors, with the humidity being one of the most important. The following table indicates some typical values:

Typical Soil Conductivity [W/(m · K)]	
Average soil	0.9
Dry sand	0.3
Sand with 8% water	0.6
Humid clay	2.5

Let's assume that the convection heat resistance between the oil and the pipe wall is negligible, so it is possible to assume that the external temperature of the pipe is 45°C. If k_g is the soil conductivity, T_c and T_g are the pipe and ground temperatures, and d is the depth at which the pipeline is buried, the heat loss will be

$$S = \frac{2\pi L k_g}{\ln\left[\dfrac{d}{r} + \sqrt{\left(\dfrac{d}{r}\right)^2 - 1}\right]} = \frac{2\pi L \times 0.9}{\ln\left[\dfrac{1.5}{0.084} + \sqrt{17.85^2 - 1}\right]} = 1.58\,L$$

$$\frac{Q}{L} = 1.58(T_1 - T_2) = 1.58(45 - 0) = 71.1\ \text{W/m} = 71.1\ \text{kW/km}$$

If W is the oil mass flow and c is its specific heat, the temperature change of the oil per kilometre will be

$$\Delta T = \frac{Q}{Wc} = \frac{71,100}{33.33 \times 2500} = 0.85\text{°C}$$

In the case of crude oils, it is necessary to avoid too high a temperature drop because the increase in viscosity may require an uneconomical pumping power. If, for example, it is defined that the minimum pumping temperature is 30°C, it would be necessary to install a crude heater every 17 km to raise the oil temperature again to 45°C.

3-4 THERMAL INSULATION

When a piece of equipment or a pipe operates at temperatures sensibly different from ambient temperature, it is usually necessary to thermally insulate it. Thermal insulation may be necessary for the following reasons:

1. *For energy savings.* The energy lost to the ambient needs to be supplied at the expense of a higher fuel or electricity consumption. In these cases, the cost of a higher insulation thickness must be economically balanced again the higher fuel consumption.

2. *To reduce the external temperature of equipment and piping to avoid burns to the operating personnel.* This is called *personal-protection thermal insulation,* and the insulation thickness is calculated so as to have external surface temperatures lower than 55 or 60°C.

3. *For process reasons.* In some cases it is necessary to keep process fluids at certain minimum temperatures to avoid undesirable reactions, solidification, precipitation, polymerization, etc. In this case, insulation thicknesses result from process calculations.

4. *In low temperature processes to avoid condensation of ambient humidity over the surface of piping and equipment.* This can lead to ice formation, with water weeping, turning the soil muddy or slippery.

3-4-1 Characteristics of a Good Thermal Insulation System

A good thermal insulation system must meet the following conditions:

1. The materials employed must be compatible with the process and must be inert when the process fluids come in contact with them.

2. The insulation system must combine a low thermal conductivity with adequate mechanical resistance. Frequently, ladders are leaned against insulated pipes, or plant operators step on insulated surfaces, which ideally must not suffer dents or damage in these situations.

3. It must have a barrier to avoid humidity penetration. The presence of water within the insulation increases the thermal conductivity dramatically.

4. In plants where combustible materials are processed, it is necessary to evaluate the behavior of the insulation in case of fire. Insulation material must present low combustibility and smoke production. Plastic foams such as polyurethane usually present this problem. Thermal insulation manufacturers are now offering special foams with low combustibility indexes.

3-4-2 Thermal Insulation Materials

Materials used in thermal insulation applications generally are grouped according to the temperature range in which they are employed. The temperatures that limit these ranges are somewhat arbitrary. When the temperature reaches an upper limit, materials may be damaged or become uneconomical because their thermal conductivity increases. A lower limit usually means that the material is not competitive because there are cheaper materials that can perform satisfactorily. Within each temperature range, the selection is made taking into consideration other properties and cost.

Cryogenic Range (–260 to –100°C).   Within this range, there are two types of insulation: vacuum and massive, the latter consisting of one or more solid phases distributed with a gas, such as dry air, to produce a very low thermal conductivity. Vacuum insulation systems, consisting typically of highly polished metal supporting walls with a vacuum space between them (sometimes with multiple metal reflective foils or specified powders inside), are usually custom-designed and installed by the insulation vendors.

The theory behind vacuum cryogenic insulation systems is that of the Dewar flask, similar to those used for cold or hot beverage conservation. It relies in part on vacuum between the walls and in part on reflection of radiant heat. Without gas inside, heat transfer will be mainly by radiation. Coating the inside hot surface facing the evacuated area reduces heat transfer to a level proportional to the emissivity of the coating (e.g., 0.01 for silver).

To reduce the heat transfer between the two walls even more, radiation shields are installed. This technique is based on the fact that radiation heat transfer often can be cut in half by installing a radiation shield between the hot and cold surfaces. Powders such as expanded perlite, diatomaceous earth, or fibers may be used for this purpose. The multilayer type, a series of reflective foil shields of aluminum separated by low-conductivity fillers such as fiberglass, comes in blanket form.

Less costly than the evacuated forms just described are foam types. Foamed polyurethane and polystyrene, either foamed into flexible sheets, foamed in situ, or foamed in rigid insulation sections may be used within this range, but they are not completely resistant to water vapor permeation. Foamed glass exhibits a somewhat better performance.

Low-Temperature Range (–100 to 0°C).   Some of the evacuated types of insulation are used in the lower end of this range, but foams are employed more. In low temperatures, the main problem is moisture permeation. Table 3-1 shows water permeability values for four generic insulation materials. It can be seen that the permeability of fibrous materials is much higher than the permeability of foams. The microstructure of a foam consists of closed cells, and moisture cannot migrate through the insulation, as in the case of fibers.

TABLE 3-1 Water Permeability Values for Four Generic Insulation Materials

Material	Permeability (perm-in)
Foamed glass	0.00
Urethane foam	0.3–0.6
Polyestirene foam	1–4
Fiberglass	100–200

Penetration by water or water vapor can increase the conductivity of insulation dramatically, so it is always necessary to protect the external side of the insulation with a moisture barrier. Water vapor permeation is not a serious problem until the moisture condenses. In high-temperature insulation, the problem is not serious because the vapor pressure of water is higher at the temperature of the internal face of the insulation than at the temperature of the external face. This tends to expel the humidity. In such cases, only protection against rain or water splashing is necessary.

When the temperature of the internal face is lower than the dew-point temperature of the ambient air, however, as in cold-temperature insulation, water vapor may penetrate and condense into the insulation owing to a driving force that is proportional to the temperatures difference between both faces of the insulation. Even when a moisture barrier is installed, it is possible to have water vapor penetration through unsealed joints or sites where the barrier may be damaged. Localized freezing of this water vapor can cause a minor damage. During plant shutdowns, however, the ice melts, and water may travel through the insulation, thus spreading the problem. To avoid water penetration, moisture barriers are used. There are three different types:

- *Structural barriers.* These are rigid sheets of reinforced plastic, aluminum, or stainless steel or galvanized iron plate, flat or corrugated.
- *Membrane barriers.* These are metal foils, laminated foils, treated papers, plastic films, and coated felt or paper. They are either part of the insulation or may be supplied separately.
- *Coating barriers.* These are in fluid form, usually as a paint. The material can be asphaltic, resinous, or polymeric. These provide a seamless coating but require time to dry.

If, for any incident, water gets into the insulation anyway, the material must be heated and dried. This can be problematic in the case of plastic foams because they usually do not resist high temperatures well.

Intermediate Range (0 to 520°C).   This is the usual range in the process industry. The commonly used materials include the following:

Calcium Silicate.   This is a mixture of lime and silica, usually reinforced with organic or inorganic fibers and molded in the desired shape, either as blocks or panels, half-round segments for pipes, or special forms for the insulation of valves and fittings. The reason why it is usually preferred over other materials is its compressive strength. It can be used at up to 900°C.

Mineral Wool.   This is obtained by melting rocks and slag in furnaces at very high temperatures. The fibers usually are bonded together with a heat-resistant binder. The maximum temperature of use is almost as high as that for calcium silicate, but its compressive strength is much lower. It is available in both rigid molded form for piping or vessels and as flexible blankets for irregular surfaces.

Fiberglass (Up to 400°C).   Molten glass can be extruded in fibers to manufacture blankets or mattresses without binders, or they can be agglomerated, forming rigid or semirigid molded elements. Its maximum use temperature is lower than that of mineral wool. Additionally, glass fibers are irritating to the skin, which makes installation work uncomfortable.

Cellular Glass or Cellular Foam (Up to 400°C).   The same material used in low-temperatures applications also can be used in this range.

High-Temperature Range (More than 550°C).   Usually in this temperatures range, not only high temperatures but also other process conditions may be severe as well. Insulation may be required to resist abrasion and erosion by molten materials, direct flame impingement, corrosive atmospheres, and severe thermal shocks. The application of these types of insulation usually requires special installation procedures and specialized personnel. The materials usually employed are

Miner fiber or calcium silicate up to 900°C

Ceramic fibers (Al_2O_3-SiO_2) up to 1400°C

Molded ceramic refractories up to 1650°C

Metallic oxides fibers such as Al_2O_3 or ZrO_2 up to 1650°C

Carbon fibers up to 2000°C

3-4-3 Factors Affecting Thermal Conductivity

Insulation materials usually present a fibrous, cellular, or porous structure in which a solid matrix enclose spaces where air is present. Within this structure, heat transmission is due to a combination of the three basic mechanisms: conduction, convection, and radiation.

Conduction takes place through the solid matrix that forms the insulation, for example, the solid fibers in a mineral wool blanket. It is clear, then, that to reduce conduction heat transfer, it is necessary that the base material be of low conductivity and that the number of fibers be as low as possible. This means a low-density material.

In the free space, air is enclosed. The temperature differences between fibers create interior natural convection streams that originate the second heat transfer mechanism. The size, shape, and orientation of these hollow spaces affect the convection heat transmission through the insulation. It is convenient to create small cells to limit air movement. This means that the higher the material density, the lower will be the convection contribution to heat transfer (the opposite of conduction). As explained previously, another way to reduce convection is vacuum insulation, but this is not practicable in most cases.

The third mechanism is radiation heat transfer between fibers through the insulation thickness. This mechanism depends on the fourth power of the absolute temperatures, so it is very important in high-temperature insulation. The radiation heat transfer is reduced with radiation shields. This means the installation in the radiation path of solid elements that absorb and re-emit energy, thus reducing its thermal level.

Thus, to create a good insulation material, it is necessary to reduce the radiation transmission with an adequate number of absorbers, to create small cells to minimize air convective movement, and to install as little mass as possible to reduce conduction but at the same time maintain good mechanical properties. All the above-mentioned mechanisms work together in the insulation, forming a complex system. To simplify examination of the heat transfer mechanisms and their combined effects, the familiar term *thermal conductivity,* or simply *k,* is used.

As we see, there are too many factors for *k* to be a description of an intrinsic property of a homogeneous material, as it purports to be. However, it is the only practical way to define the characteristics of an insulation, so it is used widely.

The relative importance of the different heat transfer mechanisms changes when the insulation temperature changes. At high temperatures, radiation between fibers is the dominant mechanism, so it is advantageous to use insulation materials with rather high density. On the other hand, at low temperatures, when the importance of radiation decreases, a high density favors conduction and is harmful to the resulting conductivity.

This means that a specific material may perform better than another at a certain temperature, but when the temperature changes, the situation may invert. Comparisons must be done with care, and ideally, thermal conductivity values have to be obtained at the temperature at which the material will be used.

Table 3-2 shows typical ranges in which the thermal conductivity of insulation materials may vary. As can be appreciated, the ranges are ample, and it is always necessary to consult the vendor for more specific information.

TABLE 3-2 Temperatures, Conductivities, and Typical Densities of Insulation Materials

Material	Temperature (°C)	k [W/(m · K)]	Density (kg/m^3)
Urethane foam	−150 to 100	0.015–0.020	25–45
Fiberglass blankets	−150 to 200	0.024–0.085	9–45
Fiberglass panels	0 to 450	0.033–0.08	35–90
Calcium silicate	25 to 650	0.031–0.12	90–150
Molded mineral fiber	up to 1000	0.051–0.13	9–45
Cellular glass	−200 to 250	0.028–0.10	100–140
Mineral fiber blankets	up to 750	0.04–0.05	110–170

3-4-4 Installation

Piping. Piping is usually insulated with flexible blankets of mineral wool or fiberglass or with rigid half-round premolded segments. Blankets are installed by wrapping the pipe and tying them with galvanized wire. Sometimes the blankets are manufactured with hexagonal wire netting in one of their faces

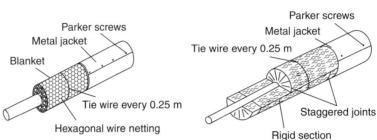

Insulation with blankets

Parker screws
Metal jacket
Blanket
Tie wire every 0.25 m
Hexagonal wire netting

Insulation with rigid sections

Parker screws
Metal jacket
Tie wire every 0.25 m
Staggered joints
Rigid section

FIGURE 3-9 Pipe insulation systems.

to facilitate installation. Sometimes they are supplied with vapor barrier incorporated, which may consist on asphaltic felt or paper in one of their faces. Once the insulation is secured, a vapor barrier (e.g., felt embedded in asphaltic paint) can be applied, and then it may be covered with an aluminum, stainless steel, or galvanized iron jacket. The jacket is installed by wrapping the insulated pipe with metal sheets with a certain overlap and fixing the sheets with Parker screws or rivets.

Valves and flanges usually are covered by insulation-filled metallic boxes with fast clamps that can be removed and reinstalled easily.

Premolded sections are rigid half segments that are supplied in different sizes to match the pipe diameters. The end joints of both half segments are staggered by beginning a length of pipe with half- and full-length sections of insulation. The segments then are fastened with wire or metal strips and covered with the metallic jacket (Fig. 3-9). Premolded sections are manufactured in a variety of materials, such as calcium silicate, different fibers with organic or inorganic binders, rigid polyurethane foam, etc.

When long vertical pipe lengths are insulated (e.g., piping connected to distillation columns), support rings are welded to the pipe to avoid insulation displacement owing to excessive weight.

Tanks and Vessels. Tanks and vessels may be insulated with either flexible blankets or rigid panels. Usually the vessel manufacturer installs pins, clips, or studs where the material is fixed. The vessel or tank is insulated in successive rings that, when completed, are fastened with wire or metal bands. Insulating cement may be applied to seal the joints. Metal jackets or another type of finishing is applied to protect the insulation from rain or moisture. Vertical vessels or tanks must be provided with support rings approximately every 2 m to avoid excessive weight. To apply insulation on hemispherical or semielliptical vessel heads, usually support rings are welded to the vessel. These rings allow building a wire net to hold the insulation in position. Joints are sealed with insulation cement, and the metal cover is site-manufactured in radial segments. Metal jacket installation is a craftsman work that must be performed with care because it greatly influences the aesthetic of the plant.

3-4-5 Economical Thickness

When the reason for the use of thermal insulation is to reduce energy expenses, optimal thickness results from an economic calculation that considers insulation and energy costs. This calculation is greatly influenced by many parameters such as ambient temperatures, process conditions, maintenance and installation cost, amortization period, etc. Companies may perform these calculations with computer programs and, with their results, construct tables of recommended thicknesses as a function of the process temperatures and pipe diameters for a given set of the remaining parameters. These tables then are adopted as standards. For example, Table 3-3 shows recommended insulation thicknesses for calcium silicate insulation in a high-temperature application for mild weathers adopted by an oil company.

When the reason for thermal insulation is personal protection, calculations are performed to have a temperature in the outer insulation face not higher than 55°C. In this case, different thicknesses are obtained, as shown in Table 3-4.

TABLE 3-3 Economical Thickness in Millimeters for Calcium Silicate Insulation (Hot Surfaces)

Diameter (in)	Temperature (°C)														
	50	75	100	125	150	175	200	250	300	350	400	450	500	550	600
½	25	25	25	25	38	38	38	38	50	75	75	75	89	89	89
¾	25	25	25	38	38	38	38	38	50	75	75	75	89	89	89
1	25	25	25	38	38	38	38	50	75	75	75	89	89	89	100
1½	25	25	38	38	38	38	50	50	75	75	75	89	89	100	100
2	25	25	38	38	38	50	50	75	75	89	89	89	100	100	100
3	25	38	38	38	50	50	50	75	89	89	89	100	100	140	140
4	25	38	38	38	50	50	50	75	89	89	100	100	100	140	140
6	25	38	38	38	50	50	75	75	89	89	100	100	100	140	140
8	25	38	38	38	50	50	75	75	89	89	100	100	140	140	140
10	25	38	38	38	50	50	75	75	89	89	100	100	140	140	140
12	25	38	38	38	50	50	75	75	89	89	100	100	140	140	140
14	25	38	38	38	50	50	75	75	89	89	100	100	140	140	140
16	25	38	38	38	50	50	75	75	89	89	100	100	140	140	140
18	25	38	38	38	50	50	75	75	89	89	100	100	140	140	140
20	25	38	38	38	50	50	75	75	89	89	100	100	140	140	140
Plane	25	38	50	50	50	63	89	100	100	114	114	140	153	165	165

TABLE 3-4 Calcium Silicate Insulation Thickness for Personal Protection

Diameter (in)	Temperature (°C)														
	50	75	100	125	150	175	200	250	300	350	400	450	500	550	600
½	0	25	25	25	25	25	25	25	25	25	38	38	38	50	50
¾	0	25	25	25	25	25	25	25	25	25	38	38	50	50	63
1	0	25	25	25	25	25	25	25	25	38	38	38	50	50	63
1½	0	25	25	25	25	25	25	25	25	38	38	50	50	63	75
2	0	25	25	25	25	25	25	25	38	38	50	50	63	63	75
3	0	25	25	25	25	25	25	25	38	38	50	63	75	75	89
4	0	25	25	25	25	25	25	38	38	50	50	63	75	89	100
6	0	25	25	25	25	25	25	38	50	50	63	75	89	100	125
8	0	25	25	25	25	25	25	38	50	63	75	75	89	100	125
10	0	25	25	25	25	25	25	38	50	63	75	89	100	114	125
12	0	25	25	25	25	25	38	38	50	63	75	89	114	125	140
14	0	25	25	25	25	25	38	50	63	75	89	100	114	125	140
16	0	25	25	25	25	25	38	50	63	75	89	100	114	140	150
18	0	25	25	25	25	25	38	50	63	75	89	100	125	140	150
20	0	25	25	25	25	25	38	50	63	75	89	114	125	140	165
Plane	0	25	25	38	38	50	63	75	100	125	150	189	216	250	300

GLOSSARY

A = area (m^2)

c = specific heat [J/(kg·K)]

h = convection film coefficient [W/(m^2·K)]

k = thermal conductivity [W/(m·K)]

L = length (m)

m = dimensionless parameter = k/hr_m

q = heat flux (W/m^2)

Q = heat flow (W)

r = radial coordinate or radius (m)

r_m = thickness or radius (m)

R = radius (m)

S = parameter in Eq. (3-2-40) (W/K)

T = temperature (K)

u = internal energy (J/kg)

U = overall heat transfer coefficient [W/(m^2·K)]

W = power per unit area (W/m^2)

W = mass flow (kg/s)

x, y, z = cartesian coordinates

REFERENCES

1. Rohsenow W. M., Hartnett J. P.: *Handbook of Heat Transfer.* New York: McGraw-Hill, 1973.
2. Kreith F.: *Principles of Heat Transfer.* New York: International Textbook Company, 1973.
3. Wong H. Y.: *Handbook of Essential Formulae and Data on Heat Transfer.* London: Longman Group, 1977.
4. McAdams W.: *Heat Transmision.* New York: McGraw-Hill, 1954.

CHAPTER 4
CONVECTION

Convection is a heat transfer mechanism that takes place at a solid-fluid interface. The fluid acts as a vehicle for energy transport, and the heat-flux density depends on the velocity with which the fluid is renewed over the surface.

Two different flow patterns can exist: laminar and turbulent. In *laminar* flow, the fluid streamlines are stable, and then, once a steady state is reached, the velocity at every point is constant. In *turbulent* flow, even though we also speak about *steady state,* meaning that an average constant velocity exists at every location, there are random fluctuations around this average value. These random fluid movements are called *eddies,* and they cannot be predicted and provoke an internal agitation that greatly influences heat transfer.

In laminar-flow systems, it is possible to write differential equations for the momentum and energy balances that can be solved to find the velocity and temperature profiles in the fluid. By knowing the temperature distribution, it is possible to calculate the heat-flux density at the interface. In turbulent-flow systems, since the temperature and velocity distributions are not continuous functions, this approach is not possible, and it is necessary to use experimental correlations. This will be explained in more detail later in this chapter.

4-1 FORCED CONVECTION OVER A FLAT PLATE

4-1-1 Boundary Layer

Let's consider the flat plate shown in Fig. 4-1 over which a fluid circulates. The fluid approaches the plate with a velocity v_∞. The plate exerts a perturbation on the flow. This perturbation takes place in a region that extends up to a certain distance from the plate, so beyond this distance the fluid velocity is also v_∞, which is called the *free-stream velocity.*

The region of the fluid where the solid surface exerts its influence on the flow is called the *boundary layer.* Since the velocity increases asymptotically to v_∞, it is usual to define the boundary layer as the fluid region where the velocity is below 99 percent of the free-stream velocity. We shall analyze how the solid perturbs the flow and how is the velocity field within the boundary layer.

4-1-2 Shear Stress in a Moving Fluid

Refering again to Fig 4-1, when the fluid comes in contact with the plate, shear stresses develop in the fluid that tend to slow down the fluid layers closer to the solid surface. We can think on this with the simple logic of the action and reaction principle. The fluid exerts a drag force over the plate, so the plate must exert an opposite force over the fluid. If the plate is fixed and cannot move, the effect is that the fluid begins to slow down. The effect appears initially in the layers of fluid closer to the solid. The layer immediately adjacent to the solid stops completely, and this layer, in turn, begins to slow down the adjacent layers, and a velocity profile, as shown in Fig. 4-2, develops in the fluid.

This effect is transmitted from one layer to another by shear stresses. Let's consider a control differential volume belonging to the fluid located at a certain distance y_1 from the solid (Fig. 4-3). We shall call this *differential volume I.*

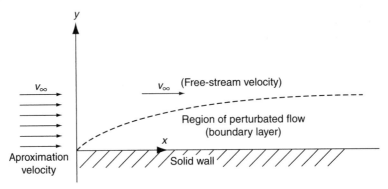

FIGURE 4-1 Fluid flowing over a flat plate.

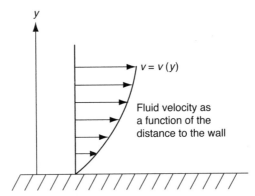

FIGURE 4-2 Velocity profile in a fluid moving over a plate.

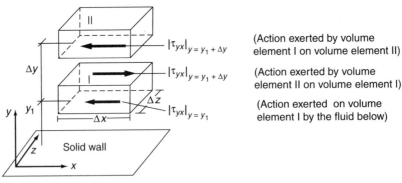

FIGURE 4-3 Shear forces in the boundary layer.

As mentioned previously, the fluid that is below this volume exerts a slowdown effect over the volume. This effect is explained by the action of a force that appears on the lower horizontal face of the control volume I.

We shall call τ (shear stress) the force exerted per unit area. Shear stresses are normally designated by two subscripts. The first one indicates the plane in which the force acts, designated by its normal. In this case, the shear stress appears in a plane whose normal is y. The second subscript indicates the direction of the force. In this case, the direction is that of the x axis. This shear stress thus will be designated as τ_{yx}.

If y_1 is the vertical coordinate of the lower face of the control volume, the shear stress acting on this face will be $|\tau_{yx}|_{y\,=\,y1}$, and the force will be $|\tau_{yx}|_{y\,=\,y1}\Delta x\Delta z$. This differential volume, in turn, exerts a similar slowdown effect on the fluid that is above it

If volume II is a differential volume immediately adjacent to volume I (even though it was represented separately for drawing clarity), we see that a shear stress appears on the lower face of volume II whose value is $|\tau_{yx}|_{y\,=\,y1+\Delta y}$. It is then obvious that differential volume II exerts over differential volume I a force of the same magnitude and opposite direction as shown in the figure.

Any differential volume such as volume I, then, is subject to the action of two forces of opposite directions. For an infinite fluid moving over a flat plate, these forces decrease in intensity as one moves away from the plate. This means that the force acting on the lower face of volume I is higher than that acting on the upper face, and the result is a net force opposing to the fluid movement

The shear stress at every fluid location is related to the velocity gradient by

$$\tau_{yx} = \mu \frac{\partial v_x}{\partial y} \qquad (4\text{-}1\text{-}1)$$

This means that as the shear stress decreases, the slope of the velocity profile also decreases, and the fluid velocity remains constant at a certain distance from the plate. This distance is the *boundary-layer thickness*. This means that beyond the boundary layer, there are no more shear stresses on the fluid. This region is called the *undisturbed free stream* or *potential flow regime*.

4-1-3 Boundary-Layer Development

Let's refer to Fig. 4-4. The x coordinate indicates the horizontal distance to the plate leading edge, and the y coordinate is the vertical distance to the plate. When the fluid comes in contact with the plate, the fluid layers closer to the plate slow down, and the boundary layer begins to develop. As long as the fluid progresses along the x axis, the perturbation extends to the fluid layers farther from the solid, and the thickness of the boundary layer increases with x. The figure shows the development of the boundary layer and the velocity profiles at different values of the x coordinate.

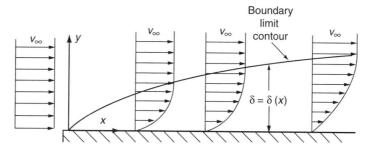

FIGURE 4-4   Velocities distribution in the boundary layer.

4-1-4 Laminar and Turbulent Boundary Layers

The boundary layer can be laminar or turbulent. However, even in a turbulent regime, there is always a region near the leading edge of the plate where the boundary layer is completely laminar (Fig. 4-5). This is so because in this region the velocity gradients are high, so the shear stresses are also high. These shear stresses exert a laminating action over the fluid movement. However, as the fluid progresses over the plate, the velocity gradients decrease, and eddy streams can penetrate into the boundary layer, which becomes turbulent.

We shall call x_c (critical distance) the value of x when the boundary layer becomes turbulent. Even in a turbulent boundary layer, the region near the solid wall, where fluid velocities are very low, is in laminar

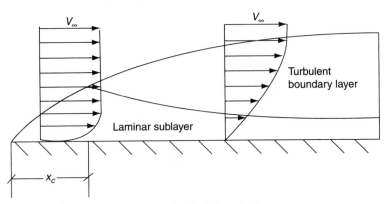

FIGURE 4-5   Laminar and turbulent regions in the boundary layer.

flow. This region is called the *laminar sublayer.* The value of x_c depends on the approximation velocity and on the roughness of the surface. It is normally accepted that for regular-roughness surfaces,

$$x_c = \frac{5 \times 10^5 \mu}{\rho v_\infty}$$

(4-1-2)

or

$$\frac{x_c \rho v_\infty}{\mu} = 5 \times 10^5$$

(4-1-3)

The force that the fluid exerts over the plate or the plate exerts over the fluid per unit area is

$$\tau_0 = \tau_{yx}\big|_{y=0} = \mu \frac{\partial v_x}{\partial y}\bigg|_{y=0}$$

(4-1-4)

For a laminar boundary layer, it is possible to develop theoretical expressions for this derivative, and it is then possible to calculate the force analytically. Theoretical solutions to this case were developed by Blasius,[1] and his results coincide remarkably well with experimental data[14] (Fig. 4-6).

However, in a turbulent regime, the momentum and energy-balance equations cannot be integrated, and the derivative cannot be calculated. The usual approach then is to define a coefficient called the *drag coefficient* or *friction factor* C_f as

$$\tau_0 = C_f \frac{1}{2} \rho v_\infty^2$$

(4-1-5)

And the friction factor is obtained empirically. Usually, experimental data are correlated as

$$C_f = C_f(\rho, v, \mu, L)$$

and a mean friction factor for the entire plate is defined. The correlations normally are expressed as

$$C_f - C_f\left(\frac{\rho v_\infty L}{\mu}\right)$$

(4-1-6)

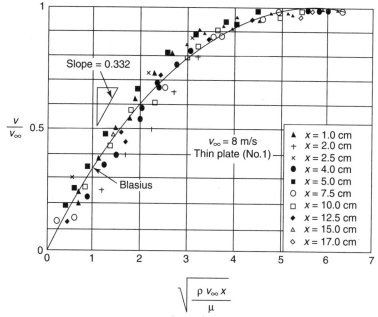

FIGURE 4-6 Velocity profile in a laminar boundary layer according to Blasius with experimental data of Hansen.[14] (*National Advisory Committee for Aeronautics NACA TM 585.*)

where $\rho v_\infty L/\mu$ is the Reynolds number for the entire plate (Re_L).

For example, for Reynolds numbers between 5×10^5 and 1×10^7 and excluding the region of boundary-layer development, it was found[2] that

$$C_f = 0.072 \mathrm{Re}^{-0.2} \qquad (4\text{-}1\text{-}7)$$

4-1-5 Heat Transfer from a Flat Plate to a Fluid in Motion

Let's consider now that the plate is at a temperature T_w, whereas the fluid approximation temperature is T_∞. When the fluid comes in contact with the plate, its temperature begins to increase owing to the heat flux. For low x values, the perturbation extends only to the region of the fluid close to the interface, but as long as the fluid moves farther, the perturbation progresses, and a temperature profile develops. For any x at enough distance from the plate, the temperature is again T_∞.

The region where the temperature changes from T_w to T_∞ is called the *thermal boundary layer*. Figure 4-7 shows the evolution of the thermal boundary layer and the temperature profiles in the system.

Since the velocity of the fluid that is in contact with the plate is nil, in that region the fluid behaves like a solid, and there will be a conduction heat flow from the plate to the fluid with a heat-flux density given by

$$q = -k \left(\frac{\partial T}{\partial y} \right)_{y=0} \qquad (4\text{-}1\text{-}8)$$

Let's consider a differential volume of fluid within the boundary layer (Fig. 4-8). There is a fluid flow entering the volume on the left side and leaving on the right side. Since the flowing fluid receives heat by conduction from the plate, it suffers a temperature increase, which means that the fluid is removing heat from the control volume.

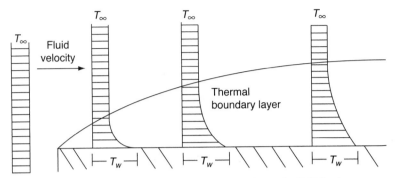

FIGURE 4-7 Temperatures distribution in the boundary layer of a fluid flowing over a flat plate.

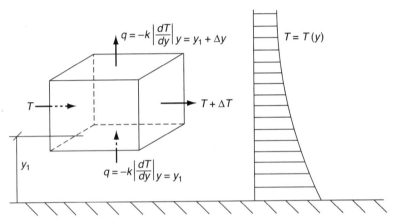

FIGURE 4-8 Heat balance on a volume element in the boundary layer.

Since part of the heat entering the volume by conduction is removed by the fluid, the heat-flux density on the upper side is smaller than that on the lower side. Since the conduction heat-flux density is related to the thermal gradient, it is clear that $\partial T/\partial y$ decreases as we move away from the plate.

The shape of the temperature profile depends on the hydrodynamic conditions and on the properties of the fluid—density, specific heat, and viscosity. In laminar-flow systems, it is possible to develop theoretical models that allow one to calculate the partial derivative of Eq. (4-1-8) and solve the problem analytically. Pohlhausen[10] used the velocity profiles calculated previously by Blasius to obtain the solution of the heat transfer problem and obtain the temperature profiles in the system. At low Reynolds numbers, the analytical solution coincides with experimental data (Fig. 4-9). However, in the majority of systems in process engineering, this approach is not possible, and an empirical coefficient called the *convection film coefficient* is defined. This coefficient plays the same role as the friction factor in fluid mechanics.

4-1-6 The Convection Film Coefficient

Since the analytical treatment has the limitations explained earlier, the usual approach to the problem consists of using empirical correlations. These correlations generally are expressed in terms of a coefficient called the *convection film coefficient,* defined as the quotient between the heat flux per unit area and the temperature difference between the wall and the bulk of the fluid.

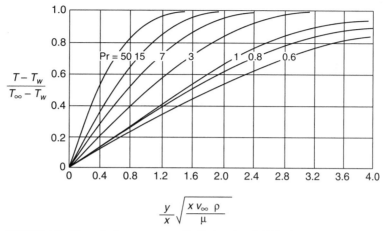

FIGURE 4-9 Dimensionless temperature distributions in a fluid flowing over a heated plate in laminar flow.

This coefficient may be defined either for the entire surface or for an area element. In the first case, a mean coefficient h_L, valid for the entire length of the plate, is defined as

$$Q/A = h_L(T_w - T_\infty) \tag{4-1-9}$$

where Q is the total heat delivered by the plate to the fluid and A is the total area of the plate.

In the second case, a local coefficient h_x is defined. This coefficient is a function of the position x, and the definition equation is

$$\frac{dQ}{dA} = h_x(T_w - T_\infty) \tag{4-1-10}$$

where dQ is the heat delivered to the fluid per unit time by a differential area dA located at a distance x from the leading edge of the plate. The total heat delivered to the fluid in the entire length of the plate can be calculated as

$$Q = \int_A h_x(T_w - T_\infty)dA = \int_0^L h_x(T_w - T_\infty)b\,dx \tag{4-1-11}$$

where b is the plate width. Then, comparing with Eq. (4-1-9), we get

$$h_L = \frac{\int_0^L h_x\,dx}{L} \tag{4-1-12}$$

It is quite evident that the film coefficients depend on the velocity distribution of the fluid, on its thermal conductivity, and on the properties that affect its capacity to transport energy (i.e., specific heat and density). As a general expression, it can be written

$$h_L = h_L(v_\infty, L, \rho, \mu, c, k) \tag{4-1-13}$$

or

$$h_x = h_x(x,\rho,c_p,\mu,k,v_\infty)$$
(4-1-14)

and these relations are found experimentally.

4-1-7   Effect of Fluid Velocity on the Film Coefficient *h*

It was explained that within the boundary layer there is always a sublayer close to the solid surface that is free of turbulences. The turbulence existing in the remaining portion of the boundary layer exerts a mixing effect and homogenizes the fluid properties. Then, in the turbulent boundary layer region, the velocity and temperature gradients are much less steep than in the laminar sublayer. This is shown in Fig. 4-10.

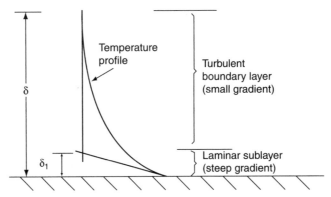

FIGURE 4-10   Temperature profile and laminar sublayer thickness.

As an approximation, we can assume that the temperature profile can be replaced by the two tangents drawn at $y = 0$ and $y = \delta$, as shown in Fig. 4-10. The intersection of both tangents defines a distance δ_1 that approximately coincides with the laminar sublayer thickness.

We have explained that the heat-flux density always can be calculated as

$$q = \frac{dQ}{dA} = -k\left.\frac{dT}{dy}\right|_{y=0}$$
(4-1-15)

And then

$$q = k\frac{T_w - T_\infty}{\delta_1}$$
(4-1-16)

But remembering the definition of the convection film coefficient, that is,

$$q = \frac{dQ}{dA} = h(T_w - T_\infty)$$
(4-1-17)

and comparing Eq. (4-1-16) with Eq. (4-1-17), we get

$$h = \frac{k}{\delta_1}$$
(4-1-18)

When the fluid velocity is high, the turbulent eddies penetrate deeper within the boundary layer, and the laminar sublayer thickness is reduced, thus increasing h. This means that as a general rule, the higher the fluid velocity, the higher the film coefficient will be.

4-1-8   Correlations for Heat Flow from a Planar Surface to a Moving Fluid

With the help of dimensional analysis, it is possible to demonstrate that the functions represented by Eqs. (4-1-13) and (4-1-14) will be of the type

$$\frac{h_x x}{k} = f\left(\frac{\rho x v_\infty}{\mu}, \frac{c_p \mu}{k}\right) \tag{4-1-19}$$

or for the mean coefficient

$$\frac{h_L L}{k} = f\left(\frac{\rho L v_\infty}{\mu}, \frac{c_p \mu}{k}\right) \tag{4-1-20}$$

The left side of Eqs. (4-1-19) and (4-1-20) are called the local and mean Nusselt numbers, respectively (Nu_x and Nu_L).

Note   Dimensional analysis is a technique that allows combining the physical magnitudes that express the solution of a problem in a smaller number of dimensional groups, thus reducing the number of variables to correlate. This subject can be examined more closely in ref. 2.

The correlations usually accepted for the case of a fluid circulating over a flat plate are the following:

1. For laminar flow ($\text{Re}_x < 5.10^5$),

$$\text{Nu}_x = \frac{x h_x}{k} = 0.33 \text{Re}_x^{1/2} \text{Pr}^{1/3} \tag{4-1-21}$$

or

$$h_x = 0.33 \frac{k}{x} \text{Re}_x^{1/2} \text{Pr}^{1/3} \tag{4-1-22}$$

According to Eq. (4-1-12), the mean Nusselt number for the entire plate, if the laminar regime exists over the complete length of the plate, can be obtained by integrating the previous expression from $x = 0$ to $x = L$ and dividing by L, that is,

$$h_L = \frac{\int_0^L h_x \, dx}{L} = \frac{0.33}{L} \int_0^L \frac{k}{x^{1/2}} \left(\frac{\rho v_\infty}{\mu}\right)^{1/2} \left(\frac{c\mu}{k}\right)^{1/3} dx$$

$$= 0.66 \frac{k}{L} \left(\frac{L v_\infty \rho}{\mu}\right)^{1/2} \left(\frac{c_p \mu}{k}\right)^{1/3} \tag{4-1-23}$$

$$\text{Nu}_L = 0.66 \text{Re}_L^{1/2} \text{Pr}^{1/3} \tag{4-1-24}$$

2. For turbulent flow $(\mathrm{Re}_x > 5.10^5)$,[2]

$$\mathrm{Nu}_x = \frac{xh_x}{k} = 0.0288\,\mathrm{Pr}^{1/3}\,\mathrm{Re}_x^{0.8} \tag{4-1-25}$$

and the mean coefficient for a plate of length L can be calculated with Eq. (4-1-12), and the following expression for Nu_L is obtained:

$$\mathrm{Nu}_L = \frac{hL}{k} = 0.036\,\mathrm{Pr}^{1/3}\,\mathrm{Re}_L^{0.8} \tag{4-1-26}$$

It is necessary to note that to obtain the preceding expression, we have neglected the existence of the laminar zone between $x = 0$ and $x = x_c$. The expression then is only valid for $L \gg x_c$. The laminar region can be included in the analysis if Eq. (4-1-21) is used between $x = 0$ and $x = x_c$ and Eq. (4-1-25) between $x = x_c$ and $x = L$. Considering $\mathrm{Re}_C = 5.10^5$, this leads to the following expression:[2]

$$\mathrm{Nu}_L = 0.036\,\mathrm{Pr}^{1/3}(\mathrm{Re}_L^{0.8} - 23{,}200) \tag{4-1-27}$$

4-2 FORCED CONVECTION INSIDE TUBES

4-2-1 Flow of Fluids Inside Tubes—Velocity Distribution

Let's consider a fluid entering into a tube with velocity v_o. At the section immediately after the entrance, almost all the fluid maintains the inlet velocity except in the region very close to the tube wall, where the fluid is slowed down owing to the shear-stress effect.

A boundary layer then starts developing. As long as the fluid progresses into the tube, the boundary layer extends toward the center of the tube, and successive fluid layers begin to slow down.

Since the total mass flow is constant from the inlet to the outlet, if the velocity of the fluid close to the tube wall decreases, the velocity near the center of the tube must increase. Then a velocity profile as shown in Fig. 4-11 begins to develop.

At a certain distance from the tube inlet, the boundary layer has reached the center of the tube, and from then on, the velocity profile remains constant. We then speak of a fully developed velocity profile. The distance extending from the inlet section to the section where the velocity profile is fully developed is called the *entrance region* L_e. The boundary layer can be in either laminar or turbulent flow. Whatever the

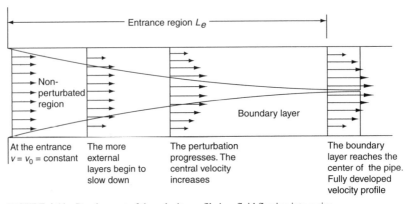

FIGURE 4-11 Development of the velocity profile in a fluid flowing into a pipe.

velocity with which the fluid enters the tube, there is always a region near the entrance where the velocity gradients are very high and the boundary layer is laminar. The central core can be in turbulent flow if the inlet velocity is high (Fig. 4-12a). As long as the influence of the wall extends to the bulk of the fluid, the shear stresses decrease. It is then possible that eddies penetrate into the boundary layer, which becomes turbulent (Fig. 4-12b). Finally, at a certain point downstream, the boundary layer reaches the center of the tube (Fig. 4-12c). There is always a region close to the tube wall where the boundary layer is in laminar flow. This is the laminar sublayer.

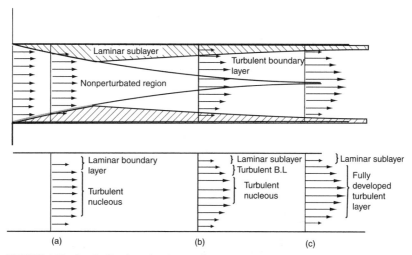

FIGURE 4-12   In-tube flow boundary layer regions.

It was found that the necessary condition for the boundary layer to remain in laminar flow is

$$\mathrm{Re} = \frac{Dv\rho}{\mu} < 2100 \tag{4-2-1}$$

where v is a mean velocity defined as

$$v = \frac{W}{\rho a_t} \tag{4-2-2}$$

where W = mass flow (kg/s)
     ρ = density (kg/m^3)
     a_t = flow area (= cross section of the tube)(m^2)

In cases when the boundary layer remains in laminar flow, it is possible to find the mathematical function of the velocity profile theoretically, and for the fully developed profile region, the well-known parabolic distribution function is obtained, that is,

$$v_z = v_{z,\max}\left[1-(r/R)^2\right] \tag{4-2-3}$$

In this equation $v_{z,\max}$ is the velocity at the center of the tube.

The total mass flow circulating through the pipe is equal to the integral of the mass flow density in the cross section of the tube. Let's consider an area element with the shape of an annulus, as shown in

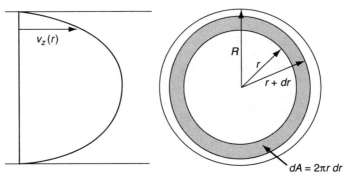

FIGURE 4-13 Parabolic velocity distribution in laminar flow.

Fig. 4-13. Since v_z is a function of the radial coordinate, in all points of dA the velocity is the same. The mass flow through dA then will be

$$dW = v_z \rho \, dA \qquad (4\text{-}2\text{-}4)$$

Integrating in the entire section,

$$W = \int dW = \int v_z \rho \, dA \qquad (4\text{-}2\text{-}5)$$

and comparing with Eq. (4-2-2), we get

$$v = \frac{\int v_z \rho \, dA}{\rho a_t} = \frac{\int v_z \, dA}{a_t} = \frac{\int_0^R v_{z,\max}\left[1 - (r/R)^2\right] 2\pi r \, dr}{\pi R^2} = \frac{v_{z,\max}}{2} \qquad (4\text{-}2\text{-}6)$$

(the integration details are left to the reader). This means that in laminar flow, the mean velocity in any section is half the velocity at the tube center.

In turbulent flow, the velocity profile is more planar. The velocity varies from 0 to about 99 percent of the maximum value in a very short distance, and a turbulent core exists where the velocity is practically uniform (Fig. 4-14). Then a velocity measurement with a Pitot tube at the center of the pipe can be considered, for all practical purposes, the mean velocity in the pipe section.

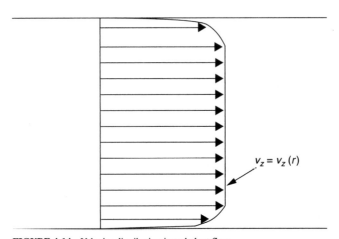

FIGURE 4-14 Velocity distribution in turbulent flow.

4-2-2 Pressure Drop

Owing to the friction at the tube wall, the fluid will suffer a pressure drop when circulating through the tube. This pressure drop can be related to the shear stress at the tube wall.

Let's consider a portion of the fluid contained between two pipe sections which are separated by a length dz. The forces acting over this portion of the fluid are shown in Fig. 4-15. Then, making a balance of forces, we get

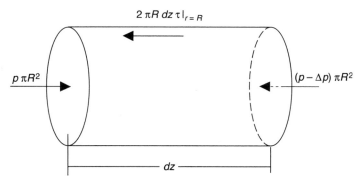

FIGURE 4-15   Forces acting on a segment of fluid.

$$dp\pi R^2 = \tau_{r=R} 2\pi R\,dz \tag{4-2-7}$$

$$dp = \tau_{r=R} \frac{2\,dz}{R} \tag{4-2-8}$$

And the shear stress at the tube wall is

$$\tau_{r=R} = -\mu \left(\frac{dv_z}{dr}\right)_{r=R} \tag{4-2-9}$$

(the minus sign is necessary because the radial coordinate has its origin at the center of the pipe, so the derivative is negative).

In laminar flow, the partial derivative in Eq. (4-2-9) can be calculated from the velocity distribution of Eq. (4-2-3). Then, combining with Eq. (4-2-6), we get

$$|\tau_{r=R}| = \mu v_{z,\max} \frac{2}{R} = \mu \frac{4v}{R} \tag{4-2-10}$$

And replacing in Eq. (4-2-8), we get

$$dp = 8\mu \frac{v}{R^2}\,dz = 32\mu \frac{v}{D^2}\,dz \tag{4-2-11}$$

This expression relates the pressure drop in a pipe with the mean fluid velocity in laminar flow.

When the flow regime is turbulent, the partial derivative of Eq. (4-2-9) cannot be calculated, and an analytical solution is not possible. The usual approach is to define a friction factor as

$$\tau_{r=R} = \frac{1}{2} f \rho v^2 \tag{4-2-12}$$

and the correlation of the friction factor as a function of the rest of the system variables is obtained experimentally.

Combining Eq. (4-2-8) and Eq. (4-2-12), we get

$$dp = 4f \frac{dz}{D} \frac{1}{2} \rho v^2 \qquad (4\text{-}2\text{-}13)$$

The friction factor is a function of the Reynolds number and the roughness of the tube surface. A well-known graph representing the friction factor as a function of the Reynolds number (Fig. 4-16) is due to Moody.[9] The curve's parameter is the relative roughness, defined as the quotient between the tube surface roughness and the pipe diameter. For commercial steel pipes in process installations, it is normally accepted that a standard roughness of 45 μm can be used in the calculations. This graph has different regions.

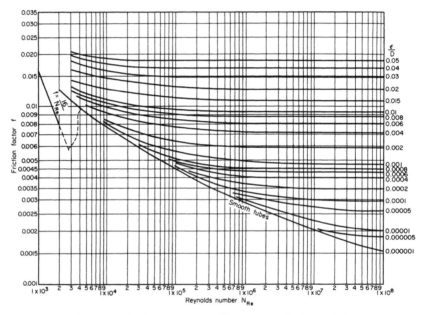

FIGURE 4-16   Fanning friction factor versus Reynolds number based on Moody.[9] (*Reproduced with permission of the American Society of Mechanical Engineers.*)

The region where 2100 < Re corresponds to the laminar flow. In that region, the mathematical expression for the friction factor can be obtained combining Eq. (4-2-11) and Eq. (4-2-13), which gives

$$f = \frac{16}{\text{Re}} \qquad (4\text{-}2\text{-}14)$$

At high Reynolds numbers, the dependence of the friction factor on the Reynolds number is very weak, and it mainly depends on the surface roughness of the tube.

The intermediate region corresponds to the transition regime, in which the friction factor depends on both the Reynolds number and the surface roughness. This graph allows calculation of the frictional pressure drop in a pipe. Knowing the mass flow W circulating in the pipe, if D is the internal diameter, we calculate

$$\text{Re} = \frac{GD}{\mu} \qquad (4\text{-}2\text{-}15)$$

where G = mass flow density or mass velocity = $4W/\pi D^2$.

With the Reynolds number it is possible to obtain the friction factor, and finally. we can calculate the pressure drop by integrating Eq. (4-2-13), and we get,

$$\Delta p = 4f\frac{L}{D}\frac{\rho v^2}{2} \tag{4-2-16}$$

4-2-3 Nonisothermal Flow Inside Tubes

Let's consider the system shown in Fig. 4-17, where a fluid at temperature T_{b_1} contained in a tank is pumped through a pipe to a second tank. If part of the pipe is heated, the fluid temperature will increase, and the fluid will arrive at the second tank at a higher temperature.

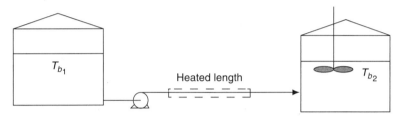

FIGURE 4-17 Fluid heating into a pipe.

Let's assume that the second tank is perfectly mixed, and we shall call T_{b_2} the temperature of the fluid contained in it. If the mass transferred from one tank to the other per unit time is W (kg/s), the heat flow (heat delivered per unit time) will be that necessary to heat up W kg/s from a temperature T_{b_1} to temperature T_{b_2}. This can be expressed as

$$Q = W(i_{b_2} - i_{b_1}) \tag{4-2-17}$$

where i_{b_1} and i_{b_2} are the specific enthalpies of the fluid at temperatures T_{b_1} and T_{b_2}.

Assuming that the enthalpies are only a function of temperature (i.e., the case of uncompressible fluids), Eq. (4-2-17) can be expressed as

$$Q = Wc(T_{b_2} - T_{b_1}) \tag{4-2-18}$$

where c is the specific heat or mass heat capacity [J/(kg · K)].

Let's analyze now what happens in the heated length. The fluid at temperature T_{b_1} enters the heated pipe. We shall assume that the internal surface of the tube is at a uniform temperature T_w, where $T_w > T_{b_1}$. Owing to this temperature difference, heat begins to flow from the tube surface to the fluid, and the regions of the fluid closer to the tube wall begin to increase their temperature.

In the same way that in fluid dynamics it is assumed that the velocity of the fluid in contact with the solid wall is zero, in heat transfer, the hypothesis normally assumed is that at the interfaces there is no discontinuity in the temperature profile, and then the temperature of the fluid in contact with the tube wall is also T_w. As long as the fluid progresses through the pipe, the perturbation extends toward the interior of the conduit, and finally, at a certain length, all the fluid has increased its temperature above the initial T_{b_1} (Fig. 4-18). Thus a temperature distribution $T = f(r, z)$ develops into the pipe, where z is the axial coordinate.

If the pipe were of an infinite length, at the outlet end, all the fluid would be heated up to the wall temperature, and a planar temperature profile again would exist. If, on the other hand, we consider that the pipe has a certain finite length, at the outlet section, the fluid will exit with a certain temperature distribution $T = f(r)$ like the profile indicated with (c) in Fig. 4-18.

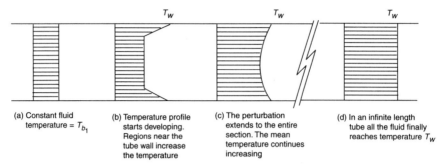

(a) Constant fluid temperature = T_{b_1}

(b) Temperature profile starts developing. Regions near the tube wall increase the temperature

(c) The perturbation extends to the entire section. The mean temperature continues increasing

(d) In an infinite length tube all the fluid finally reaches temperature T_w

FIGURE 4-18 Development of the temperature profile in a pipe.

The fluid leaving the heated tube enters the second tank, where it is perfectly mixed, resulting a uniform temperature T_{b_2}. It is said that T_{b_2} is the *mixing-cup temperature* at the outlet section. At any section of the pipe, it is possible to define the mixing-cup temperature as the temperature that a fluid passing the cross-sectional area of the conduit during a given time interval would assume if the fluid were collected and mixed in a cup. The mixing-cup temperature also can be called the *average bulk temperature* in the section under consideration.

The mixing-cup temperature corresponding to any section of the pipe can be calculated if the velocity and temperature profiles of the fluids in that section are known. If we apply the expression of the first law of thermodynamics for a circulating system to the heated pipe we are considering, we get

$$Q = \text{outlet enthalpy per unit time} - \text{inlet enthalpy per unit time} = \tilde{i}_2 - \tilde{i}_1 \qquad (4\text{-}2\text{-}19)$$

The inlet enthalpy to the system is easy to calculate. Since the fluid enters the system at a uniform temperature T_{b_1}, and selecting the reference state of the enthalpy function such that

$$i = cT \qquad \text{(enthalpy per unit mass)} \qquad (4\text{-}2\text{-}20)$$

the inlet enthalpy then will be

$$\tilde{i}_1 = WcT_{b_1} \quad \text{(J/s)} \qquad (4\text{-}2\text{-}21)$$

To calculate the outlet enthalpy, it is necessary to calculate the flow integral across the outlet section. Let's consider an annulus-shaped surface element belonging to the outlet section (Fig. 4-19) with the area

$$dA = 2\pi r dr \qquad (4\text{-}2\text{-}22)$$

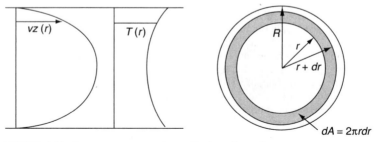

FIGURE 4-19 Temperature and velocity profiles in a tube section.

Since the fluid temperature and velocity in any pipe section are a function of the radial coordinate, they can be considered constant in that surface element. The mass flow through the surface element then will be

$$dW = 2\pi r dr \, \rho v \qquad (4\text{-}2\text{-}23)$$

And the enthalpy flow is

$$d\tilde{i}_2 = 2\pi r dr \, \rho v c \, T \qquad (4\text{-}2\text{-}24)$$

The enthalpy flow through the entire outlet section then will be

$$\tilde{i}_2 = \int_0^R 2\pi r dr \, \rho v c T \qquad (4\text{-}2\text{-}25)$$

By combining Eqs. (4-2-19), (4-2-21), and (4-2-25), we get

$$Q = \int_0^R 2\pi r dr \, \rho v c T - W c T_{b_1} \qquad (4\text{-}2\text{-}26)$$

But comparing with Eq. (4-2-18) results in

$$W c (T_{b_2} - T_{b_1}) = \int_0^R 2\pi r dr \, \rho v c T - W c T_{b_1} \qquad (4\text{-}2\text{-}27)$$

The mixing-cup or average bulk fluid temperature T_{b_2} then will be

$$T_{b_2} = \frac{\int_0^R \rho c v T (2\pi r dr)}{Wc} = \frac{\int_0^R \rho c v T (2\pi r dr)}{\int_0^R \rho c v (2\pi r dr)} = \frac{\int_0^R v T r dr}{\int_0^R v r dr} \qquad (4\text{-}2\text{-}28)$$

Equation (4-2-28) allows calculation of the average bulk temperature at any section if you know the velocity and temperature profiles corresponding to that section. This is the temperature we can measure with a thermometer inserted into the pipe downstream of the heater at a distance long enough that the fluid can reach a uniform temperature. In practice, this distance is only a few diameters.

Additionally, in the majority of cases, we shall deal with turbulent flow, and the temperature profiles are almost planar owing to the mixing effect produced by the turbulent eddies. In this case, the average bulk fluid temperature practically coincides with the temperature at the center of the pipe. Then, by inserting a thermometer into the pipe as shown in Fig. 4-20, it is possible to measure with a good approximation the mixing-cup or average bulk temperature of the fluid, and it is always a good practice to install the instrument a few diameters downstream of the outlet section of the heater.

If the total heat flow transferred to the fluid is known, the mixture temperature at the outlet section can be obtained from

$$T_{b_2} = T_{b_1} + \frac{Q}{Wc} \qquad (4\text{-}2\text{-}29)$$

From now on, every time we shall speak about the temperature of a fluid at a pipe section without any other explanation, we shall be referring to its mixing-cup or average bulk temperature.

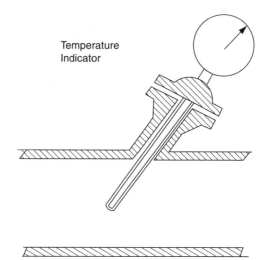

FIGURE 4-20   Average bulk temperature measurement.

4-2-4   Development of the Temperature Profile

We have explained how the temperature profile develops into the fluid as it circulates into the pipe (see Fig. 4-18). Let's define a dimensionless temperature as

$$T^*_{(r,z)} = \frac{T_w - T_{(r,z)}}{T_w - T_{b(z)}} \qquad (4\text{-}2\text{-}30)$$

where $T_{b(z)}$ is the average bulk temperature at the section under consideration. It can be observed that the temperature distribution $T^*_{(r,z)}$ becomes practically independent from z after a short length following the inlet section.

This means that when the temperature distribution in the system is represented in terms of this dimensionless temperature, we get a profile as shown in Fig. 4-21. The length of the inlet region after which the temperature profile is completely developed in laminar flow approximately equals $0.05D(\text{Re} \cdot \text{Pr})$. In turbulent flow, this thermal inlet region is much shorter, and the temperature profile develops very quickly.

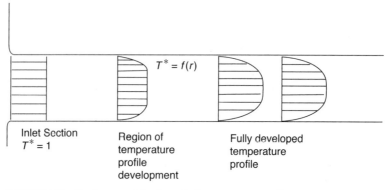

$T^* = f(r)$

Inlet Section
$T^* = 1$

Region of
temperature
profile
development

Fully developed
temperature
profile

FIGURE 4-21   Development of the dimensionless temperature profile.

4-2-5   Heat Transfer Rate from the Wall to the Fluid—Local Film Heat Transfer Coefficient

Let's consider that at a certain section of the heated pipe the average bulk temperature of the fluid is T (from now on, we shall drop the subscript b to indicate bulk temperatures). If T_w is higher than T, heat will flow from the surface of the pipe to the fluid. (we have to point out that the analysis can be done assuming that the heat transfer is in the opposite direction, and all the conclusion we shall obtain will continue to be valid.) The magnitude of the heat flux that is established depends on the hydrodynamics of the system.

Since the velocity of the fluid in contact with the solid wall is nil, the fluid in this region behaves like a solid, and the only heat transfer mechanism is conduction. Then we can always say that

$$q_r\big|_{r=R} = k\frac{dT}{dr}\bigg|_{r=R} \tag{4-2-31}$$

(We didn't include the minus sign because we want q to be positive when flowing from the wall to the fluid, in the negative direction of the radial coordinate.) This means that should we know the temperature distribution in the system, it would be possible to calculate the heat flux to the fluid with Eq. (4-2-31)

As in the case of a flat plate, in certain simple cases and in laminar flow, it is possible to obtain an analytical solution to the problem. However, the analytical procedure is limited to laminar-flow systems, which normally have little practical interest because most process equipment work in a turbulent regime.

The practical approach, then, is to define a local film convective heat transfer coefficient h_{loc} as

$$dQ = h_{\text{loc}}dA(T_w - T) \tag{4-2-32}$$

or

$$dq = h_{\text{loc}}(T_w - T) \tag{4-2-33}$$

where dA is the differential lateral area of the pipe ($=2\pi R dz$), and dQ is the heat flow transmitted through the wall.

The problem to be solved is to find an empirical correlation to calculate h_{loc} as a function of the system parameters. Before addressing this subject, we shall analyze some characteristics of the convective film heat transfer coefficients we have just defined.

4-2-6   Physical Interpretation of the Coefficient h_{loc}

We have seen that the heat flux at the tube wall can be expressed as

$$dq\big|_{r=R} = k\frac{dT}{dr}\bigg|_{r=R} \tag{4-2-34}$$

Combining this expression with Eq. (4-2-33), we get

$$k\frac{dT}{dr}\bigg|_{r=R} = h_{\text{loc}}(T_w - T) \tag{4-2-35}$$

If we write this equation as a function of the dimensionless temperature T^* defined by Eq. (4-2-30) and of a dimensionless radial coordinate $r^* = r/D$, we can write

$$-k\frac{dT^*}{dr^*}\bigg|_{r=R} = h_{\text{loc}}D \tag{4-2-36}$$

or

$$\frac{h_{loc}D}{k} = -\left.\frac{dT^*}{dr^*}\right|_{r=R} \tag{4-2-37}$$

The left side of Eq. (4-2-37) is called the *local Nusselt number* (Nu_{loc}) and is a function of the coordinate z because h_{loc} also is. Thus we see that the Nusselt number, as well as the film convection coefficient, can be interpreted as a local temperature gradient evaluated at the tube wall.

But we have seen in Sec. 4-2-3 that the dimensionless temperature profile remains constant after a certain inlet length. This means that after this inlet section, the local coefficient h_{loc} is constant and independent of the axial position.

4-2-7   Mean Temperature Difference for the Entire Tube Length

Continuing with the analysis of the system we were considering, we see that the fluid that enters the heated pipe with a temperature T_1 leaves with a temperature T_2. This means that while the fluid circulates through the pipe, its temperature increases. Figure 4-22 represents the evolution of the fluid temperature as a function of the axial coordinate z.

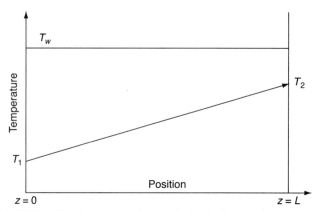

FIGURE 4-22   Temperature variation in the longitudinal direction.

We see that the temperature difference between the wall and the fluid varies from $(T_w - T_1)$ at the inlet end to $(T_w - T_2)$ at the opposite end. We want to define a mean temperature difference between the tube wall and the fluid. If we admit that the film coefficient h_{loc} is constant in the entire tube length, we get

$$dQ = WcdT = h_{loc}\pi D_o dx(T_w - T) \tag{4-2-38}$$

$$\therefore \frac{dT}{T_w - T} = \frac{h_{loc}\pi D_o dx}{Wc} \tag{4-2-39}$$

Integrating,

$$Wc = \frac{h_{loc}\pi D_o L}{\ln\dfrac{T_w - T_1}{T_w - T_2}} \tag{4-2-40}$$

Multiplying both members by

$$(T_2 - T_1) = (T_w - T_1) - (T_w - T_2) \qquad (4\text{-}2\text{-}41)$$

$$Wc(T_2 - T_1) = Q = h_{loc} A \frac{(T_w - T_1) - (T_w - T_2)}{\ln \dfrac{T_w - T_1}{T_w - T_2}} = h_{loc} A \Delta T_{ln} \qquad (4\text{-}2\text{-}42)$$

In this equation, ΔT_{ln} is the logarithmic mean temperature difference. This is a mean value between the temperature differences at both ends of the tube.

Since Eq. (4-2-42) allows calculation of the total heat flow, the local film coefficient h_{loc} will be called, in what follows, merely h on the understanding that it must be used with the logarithmic mean temperature difference ΔT_{ln}.

4-2-8 Experimental Determination of the Convection Film Coefficients—Dimensionless Numbers

We shall analyze what are the system variables influencing the film heat transfer coefficients. First, the film coefficient h will be a function of the hydrodynamic conditions and flow regime, which are characterized by the Reynolds number. Then h will be a function of all the variables included in the Reynolds number. It also will depend on the fluid properties related to the energy transport. These are the thermal conductivity k and the heat capacity c. Finally, h also will be a function of the tube length L. We then can write

$$h = f(D,\ v,\ \rho,\ \mu,\ c,\ k,\ L) \qquad (4\text{-}2\text{-}43)$$

Using the techniques of dimensional analysis, we come to the conclusion that the mathematical solution of the problem can be expressed by a relationship of the type

$$\frac{hD}{k} = f\left(\frac{Dv\rho}{\mu},\ \frac{c\mu}{k},\ \frac{L}{D} \right) \qquad (4\text{-}2\text{-}44)$$

We have assumed that the fluid properties are constant. In the case of fluids whose properties change considerably with temperature along the system under consideration, the usual approach is to evaluate the physical properties at the arithmetic mean fluid temperature between the inlet and the outlet.

The dimensionless groups included in Eq. (4-2-44) are called *Nusselt* (hD/k), *Reynolds* ($Dv\rho/\mu$), and *Prandtl* ($c\mu/k$) *numbers*. The function indicated in Eq. (4-2-44) must be obtained experimentally.

4-2-9 Influence of Wall Temperature

As was the case for a flat plate, in the proximity of the tube wall there is always a sublayer in laminar flow. This laminar sublayer represents an important resistance to the heat transfer. The thickness of this sublayer depends mainly on the fluid viscosity in the region of the fluid near the tube wall.

For example, let's consider the following two cases:

Case 1: Water circulates through a tube of diameter D with a mass flow rate W kg/s. At a certain section of the tube, the average bulk temperature of the fluid is $T = 323$ K. The tube wall temperature is 373 K. This means that the fluid is being heated.

Case 2: Through the same tube circulates an identical water flow rate. The average bulk temperature at a considered section is again 323 K, but the tube wall is at 273 K. This means that the fluid is being cooled.

We see that both cases have the same bulk temperatures, so the Reynolds and Prandtl numbers will be the same. However, in case 1, since the fluid near the tube wall is at high temperature, the thickness of the laminar sublayer will be smaller than in case 2. (This is due to the fact that the viscosity of liquids decreases with increasing temperatures.) The film coefficient h then will be higher in case 1 than in case 2.

This means that we have to add another variable to the list defined by Eq. (4-2-43), which is the viscosity at the tube wall temperature. The mathematical function expressed by Eq. (4-2-44) then is modified as

$$\text{Nu} = \frac{hD}{k} = f\left(\frac{Dv\rho}{\mu}, \frac{c_p\mu}{k}, \frac{L}{D}, \frac{\mu_w}{\mu}\right) \tag{4-2-45}$$

(It must be noted that in the case of gases, the effect is the opposite because the viscosity of gases increases with temperature.)

4-2-10 Empirical Correlations

The form of Eq. (4-2-45) is determined experimentally, and the following empirical correlations are found:

1. *Laminar flow (Re < 2100)*. One of the best known correlations is due to Sieder and Tate:[3]

$$\frac{hD}{k} = 1.86\left[\text{Re} \cdot \text{Pr}(D/L)\right]^{0.33}\left(\frac{\mu}{\mu_w}\right)^{0.14} \tag{4-2-46}$$

This equation is valid for heat transfer in laminar flow and with constant tube wall temperature. All the physical properties except μ_w must be evaluated at the average bulk temperature at the section under consideration or at a mean temperature between the average bulk temperatures of the inlet and outlet sections. Equation (4-2-46) has been reported to have a maximum error of 20 percent for $\text{Re} \cdot \text{Pr}(D/L) > 10$. For smaller mass velocities, the values of h obtained are too small.

2. *Turbulent flow (Re > 10,000)*. In the turbulent region, the following correlation, also due to Sieder and Tate,[3] is used:

$$\text{Nu} = \frac{hD}{k} = 0.023\,\text{Re}^{0.8} \cdot \text{Pr}^{0.33}\left(\frac{\mu}{\mu_w}\right)^{0.14} \tag{4-2-47}$$

It can be seen, analyzing Eq. (4-2-47), that in turbulent regime the film coefficient h is independent of the ratio D/L. This is due to the fast development of the temperature profiles in turbulent flow.

3. *Transition regime (2100 < Re < 10,000)*. In this region, the flow pattern can be unstable, and thus it is not possible to predict the Nusselt number. In certain cases, as we shall see later, a graphic interpolation between laminar and turbulent curves is performed. However, it is advisable not to design heat transfer equipment operating in this region because of the uncertainties of the results. An empirical expression for cases where the transition zone cannot be avoided is[11]

$$\frac{h}{c\rho v} = 0.116\left(\frac{\text{Re}^{0.66} - 125}{\text{Re}}\right)\left[1 + \left(\frac{D}{L}\right)^{0.66}\right]\text{Pr}^{-0.66}\left(\frac{\mu}{\mu_w}\right)^{0.14} \tag{4-2-48}$$

4-2-11 Other Dimensionless Groups

The heat transfer correlations are sometimes expressed as a function of other dimensionless groups that are combinations of the ones we have introduced. A dimensionless group that frequently appears in the literature is the *Stanton number*, which is defined as

$$\text{St} = \frac{\text{Nu}}{\text{Re} \cdot \text{Pr}} = \frac{h}{\rho cv} \tag{4-2-49}$$

Another is the dimensionless group j_H, which is defined[7] as

$$j_H = \text{St} \cdot \text{Pr}^{2/3} \left(\frac{\mu_w}{\mu}\right)^{0.14} = \frac{h}{cG} \text{Pr}^{2/3} \left(\frac{\mu_w}{\mu}\right)^{0.14}$$

(4-2-50)

where G is the mass velocity or mass flow density $= \rho v$ [kg/(s · m²)]. The dimensionless number j_H then results exclusively in a function of the Reynolds number and the ratio D/L, which according to Eqs. (4-2-46) and (4-2-47) can be expressed as follows:

For laminar flow,

$$j_H = 1.86 \text{Re}^{-2/3} (D/L)^{1/3}$$

(4-2-51)

For turbulent flow,

$$j_H = 0.023 \text{Re}^{-0.2}$$

(4-2-52)

A graph showing j_H as a function of the Reynolds number is presented in Fig. 4-23.

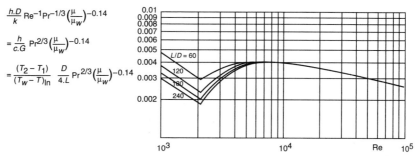

FIGURE 4-23 Graph of j_H versus Reynolds number. (*From E. Sieder and G. Tate, Ind Eng Chem 28:1429–1435, 1936. Reproduced with permission from the American Chemical Society.*)

In the transition region, the curves were drawn by graphic interpolation between the laminar and turbulent region curves. However, as was explained earlier, it is advisable to avoid this region.

Other useful expressions for j_H can be obtained by combining the definition of Eq. (4-2-50) with Eq. (4-2-42) and an enthalpy balance for the system. We can write

$$Q = \frac{h(T_w - T_1) - (T_w - T_2)}{\ln\dfrac{T_w - T_1}{T_w - T_2}} \pi DL = Wc(T_2 - T_1)$$

(4-2-53)

$$h = \frac{Wc}{\pi DL} \ln \frac{(T_w - T_1)}{(T_w - T_2)}$$

(4-2-54)

and replacing in Eq. (4-2-50), we get

$$j_H = \frac{D}{4L} \ln \frac{T_w - T_1}{T_w - T_2} \text{Pr}^{2/3} \left(\frac{\mu}{\mu_w}\right)^{-0.14}$$

(4-2-55)

This expression allows direct calculation of the outlet temperature without explicit calculation of the film coefficient h.

D. Kern[8] presented another definition of j_H as

$$j_H \text{ (Kern)} = \text{Nu} \cdot \text{Pr}^{-1/3} \left(\frac{\mu_w}{\mu}\right)^{0.14} = \frac{hD}{k} \left(\frac{c\mu}{k}\right)^{-1/3} \left(\frac{\mu_w}{\mu}\right)^{0.14} \tag{4-2-56}$$

Please note the difference with the definition of Eq. (4-2-50). The coefficient j_H defined in Eq. (4-2-50) was multiplied by the Reynolds number. The advantage of this definition is that the curves of j_H (Kern), as a function of the Reynolds number, always slopes upward, and interpolation in the transition region is easier.

Example 4-1   A natural gas stream at a pressure of 30 bar and at a temperature of 20°C must be expanded to approximately atmospheric pressure. Since the expansion of a gas through a valve produces a temperature decrease owing to Joule Thompson effect, it is necessary to preheat the gas stream to avoid too low temperatures that may produce ambient humidity condensation and icing on the exterior surfaces of the tubes. It was calculated that the gas must be heated up to a temperature of 40°C before the expansion.

The operation will be performed in an indirect heater. This is a unit consisting of a coil, into which the gas circulates, submerged in a hot-water bath (the water is maintained at high temperature with gas burners).

For a first estimation of the coil size, we assume that the water temperature can be regulated so that the tube wall temperature is maintained at 60°C. The coil will be built with 2-in Schedule 40 pipe. The gas mass flow is 2,235 kg/h, and its properties are

Temp. (°C)	30	60
Viscosity (cP)	0.0119	0.0134
Thermal conductivity [W/(m · K)]	0.0355	
Specific heat [J/(kg · K)]	2,381	
Density (kg/m³)	23.367	

It is desired to know:

a. What tube length will be necessary, and what will be the pressure drop
b. What will be the gas outlet temperature if the gas mass flow is doubled and if it is reduced to half the design value

Solution   a. The gas inlet temperature to the coil is 20°C, and the outlet temperature is 40°C. Then all the physical properties will be evaluated at the mean temperature of 30°C. The amount of heat to be delivered is

$$Q = Wc(t_2 - t_1) = (2{,}235/3{,}600) \times 2{,}381 \times (40 - 20) = 29{,}564 \text{ W}$$

The inlet diameter of a 2-in Schedule 40 pipe is 0.0525 m. Thus

Flow area $= a_t = \pi(0.0525)^2/4 = 2.164 \times 10^{-3}$

$G = \rho v = W/a_t = 2{,}235/(3{,}600 \times 2.164 \times 10^{-3}) = 287 \text{ kg/(s} \cdot \text{m}^2)$

$v = 12.2 \text{ m/s}$

$\text{Re} = D_i G/\mu = 0.0525 \times 287/(0.0119 \times 10^{-3}) = 1.266 \times 10^6$

$\text{Pr} = c\mu/k = 2{,}381 \times (0.0119 \times 10^{-3})/0.0355 = 0.798$

$(\mu/\mu_w)^{0.14} = (0.0119/0.0134)^{0.14} = 0.983$

In turbulent flow,

$$h_i = 0.023(k/D_i)\text{Re}^{0.8}\text{Pr}^{0.33}(\mu/\mu w)^{0.14} = 1{,}081 \text{ W/(m}^2 \cdot \text{K)}$$

If the entire wall temperature is maintained at 60°C, the logarithmic mean temperature difference is

$$\frac{(T_w - t_2) - (T_w - t_1)}{\ln \dfrac{T_w - t_2}{T_w - t_1}} = \frac{(60 - 40) - (60 - 20)}{\ln(20/40)} = 28.8°C$$

The necessary area then will be $A_i = Q/(h_i. \text{ LMTD}) = 29{,}564/(1{,}081 \times 28.8) = 0.95 \text{ m}^2$, which corresponds to a tube length $L = A_i/\pi D_i = 0.95/3.14 \times 0.0525 = 5.76$ m.

The roughness of a commercial steel pipe is 0.045 mm. The relative roughness, then, is 9×10^{-4}. The friction factor obtained from Fig. 4-6 is 0.005. The frictional pressure drop in the coil will be

$$\Delta P = 4f(L/D_i)\rho v^2/2 = 4 \times 0.005 \times (5.76/0.0525) \times 23.367 \times 12.2^2/2 = 3{,}815 \text{ Pa}$$

b. For a mass flow of 4,470 kg/h, the outlet temperature will be different. We do not know the mean temperature of the gas within the coil. As the physical properties of the gas do not change sensibly within the operating temperatures range, we can use the same physical properties as in the preceding case.
Since the film coefficient varies with the 0.8 power of the velocity, in this case it will be

$$h_i = 1{,}081 \times 2^{0.8} = 1{,}882 \text{ W/(m}^2 \cdot \text{K)}$$

To calculate the outlet temperature, an iterative procedure will be followed:

1. Assume t_2.
2. Calculate logarithmic mean temperature difference (LMTD).
3. Calculate $Q = h_i A_i \text{LMTD}$.
4. Verify $t_2 = t_1 + Q/Wc$.

For an outlet temperature of 38°C, we get LMTD = 30.1. Thus,

$$Q = 1{,}882 \times 0.95 \times 30.1 = 53{,}815 \text{ W}$$

$$t_2 = 20 + 53{,}815/(1.241 \times 2{,}381) = 38.2°C$$

We see that the amount of heat transferred is sensibly higher than in the preceding case, and the outlet gas temperature decreases only 2°C.

The pressure drop varies with the velocity squared (the friction factor is roughly constant), and it will then be $\Delta P = 15{,}260$ Pa

In the same way, for a gas mass flow of 1,117 kg/h, with the same procedure we obtain $t_2 = 42°C$.

4-2-12 Equivalent Diameter

The heat transfer correlations used for fluids flowing within tubes can be extended for non-circular-section conduits. In such cases, it is necessary to replace the internal diameter in Reynolds and Nusselt numbers by an equivalent diameter of the section under consideration. The equivalent diameter is defined as

$$D_{eq} = 4R_H \qquad (4\text{-}2\text{-}57)$$

where R_H is the hydraulic radius, which is

$$R_H = \frac{\text{flow area}}{\text{heat transfer perimeter}} \qquad (4\text{-}2\text{-}58)$$

Example 4-2   A double-pipe heat exchanger is formed by two concentric tubes. One fluid circulates through the inner tube, and the other one through the annular space. Calculate the heat transfer film coefficient for a gasoline stream flowing through the annular space of an exchanger of this type in the following conditions:

External diameter of the internal tube = 25.4 mm.

Internal diameter of the external tube = 39 mm.

Flow rate = 5.2 m³/h

Properties: ρ = 710 kg/m³; c = 2,442 J/(kg · K); μ = 0.4 cP; k = 0.18 W/(m · K)

Neglect the correction for tube wall temperature.

Solution   Flow area = $\pi(0.039^2 - 0.0254^2)/4 = 0.000687 \text{ m}^2$

$$\text{Heat transfer perimeter} = \pi \times 0.0254 = 0.0797 \text{ m}^2/\text{m}$$

$$D_{eq} = 4 \times 0.000687/0.0797 = 0.0344 \text{ m}$$

$$v = \frac{5.2}{3,600 \times 0.000687} = 2.1 \text{ m/s} \qquad \text{Re} = \frac{Dev\rho}{\mu} = \frac{0.0344 \times 2.1 \times 710}{0.4 \times 10^{-3}} = 128,200$$

$$\text{Pr} = \frac{2,442 \times 0.4 \times 10^{-3}}{0.18} = 5.4$$

$$h = 0.023\frac{k}{D_{eq}}\text{Re}^{0.8} \times \text{Pr}^{0.33} = 0.023\frac{0.18}{0.0344}128,200^{0.8}\,5.4^{0.33} = 2,565 \text{ W/(m}^2 \cdot \text{K)}$$

4-3 FORCED CONVECTION AROUND SUBMERGED OBJECTS

4-3-1 Fluids Flow over Submerged Bodies

Up to now, in all cases we analyzed, the fluid flow was parallel to the solid surface. However, there are important cases in process equipment where the fluid is incident over the solid surface perpendicularly or at angles to the axes of these bodies.

The most important case is that of a fluid flowing at right angles to cylinders. This situation is frequent in heat exchangers, where a fluid circulates inside the tubes of a bundle and another fluid circulates externally in a cross-flow arrangement.

Refer to Fig. 4-24. The fluid approaching the cylinder with a velocity v_∞ reduces its velocity while it moves toward the body surface. While the fluid slows, the pressure increases. At point A (called the *stagnant point*), fluid velocity is zero, and the pressure at this point rises approximately one velocity head (that is, $\rho v^2/2$) above the pressure in the incoming free stream. From that point, the fluid accelerates again while it flows along the contour of the cylinder. The maximum velocity is reached at an angle θ of approximately 90 degrees, and the pressure is at a minimum in this zone. From this point, the velocity decreases again with another increase in pressure, and a second stagnant point is reached for $\theta = 180$ degrees with maximum pressure.

This idealized behavior corresponds to fluids in laminar flow at very low fluid velocities. The pressure distribution is symmetric with respect to the plane corresponding to $\theta = 90$ degrees, which means that the pressure forces do not exert a net force over the body.

All the drag force is due to tangential viscous stress. Characterization of the flow regime is done employing a Reynolds number ($Dv_\infty\rho/\mu$), and the condition for the flow pattern described corresponds to Reynolds numbers below 10.

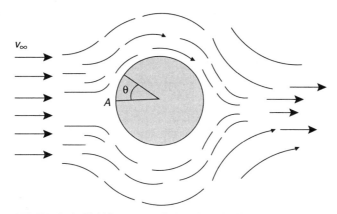

FIGURE 4-24   Fluid flow past a cylinder at low velocities.

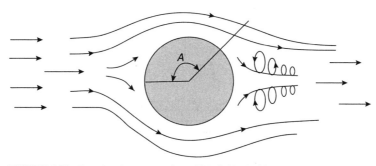

FIGURE 4-25   Boundary-layer separation at high fluid velocities.

For higher fluid velocities, the viscous forces increase, and part of the kinetic energy of the fluid is irreversibly dissipated as friction. The pressure decrease that takes place in the front half of the cylinder cannot be recovered completely in the rear half. Thus the streamlines separate from the solid surface, adopting the shape indicated in Fig. 4-25.

A turbulence region appears in the rear part of the cylinder, and the average pressure in the rear half is smaller than in the front half. The angle at which the streamlines separate from the surface also depends on the Reynolds number.

Since the pressure distribution is not symmetric, the integral of the pressure force over the solid surface is not nil and results in a net force acting on the cylinder. The total force thus is the sum of a viscous and a pressure effect.

Since the pressure and velocity distributions can only be obtained analytically in simple geometries and laminar flow, in the more general case it is necessary to find another approach to calculate the net force acting on the body. The experimental results normally are expressed by means of a drag coefficient c_D defined as

$$F = c_D \frac{1}{2} \rho v_\infty^2 A_p$$

In this equation, A_p is the body area projected in a direction normal to the flow.

The drag coefficients are correlated as a function of the Reynolds number. Figure 4-26 correlates the drag coefficients for the case of cylinders and spheres.

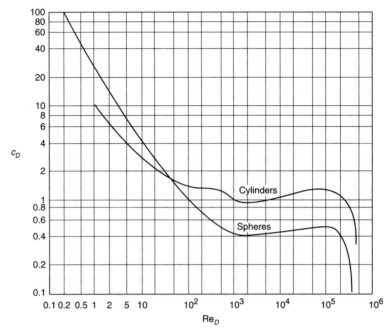

FIGURE 4-26 Drag coefficient versus Reynolds number for long circular cylinders and spheres in cross-flow. (*From C.E. Lapple and C.B. Shepherd, "Calculation of particle trajectories," IEC 32:606, (1940) Reproduced with permission from the American Chemical Society.*)

4-3-2   Convection Heat Transfer Coefficient

If a body exposed to a fluid stream exchanges heat with the fluid, a heat transfer coefficient h_m can be defined as

$$Q = h_m A(T_w - T_\infty) \qquad (4\text{-}3\text{-}1)$$

where T_w = temperature of the body surface
T_∞ = temperature of the fluid far from the surface (bulk temperature)
A = body area
h_m = average heat transfer coefficient for the entire surface

It can be easily understood that since the hydrodynamic conditions are not constant over the body surface, the heat-flux density will not be either. This means that it is also possible to define a local heat transfer coefficient as

$$dQ = h_{loc} dA(T_w - T_\infty) \qquad (4\text{-}3\text{-}2)$$

Here, h_{loc} will depend on the position of dA over the surface.

Figure 4-27 is a plot of the local Nusselt number ($h_{loc}D/k$) versus the angular coordinate θ at given parameters of the Reynolds number for air flowing in cross-flow pattern past a cylinder. It can be appreciated that the local Nusselt number is highly dependent on the position, and the location of the zone with higher heat transfer coefficients depending on the flow regime.

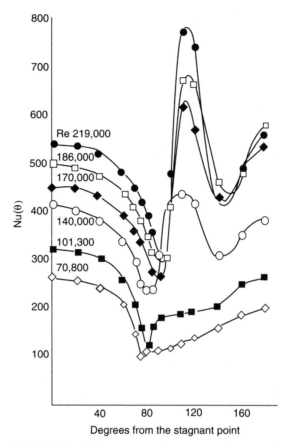

FIGURE 4-27 Unit heat transfer coefficient around a cylinder normal to an air stream. (*From W. H. Giedt, "Investigation of Variation of Point Unit Heat Transfer Coefficient around a Cylinder Normal to an Air Stream"Trans ASME 71: 375-381 (1949). Used with permission from the American Society of Mechanical Engineers.*)

However, for practical purposes, calculation of the mean heat transfer coefficient h_m defined by Eq. (4-3-1) is of much higher interest. The usual expressions are of the type

$$Nu = f(Pr, Re)$$

where Nu = average Nusselt number = $\dfrac{h_m D}{k}$

Re = Reynolds number = $\dfrac{Dv\rho}{\mu}$

Pr = Prandtl number = $\dfrac{c\mu}{k}$

Unless specifically stated, all the physical properties to be used in these correlations must be evaluated at the film temperature T_f, defined as

$$T_f = (T_w + T_\infty)/2$$

In what follows, the best-known correlations for simple geometries will be presented

Correlations for Single Cylinders in Cross-Flow. For gases flowing past cylinders, the correlations are of the type

$$\text{Nu} = B\text{Re}^n \tag{4-3-3}$$

where B and n are obtained as a function of the Reynolds number according to Table 4-1.[12] In the case of liquids, the values resulting from Eq. (4-3-3) must be multiplied by $1.1\text{Pr}^{1/3}$.

TABLE 4-1 Values of the Constants to Be Used in Eq. (4-3-3)

Re	B	n
0.4–4	0.891	0.33
4–40	0.821	0.385
40–4,000	0.615	0.466
4,000–40,000	0.174	0.615
40,000–400,000	0.0239	0.805

Correlations for Spheres. The following correlations were suggested:[4]

For liquids with $1 < \text{Re} < 70,000$,

$$\text{Nu} = 2 + 0.6\text{Re}^{1/2} \cdot \text{Pr}^{1/3} \tag{4-3-4}$$

For air with $20 < \text{Re} < 150,000$,

$$\text{Nu} = 0.33\text{Re}^{0.6} \tag{4-3-5}$$

For other gases with $1 < \text{Re} < 25$,

$$\text{Nu} = 2.2\text{Pr} + 0.48\text{Re}^{1/2} \cdot \text{Pr} \tag{4-3-6}$$

And for $25 < \text{Re} < 150,000$,

$$\text{Nu} = 0.37\text{Re}^{0.6} \cdot \text{Pr}^{1/3} \tag{4-3-7}$$

Banks of Tubes. This is the case of highest interest in process engineering because it is the geometry found in most heat exchangers. This subject was studied widely by Bergelin, Brown, and Doberstein as part of a research program on heat exchangers at the University of Delaware.[5] The experiments were performed with an ideal bank of tubes, as shown in Fig. 4-28. The half tubes located at both sides simulated the continuity of the bank.

An ample range of Reynolds numbers was covered, and different geometric arrays were investigated. The definition of the Reynolds number employed to correlate the results is

$$\text{Re} = \frac{DG_m}{\mu} \tag{4-3-8}$$

where D = external diameter of the tubes
G_m = mass velocity = W/a_m
W = mass flow (kg/s)
a_m = minimum flow area in the system. (This is the free area at the central plane of a tube's row.)

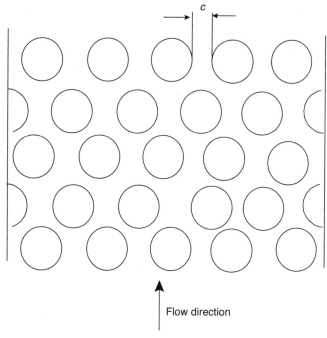

FIGURE 4-28   Ideal bank of tubes.

If c = free distance between tubes (see figure), b = width of the tubes bank in the direction normal to the plane of the drawing, and N_t = the number of tubes per row, then

$$a_m = cbN_t \ \ (\text{m}^2)$$

Square and triangular tube patterns with different tube spacing were tested. In all cases, the heat transfer coefficient and the pressure drop of the fluid across the bank were measured.

The convection heat transfer coefficient can be correlated with an expression of the type

$$j = \left(\frac{h}{cG_m}\right)\left(\frac{c\mu}{k}\right)\left(\frac{\mu}{\mu_w}\right)^{0.14} = f\left(\frac{DG_m}{\mu}\right) \tag{4-3-9}$$

And the pressure drop is expressed in terms of a friction factor, that is,

$$\Delta p = 2f\frac{G_m^2}{\rho}N_f\left(\frac{\mu}{\mu_w}\right)^{0.14} \tag{4-3-10}$$

where N_f is the number of tube rows in the bank.

The experimental correlations are shown in Fig. 7-27 (in Chap. 7). Analysis of the curves shows the existence of different flow regimes. It is generally accepted that the region of Reynolds numbers below 100 corresponds to a laminar-flow regime. At a Reynolds number around 100, a transition region begins, with the appearance of random eddies.

It has been observed that for square arrays, eddies appear simultaneously and suddenly in all the tube rows. In staggered arrays, on the other hand, the eddies appear first in the most downstream rows, and if the Reynolds number is increased, the phenomenon propagates upstream. This makes the transition from one regime to the other more gradual in staggered arrays (triangles or rotated squares) than in nonstaggered arrays (squares). (See the shape of the curves in Fig. 7-27.)

The results shown in the figure correspond to a bank with 10 tube rows. These results also can be used for a higher number of rows, but the results must be corrected if the number of rows is less than 10.

4-4   NATURAL CONVECTION

4-4-1   Heat Transfer Mechanism in Natural Convection

Let's consider a body at initial temperature T_o that at a certain time is submerged into a fluid at temperature T_∞, where $T_o > T_\infty$. Owing to the temperature difference, heat begins to flow from the body to the fluid.

The heat transfer provokes an increase in the temperature of the fluid. Thus the density of the fluid closer to the body decreases below the density of the bulk of the fluid. This density difference creates an ascendant movement of the fluid closer to the body, which is replaced by fresh fluid coming from the colder zones The shapes of the streamlines for several geometries are shown in Fig. 4-29. This fluid movement, in turn, helps to increase the heat transfer by convection because it allows the renewal of the fluid in the vicinity of the body surface.

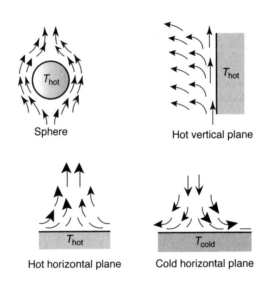

Enclosed space between walls at
different temperatures

FIGURE 4-29   Streamlines in natural convection.

The difference with respect to the other cases we studied previously is that now the heat transfer is what provokes the movement of the fluid. The fluid velocity is not imposed externally. It is not a predefined variable. The velocity profile depends on the thermal characteristics of the system.

The heat transfer coefficient in natural convection depends, as always, on the fluid properties c, ρ, μ, and k. The cause of the fluid movement is the change in density created by the applied temperature difference. The physical property that indicates how the density varies with temperature is the thermal expansion coefficient β, which is defined as

$$\beta = -\frac{1}{\rho}\left(\frac{\partial \rho}{\partial T}\right)_{p=cte} \tag{4-4-1}$$

For the case of an ideal gas,

$$\rho = \frac{Mp}{RT} \tag{4-4-2}$$

Then

$$\beta = -\frac{1}{\rho}\left(\frac{\partial \rho}{\partial T}\right)_{p} = \frac{1}{T} \tag{4-4-3}$$

We see that for an ideal gas the thermal expansion coefficient is the reciprocal of the absolute temperature. For other fluids, β must be calculated from tabulated data of densities at different temperatures.

The convection heat transfer coefficient obviously will be a function of the temperature difference between solid and fluid as well because this temperature difference is the cause of the fluid movement. We then can say

$$h = h \ [c, \ \mu, \ \rho, \ \beta, \ (T_o - T_\infty), \ k, \ L] \tag{4-4-4}$$

where L is a characteristic length of the geometry.

By applying dimensional analysis techniques, we can find a set of dimensionless groups to correlate the phenomenon. One possible solution is

$$\frac{hL}{k} = f\left[\frac{L^3 \rho^2 g \beta (T_o - T_\infty)}{\mu^2}, \frac{c\mu}{k}\right] \tag{4-4-5}$$

The first group within brackets is called the *Grashoff number* (Gr), that is,

$$\mathrm{Gr} = \frac{L^3 \rho^2 g \beta (T_o - T_\infty)}{\mu^2} \tag{4-4-6}$$

The other dimensionless groups in Eq. (4-4-12) are the Nusselt and Prandtl numbers. It is observed experimentally that the correlations are usually of the type

$$\mathrm{Nu} = a(\mathrm{Gr} \cdot \mathrm{Pr})^b \tag{4-4-7}$$

This means that the product $\mathrm{Gr} \cdot \mathrm{Pr}$ behaves as a single dimensionless group in the correlations. The physical properties included in the dimensionless groups are evaluated at the film temperature, which is the arithmetic mean value between the surface temperature and the temperature of the bulk of the fluid.

4-4-2 Correlations for Different Geometries

The correlations for the cases of higher practical interest are presented below.

Vertical Plates and Vertical Cylinders. According to Eckert and Jackson,[6]

$$\text{Nu}_L = \frac{hL}{k} = 0.555(\text{Gr}_L \cdot \text{Pr})^{1/4} \qquad \text{for Gr}_L \cdot \text{Pr} < 10^9 \qquad (4\text{-}4\text{-}8)$$

$$\text{Nu}_L = \frac{hL}{k} = 0.021(\text{Gr}_L \cdot \text{Pr})^{2/5} \qquad \text{for Gr}_L \cdot \text{Pr} > 10^9 \qquad (4\text{-}4\text{-}9)$$

In these expressions, L is the height of the plate or cylinder, and h is the average coefficient for the entire plate.

Horizontal Cylinders. According to McAdams,[4]

$$\text{Nu} = \frac{hD}{k} = 0.53(\text{Gr}_D \cdot \text{Pr})^{1/4} \qquad \text{for } 10^4 < \text{Gr}_D \cdot \text{Pr} < 10^9 \qquad (4\text{-}4\text{-}10)$$

$$\text{Nu} = \frac{hD}{k} = 0.114(\text{Gr}_D \cdot \text{Pr})^{1/3} \qquad \text{for } 10^9 < \text{Gr}_D \cdot \text{Pr} < 10^{12} \qquad (4\text{-}4\text{-}11)$$

Horizontal Planar Surfaces. According to McAdams,[4] for hot horizontal planes facing upward or cold horizontal planes facing downward,

$$\text{Nu}_L = 0.54(\text{Gr}_L \cdot \text{Pr})^{1/4} \qquad \text{for } 10^5 < \text{Gr}_L \cdot \text{Pr} < 2 \times 10^7 \qquad (4\text{-}4\text{-}12)$$

$$\text{Nu}_L = 0.14(\text{Gr}_L \cdot \text{Pr})^{1/3} \qquad \text{for } 2 \times 10^7 < \text{Gr}_L \cdot \text{Pr} < 3 \times 10^{10} \qquad (4\text{-}4\text{-}13)$$

For hot horizontal plates facing downward and cold horizontal plates facing upward in the range of $3 \times 10^5 < \text{Gr} \cdot \text{Pr} < 10^{10}$, the recommended correlation is

$$\text{Nu}_L = 0.27(\text{Gr}_L \cdot \text{Pr})^{1/4} \qquad (4\text{-}4\text{-}14)$$

In these expressions, the characteristic length is the length of one side if the surface is a square or the mean value of both sides if it is a rectangle or 0.9 times the diameter if it is a circle.

Spheres. According to Ranz and Marshall,[13] for single spheres in an infinite fluid,

$$\text{Nu}_D = 2 + 0.60\,\text{Gr}_D^{1/4} \cdot \text{Pr}^{1/3} \qquad (4\text{-}4\text{-}15)$$

Example 4-3 Calculate what must be the temperature of the bath in Example 4-1 to reach the required 60°C at the tube wall.

Solution The heat flux transferred from the external fluid (hot water) to the tube wall must be the same as that transferred from the tube wall to the process fluid. Thus

$$Q = 29{,}564 \text{ W} = h_o A_o (T - 60)$$

where T is the temperature of the water, and A_o is the external area of the coil. An iterative procedure must be employed because h_o is also a function of the water temperature. The steps are as follows:

1. Assume T.
2. Calculate h_o.
3. Verify Q.

Let's assume the $T = 90°C$. The film temperature of the water is $(90 + 60)/2 = 75°C$. At this temperature, the physical properties of the water are

Density = 974.8 kg/m³
Viscosity = 0.377 cP.
Thermal conductivity = 0.666 W/(m · K)
Specific heat = 4,190 J/(kg · K)

To calculate the thermal expansion coefficient β, we need to know the density at other temperatures. At 80°C, the density of water is 971.8; thus,

$$\beta = -\frac{1}{\rho}\frac{\Delta\rho}{\Delta T} = \frac{1}{973.3}\frac{974.8-971.8}{5} = 6.15\times10^{-4}$$

Then

$$Gr.Pr = \left(\frac{D_o^3\rho^2 g\ \beta\Delta T}{\mu^2}\frac{c\mu}{k}\right) = \left(\frac{0.0604^3\times974.8^2\times9.8\times6.15\times10^{-4}\times30}{(0.377\times10^{-3})^2}\frac{4,190\times0.377\times10^{-3}}{0.666}\right) = 6.3\times10^8$$

$$Nu = 0.53(Gr\cdot Pr)^{0.25} = 84$$

$$h_o = Nu\cdot k/D_o = 84(0.666/0.0604) = 926\ W/(m^2\cdot K)$$

$$Q = h_o A_o(T - T_w) = 926(\pi D_o L)(90 - 60) = 926(\pi 0.0604\times5.76)\times30 = 30,347\ W$$

This is approximately equal to 29,567 W.

It is important to point out that this calculation assumes that the heat transfer surfaces are clean and free from deposits or fouling. When the heat transfer surfaces get dirty, additional resistances to heat transfer must be considered in the calculations. This subject will be explained in more detail in Chap. 5.

Simplified Correlations for Air at Atmospheric Pressure. McAdams[4] presented simplified correlations that are valid for natural convection in air. These expressions can be obtained by substituting the physical properties of air in the Grashoff and Prandtl numbers so that the coefficient h is merely a function of the temperature difference and the characteristic length. These simplified correlations are shown in Table 4-2. Since they are dimensional equations, the units indicated in the table must be used for the correlations to be valid.

In Table 4-3, the physical properties of air at several temperatures are included to allow calculation of the dimensionless groups.

4-4-3 Combined Coefficient of Convection and Radiation

When a body is immersed in an infinite ambient, superimposed on the convective mechanism there exists radiation heat transfer, so the heat transfer rate is the sum of the contributions of both mechanisms, as was explained in Chap. 2. That is,

$$q = h(T_w - T_o) + \sigma\varepsilon(T_w^4 - T_o^4) \tag{4-4-23}$$

In this expression, σ is 5.67×10^{-8} W/(m² · K⁴), and ε is the surface emissivity, which depends on the type and condition of the surface. The thermal insulation of process equipment is usually finished with an aluminum or galvanized iron jacket. Table 4-4 indicates the emissivity of commonly used materials.

TABLE 4-2 Simplified Equations for Natural Convection in Air at Atmospheric Pressure and Moderate Temperatures

Geometry	Characteristic Dimension L	Flow Regime	Range of Gr · Pr	Heat Transfer Coefficient h_a [W/(m² · K)]
Vertical plates and cylinders	Height	Laminar	10^4–10^9	$h_a = 1.42(\Delta T/L)^{1/4}$ (4-4-16)
		Turbulent	10^9–10^{13}	$h_a = 1.31\Delta T^{1/3}$ (4-4-17)
Horizontal cylinders	External diameter	Laminar	10^4–10^9	$h_a = 1.32(\Delta T/L)^{1/4}$ (4-4-18)
		Turbulent	10^9–10^{13}	$h_a = 1.24\Delta T^{1/3}$ (4-4-19)
Horizontal plates, upper surface cold or lower surface hot	As defined in the text	Laminar	10^5–2×10^7	$h_a = 1.32(\Delta T/L)^{1/4}$ (4-4-20)
		Turbulent	2×10^7–3×10^{10}	$h_a = 1.52\Delta T^{1/3}$ (4-4-21)
Horizontal plates, upper surface hot or lower surface cold	As defined in the text	Laminar	3×10^5–3×10^{10}	$h_a = 0.59(\Delta T/L)^{1/4}$ (4-4-22)

Note: L in meters; ΔT in °C. These correlations can be extended to higher or lower pressures with respect to atmospheric pressure by multiplying by the following factors: $(p/1.033)^{1/2}$ for laminar flow and $(p/1.032)^{2/3}$ for turbulent flow, with p in absolute atmospheres.

TABLE 4-3 Properties of Air

Temperature (°C)	Thermal Conductivity k [W/(m · K)]	$\dfrac{\rho^2 g\beta}{\mu^2}$ [1/(m³ · K)]	Kinematic Viscosity μ/ρ (m²/s)	Prandtl Number
10	0.0250	120.3×10^6	14.4×10^{-6}	0.711
20	0.0257	102.9×10^6	15.3×10^{-6}	0.709
30	0.0264	87.4×10^6	16.2×10^{-6}	0.707
40	0.0272	75.8×10^6	17.2×10^{-6}	0.705
50	0.028	65.7×10^6	18.2×10^{-6}	0.704
60	0.0287	57.0×10^6	19.2×10^{-6}	0.702
70	0.0295	49.4×10^6	20.2×10^{-6}	0.701
80	0.0303	43.1×10^6	21.3×10^{-6}	0.699
90	0.0310	38.1×10^6	22.4×10^{-6}	0.697
100	0.0318	33.7×10^6	23.5×10^{-6}	0.695

TABLE 4-4 Emissivities

Material	ε
Aluminum plate	0.1–0.2
Aluminum-based paint	0.3–0.7
Steel plate	0.94–0.97
White paint	0.84–0.92
Asphaltic mastic	0.93

A radiation heat transfer coefficient is sometimes defined as

$$h_R = \frac{\sigma\varepsilon(T_w^4 - T_o^4)}{T_w - T_o}$$
(4-4-24)

Equation (4-4-23) then can be written as

$$Q = (h + h_R)A(T_w - T_o)$$
(4-4-25)

4-4-4 Heat Loss through an Insulated Wall in Natural Convection

In Chap. 3 we studied the equations to calculate the heat loss through an insulated wall. Let's consider a vessel containing a hot liquid whose walls are insulated with mineral wool. The heat loss through the wall can be calculated as

$$q = U(T_i - T_o) \qquad (4\text{-}4\text{-}26)$$

with

$$\frac{1}{U} = \frac{1}{h_i} + \frac{\delta}{k} + \frac{1}{h_o} \qquad (4\text{-}4\text{-}27)$$

In Eq. (4-4-27), h_i is the natural convection coefficient of the internal side (liquid-wall), and h_o is the wall-air coefficient; δ is the insulation thickness, and k is its thermal conductivity (the resistance of the metallic wall was neglected). The calculation is not straightforward because both h_i and h_o depend on the temperature of the walls in contact with the fluids, and these are not known. It is necessary, therefore, to follow an iterative procedure to solve the problem. The proposed methodology is

1. Make a first guess for U using approximate values for the h's.
2. Calculate q.
3. Estimate the internal and external temperatures of the wall with $\Delta T = q/h$.
4. With these temperatures, calculate the film temperatures and the convection coefficients with the respective correlations.
5. Recalculate U and q with the new coefficients.
6. Recalculate the wall temperatures.
7. Compare with the previous ones and go back to step 4 if necessary.

For a first guess of the heat transfer coefficients, the following values are suggested:

For water and aqueous solutions	1,000 W/(m² · K)
For light organic liquids	350 W/(m² · K)
For heavy organic liquids	75 W/(m² · K)
For gases (combined convection radiation coefficient)	10 W/(m² · K)
For steam as the heating medium	8,500 W/(m² · K)

Example 4.4 Calculate the heat loss of a vertical cylinder tank with 20-m diameter and 10-m height insulated with $1\frac{1}{2}$-in mineral wool in the vertical walls. The tank roof is not insulated. The tank contains a product whose properties are given in the following table and has to be maintained at a minimum temperature of 22°C when the external temperature is 0°C. It can be considered that the tank is full with liquid to near the roof height. The roof is in contact with the internal vapor atmosphere. The volatility of the product is low, so this internal atmosphere is mainly air. The thermal conductivity of the insulation is 0.048 W/(m · K).

Liquid Properties

Temperature (°C)	22	10
ρ (kg/m³)	802.4	809.7
μ (cP)	18	27
c [J/(kg · K)]	2,260	2,115
k [W/(m ·K)]	0.148	0.148

Solution   We shall first calculate the thermal expansion coefficient β using density data:

$$\beta = -\frac{1}{\rho}\frac{\Delta\rho}{\Delta T} = \frac{1}{806}\frac{809.7 - 802.4}{12} = 7.5 \times 10^{-4}\,\text{K}^{-1} \qquad \text{(to be assumed as constant)}$$

We shall calculate the heat loss by natural convection and radiation from the vertical wall and roof next:
Vertical wall

Calling T_1: liquid temperature (22°C)
$\qquad\quad$ T_2: metallic wall temperature
$\qquad\quad$ T_3: external temperature of the insulation
$\qquad\quad$ T_4: air temperature (0°C)
$\qquad\quad$ h_i: internal heat transfer coefficient
$\qquad\quad$ h_o: external convection-radiation combined coefficient

We shall first calculate an overall heat transfer coefficient using the estimated h's mentioned in the preceding section. This will be

$$\frac{1}{U} = \frac{1}{h_i} + \frac{1}{h_o} + \frac{\Delta x}{k_{\text{ins}}} = \frac{1}{75} + \frac{1}{10} + \frac{0.038}{0.048} = 0.905\ (\text{m}^2 \cdot \text{K})/\text{W}$$

With this coefficient, the heat loss through the wall would be

$$q = U(T_1 - T_4) = (22 - 0)/0.905 = 24.3\ \text{W/m}^2$$

And temperatures T_2 and T_3 may be calculated as

$$T_2 = T_1 - \frac{q}{h_i} = 22 - \frac{24.3}{75} = 21.6$$

$$T_3 = T_4 + \frac{q}{h_o} = 0 + \frac{24.3}{10} = 2.4$$

The film temperatures for calculation of the physical properties are
For air: $(2.4 + 0)/2 = 1.2$
For the liquid: $(22 + 21.6)/2 = 21.8$
And the properties are

	Liquid	Air
Temp (°C)	21.6	1.2
ρ (kg/m^3)	802.4	1.33
μ (cP)	18	0.018
c [J/(kg · K)]	2,260	965
k [W/(m · K)]	0.148	0.024
Pr	274	0.71
ΔT	0.4	2.4
L	10	10
Gr	5.84E+9	4.69E11
Gr · Pr	1.6E12	3.34E11
h conv $= 0.021k/L(\text{Gr} \cdot \text{Pr})^{0.4}$	23.6	2.08

The radiation coefficient for the external wall is

$$h_R = 5.67 \times 10^{-8} \varepsilon (T_3^4 - T_4^4)/(T_3 - T_4) = 5.67 \times 10^{-8} \times 0.2(275.4^4 - 273^4)/2.4 = 0.93 \text{ W/(m}^2 \cdot \text{K)}$$

The external combined coefficient then is $2.08 + 0.93 = 3.01$ W/(m^2 · K).
 Calculating U again, we get

$$\frac{1}{U} = \frac{1}{3.01} + \frac{1}{23.6} + \frac{0.038}{0.048} = 1.15 \ (\text{m}^2 \cdot \text{K})/\text{W}$$

Then

$$q = U\Delta T = \frac{22}{1.15} = 19.1 \text{ W/m}^2$$

And recalculating the wall temperatures, we get

$$T_2 = T_1 - \frac{q}{h_i} = 22 - \frac{19.1}{28} = 22.6$$

$$T_3 = T_4 + \frac{q}{h_o} = 0 + \frac{19.1}{3.01} = 6.3$$

Since these temperatures do not differ too much from the previous estimation, the physical properties of the fluids will be the same. Recalculating the coefficients:

	Liquid	Air
Temp (°C)	21.6	1.2
ρ (kg/m^3)	802.4	1.33
μ (cP)	18	0.018
c [J/(kg · K)]	2,260	965
k [W/(m · K)]	0.148	0.024
Pr	274	0.71
ΔT	0.4	6.3
L	10	10
Gr	5.84E+9	1.19E12
Gr · Pr	1.6E12	8.47E11
β		0.00364
h conv $= 0.021 k/L$ (Gr · Pr)$^{0.4}$	23.6	3.03

The radiation coefficient for the external wall results:

$$h_R = 5.67 \times 10^{-8} \varepsilon (T_3^4 - T_4^4)/(T_3 - T_4) = 5.67 \times 10^{-8} \times 0.2(279.3^4 - 273^4)/6.3 = 0.95 \text{ W/(m}^2 \cdot \text{K)}$$

The external combined coefficient then is $3.03 + 0.95 = 3.98$ W/(m^2 · K).
 Recalculating U, we get

$$\frac{1}{U} = \frac{1}{3.98} + \frac{1}{23.6} + \frac{0.038}{0.048} = 1.08 \ (\text{m}^2 \cdot \text{K})/\text{W}$$

Then

$$q = U\Delta T = \frac{22}{1.08} = 20.3 \text{ W/m}^2$$

A new iteration is not necessary. The heat loss through the wall will be $20.3 \times \pi \times D \times L = 20.3 \times \pi \times 20 \times 10 = 12{,}750$ W.

Heat Loss through the Roof

If T_i = internal temperature = 22°C
  T_w = wall temperature (equal at both sides because there is no insulation)
  T_a = air temperature
  h_i = internal coefficient
  h_o = external coefficient

then it must be $h_i(T_i - T_w) = h_o(T_w - T_o)$.

Each one of the coefficients must be calculated by iteration with the corresponding ΔT. As a first guess, we can assume that the wall temperature is the mean value between the internal temperature and the air temperature; this is 11°C. The film temperatures for the internal and external air are, respectively, 16.5 and 5.5°C. In both cases we shall use the correlations for a hot horizontal flat surface facing upward or a cold surface facing downward.

	Internal Air	External Air
Temp (°C)	16.5	5.5
ρ (kg/m³)	1.262	1.31
μ (cP)	0.019	0.018
c [J/(kg · K)]	950	957
k [W/(m · K)]	0.025	0.025
Pr	0.71	0.71
ΔT	11	11
L	20	20
Gr	1.31E13	1.59E13
Gr · Pr	9.32E12	1.13E13
β	0.00345	0.00359
h conv = 0.14(Gr · Pr)$^{0.33}$	3.39	3.5

It must be noted that even the product Gr · Pr exceeds the recommended value for the correlation we use it because we do not have a more specific correlation available. The radiation heat transfer coefficients are calculated as

$$h_{Ri} = 5.67 \times 10^{-8}\varepsilon(T_i^4 - T_w^4)/(T_i - T_w) = 5.67 \times 10^{-8} \times 0.9(295^4 - 284^4)/11 = 4.95 \text{ W/(m}^2 \cdot \text{K)}$$

$$h_{Ro} = 5.67 \times 10^{-8}\varepsilon(T_w^4 - T_o^4)/(T_w - T_o) = 5.67 \times 10^{-8} \times 0.9(284^4 - 273^4)/11 = 4.41 \text{ W/(m}^2 \cdot \text{K)}$$

The combined radiation-convection coefficients then will be

$$h_i = 3.39 + 4.95 = 8.34 \text{ W/(m}^2 \cdot \text{K)}$$

$$h_o = 3.5 + 4.41 = 7.91 \text{ W/(m}^2 \cdot \text{K)}$$

And the overall coefficient will be

$$\frac{1}{U} = \frac{1}{8.34} + \frac{1}{7.91} = 0.246 \therefore U = 4.05 \text{ W/(m}^2 \cdot \text{K)}$$

The heat-flux density through the roof then will be

$$q = U\Delta T = 4.05 \times 22 = 89 \text{ W/(m}^2 \cdot \text{K)}$$

Since the internal and external coefficients are approximately equal, the wall temperature will be very close to the assumed value that was estimated as the arithmetic mean temperature between the internal and external temperatures. Recalculation thus is not necessary.

$$\text{Heat loss through the roof} = \pi D^2/4 \times 89 = 3.14 \times 20^2/4 \times 89 = 27,946 W$$

Heat Loss through the Floor

For a plate at temperature T_w lying on a semi-infinite solid, the heat loss can be calculated as

$$Q = 2Dk_s(T_w - T_s)$$

where D is the plate diameter, K_s is the thermal conductivity of the solid, and T_s is the temperature of the solid far from the plate (see Fig. 3-8).
    If the soil conductivity is 1.5 W/(m · K), this will be

$$Q = 2 \times 20 \times 1.5(T_w - 0) = 60(T_w - 0) \text{ W}$$

To calculate h_i, we must use the expression corresponding to a cold plate facing upward, that is,

$$\text{Nu} = 0.27(\text{Gr} \cdot \text{Pr})^{0.25}$$

After some trials, we shall assume a wall temperature of 21°C. Since this value is sensibly equal to the bulk liquid temperature, we shall use the physical properties at this temperature

$$\text{Gr} \cdot \text{Pr} = \left(\frac{D^3\rho^2 g\beta\Delta T}{\mu^2} \frac{c\mu}{k}\right) = \left[\frac{(0.9 \times 20)^3 \times 802.4^2 \times 9.8 \times 7.5 \times 10^{-4} \times 1}{(18 \times 10^{-3})^2} \frac{2,115 \times 18 \times 10^{-3}}{0.148}\right] = 0.23 \times 10^{14}$$

The product Gr · Pr is again out of the range of the available correlation, but we shall use it anyway. We get Nu = 591.

$$h_i = \text{Nu}.k/0.9D = 591 \times 0.148/18 = 4.85 \text{ W/(m}^2 \cdot \text{K)}$$

Verification of the floor temperature:

$$h_i(20 - T_w)\pi D^2/4 = 60(T_w - 0)$$

$$4.85(22 - 21) \times 314 = 60(21 - 0)$$

And 1,520 = 1,260 reasonably well. Then

$$Q = (1,520 + 1,260)/2 = 1,390 \text{ W}$$

The total heat loss of the tank thus is

$$Q = 12,750 + 27,946 + 1,390 = 42,086 \text{ W}$$

GLOSSARY

A = area (m²)

A_p = projected area (m²)

a = flow area (m²)

c = heat capacity at constant pressure [J/(kg · K)]

c_D = drag coefficient

D = diameter (m)

F = force (N)

G = mass flow density [kg/(m · s)]

h = film coefficient [W/(m² · K)]

i = enthalpy per unit mass (J/kg)

$\tilde{i}$ = enthalpy flow (W)

j_H = dimensionless factor

k = thermal conductivity [W/(m · K)]

L = length (m)

M = molecular weight

Nu = Nusselt number

p = pressure

Pr = Prandtl number

Q = heat flux (W)

q = heat-flux density (W/m²)

R = ideal gas constant

R = radius (m)

r = radial coordinate (m)

Re = Reynolds number

T = temperature (K)

U = overall heat transfer coefficient [W/(m² · K)]

v = velocity (m/s)

W = mass flow (kg/s)

x, y, z = rectangular coordinates

β = thermal expansion coefficient (1/K)

δ = thickness (m)

ρ = density (kg/m³)

μ = viscosity [kg/(m · s · cP)]

τ = shear stress

Subscripts

1 = inlet

2 = outlet

b = bulk

f = friction factor
i = internal
loc = local
m = mean
o = external
t = tube
w = wall

REFERENCES

1. Blasius H.: "Grenzschleten in Flussig Keiten mit Kleiner Reibung," M: *Z. Math u Phys* 56(1), 1908.
2. Kreith F., Bohn M.: *Principles of Heat Transfer*, 6th ed., Brooks Cole, 1998.
3. Sieder E., Tate G.: "Heat Transfer and Pressure Drop of Liquids in Tubes," *Ind. Eng. Chem.* 28:1429–1435, 1936.
4. McAdams W. H.: *Heat Transmission*. New York: McGraw-Hill, 1954.
5. Bergelin O. P., Brown G. A., Doberstein S. C.: "Heat Transfer and Fluid Friction During Flow across Tube Banks," *Trans ASME*, 74:953–960, 1952.
6. Eckert E. R. G., Jackson W.: NACA Report 1015, July 1950.
7. Colburn A. P.: *Trans AICh E*, 29:174, 1933.
8. Kern D.: *Process Heat Transfer*. New York: McGraw-Hill, 1950.
9. Moody L.: "Friction Factor for Pipe Flow," *Trans ASME*, 66:671–684, 1944.
10. Pohlausen E.: "Der Warmeaustausch zwischen festen Korpern und Flüssigkeiten mit kleiner Reibung und kleiner Wärmeleitung," *Z Angew Math Mech*, 1:115, 1921.
11. Hansen M.: *Z Ver Deut Ing Beih Verfarenstech*, 4:91, 1934.
12. Hilpert R.: *Forsch Gebiete Ingenieurw*, 4:215–224, 1933.
13. Ranz W.E., Marshall W. R.: *Chem. Eng. Progr.*, 48:141–146, 173–180, 1952.
14. Hansen M.: "Velocity Distribution in a Boundary Layer of a Submerged Plate," NACA Report 582, 1930.

CHAPTER 5
FUNDAMENTALS OF HEAT TRANSFER BETWEEN FLUIDS

In almost all process industries, it is necessary to perform heat transfer unit operations over fluid streams. The purpose of these operations may be either to heat up or cool down a fluid stream or to produce a change of phase either by vaporization or condensation.

For a fluid stream to deliver heat, it is necessary to have another stream capable of receiving it. Then both streams can interchange heat, and this operation is performed in a piece of equipment called a *heat exchanger.*

The streams handled in a process plant can be divided into two groups:

1. Process streams
2. Utility streams

Process streams are the streams that participate in the mass balance of the plant. They can be feed-stocks, reaction products, fractionation products, effluents, intermediate streams, etc. *Utility streams* are the streams that do not participate in the mass balance of the plant because they do not mix with the process streams, but they can affect the energy balances. The most typical examples are the steam used as heating media and the cooling water of the plant refrigeration circuits. Usually in process industries, steam and cooling water installations are centralized, and they serve several process units.

The streams in a heat exchanger can be process or utility streams. If there is a process stream that needs to be heated up and another stream that needs to be cooled down, it may be possible to put them in contact as a way to save energy.

A typical example can be found in distillation operations. Usually the feed stream to a distillation column has to be heated up before being introduced into the column. The column bottom product, in turn, must be cooled down before being sent to storage. Since the temperature of the column bottoms is always higher that the column feed temperature, it is possible to exchange heat between them, as shown in Fig. 5-1 (unit *A*).

If it is not possible to combine process streams this way, it will be necessary to perform the heating or cooling operations with utility streams. For example, in Fig. 5-1 we can see a reboiler that is used to vaporize the liquid column bottoms, delivering the heat required by the process. This unit (*B*) receives steam as heating media. (Note: *Reboiler* is the term specifically used to designate heat exchangers used to vaporize distillation column bottoms.)

As another example, at the top of the column we can see a condenser (*C*). This piece of equipment is used to condense the light vapors coming from the column. In this case, the heat is withdrawn with cooling water.

Even though there is no universally accepted criterion, we shall normally use the term *heat exchanger* to designate units in which sensible heat is transferred between streams. When there is a change of phase of the process stream, the units will be designated as *condensers, vaporizers, reboilers,* etc. In what follows, we shall study the principles of heat transfer between fluids.

5-1 THE ENTHALPY BALANCE

Most process plants operate in a continuous mode, and we normally refer to the heat transferred per unit time. We shall call W the mass flow rate (kg/s) of a stream. The subscript h will make reference to the hot stream, and subscript c will designate the cold stream. In all cases, subscript 1 will denote heat exchanger inlet conditions, whereas subscript 2 will refer to outlet conditions of any stream.

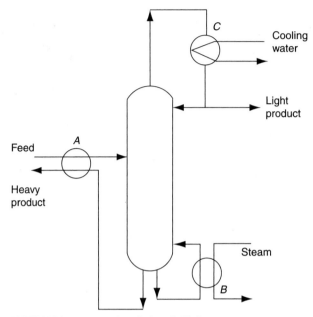

FIGURE 5-1 Heat transfer units in a distillation process

To designate the fluid temperatures, we shall use T for the hot-stream temperatures and t for the cold-stream temperatures. Q will be the heat flow or heat exchanged per unit time (J/s)

If a hot stream delivers heat to a cold medium, it will suffer an enthalpy decrease. Then

$$Q = W_h(i_{h_1} - i_{h_2}) \tag{5-1-1}$$

where i is the specific enthalpy (J/kg). If the fluid cools down without a change of phase, the enthalpy difference can be expressed as

$$i_{h_1} - i_{h_2} = c_h(T_1 - T_2) \tag{5-1-2}$$

where c is the heat capacity or specific heat [J/(kg · K)]. It then can be said that

$$Q = W_h c_h(T_1 - T_2) \tag{5-1-3}$$

If the hot fluid is a vapor of a pure substance, saturated at a certain pressure, when delivering heat, it will suffer an isothermal condensation, and

$$i_{h_1} - i_{h_2} = \lambda_h \qquad \text{(latent heat of condensation)} \tag{5-1-4}$$

Then, in this case,

$$Q = W_h \lambda_h \tag{5-1-5}$$

If the fluid is a mixture of vapors, the condensation will not be isothermal, and Eq. (5-1-1) must be used, calculating i_{h_1} as the specific enthalpy of the vapor mixture at its initial state and i_{h_2} as the enthalpy of the condensed liquid at its final state.

It is also possible that the hot fluid is a superheated vapor, which initially cools down to its saturation temperature and then undergoes an isothermal condensation. Additionally, this condensation can be followed by further cooling of the condensed liquid. In this case, Eq. (5-1-1) also must be applied, calculating the specific enthalpies of the initial and final state.

The heat delivered by the hot stream is received by the cold stream. This fluid will suffer an evolution inverse to the one just described. If the fluid heats up without a change of phase, then

$$Q = W_c c_c (t_2 - t_1) \tag{5-1-6}$$

And if it undergoes an isothermal vaporization, then

$$Q = W_c \lambda_c \tag{5-1-7}$$

where λ_c is the latent heat of vaporization.
Generically, for any situation,

$$Q = W_c (i_{c_2} - i_{c_1}) \tag{5-1-8}$$

Obviously, the heat delivered by the hot fluid is the same heat absorbed by the cold fluid. Thus

$$Q = W_c (i_{c_2} - i_{c_1}) = W_h (i_{h_1} - i_{h_2}) \tag{5-1-9}$$

This expression may assume the following particular forms:

1. Heat exchange between two fluids without change of phase:

$$W_h c_h (T_1 - T_2) = W_c c_c (t_2 - t_1) \tag{5-1-10}$$

2. Condensation of a pure saturated vapor with a refrigerant medium that exchanges sensible heat:

$$W_h \lambda_h = W_c c_c (t_2 - t_1) \tag{5-1-11}$$

3. Evaporation of a pure saturated liquid with a heating medium that exchanges sensible heat:

$$W_h c_h (T_1 - T_2) = W_c \lambda_c \tag{5-1-12}$$

4. Evaporation of a pure saturated liquid heated with steam or another condensing pure vapor:

$$W_h \lambda_h = W_c \lambda_c \tag{5-1-13}$$

5-2 HEAT TRANSFER AREA AND HEAT TRANSFER COEFFICIENT

To perform a heat transfer operation between two fluids, the following are required:

1. *A temperature difference between both streams.* The higher this difference, the higher will be the heat transfer rate.
2. *Both fluids must be separated by a surface through which the heat can be transferred.* This is called the *heat transfer area.* For example, if one of the fluids circulates through a tube, and the other is outside the tube, the heat transfer area is the surface of the tube.

It is easy to imagine that the larger the contact area between both fluids, the higher will be the amount of heat that can be exchanged per unit time. The two preceding statements can be summarized in the following expression:

$$Q = UA\Delta T \tag{5-2-1}$$

where the proportionality constant U is called the *overall heat transfer coefficient* and ΔT is the temperature difference between both streams.
We see that in problems dealing with heat transfer between fluids, there are two types of equations. Equations (5-1-1) to (5-1-13) represent the heat balances. They allow calculation of the amount of heat that

must be transferred to achieve a certain process condition in the streams that participate in the operation. These equations are associated with the thermodynamics of the process and are completely independent of the design of the equipment in which the process is performed.

On the other hand, Eq. (5-2-1) represents the kinetics of the heat transfer process. It allows calculation of the area of the heat transfer device needed to achieve a heat flux Q between two streams whose temperature difference is ΔT. This area depends on the heat transfer coefficient U, which can be modified by changing the characteristics of the apparatus. The basic objective of the design will be to reach the highest possible value for U compatible with all process restrictions.

In any equipment design problem, both types of equations are always present. This means that it will be necessary to combine a balance equation with a kinetic equation that will allow evaluating whether the heat transfer area is large enough to achieve the desired objective.

The third type of equation includes those necessary to calculate the heat transfer coefficient U. All the art and science of heat transfer are centered on this point. The heat transfer coefficient U is calculated as a function of the film coefficients, as we shall see in what follows.

5-3   EXPRESSION OF THE OVERALL HEAT TRANSFER COEFFICIENT AS A FUNCTION OF THE FILM COEFFICIENTS

5-3-1   Individual Heat Transfer Resistances

Let's assume that we have a tube with a hot fluid at temperature T circulating through it. The tube is submerged into a cold fluid at temperature t (Fig. 5-2). Let D_i and D_o be the diameters of the internal and external surfaces of the tube, respectively.

In order for heat to flow by conduction through the tube wall, from the internal to the external surfaces, there must be a temperature difference between them. Let us call T_{wi} the temperature of the internal surface and T_{wo} the temperature of the external surface of the tube wall.

We then can express the heat transfer rate between the internal fluid and the wall as

$$Q = h_i A_i (T - T_{wi}) \tag{5-3-1}$$

where h_i is the convection film coefficient for the internal fluid. Through the tube wall, heat is transmitted by conduction, and we can write

$$Q = \frac{k A_m (T_{wi} - T_{wo})}{1/2 (D_o - D_i)} \tag{5-3-2}$$

where k = thermal conductivity of the tube material

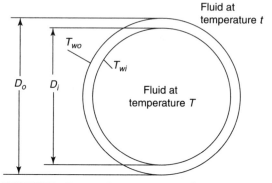

FIGURE 5-2   Heat transfer through a tube wall.

A_m = logarithmic mean area

$$A_m = \frac{\pi L(D_0 - D_i)}{\ln \dfrac{D_0}{D_i}} \tag{5-3-3}$$

The heat transfer from the external wall to the cold fluid can be expressed as

$$Q = h_o A_o (T_{wo} - t) \tag{5-3-4}$$

Equations (5-3-1), (5-3-2), and (5-3-4) can be written as

$$T - T_{wi} = Q \frac{1}{h_i A_i} \tag{5-3-5}$$

$$T_{wi} - T_{wo} = Q \frac{1/2(D_o - D_i)}{kA_m} \tag{5-3-6}$$

$$T_{wo} - t = Q \frac{1}{h_o A_o} \tag{5-3-7}$$

By summing the previous expressions, we get

$$T - t = Q \left[\frac{1}{h_i A_i} + \frac{1}{h_o A_o} + \frac{1/2(D_o - D_i)}{kA_m} \right] \tag{5-3-8}$$

Comparing with Eq. (5-2-1), we see that

$$\frac{1}{UA} = \frac{1}{h_i A_i} + \frac{1}{h_o A_o} + \frac{1/2(D_o - D_i)}{kA_m} \tag{5-3-9}$$

This expression is not completely defined because we have not yet decided which is the area A. In this case, we have an internal area A_i and an external area A_o. Either of them can be used to define a heat transfer coefficient U. Obviously, the value of this coefficient will depend on which area we choose. If we refer to the external area, the expression defines a coefficient U_o, and if we refer to the internal area, we shall be defining a coefficient U_i, and it will be

$$\frac{1}{U_o A_o} = \frac{1}{U_i A_i} = \frac{1}{h_i A_i} + \frac{1}{h_o A_o} + \frac{1/2(D_o - D_i)}{kA_m} \tag{5-3-10}$$

$$\frac{1}{U_o} = \frac{1}{h_1(A_i/A_o)} + \frac{1}{h_o} + \frac{1/2(D_o - D_i)}{k(A_m/A_o)} \tag{5-3-11}$$

or

$$\frac{1}{U_i} = \frac{1}{hi} + \frac{1}{h_o(A_o/A_i)} + \frac{1/2(D_o - D_i)}{k(A_m/A_i)} \tag{5-3-12}$$

Either of these two coefficients can be used. However, since normally the heat exchanger tubes are standardized by their external diameter, the current practice is to use the coefficient U_o defined by Eq. (5-3-11). From now on, we shall always make reference to it, unless specifically otherwise indicated. For simplicity, we shall drop the subscript and use the symbol U. For the same reason, the external area simply shall be called A.

Since Eq. (5-2-1) can be written as

$$Q = \frac{\Delta T}{(1/UA)}$$ (5-3-13)

if we remember Ohm's law of the electric circuits, that is,

$$i = \frac{\Delta V}{R}$$ (5-3-14)

where i = current intensity
ΔV = electrical potential difference
R = resistance

then, looking at Eqs. (5-3-13) and (5-3-14), we see that an analogy between both phenomena can be established, and we can consider the term $1/(UA)$ as a resistance to the heat transmission and ΔT as the driving force that provokes the heat flux. The resistance, in turn, is formed by three resistances in series, represented by the three terms of Eq. (5-3-12).

Normally, the heat exchanger tubes are made from metal, and owing to their high thermal conductivity, the third term of Eq. (5-3-12) can be neglected, resulting in

$$\frac{1}{U} = \frac{1}{h_i(A_i/A_o)} + \frac{1}{h_o}$$ (5-3-15)

It is possible to define an internal film coefficient referred to the external area h_{io} as

$$h_{io} = h_i(A_i/A_o)$$

and then Eq. (5-3-15) can be written as

$$\frac{1}{U} = \frac{1}{h_{io}} + \frac{1}{h_o}$$ (5-3-16)

5-3-2 Fouling Resistance

When heat transfer equipment has been operating for a certain time, scale, algae, suspended solids, insoluble salts, or other fouling agents can deposit over the internal or external surfaces of the tubes. This adds two additional resistances to those previously considered in the calculation of U because the heat has to flow by conduction through the two scale layers, as shown in Fig. 5-3.

We shall call the additional internal and external fouling resistances Rf_i and Rf_o, respectively. Their units are $[(m^2 \cdot K \cdot s)/J]$. Then the overall resistance to the heat transmission, once these fouling layers are formed, is given by

$$\frac{1}{U} = \frac{1}{h_{io}} + \frac{1}{h_o} + Rf_i + Rf_o$$ (5-3-17)

The combined fouling resistance is the sum

$$Rf = Rf_i + Rf_0$$ (5-3-18)

As long as the tubes foul and the resistance to heat transfer grows, the heat transfer capacity of the unit is reduced. For this reason, heat transfer equipment designers must make allowances to anticipate these deposits. It is necessary to perform an estimation of the value that this resistance can reach in the service time between two exchanger cleanings.

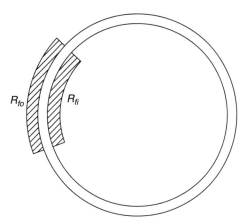

FIGURE 5-3   Fouling resistances.

This resistance can be estimated on the basis of designer knowledge and previous experience with the fluids or with the help of tables that give typical values for different applications. A fouling resistances table is included in App. F. Once the fouling resistances corresponding to each fluid have been selected, the overall heat transfer coefficient can be calculated with Eq. (5-3-17). The heat transfer area then is obtained as

$$A = \frac{Q}{U\Delta T} \qquad (5\text{-}3\text{-}19)$$

When the heat exchanger is put into service, during the initial plant startup or after a thorough cleaning operation, the fouling layer is not present, and the overall heat transfer coefficient is given by Eq. (5-3-16). We shall call U_c (clean coefficient) the term defined by

$$\frac{1}{U_c} = \frac{1}{h_o} + \frac{1}{h_{io}} \qquad (5\text{-}3\text{-}20)$$

U_c then will be higher than the design U. This means that at the beginning of operations, the exchanger will be capable to transfer a heat flux Q' given by

$$Q' = U_c A\Delta T \qquad (5\text{-}3\text{-}21)$$

with $Q' > Q$.

Usually it is always possible to adjust the performance of a unit when it is oversized, for example, by reducing the fluid flow rates (see Examples 5-6 and 5-7), so this oversizing normally is not a problem. As long as the unit fouls, the heat transfer resistance increases up to the design value. At that time, the exchanger must be cleaned because otherwise it will not be capable of transferring the required heat flow.

Example 5-1   It is necessary to exchange 500,000 W between two fluids. Their mean temperature difference is 50 K. The film heat transfer coefficients h_o and h_{io} were calculated as 1,500 and 3,000 J/(s · m^2 · K), respectively. Previous experience allows us to assume that fouling resistances of 0.0002 and 0.0003 (m^2 · K)/W will be deposited annually on the internal and external surfaces of the tubes, respectively. Calculate the required heat transfer area of the exchanger that will allow continuous operation for 1 year between cleanings.

Solution   The clean heat transfer coefficient will be

$$U_c = (1/h_{io} + 1/h_o)^{-1} = (1/1{,}500 + 1/3{,}000)^{-1} = 1{,}000 \text{ J/(s · m}^2 \cdot \text{K)}$$

The accumulated resistance after 1 year will be

$$Rf = Rf_i + Rf_o = 0.0003 + 0.0002 = 0.0005 \ (m^2 \cdot s \cdot K)/J$$

And the overall dirty heat transfer coefficient will be

$$U = (1/U_c + Rf)^{-1} = (1/1,000 + 0.0005)^{-1} = 666 \ J/(s \cdot m^2 \cdot K)$$

The required area is

$$A = Q/(U \Delta T) = 500,000/(666 \times 50) = 15 \ m^2$$

If fouling were not considered, the required area would have been

$$A = Q/(U_c \Delta T) = (500,000/1,000 \times 50) = 10 \ m^2$$

This means that fouling considerations make a 50 percent increase in the heat transfer area necessary.

5-3-3 Controlling Resistance

When calculating the overall heat transfer coefficient, it may happen that one of the three terms of the equation

$$\frac{1}{U} = \frac{1}{h_{io}} + \frac{1}{h_o} + Rf \tag{5-3-22}$$

is of a higher order of magnitude than the others. In this case, it is said that this is a *controlling resistance,* and the overall coefficient U practically coincides with the corresponding film coefficient.

For example, if $h_o \ll h_{io}$ and $h_o \ll 1/Rf$, then the term $1/h_o$ is the controlling resistance, and practically, $U = h_o$. In this case, all the efforts to improve the design will be focused on improving this coefficient because any modification that only improves h_{io} will have little influence in the value of U.

It even can be possible for Rf to be much higher than the terms $1/h$, and then fouling will be the controlling resistance. In this case, very little can be done to reduce the heat transfer area. In such a situation, it would be very important to make a good prediction of Rf, thus avoiding unnecessary safety margins that greatly affect the area calculation.

5-4 MEAN TEMPERATURE DIFFERENCE BETWEEN TWO FLUIDS

Let's suppose that it is desired to cool down a mass flow rate W_h (kg/s) of a hot fluid initially at temperature T_1 down to a temperature T_2. A mass flow rate W_c of a cooling fluid initially at temperature t_1 will be employed. The cold fluid will heat up to a temperature t_2. All these magnitudes are related by the energy balance

$$Q \ (J/s) = W_h c_h (T_1 - T_2) = W_c c_c (t_2 - t_1) \tag{5-4-1}$$

To perform this operation in a continuous mode, we need a heat exchanger that will receive both streams at the inlet temperatures and will have enough heat transfer area to allow the required heat exchange so that both streams leave the unit at temperatures T_2 and t_2. The simplest piece of equipment in which this operation can be carried out is a double-tube heat exchanger, which consists in two concentric tubes, as shown in Fig. 5-4.

One of the fluids circulates through the internal tube, and the other through the annular-section conduit enclosed by both tubes. Both fluids are separated by the wall of the internal tube, and as long as they circulate, they exchange heat through the tube wall.

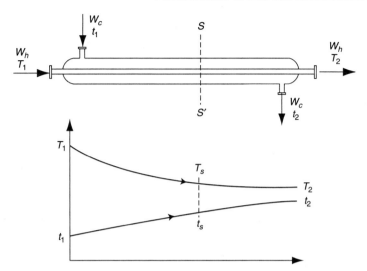

FIGURE 5-4   Parallel-currents double-tube heat exchanger.

As a consequence of this heat exchange, the temperatures of both streams change continuously along the unit from inlet to outlet, as shown in the lower diagram in Fig. 5-4, which represents the temperatures of the fluids at different sections along the exchanger. For example, in section SS' (upper diagram), located at a distance L_s from the inlet end, the temperatures of the hot and cold fluids are T_s and t_s, respectively. The driving force for heat transfer in that section then is

$$\Delta T = T_s - t_s \tag{5-4-2}$$

However, it can be appreciated in the diagram that this temperature difference changes along the unit, having a maximum value at the inlet end, given by $(T_1 - t_1)$, while the driving force is minimum at the outlet end, given by $(T_2 - t_2)$. This means that the heat transfer per unit area varies along the unit, or (which is the same) the heat-flux density is not uniform over the entire heat transfer surface. If in correspondence with section SS' we consider a differential length element of the heat exchanger, as shown in Fig. 5-5, this element, whose length is dx, is associated with a heat transfer area dA, which is

$$dA = \pi D_o dx \tag{5-4-3}$$

where D_o is the external diameter of the internal tube. (As explained previously, when the heat transfer surface is a tube, we shall always make reference to its external area.)

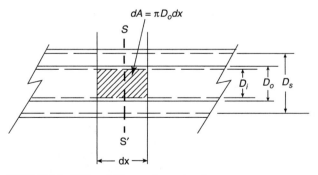

FIGURE 5-5   Differential heat exchanger length.

Since the temperature difference between both fluids changes along the tube length, the heat transfer equation must be written in differential mode, that is,

$$dQ = U dA (T - t) \qquad (5\text{-}4\text{-}4)$$

where U is calculated by Eq. (5-3-22).

However, for equipment design, we need integral expressions that allow us to calculate the total heat flux over the entire length of the tube. This means an expression of the type

$$Q = U A \Delta T_m \qquad (5\text{-}4\text{-}5)$$

where Q = total heat exchanged per unit time
$A = \pi D_o L$ = heat transfer area
U = overall heat transfer coefficient

It must be noted that if we assume that the physical properties of the fluids are constant, the overall heat transfer coefficient, which depends only on the fluid properties, velocities, and exchanger geometry, also will be constant along the heat exchanger.

ΔT_m is a mean temperature difference that must be defined so as to satisfy Eq. (5-4-5). In what follows, we shall see how this mean temperature difference must be calculated as a function of the fluid temperatures at both ends of the unit.

5-4-1   Countercurrent and Parallel-Flow Configurations

There are basically two ways in which the fluids can circulate through the unit. They are

1. Parallel-flow (also called *co-current*) configuration

2. Countercurrent configuration

The first one corresponds to the diagram in Fig. 5-4. We can see there that both fluids enter the exchanger at the same end and circulate through it in the same direction. In this way, the temperature difference is highest at the inlet end and reduces along the unit toward the outlet end.. The other circulation arrangement, called *countercurrent,* is that shown in Fig. 5-6. We see that the two streams travel through the unit in opposite directions.

Depending on the flow rates and heat capacities of both streams, the temperature difference will be higher at one end or the other. For example, if the product of the flow rate and the heat capacity of the hot fluid $W_h c_h$ is higher than that of the cold fluid ($W_c c_c$) after having exchanged heat Q, the temperature change of the hot fluid is

$$T_1 - T_2 = \frac{Q}{W_h c_h} \qquad (5\text{-}4\text{-}6)$$

whereas the temperature change of the cold fluid is

$$t_2 - t_1 = \frac{Q}{W_c c_c} \qquad (5\text{-}4\text{-}7)$$

And $T_1 - T_2 < t_2 - t_1$.

This means that the cold fluid is heated up more than the hot fluid is cooled down, and then the fluid temperatures tend to approach at the hot end of the exchanger (Fig. 5-7a).

If, on the other hand, $W_h c_h < W_c c_c$, the situation is the opposite, and the fluid temperatures tend to approach at the cold end of the exchanger (T_2 approaches more to t_1 than t_2 approaches T_1. This situation is shown in Fig. 5-7b).

One of the first conclusions that arises from the simple observation of Figs. 5-4 and 5-6 is that with a parallel-flow configuration it is not possible to cool down the hot fluid below the outlet temperature of the cold fluid because there always must be a positive ΔT for heat transfer to exist. As a maximum, with a

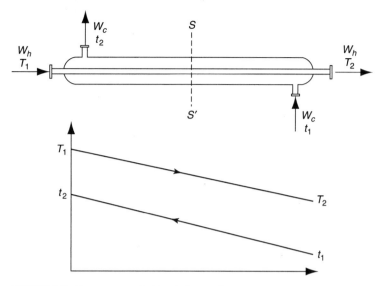

FIGURE 5-6 Countercurrent double-tube heat exchanger.

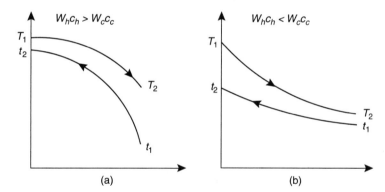

FIGURE 5-7 Temperature diagrams in countercurrent configuration.

very long unit, it is possible that both outlet temperatures could be very close, as shown in Fig. 5-8. This circumstance represents a limitation to the maximum heat transfer that can be achieved with a co-current arrangement that does not exist in countercurrent. In the case of countercurrent heat exchangers, it is possible to have $T_2 < t_2$, as shown in Fig. 5-9.

5-4-2 Thermal Diagrams of Heat Exchangers

There are two different methods to represent the thermal diagrams of a heat exchanger. The first one consists of a plot of the fluid temperatures as a function of position. This is what we used up to now. In this case, the abscissa is the x coordinate, and the graph extends from $x = 0$ to $x = L$.

Since the heat exchanged per unit area is

$$\frac{dQ}{dA} = U(T - t) \qquad (5\text{-}4\text{-}8)$$

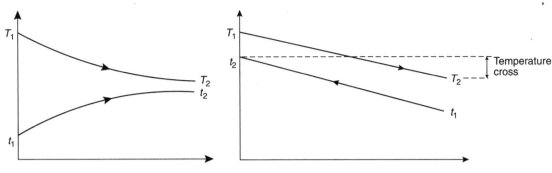

FIGURE 5-8 Temperature diagram in a parallel-flow configuration.

FIGURE 5-9 Temperature diagram in a countercurrent configuration.

this can be written as

$$\frac{dQ}{dx} = \pi D_o U(T - t) \tag{5-4-9}$$

It then can be seen that the heat exchanged per unit length decreases when the temperature difference between the fluids decreases. Since the temperature change of each stream is proportional to the heat received or yielded, the temperature change per unit length of each fluid also decreases with a reduction in ΔT. This results in a reduction in the slopes of the curves in the T-x diagram, as shown in Fig. 5-10 for a parallel-current-configuration exchanger. It also can be seen that in the zone near the outlet end, the temperatures change per unit length is small as a consequence of the reduction in ΔT.

In the case of a countercurrent heat exchanger, the shapes of the curves will be as shown in Fig. 5-11.

The second method of representation, which is usually more appropriate, is the T-Q diagram. The difference is that the abscissa represents the heat exchanged. If the heat capacities of the fluids are constant, the temperature change for each fluid is proportional to the heat exchanged, so the evolutions of the temperatures are straight lines when represented in a T-Q diagram.

Figure 5-12 shows this type of representation for parallel-current and countercurrent heat exchangers. In relation to Fig. 5-12a, we can see that even though the amount of heat transferred between sections A

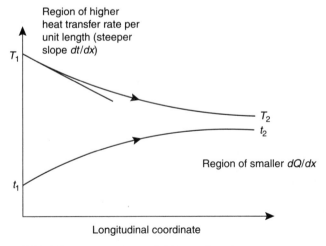

FIGURE 5-10 T-x curves in a parallel-flow configuration.

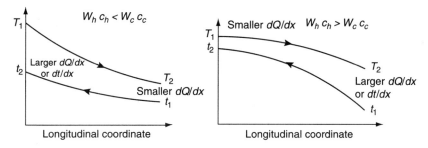

FIGURE 5-11   *T-x* curves in countercurrent configuration.

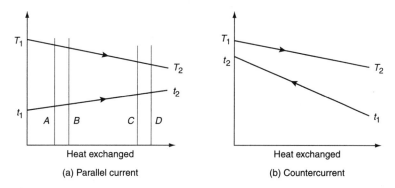

FIGURE 5-12   *T-Q* curves in parallel-current and countercurrent configurations.

and B is the same as that transferred between sections C and D, the heat transfer area existing between sections A and B is smaller than that between sections C and D because a higher temperature difference exists in the first case.

5-4-3   Logarithmic Mean Temperature Difference

Figure 5-13 represents a countercurrent heat exchanger. At the section where the axial coordinate is x, the temperature of the hot fluid is T, and the temperature of the cold fluid is t. At the section corresponding to coordinate $x + dx$, these temperatures are $T + dT$ and $t + dt$. In this case, both increments are positive because both temperatures increase with coordinate x.

A heat balance is

$$dQ = W_h c_h dT = W_c c_c dt \tag{5-4-10}$$

but

$$dQ = UdA(T - t) = U\pi D_o dx(T - t) \tag{5-4-11}$$

From Eq. (5-4-10), we obtain

$$\frac{dQ}{W_h c_h} = dT \tag{5-4-12}$$

$$\frac{dQ}{W_c c_c} = dt \tag{5-4-13}$$

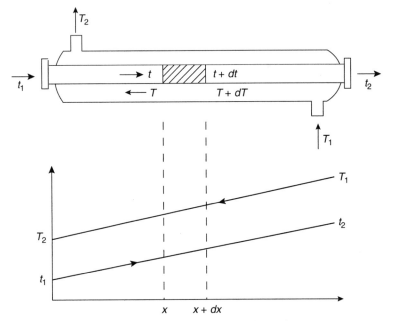

FIGURE 5-13 Temperatures evolution in a countercurrent heat exchanger.

Substracting, we get

$$dQ\left(\frac{1}{W_h c_h} - \frac{1}{W_c c_c}\right) = d(T - t) \tag{5-4-14}$$

Additionally, considering Eq. (5-4-11),

$$\frac{dQ}{U\pi D_o dx} = (T - t) \tag{5-4-15}$$

Dividing Eq. (5-4-14) by Eq. (5-4-15), we get

$$\pi D_o dx U\left(\frac{1}{W_h c_h} - \frac{1}{W_c c_c}\right) = \frac{d(T - t)}{T - t} \tag{5-4-16}$$

This differential equation can be integrated with the following limits:

For $x = 0$, $T - t = T_2 - t_1$.
For $x = L$, $T - t = T_1 - t_2$.

And we get

$$\pi D_o L . U\left(\frac{1}{W_h c_h} - \frac{1}{W_c c_c}\right) = \ln\frac{T_1 - t_2}{T_2 - t_1} \tag{5-4-17}$$

The total heat exchanged in the unit can be expressed as

$$Q = W_h c_h (T_1 - T_2) \tag{5-4-18}$$

Or

$$Q = W_c c_c (t_2 - t_1) \tag{5-4-19}$$

Then

$$\frac{1}{W_h c_h} = \frac{T_1 - T_2}{Q} \tag{5-4-20}$$

$$\frac{1}{W_c c_c} = \frac{t_2 - t_1}{Q} \tag{5-4-21}$$

Substituting Eqs. (5-4-20) and (5-4-21) into Eq. (5-4-17) and reordering gives

$$Q = (\pi D_o L) U \left[\frac{(T_1 - t_2) - (T_2 - t_1)}{\ln \dfrac{T_1 - t_2}{T_2 - t_1}} \right] \tag{5-4-22}$$

The term within brackets is the logarithmic mean temperature difference between both fluids. This means the logarithmic mean between the ΔT values at both ends of the exchanger.

By comparing with Eq. (5-4-5), we see that Eq. (5-4-22) tells us that the mean temperature difference that must be used to establish a relationship between the total area of the unit and the total heat transferred is this *logarithmic mean temperature difference* (LMTD). Equation (5-4-22) has been demonstrated for a countercurrent heat exchanger but is also valid for a co-current configuration. In this case, the expression for LMTD is

$$\text{LMTD} = \frac{(T_1 - t_1) - (T_2 - t_2)}{\ln \dfrac{(T_1 - t_1)}{(T_2 - t_2)}} \tag{5-4-23}$$

We must remark that LMTD can be used as the mean temperature difference for the unit as long as all the hypotheses made in the deduction are fulfilled. These are

- The overall heat transfer coefficient is constant.
- The fluid heat capacities are constant.

These hypotheses are not always satisfied, particularly in cases in which in addition to sensible heat, the fluids also exchange latent heat. In such cases, it is necessary to develop special expressions or to divide the unit into area elements and perform a numerical integration.

Example 5-2 It is desired to cool down 0.34 kg/s of an aqueous solution with a heat capacity similar to water from 60 to 50°C. The cooling fluid will be 0.3 kg/s of water at 25°C. A double-pipe heat exchanger will be used. The external diameter of the interior tube is 0.025 m. The overall heat transfer coefficient is estimated as 1,600 W/(m^2 · K). Calculate the necessary tube length if

a. A countercurrent arrangement is used.
b. A co-current arrangement is used.

Solution   Heat balance:

$$Q = W_h c_h (T_1 - T_2) = 0.34 \times 4{,}180 \times (60 - 50) = 14{,}210 \text{ J/s}$$

Cooling-water outlet temperature:

$$t_2 = t_1 + Q/W_c c_c = 25 + 14{,}210/(0.30 \times 4{,}180) = 36.3°\text{C}$$

Calculation of LMTD:

a. Countercurrent:

$$\text{DMLT} = \frac{(T_1 - t_2) - (T_2 - t_1)}{\ln \dfrac{(T_1 - t_2)}{(T_2 - t_1)}} = \frac{(60 - 36.5) - (50 - 25)}{\ln \dfrac{(60 - 36.5)}{(50 - 25)}} = 24.3°\text{C}$$

b. Co-current:

$$\text{DMLT} = \frac{(T_1 - t_1) - (T_2 - t_2)}{\ln \dfrac{(T_1 - t_1)}{(T_2 - t_2)}} = \frac{(60 - 25) - (50 - 36.3)}{\ln \dfrac{(60 - 25)}{(50 - 36.3)}} = 22.7°\text{C}$$

The required area in each case is

$$A = \frac{Q}{U \cdot \text{DMLT}}$$

This means that

$$14{,}210/(1{,}600 \times 24.3) = 0.36 \text{ m}^2 \qquad \text{for countercurrent}$$

and

$$14{,}210/(1{,}600 \times 22.7) = 0.391 \text{ m}^2 \qquad \text{for co-current}$$

The length of the unit in each case is

$$L = \frac{A}{\pi D_o}$$

This means that

$$L = 0.365/(\pi \times 0.025) = 4.64 \text{ m} \qquad \text{for countercurrent}$$

and

$$L = 0.391/(\pi \times 0.025) = 4.97 \text{m} \qquad \text{for co-current}$$

5-5   CALCULATION OF THE OVERALL HEAT TRANSFER COEFFICIENT AND PRESSURE DROP FOR DOUBLE-TUBE HEAT EXCHANGERS

The overall heat transfer coefficient U can be calculated from the film coefficients according to Eq. (5-3-22). To calculate these coefficients, the usual correlations for heat transfer within tubes must be employed.

5-5-1 Internal Coefficient h_i

h_i is calculated as a function of the Reynolds number, defined as

$$\text{Re} = \frac{D_i \rho v}{\mu} = \frac{D_i G}{\mu} \tag{5-5-1}$$

where G = mass flow density = ρv. The correlations are

1. For a laminar regime (Re < 2,100),[2]

$$\text{Nu} = 1.86[\text{Re} \cdot \text{Pr}(D_i/L)]^{0.33}(\mu/\mu w)^{0.14} \tag{5-5-2}$$

where Nu = Nusselt number = $h_i D_i / k$
Pr = Prandtl number = $c\mu/k$
μ/μ_w = ratio between viscosity at the mean fluid temperature and viscosity at the mean tube wall temperature

In Eq. (5-5-2), L is the length of the fluid path before a mixing or homogenization takes place. We can think that for a double-tube heat exchanger such as that shown in Fig. 5-14, the temperature of the fluid in the annulus becomes uniform when the fluid passes from one tube to the following through the

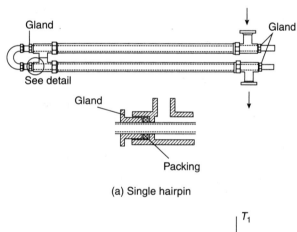

(a) Single hairpin

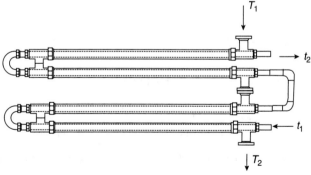

(b) Hairpins in series

FIGURE 5-14 Hairpin heat exchanger.

union tees. Then L should be taken as the length of one pass. For the fluid circulating inside the internal tube, Kern[1] suggests adopting the total length for L. This means the sum of the lengths of all passes. All the physical properties of Eq. (5-5-2), with the exception of μ_w, must be taken at the mean fluid temperature in the tube.

2. For a turbulent regime (Re > 10,000),[2]

$$\text{Nu} = 0.023\text{Re}^{0.8} \cdot \text{Pr}^{0.33}(\mu/\mu_w)^{0.14} \tag{5-5-3}$$

where the physical properties also should be evaluated at the mean fluid temperature, with the exception of μ_w, which is evaluated at the mean wall temperature.

3. The transition regime (2,100 < Re < 10,000) is a highly unstable zone, and all the correlations that have been suggested present important deviations. Some authors[1] perform graphic interpolations between the correlations corresponding to the other two regimes, but the best recommendation is to avoid this zone in the design. If this is not possible, Eq (4-2-48) should be used.

4. For water,[3] the following dimensional correlation has been suggested:

$$h_i = 1,423(1 + 0.0146t)v^{0.8}/D_i^{0.2} \tag{5-5-4}$$

where $h_i = \text{J}/(\text{m}^2 \cdot \text{s} \cdot \text{K})$
$\quad\quad t = $ mean temperature of the water (°C)
$\quad\quad v = $ velocity (m/s)
$\quad\quad D_i = $ internal diameter (m)

Kern[1] presents a graph whose values have good agreement with this correlation, and he extends the application range up to water velocities of 0.3 to 3 m/s, diameters between 0.01 and 0.05 m, and temperatures between 5 and 95°C.

5-5-2 Calculation for the Annulus Fluid

The same correlations used for the tube fluid are still valid, but the internal diameter must be substituted by the annulus equivalent hydraulic diameter. This is defined as

$$D_{eq} = 4 \times \frac{\text{flow area}}{\text{wetted perimeter}} \tag{5-5-5}$$

For heat transfer, the wetted perimeter is that corresponding to the internal tube. This is πD_o. Then

$$D_{eq} = 4 \times \frac{\pi\left(D_s^2 - D_o^2\right)}{4\pi D_o} \tag{5-5-6}$$

where D_s is the internal diameter of the external tube.

5-5-3 Pressure Drop in Double-Tube Heat Exchangers

Each time a heat exchanger must be designed or rated, it is necessary to predict the pressure drop of the fluids through the heat exchanger. This can be calculated with the usual expressions for fluid flow using a friction factor that is obtained as a function of the Reynolds number.

Calculations for the Internal-Tube Fluid. The usual expression to calculate pressure drop through a pipe is

$$\Delta p = 4f\frac{L}{D_i}\rho\frac{v^2}{2} \tag{5-5-7}$$

This equation is valid for isothermal fluids. It is usually corrected for heating or cooling by multiplying by a factor $(\mu/\mu_w)^a$, where $a = -0.14$ for turbulent flow and -0.25 for laminar flow. The friction factor that must be used in Eq. (5-5-7) is the *Fanning factor*. The graphs of the friction factor as a function of the Reynolds number can be obtained from Chap. 4, or the following expressions can be used:

1. For the zone of Re < 2,100, the friction factor can be calculated with the following expression derived from the Hagen Pouiseuille equation:

$$f = 16/\text{Re} \qquad (5\text{-}5\text{-}8)$$

2. In the turbulent region (Re > 2,100), the friction factor depends on the roughness of the tube material.

However some simplified correlations that are valid for particular situations have been suggested. For ¾- or 1-in smooth tubes, a recommended expression[4] is

$$f = 0.0014 + 0.125/\text{Re}^{0.32} \qquad (5\text{-}5\text{-}9)$$

And for commercial steel heat exchanger tubes,[5]

$$f = 0.0035 + 0.264/\text{Re}^{0.42} \qquad (5\text{-}5\text{-}10)$$

Calculations for the Annulus Fluid. In this case, the same expressions are valid, but again, the internal diameter must be substituted by the equivalent hydraulic diameter. The equivalent hydraulic diameter is also defined by Eq. (5-5-5), but the wetted perimeter is, in this case, the friction perimeter, which is the sum of the perimeters of both tubes. This equivalent hydraulic diameter will be called D'_{eq} and is

$$D'_{eq} = 4 \times \frac{\pi\left(D_s^2 - D_o^2\right)/4}{\pi(D_o + D_s)} \qquad (5\text{-}5\text{-}11)$$

5-5-4 Double-Tube Heat Exchanger Applications

The double-tube heat exchanger is one of the simplest pieces of equipment in which perform heat exchange in a continuous mode between two fluids. Figure 5-14 represents a hairpin of a small home-made heat exchanger. A hairpin is formed by two sets of concentric tubes with the corresponding connection pieces.

In cases where a higher heat transfer area is required, it is possible to add several hairpins, one after the other, in a series configuration (see Fig. 5-14b). It can be seen that all parts of the unit are standard pipe fittings, so the assembly of this type of equipment does not required highly skilled labor and can be performed in a modest shop by a pipe fitter.

It also can be appreciated in the figure that in order to avoid fluid leaks, it is necessary to install packing elements at both ends with their corresponding glands. This is necessary because the unit must be disassembled for cleaning, so welded unions are not allowed. In packed unions, it is usual to have some leakage, so the glands must be adjusted periodically. Additionally, disassembling the unit is complicated and time-consuming.

These are the reasons why this type of unit is not used very much in industry. Additionally, the maximum tube length that is usually employed is 6 m. Longer tubes can present too high a deflection and distortion of the annular space, which can cause poor flow distribution. The maximum heat transfer area of a hairpin thus is quite small, and it would be necessary to use a great number of hairpins for most industrial applications.

The unit thus is not compact, and maintenance labor is high. The tubes employed in the construction of these heat exchangers are usually any of the combinations indicated in Table 5-1.

Even though an increase in tube diameter increases the cross sections and makes it possible to handle higher flow rates, it also results in an unfavorable heat transfer-area/flow-area ratio. For example, a 3-cm-diameter tube has a flow area equal to $0.03^2\pi/4 = 0.0007$ m^2. The lateral area per meter of tube is $\pi \times D = \pi \times 0.03 = 0.094$ m^2/m. If, instead, two 2-cm-diameter tubes were employed, the flow area would

TABLE 5-1   Standard Combinations for Double-Tube Heat Exchangers

	Dimensions in Inches					
φ External Tube			φ Internal Tube			
2	¾	1	1¼			
2½	¾	1	1¼			
3	¾	1	1¼	1½	2	
4	¾	1	1¼	1½	2	3

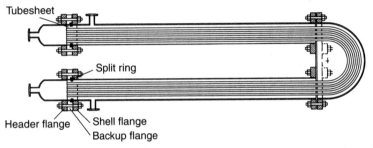

FIGURE 5-15   Multitube hairpin heat exchanger.

be $2 \times \pi \times 0.02^2/4 = 0.00063$ m^2 (a little smaller than before). The lateral heat transfer area, however, is $2\pi D = 2 \times \pi \times 0.02 = 0.125$ m^2/m. This means that by using two 2-cm-diameter tubes, we get a somewhat higher fluid velocity (which means better heat transfer coefficients) and higher heat transfer area than with a single 3-cm-diameter tube. Thus, from the heat transfer point of view, the first option is more convenient. This is the reason why the diameter of the tubes cannot be increased indefinitely. If the flow rates of the fluids require higher flow areas than those indicated in Table 5-1, additional units in parallel should be used.

It is also possible, when it is necessary to increase the flow area for the internal-tube fluid, to manufacture heat exchangers with more than one internal tube inside a single external tube. In this case, the construction is more complex because it is always necessary to be able to disassemble the unit for cleaning.

These units are specifically called *hairpin heat exchangers*. Figure 5-15 shows a sectional view of one hairpin heat exchanger model. The tubes are inserted and welded into a tubesheet. The tubesheet has a slot where a split metallic ring is inserted. This ring is split into two halves so that it can be inserted and withdrawn from the slot. The backup flange compresses the split ring against the body flange, and the unit thus is closed. To disassemble the unit, both end heads are withdrawn, and after removing the split ring, the tubes and tubesheet assembly can be extracted to the right.

These units are usually constructed by specialized companies, and there are several patent-protected closure systems. They can be a cost-effective solution in cases where the flow rates are small and it is necessary to keep a countercurrent configuration. They are limited to a few tubes because for bigger sizes this type of construction becomes difficult. They are not a competitive solution against the shell and tube heat exchangers that will be studied in later chapters. We can then say that hairpin heat exchangers are limited to applications where the required heat transfer area is less than 10 or 15 m^2.

5-6 HEAT EXCHANGER PROCESS SPECIFICATIONS

The thermal design of a heat exchanger consists of the definition of its geometric configuration to perform a certain process service. This service is determined by the process conditions that must be achieved, and the designer must look for a unit capable of satisfying them at a minimum cost.

The process data consist of

- Specifications related to thermal performance
- Specifications related to pressure drop of the streams
- Fouling factors
- Geometric restrictions

We shall see how these specifications must be considered in the design.

5-6-1 Specifications Related to Thermal Performance

The process variables that define the performance of a heat exchanger are related by the thermal balance equations:

$$Q = W_h c_h (T_1 - T_2) \tag{5-6-1}$$

$$Q = W_c c_c (t_2 - t_1) \tag{5-6-2}$$

In this system of equations, there are seven process variables: W_h, W_c, T_1, T_2, t_1, t_2, and Q. For the system to be defined, it is necessary to specify five, and the two remaining can be calculated with Eqs. (5-6-1) and (5-6-2). For example, a usual situation is to know the process flow rates, both inlet temperatures, and the outlet temperature of one of the streams. Then the other outlet temperature and the heat duty Q can be calculated from Eqs. (5-6-1) and (5-6-2).

In other cases, the process conditions require to cool-down a hot stream with flow rate W_h from a certain inlet temperature T_1 to an outlet temperature T_2 using cooling water available at a certain t_1. The design conditions for industrial cooling systems usually define an upper limit for the cooling-water return temperature. In this case, t_2 is also part of the process data, and with Eqs. (5-6-1) and (5-6-2), W_c and Q can be calculated.

When the required five variables are not defined in the process specifications, they must be adopted. For example, it is possible that neither the cooling-water flow rate nor the return temperature were defined to the designer. In this case, the designer has to adopt them. To adopt these values, it must be considered that the greater the water-mass flow to remove a certain heat flow, the smaller is the temperature increase $t_2 - t_1$ that this water will suffer.

This is illustrated in Fig. 5-16. Curve h represents the evolution of the hot fluid. Curve a shows the evolution of the cooling water if a mass flow W_{ca} is employed. If a cooling-water mass flow W_{cb} is used,

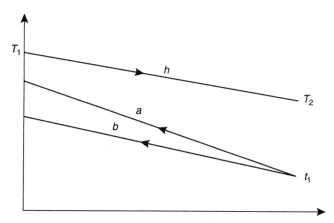

FIGURE 5-16 Effect of the cooling-water flow rate on the temperature difference.

the evolution of the cooling-water temperature is represented by curve b. In this case, $W_{cb} > W_{ca}$. Thus we can see that the mean temperature difference between the hot fluid (curve h) and the cold fluid (curve a or curve b) is higher if a higher amount of cooling water is used (curve b). This higher ΔT means that a smaller heat exchanger area will be necessary to transfer the same Q.

The problem therefore is to economically balance a smaller heat exchanger against an increase in the operational cost owing to higher water consumption (pumping power, water treatment, etc.). If the mean temperature difference between both streams is considerably higher than the temperature change of each stream, the area reduction that can be achieved by increasing the flow rate may not be significant, as illustrated in the following examples.

Example 5-3   Here, 20 kg/s of benzene [$c = 1{,}650$ J/(kg · K)] must be cooled down from 35 to 30°C. It is intended to use 4 kg/s of cooling water, available at 20°C. What would be the reduction in the heat exchanger area if the water flow rate were doubled?

Solution   For the original case, with a cooling-water flow of 4 kg/s, it will be

$$Q = W_h c_h (T_1 - T_2) = 20 \times 1{,}650 \times (35 - 30) = 165{,}000 \text{ J/s}$$

The cooling-water outlet temperature will be

$$t_2 = t_1 + Q/W_c c_c = 20 + 165{,}000/(4 \times 4{,}180) = 29.86°C$$

For a countercurrent configuration, the LMTD is

$$LMTD = \frac{(35 - 29.86) - (30 - 20)}{\ln \dfrac{(35 - 29.86)}{(30 - 20)}} = 7.3°C$$

The required area will be

$$A_1 = Q/(U \cdot \text{LMTD}) = Q/(7.3U)$$

If the water-mass flow is doubled, for the same duty, the new outlet temperature must be

$$t_2' = 20 + 165{,}000/(8 \times 4{,}180) = 24.93°C$$

and then

$$\text{DMLT} = \frac{(35 - 24.93) - (30 - 20)}{\ln \dfrac{(35 - 24.93)}{(30 - 20)}} = 10.03°C$$

The required area thus will be

$$A_2 = Q/(U \cdot \text{DMLT}) = Q/(10.03U)$$

We see that even if the heat transfer coefficients are the same, the higher water flow allows a reduction of about 30 percent in area. This difference can be even higher if we consider that with a higher mass flow, the velocity will be higher and a better heat transfer coefficient probably can be achieved.

Example 5-4   Repeat Example 5-3 considering that the benzene now must be cooled down from 65 to 60°C.

Solution   The heat duty Q is the same as in the preceding case. Then, if 4 kg/s of water is employed, the outlet temperature again will be 29.86°C. Then

$$DMLT = \frac{(65-29.86)-(60-20)}{\ln\dfrac{(65-29.86)}{(60-20)}} = 37.51°C$$

The required area thus will be

$$A_1 = Q/(37.51U)$$

If the water flow rate is doubled, the outlet temperature will be 24.93°C, and then

$$DMLT = \frac{(65-24.93)-(60-20)}{\ln\dfrac{(65-24.93)}{(60-20)}} = 40.03°C$$

And the new area will be

$$A_2 = Q/(40.03U)$$

We see that the area reduction achieved in this case is only 6 percent. This is so because with the higher water flow rate, even though its outlet temperature is 5°C lower, this has little effect on the LMTD value.

5-6-2 Specifications Concerning Pressure Drop

To improve the heat transfer coefficients, it is necessary to increase the fluid velocities. It is better, from the heat transfer point of view, to use a small-diameter, high-length tube rather than a shorter one with a higher diameter, both having the same heat transfer area.

However, the increase in velocity also means higher pressure drops. If the heat exchanger has to be installed in an existing process, the designer must adhere to the maximum allowable pressure drop. If the heat exchanger is for a new process, it may happen that the designer defines the heat exchanger pressure drop, and then the required pumps can be specified to overcome this Δp. In these cases, the problem consists of balancing a higher heat exchanger cost against a higher pumping power, and the most cost-effective solution must be adopted.

5-6-3 Fouling Factors

Adoption of the fouling factor should not be the designer's responsibility. The fouling factor is a datum that belongs to the process technology, and it must be supplied to the design engineers just like the physical properties of the fluids handled.

In cases where the thermal design is performed by the equipment vendor, it is essential that the purchaser include the R_f value in the technical specifications because in this way the proposals received from different vendors will be homogeneous and comparable. In certain cases, the designer can suggest, based on his or her expertise, fouling factors that can be accepted or not by the purchaser, but in any case, both parties must agree as to what the design fouling factor will be.

Tables in App. F can be used as a guide in the absence of more specific information.

5-6-4 Geometric Constraints

Usually there exist certain design constraints based on mechanical and layout considerations. They may be a maximum tube length, position (horizontal or vertical), relative nozzle orientation, etc. It is convenient for the designer to know the characteristics of the place where the heat exchanger has to be installed. It must be noted that a design problem admits more than one solution, and the designer must choose the most suitable for the particular case.

5-7 DESIGN PROCEDURE FOR A DOUBLE-TUBE HEAT EXCHANGER

We shall describe the necessary steps to design a double-tube heat exchanger. We assume that all the process conditions previously explained are completely defined. This means that we know the flow rates, inlet and outlet temperatures, allowable pressure drops in both streams, and fouling resistances.

We also assume that a maximum tube length L, also has been defined. The heat exchanger thus will be designed using this tube length, and we shall calculate how many hairpins in series are necessary to achieve the required heat duty. The unit configuration will be similar to that shown in Fig. 5-14b.

The design procedure consists of the following steps:

1. Selection of Tube Diameter.   The usual combinations between internal and external tubes diameters that can be adopted are shown in Table 5-1. The smaller the flow area for both fluids, the higher are the velocity and heat transfer coefficients, but this also results in a higher pressure drop. As a first approach, it is suggested to select the tube diameter considering the fluid velocities. For example, in case of low-viscosity liquids, it is advisable to work with velocities of about 1–2 m/s. For highly viscous fluids, this value must be lower.

The flow area for each fluid then will be

$$a = \frac{W}{\rho v} \tag{5-7-1}$$

Once the flow area for the annulus and tube have been calculated, it is possible to select the required tube diameters. If the flow rates are high, it may be necessary to add units in parallel.

At the end of the calculations, when the final length is known, the pressure drop for each fluid can be calculated, and it is possible that this may require a change in the selected diameters. This velocity criterion thus must be considered only as a first approximation to design.

2. Calculation of the Film Coefficients.   Once the fluid velocities are known, it is possible to calculate the Reynolds numbers for the tube and annulus. Then it is possible to obtain the film heat transfer coefficients with the correlations presented in Sec. 5-5.

The physical properties of both fluids must be obtained at the average temperature between inlet and outlet. The viscosities at the wall temperature cannot be calculated from the beginning because this temperature is unknown. Then, as a first approximation, it is assumed that the factors $(\mu/\mu_w)^{0.14}$ are unity.

With this simplification, a first value of the film coefficients h_o and h_{io} is obtained. Then, by equating the heat transfer rates at both sides of the tube wall, we get

$$h_o(T - T_w) = h_{io}(T_w - t) \tag{5-7-2}$$

if the annulus fluid is the hot fluid or

$$h_{io}(T - T_w) = h_o(T_w - t) \tag{5-7-3}$$

if the tube fluid is the hot fluid.

In Eqs. (5-7-2) and (5-7-3), T and t are the average temperatures of the hot and cold fluids, respectively, whereas T_w is a mean temperature of the wall, not necessarily coincident with the average of the wall temperatures at both ends of the exchanger. From Eqs. (5-7-2) and (5-7-3), T_w can be obtained. With T_w, it is possible to calculate the correction factors $(\mu/\mu_w)0.14$, and the previous values for h_o and h_{io} can be corrected. An iterative procedure could be proposed, but usually the first correction is good enough.

3. Calculation of the Overall Heat Transfer Coefficient.   With the individual film coefficients and the fouling resistance available, the overall heat transfer coefficient can be calculated:

$$\frac{1}{U} = \frac{1}{h_o} + \frac{1}{h_{io}} + Rf \tag{5-7-4}$$

4. Calculation of the LMTD.   With the inlet and outlet temperatures available, the LMTD can be calculated by Eqs. (5-4-22) or (5-4-23) depending on the flow configuration.

5. Calculation of the Heat Transfer Area.   The heat transfer area can be calculated as

$$A = \frac{Q}{U \cdot \text{DMLT}} \tag{5-7-5}$$

6. Calculation of the Total Tube Length and Number of Tubes.   The required tube length then will be

$$L = \frac{A}{\pi D_o} \tag{5-7-6}$$

And the number of tubes in series is obtained as

$$n_t = L/L_t \tag{5-7-7}$$

7. Calculation of the Pressure Drop.   The friction factors for both streams can be calculated with the correlations presented in Sec. 5-5, and the pressure drop for each fluid will be

$$\Delta p_t = 4f \frac{L}{D} \rho \frac{v^2}{2} \left(\frac{\mu}{\mu_w} \right)^a \tag{5-7-8}$$

where $a = -0.14$ for Re > 2,100 and $a = -0.25$ for Re < 2,100. In Eq. (5-7-8), D must be replaced by D_i or D_{eq} as required.

For heat exchangers with more than one tube, the annulus fluid suffers an additional pressure drop when passing from one tube to the next through the connecting tees. This pressure drop can be calculated as

$$\Delta p_r = \frac{n_t}{2} \frac{\rho v^2}{2} \tag{5-7-9}$$

Then, for the annulus fluid, the total pressure drop is

$$\Delta p = \Delta p_t + \Delta p_r \tag{5-7-10}$$

8. Changes to the Original Design.   If the calculated pressure drop is excessive, it will be necessary to increase the flow area, either by increasing the tube diameters or installing more branches in parallel. If the calculated pressure drop is smaller than allowable, then a reduction in the flow area can be attempted. In both cases, the design procedure must be reinitiated.

Since it is not possible to modify the flow area of the internal tube without affecting the annulus, the design of this type of equipment is difficult to optimize. Many times it is necessary to accept a poor utilization of the allowable pressure drop in one stream in order to satisfy the requirement in the other.

In any case, as was explained, this type of heat exchanger is limited to low-area and low-cost applications. Later on in this text we shall study other types of heat exchangers where, owing to the existence of a greater number of design variables, it is possible to optimize the allowable pressure drops in both streams.

Once the thermal design of a heat exchanger is completed, the mechanical design must be performed. This means verifying tubes thickness, selecting materials, rating nozzles, choosing gaskets, etc. and elaborating the mechanical drawings as well. These subjects are under the scope of the mechanical engineering and will not be treated in this text.

Example 5-5   Here, 0.8 kg/s of a solvent whose properties are indicated below must be cooled down from 40 to 30°C. A stream of ethylene glycol at 5°C will be used as coolant. The outlet temperature of ethylene glycol is limited to 25°C. Design a suitable heat exchanger for this service. The pressure drop for both streams must not be higher than 110,000 N/m². The combined fouling resistance must be 5×10^{-4} (s · m² · K)/J. The properties of the solvent at 35°C are

$\rho = 790$ kg/m³

$c = 1,922$ J/(kg · K)

$\mu = 0.95$ cP $= 0.95 \times 10^{-3}$ kg/ms

$k = 0.187$ J/(s · m · K)

Solution The properties of ethylene glycol are

$\rho = 1,010$ kg/m³

$c = 2,340$ J/(kg · K)

$k = 0.264$ J/(s · m · K)

Viscosity (kg/ms):

T, °C	10	20	30	35
μ	0.028	0.020	0.014	0.012

Ethylene glycol is used as intermediate cooling medium in refrigeration applications owing to its low freezing point. These substances are called *brines*. In this case, the operating condition of the refrigeration cycle establishes a maximum return temperature for the hot brine, and the brine mass flow must be selected not to surpass this limit.

If we select the outlet temperature as the allowable maximum, the mass flow will be at the minimum. The heat duty is

$$Q = W_h C_h (T_1 - T_2) = 0.8 \times 1,922 \times (40 - 30) = 15,376 \text{ J/s}$$

and the brine mass flow is

$$W_c = Q/[c_c(t_2 - t_1)] = 15,376/[2,340(25 - 5)] = 0.32 \text{ kg/s}$$

1. Selection of Diameters

For a preliminary selection of diameters, we shall adopt a solvent velocity of 1 m/s. The required flow area thus will be

$$a_t = W_h/\rho v = 0.80/(790 \times 1) = 1.01 \times 10^{-3} \text{ m}^2$$

If the solvent circulates through the internal tube, this section corresponds to a diameter of

$$D_i = \sqrt{\frac{4a_t}{\pi}} = 0.036 \text{ m}$$

This is roughly the diameter of a 1¼-in Schedule 40 pipe with dimensions

$$D_i = 0.035 \text{ m}$$

$$D_o = 0.0421 \text{ m}$$

$$a_t = \pi D_i^2/4 = 9.62 \times 10^{-4} \text{ m}^2$$

If the exchanger is built with this pipe, the solvent velocity will be

$$v = W/(\rho\, a_t) = 0.80/(790 \times 9.62 \times 10^{-4}) = 1.05 \text{ m/s}$$

It can be seen in Table 5-1 that a possible standard combination consists in using a 2-in external pipe. Let's select a 2-in Schedule 40 pipe. Its internal diameter is

$$D_s = 0.0525 \text{ m}$$

The annular flow area thus is

$$a_S = \pi \frac{D_s^2 - D_o^2}{4} = \pi \frac{0.0525^2 - 0.0421^2}{4} = 7.75 \times 10^{-4} \text{m}^2$$

The brine velocity into the annulus thus is

$$v = W_c/(a_s \rho) = 0.32/(7.75 \times 10^{-4} \times 1,010) = 0.408 \text{ m/s}$$

This is acceptable for a viscous fluid.

If too low a velocity were obtained in the preceding step, using a higher brine mass flow may be considered to get better heat transfer coefficients.

2. Calculation of the Heat Transfer Coefficients

Calculations for the solvent:

$$\text{Re}_t = \frac{D_i v \rho}{\mu} = \frac{0.035 \times 1.05 \times 790}{0.95 \times 10^{-3}} = 30,560$$

$$\text{Pr} = \frac{c\mu}{k} = \frac{1,922 \times 0.95 \times 10^{-3}}{0.187} = 9.76$$

$$h_i = 0.023 \text{Re}_t^{0.8} \text{Pr}^{0.33} \left(\frac{\mu}{\mu_w}\right)^{0.14} \frac{k}{D_i} = 0.023 \times 30,560^{0.8} \times 9.76^{0.33} \times \left(\frac{\mu}{\mu_w}\right)^{0.14} \times \frac{0.187}{0.035}$$

$$= 1011 \left(\frac{\mu}{\mu_w}\right)^{0.14}$$

As a first approximation, we neglect the viscosity correction factors. Then

$$h_{io} = h_i \frac{D_i}{D_o} = 1,011 \frac{0.035}{0.0421} = 839 \text{ J/(s} \cdot \text{m}^2 \cdot \text{K)}$$

Calculations for the brine:

$$\text{Equivalent diameter} = 4R_H$$

$$R_H = \frac{\text{flow area}}{\text{internal tube perimeter}} = \frac{7.75 \times 10^{-4}}{\pi \times 0.0421} = 5.85 \times 10^{-3}$$

$$\therefore D_{eq} = 4 \times 5.85 \times 10^{-3} = 0.023 \text{ m}$$

The mean temperature of ethylene glycol is 15°C. At this temperature, the viscosity is 0.024 kg/(m · s). The Reynolds number for heat transfer is

$$\text{Re}_s = \frac{D_{eq} v \rho}{\mu} = \frac{0.023 \times 0.408 \times 1,010}{0.024} = 394$$

The heat transfer coefficient can be calculated from

$$\frac{h_o D_{eq}}{k} = 1.86 \left(Re_s \, Pr \frac{D_{eq}}{L} \right)^{0.33} \left(\frac{\mu}{\mu_w} \right)^{0.14}$$

Let's assume that the exchanger shall be built with 6-m pipes. Considering L as the length of each tube and neglecting as a first approximation the viscosity correction, we get

$$h_o = 1.86 \frac{k}{D_{eq}} \left(Re_s \, Pr \frac{D_{eq}}{L} \right)^{0.33} = 1.86 \frac{0.264}{0.023} \left[394 \times \left(\frac{2340 \times 0.024}{0.264} \right) \times \frac{0.023}{6} \right]^{0.33} = 144 \, J/(s \cdot m^2 \cdot K)$$

Calculation of the wall temperature:

$$h_o(T_w - t) = h_{io}(T - T_w)$$

where T = mean temperature of the hot fluid = 35°C
$\qquad$ t = mean temperature of the cold fluid = 15°C
$\qquad$ T_w = mean wall temperature

$\qquad \therefore \; 144(T_w - 15) = 839(35 - T_w)$ $\qquad$ from where T_w = 32°C

At this temperature, the viscosity of ethylene glycol is 0.013 kg/(m · s). The correction factor for ethylene glycol is

$$\left(\frac{\mu}{\mu_w} \right)^{0.14} = \left(\frac{0.024}{0.013} \right)^{0.14} = 1.08$$

Then

$$h_o = 144 \times 1.08 = 156$$

The correction factor for the solvent is negligible because the wall temperature is very close to the mean fluid temperature.

3. Calculation of the Overall Heat Coefficient

$$U = \left(\frac{1}{h_o} + \frac{1}{h_{io}} + Rf \right)^{-1} = \left(\frac{1}{155} + \frac{1}{389} + 5 \times 10^{-4} \right)^{-1} = 124 \, J/(s \cdot m^2 \cdot K)$$

4. Calculation of the LMTD

$$LMTD = \frac{(T_1 - t_2) - (T_2 - t_1)}{\ln \frac{(T_1 - t_2)}{(T_2 - t_1)}} = \frac{(40 - 25) - (30 - 5)}{\ln \frac{(40 - 25)}{(30 - 5)}} = 19.5°C$$

5. Calculation of the Heat Transfer Area
The required area thus will be

$$A = \frac{Q}{U(LMTD)} = \frac{15,376}{124 \times 19.5} = 6.36 \, m^2$$

The area of each tube is

$$\pi D_o L = \pi \times 0.0421 \times 6 = 0.794 \, m^2$$

Then the number of tubes in series will be

$$6.36/0.794 = 8 \text{ tubes}$$

Then the proposed unit consists in four hairpins (eight tubes) 6 m in length made with 1¼-in internal tubes and 2-in external tubes connected in series. Now it is necessary to check pressure drops.

6. Calculation of Pressure Drops

Calculations for ethylene glycol: The equivalent diameter of the annulus for pressure-drop calculation is

$$D'_{eq} = D_s - D_o = 0.0525 - 0.0421 = 0.0104 \text{ m}$$

The Reynolds number is

$$\text{Re}'_s = \frac{D'_{eq} v \rho}{\mu} = \frac{0.0104 \times 0.408 \times 1010}{0.024} = 178$$

For laminar flow, it will be

$$f = 16/\text{Re}'_s = 0.090$$

$$\Delta p = 4f \frac{L}{D'_{eq}} \frac{\rho v^2}{2} \left(\frac{\mu}{\mu_w}\right)^{0.25} = 4 \times 0.090 \times \frac{8 \times 6}{0.0104} \times 1{,}010 \times \frac{0.408^2}{2} \times \left(\frac{0.013}{0.024}\right)^{0.25} = 119{,}000 \text{ N/m}^2$$

The pressure-drop terms, corresponding to the connections between hairpins, $(4 \times \rho v^2 /2)$ are in this case negligible.

We see that the brine pressure drop slightly exceeds the allowable limit, and we can assume that it will be tolerated by the process. If it were desired to reduce this Δp, an option would be to slightly reduce the brine mass flow. Even though this option would result in a brine outlet temperature higher than the allowable 25°C, the excess also would be very small.

If neither of these options is acceptable, the only possibility is to increase the tube diameter to have a higher flow area.

Calculations for the solvent: The friction factor is

$$f = 0.0035 \times \frac{0.264}{\text{Re}_t^{0.42}} = 0.0035 \times \frac{0.264}{30{,}560^{0.42}} = 6.9 \times 10^{-3}$$

and it results in

$$\Delta p = 4f \frac{L}{D_i} \frac{\rho v^2}{2} = 4 \times 6.9 \times 10^{-3} \times \frac{6 \times 8}{0.035} \times 790 \times \frac{1.05^2}{2} = 16{,}900 \text{ N/m}^2$$

This is much smaller than the allowable maximum. This means that the velocity of the solvent could be increased, which would improve h_i. However, since the controlling resistance is the ethylene glycol film, this would have little effect on the overall heat transfer coefficient U.

Additionally, if the internal tube diameter is reduced, the annulus fluid velocity and h_o also would be reduced . It would not be possible to compensate for this effect with a reduction in the external tube diameter because the annulus fluid pressure drop is very close to the maximum. Thus this design can be considered definitive.

Example 5-6   Calculate the necessary tube length for the conditions of Example 5-6 if, instead of ethylene glycol, water at the same inlet temperature (5°C) and with the same pressure-drop limitations is employed.

Solution Since a lower-viscosity fluid is employed, a higher velocity could be adopted. If we choose 2 m/s for the water, its mass flow would be

$$W = \rho v a_s = 1{,}000 \times 2 \times 7.75 \times 10^{-4} = 1.55 \text{ kg/s}$$

The temperature change for the water then will be

$$t_2 - t_1 = \frac{Q}{W_c c_c} = \frac{15{,}376}{1.55 \times 4{,}180} = 2.4°\text{C}$$

Then

$$t_2 = 5 + 2.4 = 7.4°\text{C}$$

The LMTD then will be

$$\text{LMTD} = \frac{(40 - 7.4) - (30 - 5)}{\ln \dfrac{(40 - 7.4)}{(30 - 5)}} = 28.6°\text{C}$$

To calculate the film heat transfer coefficient, we shall use Eq. (5-5-4) with the annulus equivalent diameter

$$h_o = 1{,}423(1 + 0.0146t)\frac{v^{0.8}}{D_{eq}^{0.2}} = 1{,}423\,(1 + 0.0146 \times 6.2)\left(\frac{2^{0.8}}{0.023^{0.2}}\right) = 5{,}750 \text{ J/(s} \cdot \text{m}^2 \cdot \text{K)}$$

If we neglect the change in the physical properties of the solvent with temperature, the film heat transfer coefficient for the internal tube would be the same as before. Then

$$U = \left(\frac{1}{5{,}750} + \frac{1}{839} + 5 \times 10^{-4}\right)^{-1} = 535 \text{ J/(s} \cdot \text{m}^2)$$

Thus

$$A = \frac{Q}{U(\text{LMTD})} = \frac{15{,}376}{535 \times 28.6} = 1 \text{ m}^2$$

The required length is

$$\frac{A}{\pi D_o} = \frac{1.0}{\pi \times 0.042} = 7.6 \text{ m}$$

A hairpin with two tubes 4 m in length can be used (area = 1.05 m²).

Calculation of the Water Pressure Drop
The viscosity of water at the mean temperature of 6°C is 1.5×10^{-3} kg/(m · s). Then

$$\text{Re}_s = \frac{D_{eq} v \rho}{\mu} = \frac{0.0104 \times 2 \times 1{,}000}{1.5 \times 10^{-3}} = 13{,}866$$

For turbulent flow,

$$f = 0.0035 + \frac{0.264}{\text{Re}^{0.42}} = 0.0035 + \frac{0.264}{13{,}866^{0.42}} = 8.3 \times 10^{-3}$$

$$\Delta p = 4f \frac{L}{D_{eq}} \frac{\rho v^2}{2} = 4 \times 8.3 \times 10^{-3} \frac{8}{0.0104} \times 1{,}000 \times \frac{2^2}{2} = 51{,}076 \text{ N/m}^2$$

Again, the pressure drop for the connecting tees can be neglected.

We see that it is possible to cool down the solvent with a heat exchanger whose area is six times smaller than before and with a lower pressure drop. This is due to (1) a much higher heat transfer coefficient because a lower-viscosity fluid is employed and (2) a higher LMTD because a higher coolant flow rate is employed.

It can be noticed that the controlling resistance is now in the solvent film. This means that an important reduction in the heat transfer area could be obtained by increasing its velocity. It then would be possible to improve the design by adopting a smaller tube diameter. Also notice that the reduction in the tube length results in a reduction in the pressure drop, so there is a higher margin to increase the velocities.

Example 5-7 Calculate what will be the solvent outlet temperature in the exchanger of Example 5-6 at the beginning of operation when the fouling resistance is nil.

Solution If the water flow rate is maintained, the overall heat transfer coefficient now will be

$$U_c = \frac{1}{5,750} + \frac{1}{839} = 732 \text{ J/(s} \cdot \text{m}^2 \cdot \text{K)}$$

With this coefficient, a higher heat flux will be achieved. This will result in a lower solvent outlet temperature and a higher water outlet temperature. In turn, this means a lower LMTD than in the preceding example.

The equilibrium point is reached for the values of t_2, T_2, and Q that satisfy the following set of equations:

$$Q = W_c c_c (t_2 - t_1) \tag{1}$$

$$W_c c_c (t_2 - t_1) = W_h c_h (T_1 - T_2) \tag{2}$$

$$Q = UA \frac{(T_1 - t_2) - (T_2 - t_1)}{\ln \dfrac{(T_1 - t_2)}{(T_2 - t_1)}} \tag{3}$$

This system can be solved by a trial-and-error procedure.

We shall assume t_2 and calculate Q with Eq. (1). Then, with the same t_2, we calculate T_2 with Eq. (2), and by substituting these values in Eq. (3), we calculate a new Q. The t_2 value for which the same Q is obtained by both calculations is the solution of the system (see Fig. 5-17).

Calculation Progress

Iteration	1	2	3	4
t_2	10	9	8	8, 5
Q	32,300	25,916	19,437	20,084
T_2	18.9	23.14	27.36	26.93
Q	16,081	18,451	20,660	20,440

A water outlet temperature of 8.1°C is obtained. This corresponds to a solvent outlet temperature of 27°C, which means that it is possible to cool down the solvent to a temperature 8°C lower than with a dirty unit.

Example 5-8 What should be the cooling-water mass flow for the unit in Example 5-7 (clean conditions) to have the design (30°C) outlet temperature in the solvent?

Solution If the water flow rate is reduced, the water will heat up to a higher temperature. This would reduce the LMTD, and the solvent outlet temperature also would increase. Additionally, the reduction in the water flow will result in a lower heat transfer coefficient h_o.

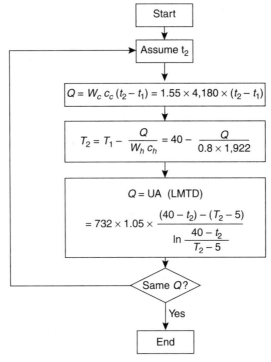

FIGURE 5-17 Flow diagram for Example 5-7.

The problem can be solved by assuming different W_c values. For each W_c, the water outlet temperature will be calculated assuming that the heat transferred is the same as in Example 5-6 (15,376 J/s). We then shall calculate the LMTD. The required heat transfer coefficient to transfer 15,376 W with that temperature difference will be

$$U = \frac{Q}{A(\text{LMTD})}$$

This value is compared with the heat transfer coefficient that can be attained in the exchangers with the assumed flow rates, which is

$$U = \left(\frac{1}{h_o} + \frac{1}{h_{io}} \right)^{-1}$$

The water flow rate that makes both values equal is the one that allows the desired heat duty. The calculation flow diagram is shown in Fig. 5-18, and the results are listed in the table below.

Trial	1	2	3	4
W_c	0.5	0.4	0.35	0.38
t_2	12.35	14.19	15.5	14.68
LMTD	26.3	25.4	24.26	25.15
$U(1)$	556	576	603	582
t	8.6	9.6	10.2	9.8
h_o	2350	2000	1830	1950
$U(2)$	620	590	570	586

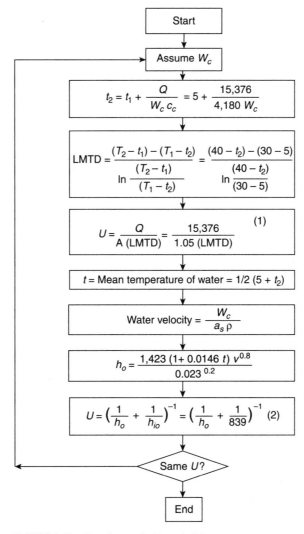

FIGURE 5-18   Flow diagram for Example 5-8.

We see that it is possible to reach the desired temperature by reducing the flow rate to one-third the design value.

GLOSSARY

a = flow area (m^2)

A = heat transfer area (m^2)

A_i = internal tube area (m^2)

A_o = external tube area (m^2)

A_m = mean tube area (m^2)

c = specific heat [J/(kg · K)]

D = diameter (m)

D_i = internal diameter of the internal tube (m)

D_o = external diameter of the internal tube (m)

D_s = internal diameter of the external tube (m)

D_{eq} = equivalent diameter for h calculations (m)

D'_{eq} = equivalent diameter for Δp calculations (m)

f = friction factor

G = mass velocity (kg/s) = ρv

h = film coefficient [J/(s · m^2 · K)]

h_o = external film coefficient [J/(s · m^2 · K)]

h_i = internal film coefficient [J/(s · m^2 · K)]

h_{io} = internal film coefficient referred to the external area [J/(s · m^2 · K)]

i = specific enthalpy (J/kg)

k = thermal conductivity [J/(s · m · K)]

L = heat transfer length (m)

L_t = tube length (m)

P = pressure (N/m^2)

Pr = Prandtl number

Q = heat exchanged (J/s)

R_f = fouling resistance

Re = Reynolds number

T = hot-fluid temperature (K or °C)

t = cold-fluid temperature (K or °C)

v = velocity (m/s)

W = mass flow (kg/s)

x = length coordinate (m)

λ = heat of condensation (J/kg)

μ = viscosity [kg/(m · s)]

ρ = density (kg/m^3)

Subscripts

1 = inlet

2 = outlet

c = cold

h = hot

i = internal

o = external

w = wall

REFERENCES

1. Kern D.: *Process Heat Transfer.* New York: McGraw-Hill, 1954.
2. Sieder E.N., Tate G.E.: "Heat Transfer and Pressure drop of Liquids in Tubes" *Ind Eng Chem* 28:1429–1436, 1930.
3. Perry R., Chilton C.: *Chemical Engineers Handbook,* 5th ed., New York: McGraw-Hill, 1973.
4. Drew T. B., Koo E. C., McAdams W.H.: *Trans AlCh E* 28:56–72, 1932.
5. Wilson R. E., McAdams W. H., Seltzer H.: *Ind Eng Chem* 14:105–119, 1922.

CHAPTER 6
SHELL-AND-TUBE HEAT EXCHANGERS

6-1 DESIGN AND CONSTRUCTION STANDARDS

The quality of a heat exchanger or any other piece of process equipment depends on a great number of constructive details ranging from the quality of the construction materials to the way in which it is packed for shipment, passing through the mechanical design of its components, constructive techniques, tolerances, tests, and many others. It would be impossible for someone requesting a quotation for the supply of process equipment or any other industrial material to specify all the necessary details to ensure the quality of the supply. However, the specification of all these details is necessary to be sure that the proposals coming from different vendors will be comparable and will accommodate the quality level intended.

To solve this problem, engineers make use of standards prepared by standards development organizations. These are governmental or private associations of companies related to a certain type of activity, either users or vendors that develop, publish, and maintain constructive standards and recommendations for almost any conceivable product.

These standards are prepared by specialist in the particular subject, taking into consideration the interests of purchasers and vendors, and they establish all the requirements to which the construction of the product must adhere in its different stages. The standards then are made public for use. Their principal objective is to provide a common reference point for purchasers and vendors, thus avoiding the necessity of specifying all the aspects previously mentioned by just specifying that the provision has to comply with the prescriptions of a certain standard.

Usually adherence to these standards is not mandatory by law. However, the use of standards is beneficial for vendors and purchasers because standards make it possible to clearly define similar and uniform conditions for all the vendors participating in a bid process and to prevent further discussions regarding the quality of the provision.

A good standard must specify all the details that may result in conflicts between vendors and purchasers, thus avoiding ambiguous terms such as *good engineering practice, usual practice,* etc. Frequently, different standards development organizations produce sector-based standards that respond to the particular requirements of a specific type of industry. For example, the American Petroleum Institute (API) is an association that normally develops standards for the severe service conditions of the oil-processing industry. Often the quality requirements for this industry are not necessary in other applications, and other standards development organizations such as the Tubular Exchanger Manufacturers Association (TEMA) or the American National Standards Institute (ANSI) may have different standards for the same type of product. The engineer has to choose the standard that best adapts to the particular requirements of the project. The adoption of an unnecessarily stringent standard will increase the cost of the project without any benefit.

Heat Exchanger Manufacturing Standards. In what follows, the main features and characteristics of shell-and-tube heat exchangers will be explained. This is by far the most common heat exchanger type employed in the process industry. The design and construction of these units, as with any other pressure vessel, normally follow the prescriptions of the American Society of Mechanical Engineers (ASME) *Boiler and Pressure Vessel Code,* Section VII, Division 1. In U.S.A this code is mandatory in many States, and because it is a recognized code, it is adopted in many projects around the world.

TEMA issues the broadly known TEMA Standards.[1] They are prepared and updated by a technical committee of association members. These standards complement the ASME Code in the particular subjects related to heat exchangers. The ASME stamp is required for all the exchangers manufactured and designed according to TEMA Standards.

Three classes of mechanical standards are defined:

- *Class R*: designates heat exchangers for the severe requirements of petroleum and other related processing applications.
- *Class C*: indicates generally moderate requirements of commercial and general process applications.
- *Class B*: specifies design and application for chemical process service.

Heat exchanger Class C and B designs are more compact and economical than Class R designs. The users must select the design/fabrication code designation for their individual application. The standards do not include recommendations for the selection of a particular class. Many chemical plants select the most severe class, Class R, because they prefer a more robust construction.

In design or construction aspects that are general for any pressure vessel, TEMA Standards refer to the ASME code (e.g., construction materials, hydraulic tests, tensile strengths, etc.), but the TEMA Standards include specific details for heat exchangers, such as tubesheet design, tube-to-tubesheet joints, tolerances, etc.

There are other heat exchangers design and construction codes than the TEMA Standards. For example, API Standard 660[7] establishes additional requirements for the severe service of the petroleum industry. However, since the TEMA Standards are the best-known heat exchanger standards, We shall make reference to them in the following topics. It must be understood that adherence to these standards is voluntary, and another standard or no standard can be adopted based on the particular characteristics of the project.

6-2 PRINCIPAL COMPONENTS OF A SHELL-AND-TUBE HEAT EXCHANGER

The double-tube heat exchanger described in Chap. 5 is limited to applications in which the heat flux Q and the fluids flow rates are small. If higher flow rates have to be processed, it would be necessary to install several units in parallel, and the resulting design would not be compact, and maintenance would be difficult.

To avoid these inconveniences, the shell-and-tube heat exchanger is normally used. The basic idea consists of installing several internal tubes into another tube of much bigger diameter, designated as the *shell*. This construction is illustrated in Fig. 6-1, and Table 6-1 provides the nomenclature for the figures.

The shell is closed at both ends by the tubesheets. These normally have a considerable thickness and are penetrated by the tubes. The tube-to-tubesheet joints must be hermetically sealed to avoid fluid leakage from one side of the tubesheet to the other. One possible joint method is the welded construction indicated in the detail of Fig. 6-1.

The tubesheets are bolted to the heads, which act as collectors and distributors of the fluid circulating inside the tubes. This fluid enters the exchanger through a head nozzle and goes into and through the tubes up to the opposite head, where it exits. The other fluid goes into the shell through an inlet nozzle, fills all the

TABLE 6-1 Component Nomenclature

1 Channel cover	14 Tubes	27 Stuffing box
2 Channel	15 Shell	28 Gland
3 Channel nozzle	16 Expansion joint	29 Floating tubesheet and skirt
4 Bonnet	17 Transversal baffle	30 Loose flange
5 Bonnet nozzle	18 Fixed tubesheet flange	31 Split ring
6 Shell flange, stationary head side	19 Pass partition plate	32 Reversing bonnet
7 Bolting	20 Tie rod and spacers	33 Floating head coverplate
8 Gasket	21 Vent connection	34 Shell cover
9 Stationary tubesheet	22 Split backing flange	35 Lifting lug
10 Fixed tubesheet	23 Floating tubesheet	36 Drain plug
11 Shell nozzle	24 Floating head	37 Shell cover flange
12 Shell flange rear end side	25 Lantern ring	38 Floating head cover flange
13 Impingement plate	26 Packing	39 Return bonnet flange

Note: Reference numbers correspond to Figs. 6-1, 6-13, 6-22, 6-26, 6-28, 6-29, and 6-31.
Source: Used with permission from the TEMA Standards.

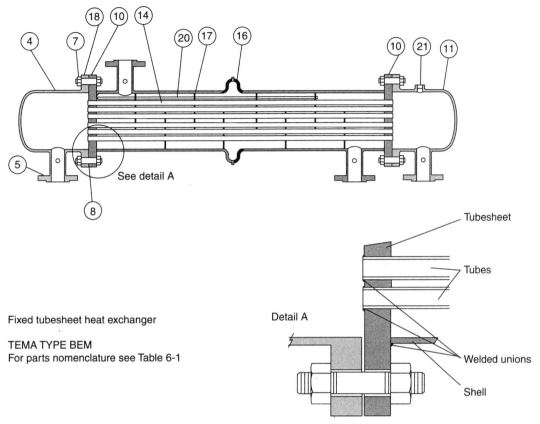

Fixed tubesheet heat exchanger

TEMA TYPE BEM
For parts nomenclature see Table 6-1

Detail A

Tubesheet

Tubes

Welded unions

Shell

FIGURE 6-1   Fixed tubesheet heat exchanger.

space surrounding the tubes, and moves toward the outlet nozzle, where it exits. So both fluids are separated by the tubes wall, which constitute the heat transfer area . If N is the number of tubes of the heat exchanger, L is the length of the tubes, and D_o is the external diameter of the tubes, the heat transfer area will be

$$A = \pi D_o N L$$

6-3 BAFFLES

We know that the heat transfer coefficients increase when fluids velocity and turbulence increase. The shell-side fluid velocity can be modified by installing a set of baffles that force the fluid movement in a direction perpendicular to the axis of the tubes. By changing the separation between these baffles, it is possible to change fluid velocity.

The most common type of baffle is the segmental baffle. Segmental baffles are circular plates, about the same diameter as the shell, to which a horizontal or vertical cut has been made. The baffles are installed into the shell in such a way that the cuts of two consecutive baffles are rotated 180 degrees. The baffles obviously must be perforated to allow the tubes to pass through them. This type of exchanger is illustrated in Fig. 6-2. The shell fluid must follow the illustrated path to reach the outlet nozzle.

It then can be seen that the shell fluid velocity has a component in the direction perpendicular to the tubes and a component in the direction parallel to the axis of the tubes. The velocity component in the direction perpendicular to the tubes is usually the most important for heat transfer. The fluid velocity

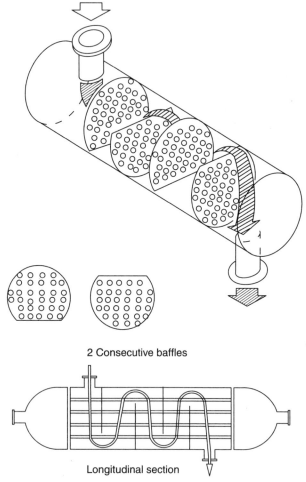

2 Consecutive baffles

Longitudinal section

FIGURE 6-2   Segmental baffles.

in the direction perpendicular to the tubes depends on the separation between the baffles. The closer the baffles are to each other, the smaller is the flow area, and the higher is the fluid velocity. This has a direct influence on the heat transfer coefficient, which increases with velocity. Of course, this increment in the velocity represents an increase in the frictional pressure drop.

The exchanger designer has to choose the baffle separation to attain the maximum possible value of the heat transfer coefficient without exceeding the allowable pressure drop of the shell-side fluid. We shall later study the available correlations to calculate the heat transfer coefficients and friction factor for a heat exchanger shell[3].

The most common baffle cut is about 25 percent of the diameter. This means that the height of the baffle window is 25 percent of the shell diameter. Figure 6-3 shows the effects in the shell fluid streamlines corresponding to different baffle cuts. For low baffle cuts (Fig. 6-3*a*), the velocity of the fluid at the window is high, with further reconversion in pressure with high turbulence and eddy formation. A large amount of the fluid energy is spent in the window area, where there are few tubes, resulting in an inefficient conversion of pressure drop in heat transfer. If, on the other hand, the baffle cut is high, there may be short circuits between the baffle tips, as shown in Fig. 6-3*c,* and there will be important zones of the fluid with low and erratic velocities and low heat transfer coefficients.[6] The optimum seems to be in an intermediate situation such as that illustrated in Fig. 6-3*b.*

Other types of baffles, not as common as segmental baffle, are shown in Fig. 6-4. These are called *disk* and *doughnut baffles.*

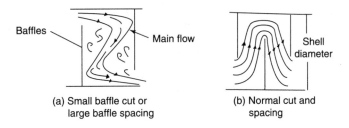

(a) Small baffle cut or
    large baffle spacing

(b) Normal cut and
    spacing

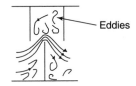

(c) Large cut or small spacing

FIGURE 6-3   Effect of the baffle cut.

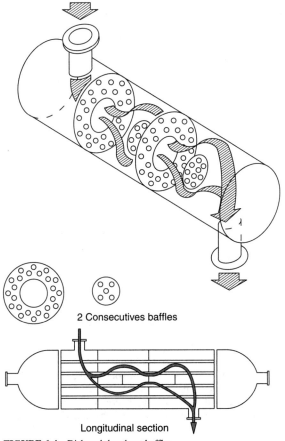

2 Consecutives baffles

Longitudinal section

FIGURE 6-4   Disk and doughnut baffles.

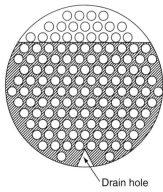

Drain hole

FIGURE 6-5   Drain holes.

Drain Holes.   Precautions must be taken to allow complete drainage of the shell side when the heat exchanger is taken out of operation. For this reason, drain holes are provided in baffles with horizontal cuts to avoid trapping liquid between consecutive baffles, as shown in Fig. 6-5.

Baffle Spacing.   As was explained previously, an increase in the heat transfer coefficient is achieved by reducing the baffle spacing. However, there is a lower limit for the baffle spacing, set by the TEMA Standards, of one-fifth the shell diameter and never less than 2 in (51 mm).

Spacing higher than the shell diameter is not common. If higher spacing is employed, the effect may be similar to that seen in Fig. 6-3*c* with poor heat transfer coefficients.

The TEMA Standards also define a maximum spacing for baffles, which is defined by mechanical considerations. This is so because the purpose of baffles is not only to improve the heat transfer but also to support the tubes. If a heat exchanger tube is not supported adequately, it can vibrate owing to the effect of the externally circulating fluid. In many cases this has contributed to tube destruction. This topic will be explained later in more detail .

The TEMA Standards specify the maximum unsupported span that heat exchanger tubes may have. Some tubes are supported at every baffle hole, with the unsupported span equal to the baffle spacing. But the tubes passing through the window are only supported every second baffle, so the unsupported span is equal to twice the baffle spacing. The maximum unsupported spans allowed by the standards for different tube materials and diameters are shown in Table 6-2.

TABLE 6-2   Maximum Unsupported Span (m)

External Tube Diameter, (mm)	Tube Material and Temperature Limit, °C	
	Carbon Steel and Low-Alloy Steels (399) High-Alloy Steels (454) Nickel-Copper (315) Nickel (454) Nickel-Chromium-Iron (537)	Aluminum and Aluminum Alloys Copper and Copper Alloys, Titanium at the Maximum Temperature Allowed by the ASME Code
¾ (19)	1.52	1.32
1 (25)	1.88	1.62
1¼ (32)	2,23	1.93
1½ (38)	2.54	2.21
2 (50)	3.17	2.79

Source: Extracted with permission from the TEMA Standards.

The baffles must be maintained firmly in their position because any vibration with respect to the tubes may wear and eventually destroy the tube at the baffle location. This is achieved with tie rods (see Fig. 6-1). Tie rods may be welded to the baffles or installed as shown on Fig. 6-6. In this construction, the tie rods pass through the baffle holes, and concentric spacers with higher diameters than the baffle holes are inserted in the tie rods to maintain the baffles in position.[4]

Provision must be made in the tubesheet layout for these rods, which is usually accomplished by omitting a tube at the selected location. This must be taken into consideration in defining the number of tubes that may be installed in a shell. Tie rods usually are threaded into the back of only one tubesheet, being free at the other end, and terminate with the last baffle by means of a lock washer.

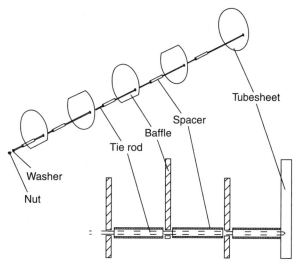

FIGURE 6-6 Tie rod and baffle spacers.

The minimum number of tie rods that may be installed is also defined in the TEMA Standards as follows:

Shell Diameter (m)	Number of Tie Rods
Up to 0.4	4
0.4–0.83	6
0.83–1.22	8
More	10

Tolerances.   The tube holes must be drilled in the baffles with a certain tolerance over the tube outside diameter. When the maximum unsupported length is 914 mm or less, this tolerance is 1/32 in (0.8 mm) in diameter. For larger unsupported lengths, it must be 1/64 in (0.4 mm).

The maximum diametral clearance between the baffle and the interior shell diameter is defined by TEMA as follows:

Nominal Shell Diameter, m	Shell Internal Diameter Minus Baffle Diameter, mm
0.15–0.43	3.2
0.45–0.99	4.8
1.016–1.37	6.4
1.39–1.75	7.9
1.78–2.1	9.5
2.15–2.54	11.1

6-4 TUBES AND TUBE DISTRIBUTION

The dimensions of the tubes used in the construction of heat exchangers usually respond to Birmingham Wire Gage (BWG) Standards. The tubes are designated by their outside diameters and a code that is related to the wall thickness. The table included in App. H shows the dimensional characteristics of tubes according to BWG Standards.

Tube diameters ranging from ¼ in (6.35 mm) to 1.5 in (38 mm) may be used. However, almost all heat exchangers are constructed with ¾-in (19-mm) or 1-in (25.4-mm) tubes.

The length of the tubes is defined by the designer. It is not advisable to use tube lengths higher than 6 m because the construction is more difficult. Cleaning operations are also more difficult with long tubes. However, there are certain instances in which it is imperative to maintain the countercurrent configuration, and large heat transfer areas are necessary (e.g., in natural gas conditioning plants to exchange heat between the inlet and outlet gas). In such cases, tubes up to 20 m long can be used.

6-4-1 Materials

A number of materials can be used in heat exchanger tubes. The material is selected based on the characteristics of the fluids from the corrosion point of view. The material of the tubes is in contact with the tube-side and shell-side fluids and must be resistant to both.

Wall thickness is selected based on a mechanical calculation to resist the design pressures. The most commonly used materials are carbon steel, alloy steel, stainless steel, bronze, nickel, and their alloys (e.g., Monel, Inconel, and Hastelloy). For very corrosive fluids, more exotic materials such as titanium or even glass can be used.

6-4-2 Impingement Protection

If the fluids are erosive, or when inlet velocities are high, it is necessary to install shield plates under the inlet nozzle to avoid the direct impact of the fluid on the surfaces of the tubes. These plates are called *impingement protection plates,* and their location can be appreciated in Fig. 6-29.

The installation of impingement plates is mandatory if the inlet fluid is a gas with abrasive particles and in heat exchangers that receive liquid-vapor mixtures or vapor with entrained liquid, where the erosive effects are intense. For single-phase fluids, TEMA Standards define when the installation of impingement protection is required, depending on the kinetic energy of the fluid entering the shell. For liquids with abrasive particles, impingement protection is required if the inlet-line ρv^2 exceeds 740 kg/(m · s²). For clean fluids, impingement protection is required if the inlet-line ρv^2 exceeds 2,232 kg/(m · s²).

TEMA defines two other important areas that also limit the inlet or outlet velocities—the shell entrance or exit area and the bundle entrance or exit area (Fig. 6-7).

Shell Entrance or Exit Area. When an impingement plate is provided, this area is the unrestricted radial flow area between the inside diameter of the shell at the nozzle and the face of the impingement plate. If no impingement plate is provided, the shell entrance or exit area also includes the flow area between the tubes within the projection of the nozzle.

Bundle Entrance or Exit Area. This is the flow area between the tubes within the compartments between first baffle and tubesheet. The area of the impingement plate, if any, must be substracted.

The TEMA Standards establish that in no case should the shell or bundle entrance or exit areas produce a ρv^2 in excess of 5953 kg/(m · s²).

6-4-3 Tube Patterns

The *tube pattern* is the geometric disposition of tubes in the tubesheet. Figure 6-8 shows the most common tube pattern types. The *tube pitch* P_t is the minimum separation between tube axes. A tube pattern is defined by the type (e.g., square, rotated square, triangle, or rotated triangle), by the tube diameter, and by the tube pitch.

Usually, triangular patterns have higher heat transfer coefficients and pressure drops than square arrays. For removable bundles, the square patterns have the advantage of allowing mechanical cleaning of the

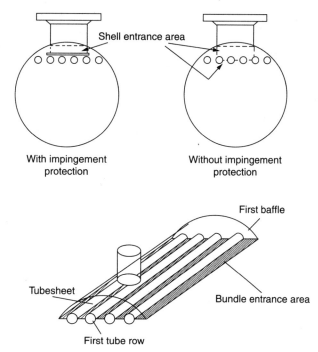

FIGURE 6-7   Shell entrance area and bundle entrance area.

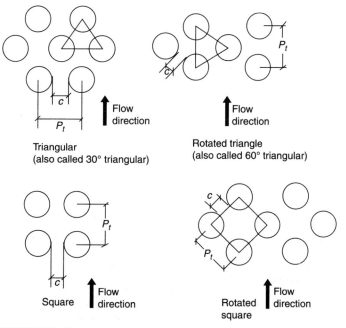

FIGURE 6-8   Tube patterns.

TABLE 6.3

Square Arrays		Triangular Arrays	
D_i, Tube Diameter	P_t, Pitch	D_i, Tube Diameter	P_t, Pitch
¾ in (19 mm)	1 in (25 mm)	¾ in (19 mm)	15/16 in (24 mm)
1 in (25 mm)	1¼ in (32 mm)	¾ in (19 mm)	1 in (25 mm)
1¼ in (32 mm)	1⁹∕₁₆ (40 mm)	1 in (25 m)	1¼ in (32 mm)
1½ in (39 mm)	1⅞ in (48 mm)	1¼ in (32 mm)	1⁹∕₁₆ in (40 mm)
		1½ in (39 mm)	1⅞ (48 mm)

outside surfaces of the tubes. When ¾- or 1-in tubes are used, the clearance between tubes c (see Fig. 6-8) is about 5 or 6 mm. The most common tube patterns are indicated in Table 6-3.

6-5 TUBES TO TUBESHEETS JOINT

There are basically two tube-to-tubesheet joining methods:

1. Expanded joint

2. Welded joint

6-5-1 Expanded Joints

When a tube is submitted to internal pressure, it suffers an expansion and increases its diameter. When the stress to which the material is subjected is lower than the yield strength, the material will recover its original shape when the applied pressure disappears. However, if the stress exceeds the yield strength, the recovery is only partial, and a permanent deformation remains.

The stress distribution as a function of the radial coordinate in the wall of a tube submitted to internal pressure is shown in Fig. 6-9. It can be seen that the stress is inversely proportional to the radius. This means that the material near the internal wall is subject to a higher stress than the more external material.

When the heat exchanger is assembled, the tube is introduced in the tubesheet hole with a clearance between them. Then internal pressure is applied, and the tube bulges out. As the pressure is increased, the material of the tube can reach its elastic limit. We can imagine the tube and tubesheet assembly as a large plate with a hole in its center and pressure acting on the hole (Fig. 6-10). The stress distribution is continuous in both materials, as shown in the figure.

With increasing pressure, enlargement continues, and the plastic zone spreads outward, including part of the tubesheet material. When the pressure is released, the tubesheet material that had not reached the elastic limit tries to recover its original size, but the material of the tube, which had reached the elastic limit, does not spring back and is compressed by the recovery of the tubesheet. This creates a tight bond between tube and tubesheet. The resulting joint has good mechanical resistance and is the most common joint method employed in heat exchanger construction.

There are several techniques to achieve this effect. The most usual is roller expanding. A tube roller is a special tool that is introduced into the tube and can be expanded against the tube wall by the action of a mandrel. Rotation of the tool does the job by squeezing the tube wall. Other techniques are based on the application of hydraulic pressure directly to the tube end or even detonation of explosive charges consisting of Primacord contained in a polyethylene cylinder inserted into the tube.

The strength and tightness of an expanded joint improve if annular grooves are provided in the tubesheet holes. Then the tube material expands into the grooves, as shown in Fig. 6-11a. The TEMA Standards define the number and dimensions of these grooves for the different heat exchanger classes.

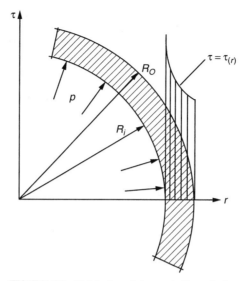

FIGURE 6-9 Distribution of stresses in the wall of a tube submitted to internal pressure.

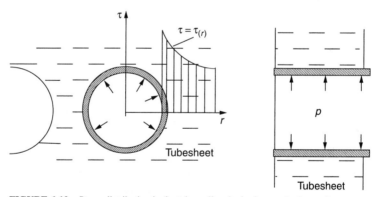

FIGURE 6-10 Stress distribution in the tube wall and tubesheet under internal pressure.

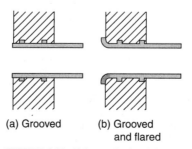

(a) Grooved   (b) Grooved
                       and flared

FIGURE 6-11 Tube-to-tubesheet joint.

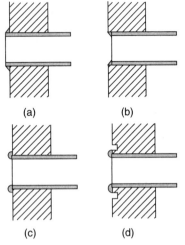

(a)　(b)

(c)　(d)

FIGURE 6-12　Welded joints

When the shell-side pressure is higher than the pressure into the exchanger heads (which means that there is an outward force acting on the tubesheet), the strength of the joint can be improved by flaring or beading the end of the tubes, as shown on Fig. 6-11*b*.

6-5-2　Welded Joints

There are several techniques to weld the tubes to the tubesheet.[5] Some of them are represented in Fig. 6-12. A welded joint is tighter than an expanded joint, but the possibility of using this technique depends on the tube and tubesheet materials.

When materials of different thicknesses are welded, the material with the higher mass can dissipate heat more easily than the thinner material, and it may be difficult to bring both pieces to the welding temperature simultaneously. For this reason, sometimes the alternative shown in Fig. 6-12*d* is adopted. This allows one to weld materials of similar thicknesses. Welding tubes to the tubesheet is a difficult operation and requires that the weld procedures be carefully prepared and tested before being applied.

Another technique that may be used is brazing. In brazing, coalescence is produced by heating the tube-tubesheet assembly to a temperature above 430°C. This melts a nonferrous filler metal that flows into the space between the tube and hole by capillary action. The base metals have higher melting points than the filler and do not melt. This joining method does not produce joints of high strength and normally is used in combination with expansion.

6-6　MULTIPASS HEAT EXCHANGERS

The heat exchanger of Fig. 6-1 can be modified as indicated in Fig. 6-13 by changing both heads. With this configuration, the tube-side fluid, which enters into the exchanger through the left head (front head), circulates through half the tubes up to the opposite head (rear head). Then the fluid inverts its direction and circulates back through the other half of the tubes up to the front head, where it exits. Since the fluid circulates twice through the exchanger, this is called a *two-pass heat exchanger.*

Please note that the front (left) head must be partitioned to avoid a fluid short circuit from inlet to outlet nozzles. This is achieved with installation of a pass-partition plate, as indicated in the figure.

If we compare the heat exchangers in Figs. 6-1 and 6-13, if both have the same shell diameter and length, the heat transfer area will be the same. However, in the case of Fig. 6-13, because the incoming fluid is distributed into a smaller number of tubes, the tube-side flow area will be half that of Fig. 6-1. Thus, if both units have to process the same flow rate, the fluid velocity in the two-pass heat exchanger will be double that in the single-pass exchanger.

The tube-side-fluid heat transfer coefficient thus will be higher in the two-pass unit. We could have reached the same result with a heat exchanger with half the number of tubes but twice the length. However, the length of a heat exchanger usually is limited by the available plot area, and additionally, as was mentioned earlier, for construction and maintenance reasons, tube lengths higher than 6 m are seldom used. Thus the alternative of increasing the number of tube passes leads to more compact designs.

By partitioning the heat exchanger heads adequately, it is possible to increase the number of tube passes, and heat exchangers with two, four, six, eight, or more passes can be constructed, as shown in Fig. 6-14.

Usually an even number of passes is adopted. It is possible to use an odd number of passes, but if an even number of passes is employed, the inlet and outlet nozzles are at the same end, which usually makes the piping layout and cleaning operations more simple.

Figure 6-15 shows possible pass-partition configurations for different numbers of passes. Since in each pass the temperature of the fluid changes, different sectors in each head will contain fluid at different

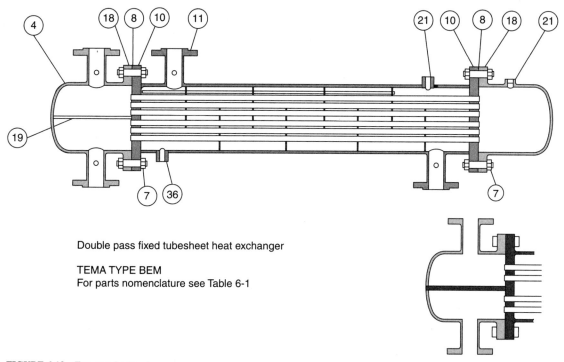

Double pass fixed tubesheet heat exchanger

TEMA TYPE BEM
For parts nomenclature see Table 6-1

FIGURE 6-13   Two-pass heat exchanger.

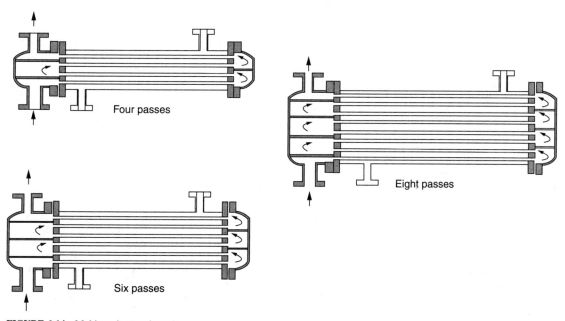

Four passes

Eight passes

Six passes

FIGURE 6-14   Multipass heat exchangers.

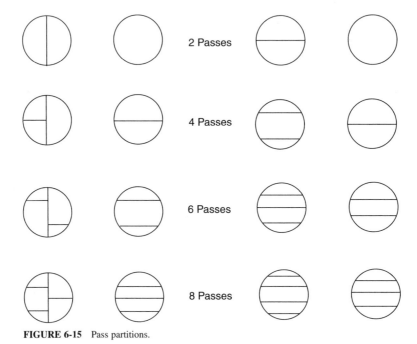

FIGURE 6-15 Pass partitions.

temperature. Since high temperature gradients can cause undesirable thermal stress on the tubesheet, it is recommended that the pass-partition configuration be such that the fluid temperature difference between adjacent sectors is less than 28°C (50°F).[6]

The pass-partition plates reduce the tubesheet area available for tubes allocation, as can be appreciated in Fig. 6-13. This normally makes it necessary to eliminate some tube rows. The tables in App. D indicate the number of tubes that can be allocated on a tubesheet for different shell diameters and tube passes. It can be seen that for the same shell diameter, the higher the number of tube passes, the smaller will be the number of tubes that may be allocated. Also, in App. E, several tube layouts for the most common tube patterns and shell diameters are included

It is left to the discretion of each manufacturer to establish a system of standard shell diameters within TEMA Standards to achieve the economies peculiar to his individual design and manufacturing facilities. However, the diameters indicated in the graphs and tables in the appendices are quite typical for all manufacturers.

6-7 HEAT EXCHANGERS WITH MULTIPLE SHELL PASSES

Let us consider a heat exchanger with four tube passes, as shown in Fig. 6-16. Let's assume that the tube-side fluid is the cold fluid. This fluid enters the exchanger at temperature t_1. In each pass, its temperature increases. The temperature evolution is represented in the lower part of the figure.

Since in every pass the temperature of the cold fluid gets closer to that of the hot fluid, the ΔT available for heat transfer is reduced, and the amount of heat transferred in each successive pass is lower. This means that the increase in temperature of the cold fluid is higher in the first passes than in the later ones.

It can be seen in the figure that the inlet temperature at the last tube pass is t'''. This stream must receive heat from the shell-side fluid, which at this section of the exchanger is at its lowest temperature T_2. In order for heat transfer to the cold fluid in the fourth tube pass to be possible, it is necessary that $T_2 > t'''$. We have explained that the temperature change experienced by the cold fluid in the last tube pass is small, so the cold

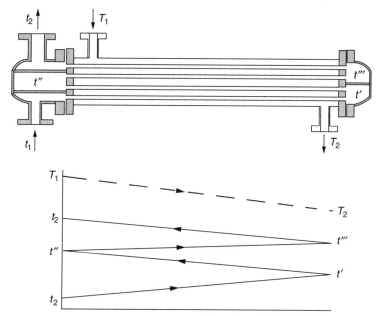

FIGURE 6-16 Temperatures evolution in a four-pass heat exchanger.

stream outlet temperature t_2 cannot be much higher than T_2 either. This means that with this type of heat exchanger, it is not possible to achieve an important temperature cross between both fluids. This limitation does not exist in true countercurrent units. However, the length of a countercurrent heat exchanger would be four times that of the four-pass exchanger with the same fluid velocity and heat transfer area. Then the tubes would be excessively long.

One way to improve the heat transfer characteristics of multipass heat exchangers is to install longitudinal baffles. These are plates whose width is similar to the shell diameter and whose length is somewhat smaller than the tube length so as to allow the shell-side fluid to cross from the upper to the lower part or vice versa at one of the exchanger ends. Above and below the longitudinal baffle, transversal baffles are installed. The resulting configuration is illustrated in the upper part of Fig. 6-17.

If the tube-side fluid performs four passes, the schematic of the unit and the evolution of the streams temperatures are those shown in Fig. 6-18. In the lower diagram, the dashed curve corresponds to the evolution of the shell-side fluid.

It can be appreciated that in this case it is simpler to achieve the temperature cross with $T_2 < t_2$. This is so because the hot fluid that exits the unit at T_2 is in contact only with the two coldest passes of the tube-side fluid. We can see that in this case $t''' > T_2$, which could not have been possible without the longitudinal baffle. This means that the longitudinal baffle improves the heat transfer characteristics in units operating with difficult temperature programs (high-temperature cross between both streams).

The heat exchanger shown in Fig. 6-18 has two shell passes. It is also possible to build units with three or more shell passes. In theses cases, it is necessary to install additional longitudinal baffles, as shown in Fig. 6-19.

The higher the number of shell passes, the higher is the temperature cross between both streams that can be achieved. We shall explain later how to decide if a particular temperature program makes it necessary to install longitudinal baffles.

A heat exchanger configuration is designated with two digits. The first digit indicates the number of shell passes, and the second is the number of tube passes. Thus, for example, the heat exchanger in Fig. 6-17 is a 1-2 configuration, whereas that in Fig. 6-18 is a 2-4 configuration because it has two shell passes and four tube passes

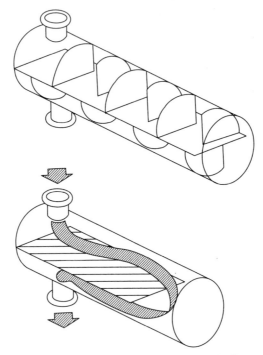

FIGURE 6-17   Heat exchanger with longitudinal baffle.

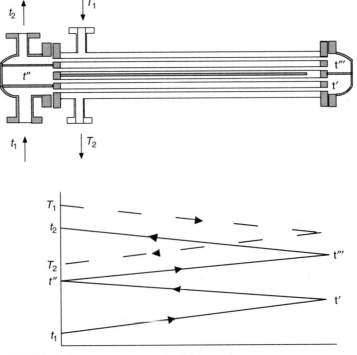

FIGURE 6-18   Temperatures evolution in a 2-4 heat exchanger.

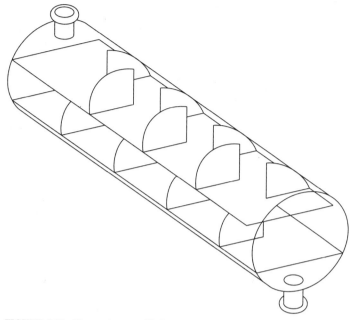

FIGURE 6-19   Heat exchanger with three shell passes.

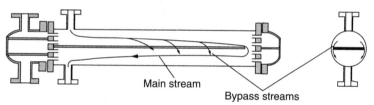

FIGURE 6-20   Leakage between shell and longitudinal baffle.

In units constructed with longitudinal baffles, it is important to achieve a tight seal in the baffle-shell joint. Otherwise, the shell fluid may short circuit the tube bundle and reach the outlet nozzle without performing a good heat exchange with the other fluid. This is shown in Fig. 6-20.

To achieve a tight seal, it is necessary to use specially designed gaskets that can be adjusted from the outside before proceeding to the final closure of the unit. In large-diameter heat exchangers, the longitudinal baffle can be welded to the shell internally. However, this is not possible in the removable-bundle units that we shall describe later.

This is why it is usually preferred to avoid the installation of longitudinal baffles. This is possible because, as we shall see, a 2-4 configuration can be achieved by adequately combining two 1-2 units. It also must be noted that a heat exchanger with two shell passes and two tube passes is equivalent to a countercurrent heat exchanger with half the tube numbers and double the tube lengths.

6-8   FRONT HEADS (INLET HEADS)

There are several types of front-head designs, as illustrated in Fig. 6-21. Two of the major considerations in the selection of heads are accessibility to the tubes and piping convenience. If a process will foul the exchanger repeatedly so that cleaning is a major factor, a head or coverplate that can be removed easily

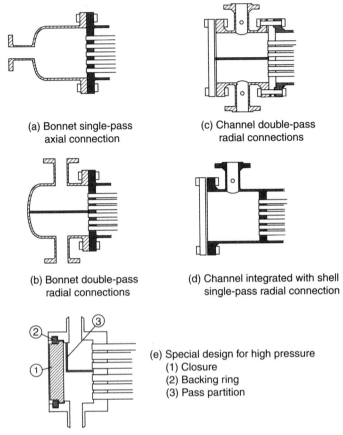

(a) Bonnet single-pass
axial connection

(c) Channel double-pass
radial connections

(b) Bonnet double-pass
radial connections

(d) Channel integrated with shell
single-pass radial connection

(e) Special design for high pressure
(1) Closure
(2) Backing ring
(3) Pass partition

FIGURE 6-21 Front head types. *(Used with permission from Standards of the Tubular Exchangers Manufacturers Association.)*

is an obvious choice. One step in this direction is to locate connections on the sides, not on the ends, of removable heads. The radial flange connections shown in Fig. 6-21b permit easy removal of the head with minimum disturbance to the piping of the whole installation. The next step is to use a channel, as in Fig. 6-21c and d.

Open-end heads that are fabricated from cylindrical sections are called *channels*. These are fitted with easily removable coverplates so that tubes can be cleaned without disturbing piping. They are used widely where fouling conditions are encountered or where frequent access for inspection is desired. The heads shown in Fig. 6-21a and b are called *bonnets*. They can be either cast or fabricated. Sometimes the channel is integral with the shell, as in Fig. 6-21d. In this case, only the coverplate can be removed.

The design shown in Fig. 6-21e is used in very-high-pressure applications. The head and tubesheet are built in a single piece. The internal pressure compresses the coverplate against the seat, and longitudinal efforts are absorbed by the backup ring, which is stressed only in shear. Pass-partition plates (3), if required, are built with rather thin plate because the differential pressure to which they are submitted is only given by the pressure drop of the fluid through the tubes.

All these designs can be used as front or rear heads. However, for removable-bundle designs, special types of rear heads must be employed, as we shall consider later.

6-9 FIXED-TUBESHEET AND REMOVABLE-BUNDLE HEAT EXCHANGERS

6-9-1 Fixed-Tubesheet Heat Exchangers

The heat exchanger shown in Fig. 6-1 has the tubesheets welded to the shell, as can be appreciated in the detail included in the figure. This means that once the unit is built, it is not possible to have access to the shell side for cleaning or inspection. This type of heat exchanger is called a *fixed-tubesheet exchanger.*

The construction sequence is as follows:

1. Tubesheet and baffle holes are drilled.
2. The tubes are installed in one of the tubesheets.
3. The baffles and spacers are introduced through the tubes and tie rods.
4. The shell is placed in position.
5. The second tubesheet is installed.
6. The shell is welded to the tubesheets.
7. The tubes are rolled into the tubesheets.

The two main disadvantages of this type of unit are

1. It cannot be disassembled for cleaning or inspection.
2. If the temperature difference between the fluids is high or the linear thermal expansion coefficients of the tube and shell materials are very different, when the exchanger comes into operation, differential expansion between shell and tubes creates forces acting on the tube-to-tubesheet joints that can damage the unit.

The first of these problems cannot be solved with this type of heat exchanger. This means that this design is not suitable for cases in which both fluids have a fouling tendency (if only one fluid is fouling, it can be allocated to the tube side because the tube interiors can be cleaned mechanically). In some cases even though a mechanical cleaning of the tube exteriors is not possible, they can be cleaned chemically by circulating a solvent or detergent. However, this alternative is not feasible in many applications where the fouling characteristics of the fluid require mechanical removal procedures.

The second problem, differential expansion between the shell and tube bundle, can be solved by installing a shell expansion joint such as that shown in Fig. 6-1. Such expansion joints act as elastic bellows absorbing the differential expansion without transmitting forces to the tubesheets. During mechanical design of the heat exchanger, it is determined if the installation of an expansion joint is necessary.

However, the most common approach to solving the two previously mentioned limitations of the fixed-tubesheet design is to adopt a removable-bundle construction, as will be explained below.

6-9-2 Removable-Bundle Heat Exchangers

In these units, the tube bundle can be removed, and this allows access to the tube exteriors to perform mechanical cleaning. There are several types of removable-bundle constructions that will be analyzed below.

Pull-Through Floating Head. In the design shown in Fig. 6-22, the right head is not fixed to the shell but is free to move longitudinally into the shell as the differential expansion requires. This is called a *floating head.* The opposite tubesheet is bolted to the shell and to the left head, which is called a *fixed head.* This tubesheet is called a *stationary tubesheet* because it does not have relative movement with respect to the shell.

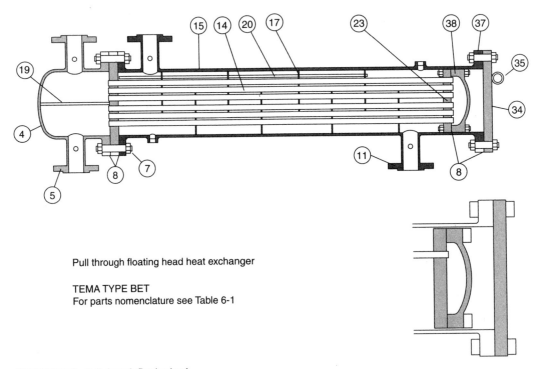

Pull through floating head heat exchanger

TEMA TYPE BET
For parts nomenclature see Table 6-1

FIGURE 6-22   Pull-through floating head.

The assembly constituted by this tubesheet, tubes, baffles, and floating head is the removable bundle. To remove the bundle, the procedure is

1. The flanged joint of the fixed head is disassembled.

2. The removable bundle is withdrawn from the left side. During this operation, the floating head passes through the shell. This is why it is called a *pull-through floating head.* Access to the exterior of the tubes then is possible.

If the tube layout pattern is square, then it is possible to perform mechanical cleaning of the exterior of the tubes. In a triangular pattern, this cleaning is not possible even if the bundle is removed. This design does not present the problem of differential expansion stress because the bundle may expand without creating any stress.

The main disadvantages of this design are

1. This design is not very suitable for single-pass heat exchangers. In these cases, it is necessary to install special flanges through the shell cover. Figure 6-23 illustrates one possible design. The flange *a* is screwed to the outlet nozzle and can be removed. Then the packing gland is loosened, and the bundle can be pulled through to the left. This can add considerably to the cost and adds a potential source of leaking. Also, the advantage of free expansion is lost.

2. The location of the floating-head flanges requires that the tube circle diameter d_1 be considerably smaller than the shell

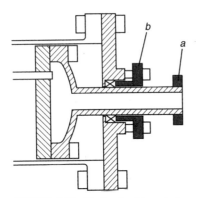

FIGURE 6-23   Floating head in a single-pass heat exchanger.

diameter D_s. This can be easily appreciated in Fig. 6-24. This leaves an empty peripheral lane where the shell-side fluid may canalize, as shown in Fig. 6-25. This allows part of the fluid to circulate between baffle windows without penetrating into the bundle. This stream is called a *bypass stream*. Bypass streams can be eliminated by installing sealing strips, as will be discussed below. The installation of sealing strips does not eliminate the need to increase the shell diameter with respect to the tube circle, which increases the cost of the unit.

3. A third problem is that the floating-head flange gasket is a hidden gasket. Any damage to this gasket will not be detected and may cause the higher-pressure fluid to contaminate the other fluid. If this cannot be tolerated because of chemical compatibility, toxicity, etc., other designs sometimes are preferred.

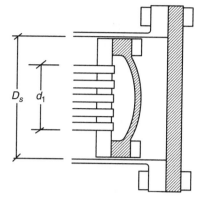

FIGURE 6-24 In a TEMA T heat exchanger, the tube circle is considerably smaller than the shell diameter.

Split-Ring Floating-Head Heat Exchanger.   This design aims to solve the second of the problems pointed out for the pull-through floating-head heat exchanger: the existence of a bypass region between the shell and the tube circle. As was explained, this diametral clearance is necessary for allocation of the removable-head flange. In the split-ring floating-head design, the term $(D_s - d_1)$, although larger than for the fixed-tubesheet design, is a good deal smaller than for the pull-through floating-head design. The construction is shown in Fig. 6-26.

It can be seen that the floating head is not bolted to the tubesheet. It is bolted to a backup ring, and the tubesheet is tightened between the ring and the head flange. This construction makes it possible for the tubesheet diameter to be smaller than the bolt circle of the flanged joint.

The backup ring is not a single piece. It is split into two halves. This makes it possible for both parts to be removed from the unit (once unbolted from the head flange) before pulling the bundle through the shell.

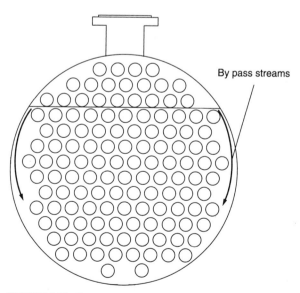

By pass streams

FIGURE 6-25 Bypass streams.

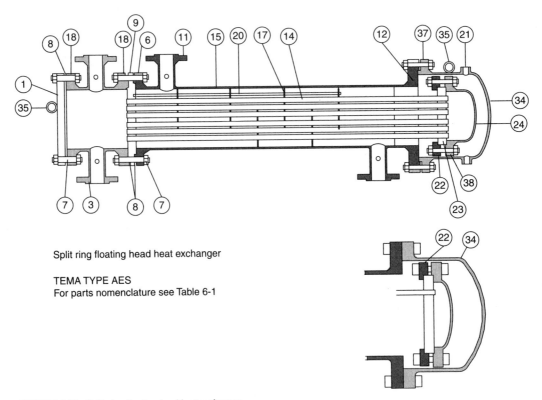

Split ring floating head heat exchanger

TEMA TYPE AES
For parts nomenclature see Table 6-1

FIGURE 6-26   Split-ring floating-head heat exchanger.

The disassembly procedure is as follows:

1. The shell cover (34) is removed.

2. The flanged joint is unbolted.

3. The ring is split, and both halves are removed.

4. The stationary head in the opposite end is removed.

5. The tube bundle is pulled out, passing the tubesheet through the shell.

Figure 6-27 is a comparison between a pull-through design and a split-ring design. It can be seen that the second alternative allows a smaller shell diameter.

U-Tube Heat Exchanger   Another way to solve the problem of the differential expansion is with the U-tube construction shown in Fig. 6-28. This design has the advantage of a lower cost because it eliminates one head, but its principal limitations are

1. It is not possible to clean the interior of the tubes because it is not possible to pass a cleaning rod through them.

2. This construction cannot be used for single-pass exchangers.

3. With the exception of the outermost tubes, individual tubes cannot be replaced. Any leaking tube must be plugged.

4. In very large diameters, support of the tubes is difficult. (The U-tube bundle becomes susceptible to vibration hazards.)

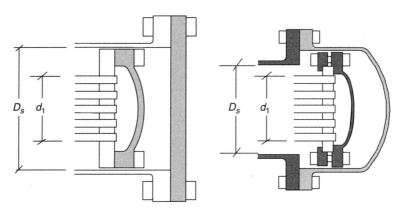

FIGURE 6-27 Comparison between TEMA T and TEMA S floating heads for the same tube-circle diameter. It can be appreciated that a smaller shell diameter is possible with the TEMA S type.

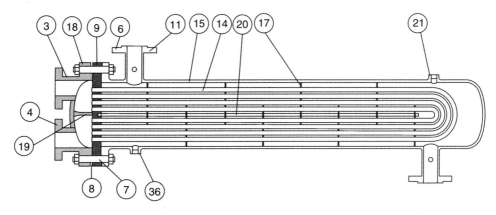

U-tube heat exchanger

TEMA TYPE BEU
For parts nomenclature see Table 6-1

FIGURE 6-28 U-tube heat exchanger.

As a counterpart, the number of gaskets is minimal. This makes this design particularly attractive in high-pressure service.

Outside-Packed Floating-Head Design.   Figure 6-29 shows this design. The floating head can move axially, and the shell side is sealed by a packing (26) that is compressed by means of a packing gland (28). The floating tubesheet skirt (29) diameter is smaller than that of the shell. It thus can be removed to the left, passing through the shell, when the unit is disassembled. The slip-on backing flange (30) is a loose flange that can be removed to the right after removal of the split shear ring (31).

Any leakage at the floating-head joint is easily detectable, but this type of packed joint should not be used when lethal, toxic, or flammable fluids are to be contained. Leakage through a stuffing box is more likely than failure of a gasketed joint.

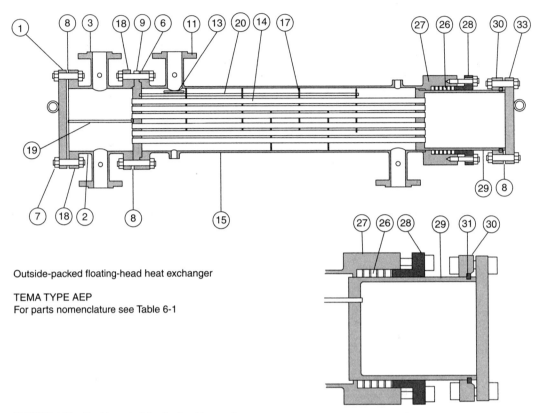

Outside-packed floating-head heat exchanger

TEMA TYPE AEP
For parts nomenclature see Table 6-1

FIGURE 6-29 Outside-packed floating-head heat exchanger.

This unit is very limited in its design pressure and temperature ranges. Design pressure must be lower than 40 bar, and design temperature must be lower than 590 K. API Standard 660, which is used in the petroleum industry, does not allow this type of construction.

Figure 6-30 shows a double-tubesheet design. This design can be used where a contamination of one fluid with the other because of tube-to-tubesheet joint leakage cannot be tolerated. This double-tubesheet

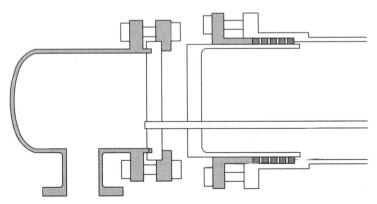

FIGURE 6-30 Double-tubesheet design.

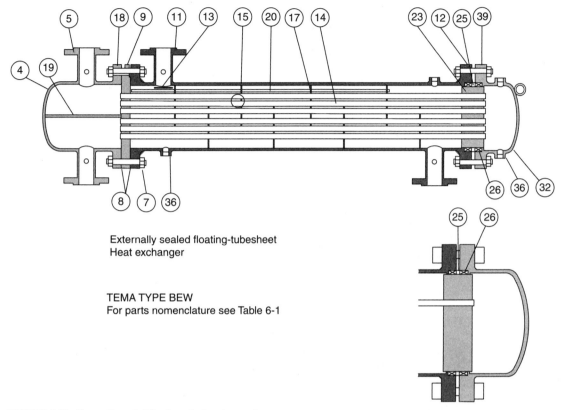

Externally sealed floating-tubesheet
Heat exchanger

TEMA TYPE BEW
For parts nomenclature see Table 6-1

FIGURE 6-31 Externally sealed floating-tubesheet heat exchanger.

construction also can be used in other types of heat exchangers, such as the fixed-tubesheet or U-tube designs, but it is not feasible in the other floating-head designs.

Externally Sealed Floating-Tubesheet or Floating-Head Outside-Packed Lantern Ring. In this construction (Fig. 6-31), an outside-packed lantern ring provides the seal between the shell-side and tube-side fluids as the tubesheet moves back and forth with expansion and contraction of tubes. Weep holes in the ring vent any seepage to the atmosphere.

The construction is relatively simple, and there is no bypass area. Maintenance is also relatively simple. However, it is even more limited than the outside-packed floating-head heat exchanger. The TEMA Standards allow this type of construction only for water, steam, air, lubricating oil, or similar services in conditions less severe than 2106 N/m² and 423 K.

6-9-3 Comparison among the Different Heat Exchangers Types

Table 6-4 lists the advantages and disadvantages of the different types of heat exchangers we have analyzed.

6-9-4 Standard Types and TEMA Designation

The several types of rear heads (described previously) can be combined with different types of front heads and shells, giving origin to a number of possible combinations. The TEMA Standards include a code to describe the heat exchanger type using three letters. The first letter designates the inlet (front) head. The

TABLE 6-4 Features of Principal Heat Exchangers Designs[1,2]

Type of Design	U-Tube (Type U)	Fixed Tubesheet (Types L, M, and N)	Pull-Through Floating Head (Type T)	Floating-Head Outside-Packed Lantern Ring (Type W)	Split-Backing-Ring Floating-Head (Type S)	Floating-Head Outside-Packed Stuffing Box (Type P)
Relative cost increases from A (least expensive) through E (most expensive)	A	B	C	C	D	E
Provision for differential expansion	Individual tubes free to expand	Expansion joint in shell	Floating head	Floating head	Floating head	Floating head
Removable bundle	Yes	No	Yes	Yes	Yes	Yes
Individual tubes replaceable	Only those in outside rows	Yes	Yes	Yes	Yes	Yes
Tube interiors cleanable	Difficult to do mechanically; can do chemically	Yes, mechanically or chemically	Yes, mechanically or chemically	Yes, mechanically or chemically	Yes, mechanically or chemically	Yes, mechanically or chemically
Tube exteriors with triangular pitch cleanable	Chemically only	Chemically only	Chemically only	Chemically only	Chemically only	Chemically only
Tubes exterior with sqare pitch cleanable	Yes, mechanically or chemically	Chemically only	Yes, mechanically or chemically	Yes, mechanically or chemically	Yes, mechanically or chemically	Yes, mechanically or chemically
Number of tube passes	Any practical even number possible	No practical limitations	No practical limitations (single-pass floating-head requires packing joint)	Limited to single- or double-pass	No practical limitations (single-pass floating-head requires packing joint)	No practical limitations
Internal gaskets eliminated	Yes	Yes	No	Yes	No	Yes

second letter refers to the shell type. The third letter designates the rear (outlet or return) head. For example, a heat exchanger of type AEL is a heat exchanger with front-head type A, shell type E, and rear-head type L. The meaning of these codes can be seen in Fig. 6-32.

Shell type K is used as a evaporator or reboiler. The rest of the elements in the figure already have been analyzed. Figures 6-1, 6-13, 6-22, 6-26, 6-28, 6-29, and 6-31 indicate the corresponding code for each type of exchanger represented.

6-10 TUBE VIBRATION

6-10-1 Vibration Damage

The tubes of a heat exchanger are supported at both ends by the tubesheets and in intermediate positions by baffles. Each tube section between two consecutive supports is a flexible element and can vibrate like the strings of a guitar. Like the strings of a musical instrument, each tube section has a natural frequency

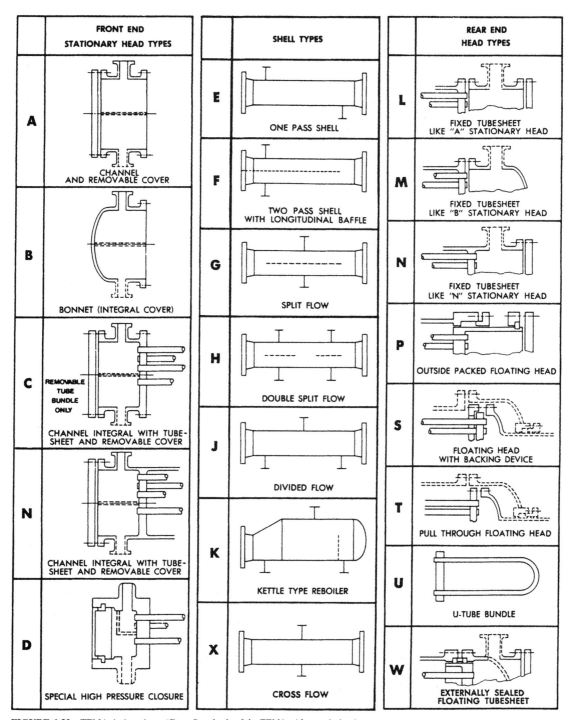

FIGURE 6-32   TEMA designations. *(From Standards of the TEMA with permission.)*

of vibration that depends on its mass (both that of the tube and that of the fluid into it), on its moment of inertia, on the way it is supported, and more important, on the distance between supports. For example, a tube section included between two consecutive baffles has a different frequency of vibration than another one of the same length included between the tubesheet and a baffle or the end of a U-tube bend.

The natural frequency of vibration also depends on the density of the shell-side fluid and the tensile strength at which the tube is submitted (the same way as tightening the strings of a guitar changes the frequency of vibration). The flow of the shell-side fluid past the tubes may produce certain periodic unbalanced forces over the tubes. When the frequency of these forces coincides with the natural frequency of the tubes, resonance occurs, and the tubes may start vibrating with large amplitude. Under certain conditions, the vibration amplitude may be large enough to produce severe damage to the tubes.

Damage can result from any of the following independent conditions or combination thereof:

1. *Collision damage.* Impact of the tubes against each other or against the shell can result in failure. The impacted area of the tube develops the characteristic flattened, boat-shaped spot, generally at the midspan of the unsupported length. The tube wall eventually wears thin, causing failure.

2. *Baffle damage.* Baffle tube holes require a manufacturing clearance over the tube outer diameter to facilitate fabrication. When large fluid forces are present, the tube can impact the baffle hole, causing thinning of the tube wall in an uneven circumferential manner, usually the width of the baffle thickness. Continuous thinning over a period of time results in tube failure.

3. *Tubesheet clamping effect.* Tubes are normally expanded into the tubesheet to minimize the crevice between the outer tube wall and the tubesheet hole. The natural frequency of the tube span adjacent to the tubesheet is increased by the clamping effect. However, the stresses owing to any lateral deflection of the tube are also maximal at the location where the tube emerges from the tubesheet, contributing to possible tube breakage.

4. *Material defect propagation.* There are some vibration frequencies that cannot produce damage to a homogeneous material. However, if a material contains flaws oriented with respect to the stress field, they can propagate, and the tube may fail.

5. *Acoustic vibration.* This phenomenon may appear in heat exchangers working with gas in the shell side. Acoustic resonance is due to gas-column oscillation. This oscillation creates an acoustic vibration and generates a sound wave. This is not usually associated with tube damage unless the acoustic resonance frequency approaches the tube natural frequency, but the heat exchanger and even the associated piping may vibrate with loud noise.

6-10-2 Vibration Mechanisms

There are four basic flow-induced vibration mechanisms that can occur in a tube bundle:

1. *Fluidelastic instability.* This is the most damaging in that it results in extremely large amplitudes of vibration. This mechanism is associated with high fluid velocities. A tube exposed to a fluid circulating at high velocity in a direction perpendicular to the tube starts vibrating, describing oval orbits around its axis.

2. *Vortex shedding.* This is also known as *Von Karman turbulence.* A tube exposed to an incident cross-flow provokes instability in the flow and the simultaneous shedding of discrete vorticities alternately from the sides of the tube. This phenomenon is referred to as *vortex shedding.* Alternate shedding of the vorticities produces harmonically varying lift and drag forces that may cause movement of the tube. When the tube oscillation frequency approaches the tube natural frequency within about 20 percent, the tube starts vibrating at its natural frequency.

3. *Turbulent buffeting.* This refers to unsteady forces developed on a tube exposed to a highly turbulent flow. The turbulence has a wide spectrum of frequencies around a dominating frequency that increases as the cross-flow increases. The effect is similar to the wind shaking or buffeting a canvas tent. This effect is observed only in heat exchangers handling high-velocity gases.

4. *Acoustic vibration.* As explained previously, this phenomenon consists on the vibration of a certain mass of gas, which can be excited by any of the preceding mechanism and provokes the appearance of sound waves.

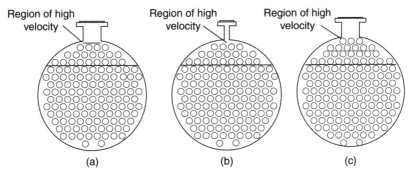

FIGURE 6-33   Restricted entrance areas.

6-10-3 Failure Regions

Tube failures have been reported in nearly all locations within a heat exchanger. But those of primary concern are

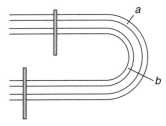

FIGURE 6-34   U-tubes bends.

1. *Nozzle entrance and exit area.* In these regions, impingement plates (Fig. 6-33a), small nozzle diameter (Fig. 6-33b), or large outer-tube limits (Fig. 6-33c) can contribute to restricted entrance or exit areas. These restricted areas usually create high local velocities that can result in damaging flow-induced vibration.

2. *U-tubes bends.* Outer rows of U-bends (like tube *a* in Fig. 6-34) have a lower natural frequency of vibration and therefore are more susceptible to flow-induced vibration failures than the inner rows.

3. *Tubesheet region.* Unsupported tube spans adjacent to the tubesheets are frequently longer than those in the baffled region of a heat exchanger owing to the space required to allocate inlet and outlet nozzles (Fig. 6-35). The higher unsupported span results in lower natural vibration frequencies of the tubes in this region. The possible high local velocities, in conjunction with the lower natural frequency, make this a region of primary concern in preventing damaging vibrations.

4. *Baffle window region.* Tubes located in baffle windows have unsupported spans equal to multiples of the baffle spacing. Long, unsupported tube spans result in a reduced natural frequency of vibration, and tubes have a greater tendency to vibrate and suffer damage.

6-10-4 Treatment of the Problem

The evaluation of potential flow-induced vibration problems in heat exchanger design is not part of the heat transfer field, so it will not be presented here in detail. However, in many cases, to solve vibration problems, it is necessary to make changes in the exchanger geometry, and this affects the thermal design.

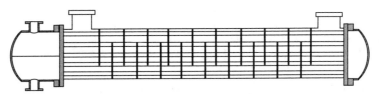

FIGURE 6-35   Unsupported span is higher at exchanger ends.

The tube vibration phenomenon is complex. The TEMA Standards provide a methodology to predict vibration damage, and the calculation procedure may be consulted there. However, it is made clear that the TEMA guarantee does not cover vibration damage.

Commercial heat exchanger design programs usually include a vibration-evaluation module based on this methodology, that alerts the designer when vibration may be expected with a certain design. The method proposed by TEMA consists of calculating the natural frequency of vibration for each tube section within the exchanger. This frequency, as was explained, depends on the unsupported span, type of support, tensile or compression stress, and physical properties of tube material and fluids. The method includes criteria to avoid tube vibration caused by all the mechanisms explained previously.

Vibration owing to fluidelastic instability is avoided if the fluid velocity at each section of the heat exchanger is below a certain critical velocity. This critical velocity is a function of the tube geometry, its natural frequency of vibration, and the viscous damping effect of the shell-side fluid. If the fluid velocity at any location is higher than the critical velocity calculated for that location, potential tube damage can be expected.

It is important to note that this velocity must be calculated point by point at every location in the shell. This requires considering internal bypass streams, leakage streams between tubes and baffles and between baffles and the shell, the effect of impingement plates, pass-partition lanes, sealing strips, etc. The calculation equations are included in the standards.

Regarding vortex shedding vibration, the TEMA Standards include equations and graphs that allow calculation of the vibration frequency resulting from this mechanism. If the tube natural frequency is less than double the vortex shedding frequency, there is a potential risk of vibration damage. In this case, it is necessary to calculate the amplitude of the tube vibration, and this must not exceed 2 percent of the tube diameter.

The standards also include the necessary equations to calculate the amplitude of the vibrations associated with the turbulent buffeting mechanism, and it also must be checked that this amplitude is smaller than 2 percent of the tube diameter.

Regarding acoustic vibration, it is necessary first to calculate the acoustic vibration frequency of the shell. This depends on the physical properties of the circulating fluids. The TEMA Standards then define criteria to compare this frequency with the vortex shedding and turbulent buffeting frequencies and decide if the acoustic resonance effect will be present. This would be a noise source.

6-10-5   Corrective Measures

When the conclusion of a vibration analysis is that there exists a potential risk of tube damage, the design needs to be changed. The design normally is corrected by

1. *Reducing the unsupported tube span.* This can be achieved by using baffles with no tubes in the window area (NTIW type = no tube in window baffles).

2. *Reducing the fluid velocity in the critical regions.* For example, by changing the tube pattern, by increasing the baffle spacing (although the counterpart is that the unsupported span increases), by increasing the nozzle diameters, or by eliminating some tubes.

The problem of the acoustic resonance may be corrected by adding a deresonating baffle parallel to the cross-flow direction to increase the shell acoustic frequency.

6-11   SPECIFICATION SHEET

The TEMA Standards include a template form to specify the characteristics of a heat exchanger. This is called a *specification sheet* or *data sheet,* and it is used as a basic engineering document in the design, purchasing, and construction stages of a heat exchanger. This document is shown in App. J.

The terminology included in the specification sheet was explained earlier, with the exception of some terms that will be defined next:

Operating pressure   This is the pressure to which the unit is submitted in normal operation.

Design pressure This is the maximum pressure at which the unit still needs to continue operating, resulting from unusual conditions, for example, during special operating procedures or as consequence of process excursions that must be tolerated. This is the pressure used for the mechanical designer to calculate thicknesses of plates, tubes, flanges, etc.

Hydraulic test pressure Any pressure vessel designed per ASME code can withstand at ambient temperature, in the absence of dynamic loads and for a limited time, pressures higher than the design pressure. Thus, if a hydraulic test is specified, the test pressure will be 1.3 times the design pressure.

Design temperature This is the maximum temperature that may coexist with the design pressure.

6-12 DESIGN AND CONSTRUCTION OF HEAT EXCHANGERS

The stages in the project of a heat exchanger are

1. Process specification
2. Thermal design
3. Mechanical design
4. Construction

We shall discuss which heat exchanger aspects must be defined in each of these stages and which are the engineering documents resulting from them.

1. *Process specification.* As we discussed for double-tube heat exchangers, in this stage the process conditions and design constrains are defined. The following data must be specified: flow rates, inlet and outlet temperatures, physical properties, fouling factors, allowable pressure drops, and design and operating pressures and temperatures. Also, this stage includes the geometric specifications that must be adopted during design, such as maximum tube length, shell and tubes fluid allocation, TEMA type and class, and sometimes construction materials. In this stage, the process engineer will fill in all these fields in the specification sheet, thus defining the design basis.

2. *Thermal design.* The methods employed to perform the thermal design will be studied in Chap. 7. The result of this thermal design is the definition of the heat exchanger geometry, including tube diameter and length, tube pattern and pitch, number of tubes and shell passes, shell diameter, number and type of baffles, and nozzle size. This is normally a process engineering activity, and further design activities correspond to the field of mechanical engineering. At the end of the thermal design, the specification sheet will be almost complete. It will be possible to sketch an outline drawing of the exchanger, indicating location of nozzles, main dimensions, types of supports, or any other information the process engineer wants to define.

3. *Mechanical design.* In this stage, all the heat exchanger components are mechanically designed, the materials specification is completed, mechanical tests are specified, and detailed drawings are prepared.

4. *Construction.* Constructive and shop drawings usually are left to the manufacturer. In this stage, all the necessary details for the construction and welding procedures and methods are defined.

Technical Specification. All the previously mentioned stages in a heat exchanger project can be carried on by different companies or different departments of the same company. Sometimes, the exchanger manufacturer performs the mechanical design or even the thermal design, depending on the type of contract and the engineering capacity of the purchaser. Usually, and especially if the heat exchanger is part of a complete project of a process unit, an engineering contractor performs the exchanger design and prepares the request for tender documentation. The advantage of this approach is that all vendors will quote on the basis of the same design, which will make proposal evaluation easier.

Usually, the purchaser of process equipment or its engineering contractor wants to supervise and monitor the different stages of the engineering and construction according to a quality plan. A document called the *material requisition* defines the scope of the supply, the drawing approval procedure, and the shop inspection and test procedures to be followed. It also specifies the technical documentation that must be delivered, such as detailed drawings, materials quality certificates, welding procedures, etc. The material requisition also includes a technical specification with specific details and standard drawings of the different components (e.g., nameplates, saddles, lifting lugs, insulation, painting, etc.) adopted by the purchaser. All these documents are included with the request for quotation.

REFERENCES

1. TEMA: *Standards of the Tubular Exchangers Manufacturers Association.*
2. Perry R. H., Chilton C.H.: *Chemical Engineer's Handbook.* New York: McGraw-Hill, 1973.
3. Ludwing: *Applied Process Design for Chemical and Petrochemical Plants.* Gulf Publishing, 1964.
4. Kern D.: *Process Heat Transfer.* New York: McGraw-Hill, 1954.
5. Yorkell: "Heat Exchanger Tube to Tubesheet Connections," *Chem. Eng.,* Feb 8:78–94, 1982.
6. Gardner K.: in *Heat Exchanger Design and Theory Sourcebook.* New York: McGraw-Hill, 1974.
7. ANSI/API Standard 660- Shell and Tube Heat Exchangers, 7th Ed., American Petroleum Institute, 2003.

CHAPTER 7

THERMAL DESIGN OF SHELL-AND-TUBE HEAT EXCHANGERS

7-1 BASIC PRINCIPLES

7-1-1 Heat Exchangers Design Techniques

The thermal design of a heat exchanger can be performed by the supplier, the purchaser, or an engineering company. Nowadays, there are many commercial design programs developed by research institutes or software suppliers that can be purchased or licensed usually on an annual basis.

The task of heat exchanger design consists of the definition of a large number of geometric parameters, such as length, diameter, and thickness of the tubes; tube pattern and pitch; baffle spacing; shell diameter and number of tubes; and number of passes in tube-side and shell-side. Some of these parameters may have been defined in advance by project specifications or designer preferences (typically tube type and pattern). Regarding tube length, usually the cost of a heat exchanger decreases if longer tubes are used, but sometimes, tube length is limited by layout restrictions. Once these parameters are defined, the remaining geometric characteristics (e.g., number of tubes, number of passes, baffle spacing, etc.) are defined by the thermal design program, following an optimization path that results in the minimum cost or minimum area.

In some cases, the thermal design program interacts with mechanical design modules and computer-aided design (CAD) software to prepare the engineering drawings of the heat exchanger. Obviously, access to these tools depends on the size of the company, the degree of specialization, and the volume of heat exchanger design jobs usually performed and is normally limited to engineering companies or process equipment suppliers. In small companies that have their own design department, heat exchanger design can be performed using simpler programs or with the help of manual calculations.

Before starting a heat exchanger design, it is important to analyze the amount of effort that should be devoted to the task. This decision must take into consideration the following factors:

1. Cost of the engineering work
2. Heat exchanger cost (The design of a large and high-cost unit obviously will deserve more attention than that for a small unit.)
3. The effect that a malfunction of the unit may have on the entire process (Equipment that is critical for the process justifies more effort in the design stage.)

The accuracy of the calculated results is another point to be considered. The two most frequent sources of uncertainty are the physical properties of the fluids and fouling factors, and quite often, because of these uncertainties, it is not worth using highly sophisticated design tools for the prediction of convection film coefficients.

Another important aspect to be considered is which part of the exchanger deserves more attention. The concept of *controlling resistance* always must be kept in mind. For example, if fouling is the controlling resistance in a heat exchanger, it is more important to have a good prediction of the fouling factor than to have a very accurate method to calculate the film coefficients.

Fouling deserves a special paragraph. Caution needs to be exerted in selecting fouling factors. An excessively large fouling factor will result in an oversized heat exchanger. A large fouling factor is sometimes adopted as a safety margin to cover uncertainties in fluid properties and even in process knowledge. This is not an advisable procedure for many reasons. First, the effect of fouling resistance on heat exchanger design depends on the relative magnitude of fouling with respect to the fluid film resistances. It is thus not possible to know in advance which will be the resulting oversizing. If fouling is the controlling resistance, too large a fouling factor may result in a unit with two or three times more area than really necessary. Additionally, a numerical value for the fouling factor written in a specification sheet usually has the effect of setting a precedent for future designs that may result in unnecessary penalization.

Unfortunately, the experimental determination and validation of fouling factors is not a usual practice in process industries. This is due to the fact that to get reliable results, long- and medium-term programs must be established, and dedicated instrumentation installation may be necessary Thus, in the absence of more specific information, the fouling resistances included in App. F can be adopted

This chapter will examine the methods of Kern[1] and Bell[9] for heat exchanger design. Both approaches require only a small calculation capacity and can be performed either manually or with simple spreadsheets. Whenever possible, the correlations are presented numerically and not graphically so that they can be easily programmed. When the original sources did not include numerical expressions, regressions of the graphic correlations were performed. This chapter deals only with heat exchangers in which there is no phase change of the streams. Other types of units, such as condensers and reboilers, will be treated in later chapters.

7-1-2 Cross Mixing Hypothesis and Mean Temperature Difference

When two circulating fluids exchange heat in a co-current or countercurrent pattern, as in the double-pipe heat exchanger shown in Fig. 7-1, the heat exchanged can be calculated as

$$Q = UA(\text{LMTD}) \tag{7-1-1}$$

LMTD is the logarithmic mean temperature difference, which can be calculated with the expressions included in Fig. 7-1 corresponding to each one of the two cases.

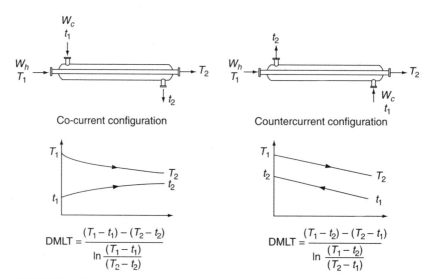

$$\text{DMLT} = \frac{(T_1 - t_1) - (T_2 - t_2)}{\ln \dfrac{(T_1 - t_1)}{(T_2 - t_2)}}$$

$$\text{DMLT} = \frac{(T_1 - t_2) - (T_2 - t_1)}{\ln \dfrac{(T_1 - t_2)}{(T_2 - t_1)}}$$

FIGURE 7-1 LMTD in co-current and countercurrent configurations.

The overall heat transfer coefficient can be calculated from the film coefficients of both fluids as

$$\frac{1}{U} = \frac{1}{U_c} + R_f \tag{7-1-2}$$

$$\frac{1}{U_c} = \frac{1}{h_o} + \frac{1}{h_{io}} \tag{7-1-3}$$

$$h_{io} = h_i \frac{D_i}{D_o} \tag{7-1-4}$$

where U = overall heat transfer coefficient
U_c = clean overall heat transfer coefficient
R_f = combined fouling resistance
h_{io} = convection heat coefficient of the tube-side fluid referred to the external surface
h_o = annulus fluid convection film coefficient
D_i = internal diameter of the inner tube
D_o = external diameter of the inner tube

In shell and tubes heat exchangers with only one tube pass, the inlet and outlet nozzles can be arranged in a countercurrent or co-current configuration, as shown in Fig. 7-2. However, in this type of heat exchanger, the shell fluid follows the path indicated in Fig. 7-3 owing to the presence of the baffles. We see that there is a component of the flow perpendicular to the tubes (cross-flow), and at the same time, there is a velocity component in the longitudinal direction, mainly at baffle windows.

This means that the fluid circulation is not truly countercurrent or co-current because of this cross-flow component. However, in the design of this type of heat exchanger, it is usually assumed that at any cross section of the heat exchanger, the shell fluid is completely mixed, so a uniform shell fluid temperature can be considered in the entire cross section.

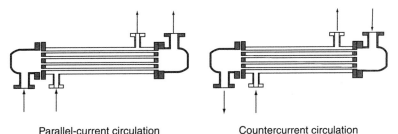

Parallel-current circulation   Countercurrent circulation

FIGURE 7-2   Nozzle arrangements in a shell-and-tube heat exchanger.

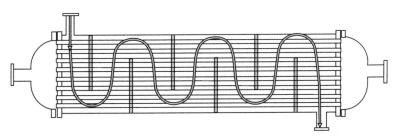

FIGURE 7-3   Shell-fluid path in a baffled heat exchanger.

Then, if it is admitted that the film coefficients for all the tubes are the same, the tube-side fluid also will heat up or cool down uniformly in all the tubes because they will be exposed to the same external temperatures.

It is then possible to consider that at each cross section the temperature of each fluid is uniform, and the temperature evolution can be represented by diagrams such as those in Fig. 7-1, and it is valid to use the logarithmic mean temperature difference to calculate the heat transferred. Equations (7-1-1) through (7-1-4) thus are valid. In these equations, A is the total area of the unit, which can be calculated by means of the expressions

$$A = Na_1 L \tag{7-1-5}$$

$$a_1 = \pi D_o \tag{7-1-6}$$

where A = heat exchanger area
   N = number of tubes
   a_1 = tube surface area per unit length = πD_o
   L = heat exchanger tube length
   D_o = external diameter of the tubes

The overall heat transfer coefficient can be calculated by means of Eqs. (7-1-2) through (7-1-4). In those equations, h_o is the film convection coefficient for the fluid circulating through the shell of the heat exchanger. The calculation of this coefficient will be explained later.

Example 7-1   *Calculation of the LMTD*
It is desired to cool down 14 kg/s of an oil having a heat capacity of 1,964 J/(kg · °C) from an inlet temperature of 107°C to 40°C. The operation will be performed using 50 kg/s of cold water available at 20°C. Calculate the LMTD using

a. A countercurrent heat exchanger
b. A parallel-flow heat exchanger

Solution   The heat to be exchanged is

$$Q = W_h c_h (T_1 - T_2) = 14 \times 1,964 \times (107 - 40) = 1.842 \times 10^6 \text{ J/s}$$

The cold water outlet temperature will be

$$t_2 = t_1 + Q/W_c c_c = 20 + 1.842 \times 10^6/50 \times 4,180 = 28.8°C$$

This value is independent on the heat exchanger configuration. For a countercurrent unit,

$$\text{LMTD} = \frac{(T_1 - t_2) - (T_2 - t_1)}{\ln \dfrac{T_1 - t_2}{T_2 - t_1}} = \frac{(107 - 28.8) - (40 - 20)}{\ln \dfrac{107 - 28.8}{40 - 20}} = 42.6°C$$

For a parallel-current unit,

$$\text{LMTD} = \frac{(T_1 - t_1) - (T_2 - t_2)}{\ln \dfrac{T_1 - t_1}{T_2 - t_2}} = \frac{(107 - 20) - (40 - 28.8)}{\ln \dfrac{107 - 20}{40 - 28.8}} = 36.9°C$$

We see that the LMTD in the countercurrent configuration is higher than that in a co-current configuration. This means that if a co-current heat exchanger is used, a larger heat transfer area will be necessary. The ratio of the required areas for both cases is

$$\frac{A_{pc}}{A_{cc}} = \frac{\text{LMTD}_{cc}}{\text{LMTD}_{pc}} = \frac{42.6}{36.9} = 1.15$$

7-2 LMTD CORRECTION FACTORS FOR OTHER FLOW CONFIGURATIONS

In a heat exchanger having more than one tube pass, the situation is different. It is no longer possible to consider a single temperature for each fluid at every cross section of the heat exchanger.

Let's consider a heat exchanger such as that shown schematically in Fig. 7-4. In a section such as AA' we can accept that the shell fluid is uniformly mixed. However, the fluid circulating through the tubes increases its temperature from t_1 to t' in the first pass and then from t' to t_2 in the second pass, so it crosses section AA' twice, with different temperatures each time. In this way, the shell fluid is in co-current flow with the first tube pass and in countercurrent flow with the second.

We have seen in Example 7-1 that a countercurrent heat exchanger requires less heat transfer area than a co-current unit with the same overall heat transfer coefficient. In a 1-2 heat exchanger, the heat transfer area will be higher than in a countercurrent unit and lower than in a co-current unit. This problem is normally handled by keeping the form of the equation

$$Q = UA\Delta T \tag{7-2-1}$$

where ΔT is an effective mean temperature difference between both streams, which must be higher than the LMTD corresponding to a parallel-flow arrangement but lower than the LMTD of a countercurrent configuration. If LMTD_{cc} is the logarithmic mean temperature difference for a countercurrent configuration, it is possible to define a factor F_t, called the LMTD *correction factor*, as

$$\Delta T = \text{LMTD}_{cc} F_t \tag{7-2-2}$$

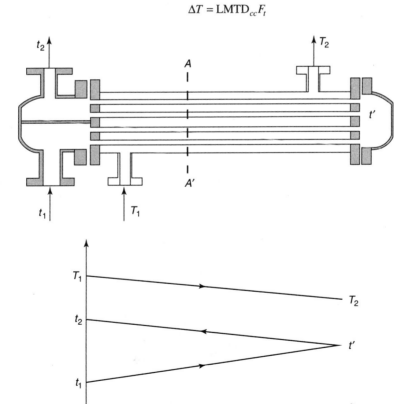

FIGURE 7-4   Temperature diagram in a 1-2 heat exchanger.

F_t is always lower than 1 because $\Delta T < \text{DMLT}_{cc}$. Equation (7-2-1) then can be written as

$$Q = UA \cdot \text{LMTD}_{cc}F_t \tag{7-2-3}$$

Then the mean temperature difference for any heat exchanger is calculated as if the configuration were countercurrent and then multiplying by the correction factor, as shown in Eq. (7-2-3). For this reason, the ΔT defined by Eq. (7-2-2) is called the *effective mean temperature difference* to distinguished it from the LMTD_{cc}, which is not the real temperature difference of the unit. From now on, the logarithmic mean temperature difference for a countercurrent configuration, LMTD_{cc} will be designated simply as LMTD, dropping the subscripts.

7-2-1 Calculation of F_t Factors

F_t factors can be calculated theoretically.[2,13] The mathematical deduction of the final expressions is tedious and will not be presented here.

The value of this factor depends on the type of unit (number of shell and tubes passes) and the inlet and outlet temperatures. A heat exchanger with one shell pass and two tube passes (1-2) has approximately the same F_t as a 1-4, 1-6, or 1-8 unit with the same inlet and outlet temperatures. This means that F_t is only a weak function of the number of tube passes (as long as it is even). Thus only the F_t values for 1-2, 2-4, 3-6, 4-8, etc. configurations are usually shown. For any other case, the value corresponding to the same number of shell passes must be used independently of the number of tube passes.

F_t values are calculated as a function of two dimensionless parameters defined as

$$R = \frac{T_1 - T_2}{t_2 - t_1} \quad \text{and} \quad S = \frac{t_2 - t_1}{T_1 - t_1}$$

Appendix B provides curves to calculate F_t factors for several flow arrangements. The analytical expressions corresponding to these curves, which can be used in computer programs, are

$$F_t = \frac{\sqrt{R^2+1}}{R-1} \frac{\ln\left[(1-P_x)/(1-RP_x)\right]}{\ln\left[\dfrac{(2/P_x)-1-R+\sqrt{R^2+1}}{(2/P_x)-1-R-\sqrt{R^2+1}}\right]} \tag{7-2-4}$$

where

$$P_x = \frac{1-\left(\dfrac{RS-1}{S-1}\right)^{1/N_s}}{R-\left(\dfrac{RS-1}{S-1}\right)^{1/N_s}}$$

N_s is the number of simple shells or number of shell passes. If $R = 1$, the preceding expression is undefined, and the following equation must be applied:

$$F_t = \frac{(P_x\sqrt{R_2+1})/(1-P_x)}{\ln\left[\dfrac{(2/P_x)-1-R+\sqrt{R^2+1}}{(2/P_x)-1-R-\sqrt{R^2+1}}\right]} \tag{7-2-5}$$

where

$$P_x = \frac{S}{(N_s - N_sS + S)}$$

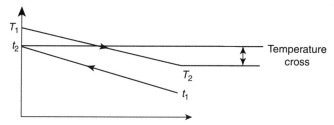

FIGURE 7-5   Temperature cross in a countercurrent heat exchanger.

Note: A downloadable interactive spreadsheet to perform these calculations is available at http://www. mhprofessional.com/product.php?isbn=0071624082

7-2-2 F_t Factor as Configuration-Selection Criterion

In a co-current heat exchanger, the outlet temperature of the hot fluid can never be lower than the outlet temperature of the cold fluid. In the limit case, where both temperatures are equal, the $LMTD_{pc}$ becomes zero, and an infinite area would be required. In a countercurrent heat exchanger, on the other hand, it is possible to have a temperature cross as shown in Fig. 7-5, obtaining a $t_2 > T_2$.

The countercurrent heat exchanger is, among all heat exchangers, the one which allows the higher temperature-cross, because, theoretically, with an infinite area, it would be possible to withdraw one of the fluids at the inlet temperature of the other, resulting in any of the situations shown in Fig. 7-6. If the product of the mass flow rate times the specific heat of the hot fluid ($W_h c_h$) is higher than that of the cold fluid ($W_c c_c$), the situation is that of Fig. 7-6a, and in the opposite case, we would have the situation of Fig. 7-6b.

In the case of multipass heat exchangers, it is possible to have some temperature cross, but there is a maximum temperature cross that is possible for each configuration and inlet temperatures. This fixes an upper limit to the amount of heat that may be transferred between both streams for each configuration. In this situation, the F_t factor is 0, so an infinite area would be necessary to reach this limit.

Even though a heat exchanger with $F_t > 0$ theoretically could do the job, if we observe the shape of the F_t curves, we can appreciate that for F_t values lower than 0.7 or 0.8, the slope of the curves is almost vertical, and F_t decreases drastically with a minor change in the parameters. So it is not safe to work in these zones, and it is recommended never to design with an F_t lower than 0.75. It must be kept in mind that the F_t curves have been developed theoretically, and in their deduction, certain hypotheses corresponding to the ideality of the heat exchange configuration were assumed. These hypotheses may not be true in a real heat exchanger, so it is not recommended to approach the steepest region of the curves in which the true F_t may be quite different from the theoretical one.

By examining the F_t curves, we see that for the same process conditions (e.g., the same inlet and outlet temperatures), when the number of shell passes increases, the F_t factor also increases. This means that if a

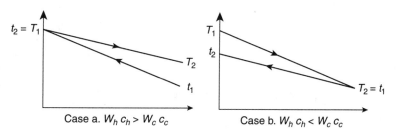

Case a. $W_h c_h > W_c c_c$ Case b. $W_h c_h < W_c c_c$

FIGURE 7-6   Temperature diagrams in a countercurrent heat exchanger with infinite area.

1-2 heat exchanger has an F_t factor lower than 0.75 for the required process conditions, it may be possible to perform the service adopting a configuration 2-4, 3-6, 4-8, etc.

Example 7-2   It is required to cool down 20 kg/s of an aqueous solution whose specific heat is 4,180 J/(kg · K) from an inlet temperature of 67°C to 37°C. The operation will be performed with cooling water available at 17°C. Determine the number of shell passes of the heat exchanger using a water mass flow of (a) 40 kg/s and (b) 20 kg/s.

Solution

a. With a water flow rate of 40 kg/s, the water outlet temperature can be calculated with the following heat balance:

$$4{,}180 \times 20 \times (67-37) = 4{,}180 \times 40 \times (t_2 - 17)$$

From which we get $t_2 = 32°C$. Calculating the parameters we have

$$R = \frac{67-37}{32-17} = 2 \quad \text{and} \quad S = \frac{32-17}{67-17} = 0.3$$

For a 1-2 heat exchanger with these parameters, we get $F_t = 0.85$. Thus it is possible to use this configuration for the required service.

b. If a water mass flow of 20 kg/s is employed, the water outlet temperature calculated with a mass balance is 47°C and then

$$R = \frac{67-37}{47-17} = 1 \quad \text{and} \quad S = \frac{47-17}{67-17} = 0.6$$

For a 1-2 configuration, a too low F_t factor is obtained with these parameters. This means that a 1-2 heat exchanger is not possible for these process conditions. With $R = 1$ and $S = 0.6$, if we move to a 2-4 configuration, we get $F_t = 0.9$, and it is thus the minimum configuration with this water flow rate. If the water mass flow is reduced further, it may not be possible to use a 2-4 configuration, and a 3-6 unit may be necessary.

In all cases, a countercurrent configuration is always possible because then $F_t = 1$. The reason for using multipass configurations is to increase the in-tube fluid velocity, improving the heat transfer coefficient, because in order to have good fluid velocity with only one tube pass, an excessively long heat exchanger may be necessary in some cases. This will be better explained in Example 7-3.

Additionally, in services requiring floating heads, the countercurrent configuration presents construction issues and is normally avoided. Obviously, U-tubes cannot be used in a one-pass configuration.

Example 7-3   It is known that to achieve a reasonable value of the heat transfer coefficient when cooling a solution with the characteristics of that in Example 7-2, a minimum fluid velocity of 1 m/s is necessary. In this case, the overall heat transfer coefficient will be about 1,300 W/(m² · K). For cleaning considerations, the solution must circulate into the tubes. It is desired to design a heat exchanger to cool 20 kg/s of solution using 20 kg/s of cooling water. The process conditions are those of case (b) in Example 7-2. The heat exchanger must be constructed with ¾-in BWG 16 tubes that have an internal diameter of 0.0157 m and an external diameter of 0.019 m. Suggest a preliminary design.

Solution   Let's consider as a first choice the use of a pure countercurrent configuration for which $F_t = 1$. The heat duty is

$$Q = W_h c_h (T_1 - T_2) = 20 \times 4{,}180 \times (67 - 37) = 2.508 \times 10^6 \text{ J/s}$$

The flow area of each tube is

$$\pi \times 0.0157^2/4 = 1.94 \times 10^{-4} \text{ m}^2$$

Assuming a 1,000 kg/m^3 density for the solution, if we want to get a velocity of 1m/s, the number of tubes in parallel will be

$$N_t = \frac{0.020 \text{ m}^3/\text{s}}{1 \text{ m/s} \times 1.94 \times 10^{-4} \text{ m}^2/\text{tube}} \simeq 102 \text{ tubes.}$$

This is the maximum number of tubes in parallel that can be used because otherwise the fluid velocity would be too low.

For a countercurrent unit, the effective mean temperature difference coincides with the LMTD. However, since $T_1 - t_2 = T_2 - t_1 = 20°C$, the LMTD is undetermined. Then $\Delta T = T_1 - t_2 = T_2 - t_1 = 20°C$.

To transfer 2,508, kW with a coefficient equal to 1,300 W/(m$^2 \cdot$ K) and a temperature difference of 20°C, the required heat transfer area is

$$A = \frac{Q}{U \cdot \text{DMLT}} = \frac{2.508 \times 10^6}{1,300 \times 20} = 96.5 \text{ m}^2$$

The heat transfer area of the 102 tubes per meter of exchanger length is

$$\pi \times 0.019 \times 102 = 6 \text{ m}^2/\text{m of exchanger length}$$

The required heat exchanger length thus would be

$$L = \frac{96.5}{6} = 16 \text{ m}$$

This is an excessive length for a single-shell heat exchanger. To achieve this length, it would be necessary to use two or three shells connected in series, as shown in Fig. 7-7, to keep the countercurrent configuration. If a more compact design is desired, a multipass heat exchanger must be used.

The alternative of a 1-2 configuration must be precluded according to the results of Example 7-2. We shall consider the possibility of using a single-shell 2-4 heat exchanger with longitudinal baffle (TEMA type F). As analyzed in Example 7-2, for a 2-4 configuration, $F_t = 0.9$; then $\Delta T = 20 \times 0.9 = 18°C$.

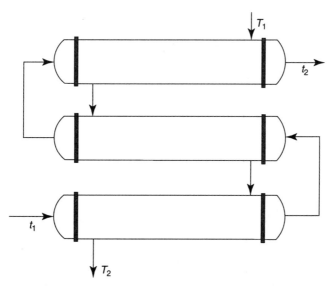

FIGURE 7-7   Countercurrent arrangement with three shells.

To keep the velocity of 1 m/s, the unit must have no more than 102 tubes in each pass. This means 408 tubes total. The required heat transfer area is

$$A = Q/U\Delta T = 2.508 \times 10^6/(1,300 \times 18) = 107 \text{ m}^2$$

The heat transfer area of the 408 tubes per meter length is

$$408 \times \pi \times 0.019 = 24 \text{ m}^2/\text{m of heat exchanger}$$

The total length will be

$$L = 107/24 = 4.5 \text{ m}$$

An acceptable decision thus will be to use a heat exchanger with 408 tubes arranged in four passes with 4.5 m length with two shell passes. The designer now must decide which is the required shell diameter to contain the 408 tubes and choose the baffle spacing to have a good heat transfer coefficient in the shell side. Of course, this is a preliminary design that must be confirmed with the design methods explained later on in this chapter.

7-2-3   Combination of Heat Exchangers

F_t factors are characteristic of a heat exchange configuration rather than of a particular type of heat exchanger. For example, with two 1-2 heat exchangers connected as shown in Fig. 7-8, we obtain the same heat exchange configuration as using a 2-4 exchanger. Both shells constitute a unit.

In this case, if both shells are identical, a single overall heat transfer coefficient can be used for the whole unit, and the F_t correction factor can be calculated from temperatures T_1, T_2, t_1, and t_2 as if it were a single 2-4 heat exchanger.

The heat exchanged thus is

$$Q = UA \times \text{LMTD} \times F_t \tag{7-2-6}$$

where

$$\text{LMTD} = \frac{(T_1 - t_2) - (T_2 - t_1)}{\ln \dfrac{T_1 - t_2}{T_2 - t_1}} \tag{7-2-7}$$

In this case, A is the total area of the two shells.

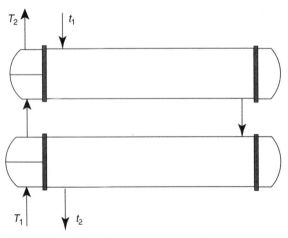

FIGURE 7-8   A 2-4 configuration using two shells.

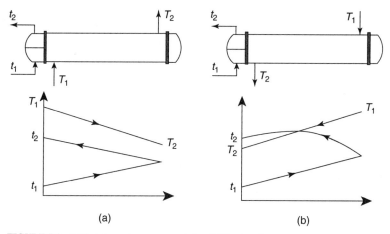

FIGURE 7-9   Different nozzle arrangements in a 1-2 heat exchanger.

7-2-4   Inlet Nozzle Orientation

A multipass heat exchanger may have the inlet nozzles of both fluids at the same end of the unit or at opposite ends. This is shown in Fig. 7-9 for the case of a 1-2 heat exchanger. It can be demonstrated[2] that the F_t values for both cases are identical.

It must be noted in Fig 7-9b that close to the cold fluid outlet, the direction of the heat flow in the second tube pass is inverted because the temperature of the hot fluid has decreased below that of the cold fluid owing to the effect of the first pass of the tube-side fluid.

7-3   BASIC DESIGN PARAMETERS OF A HEAT EXCHANGER

To design a heat exchanger means to determine the principal geometric characteristics of the unit. These are

1. Heat exchanger type (e.g., fixed tubesheet, U tubes, floating heads, etc.)
2. Diameter and array of tubes (e.g., square, triangular, rotated square and tube pitch)
3. Fluid allocation (which will be the tube fluid and which will be the shell fluid)
4. Number of shells in series, number of shell passes
5. Number of tubes, number of tube passes and shell diameter
6. Tubes length
7. Baffles type and baffles spacing

The thermal design of a heat exchanger is always a compromise between higher heat transfer coefficients and lower pressure drop in the fluids. We know that in order to improve the heat transfer coefficient, it is necessary to increase the fluid velocities. This always produces an increase in frictional pressure drops. Sometimes this can be solved by increasing the pumping power. Thus the goal is to find the best solution considering the investment cost (i.e., heat exchanger area) and operative cost (i.e., pumping power).

In other cases, the allowable pressure drops are defined by process considerations, and the designer must accept these limits at any resulting cost. He then must find the design parameter combination that results in the highest heat transfer coefficient within the pressure drop limitations. We shall briefly analyze what factors must be considered to adopt the design parameters.

Heat Exchanger Type.   The advantages and disadvantages of each type of heat exchanger were analyzed in Chap. 6 and will not be repeated here. The selection takes into consideration the cleaning facility, relative cost, thermal expansion, fluids toxicity, etc.

Tube Diameter and Tube Pattern.   The tubes used in heat exchanger manufacture are usually standardized by their external diameter. The most used standard is BWG. The tube thickness must be selected according to the internal and external pressures. Despite the fact that the thickness selection corresponds to the mechanical design, this is information that affects the thermal design and must be selected from the beginning. Appendix H shows the tube dimensions based on BWG standards. It can be seen that the higher the BWG designation, the lower is the tube thickness for the same internal diameter.

The most popular diameters are 1 or 3/4 in. Smaller tubes are difficult to clean, whereas larger diameters result in an unfavorable flow-area:transfer-area ratio.

Square tubes pattern allow mechanical cleaning of the external tubes surfaces, which is not possible with triangular arrays. On the other hand, triangular arrays allow more tubes for the same shell diameter and tube pitch, and heat transfer coefficients are somewhat higher.

At any rate, these variables do not affect the design significantly, and the tube diameter, as well as the array and tube pitch, is adopted beforehand based on design preferences, and they are seldom changed during the design process.

Fluid Allocation.   It is usually preferred to allocate the most fouling fluid to the tube side. The reason is that it is easier to clean the tube interiors than the exteriors. If one of the fluids is more corrosive, it also may be convenient to send it through the tube side because the shell then can be built with a lower-quality and cheaper material. The tube material must be resistant to both fluids.

If the flow rates of both streams are very different, it may be that the decision as to which will be the tube-side and shell-side fluids will be based on the possibility of reaching a good velocity in both fluids. However, there are other ways to act on the fluid velocities, and normally, the fluid allocation is based on the other considerations mentioned previously.

Number of Shells and/or Shell Passes.   Use of the F_t factor as a criterion to decide the number of shell passes was explained and analyzed in Examples 7-2 and 7-3. The use of shells with longitudinal baffles (TEMA type F) results in construction complications because the design must guarantee a good closure between the shell and the longitudinal baffle. Thus it is normally preferred to avoid this type of construction, and the multipass configuration is obtained combining adequately connected shells (see Fig. 7-8).

Number of Tubes and Number of Tube Passes.   To obtain a specific heat transfer area, a combination of number and length of tubes must be selected. The higher the number of tubes, the smaller will be the length required.

The cost of a heat exchanger with long tubes is lower than that of a unit with shorter tubes and the same area because the labor cost increases with the number of tubes to be welded. In addition, the cost of tubesheets and heads increases significantly with shell diameter. This is why the usual practice is to adopt the greatest length allowed by the equipment layout.

The higher the number of tubes, the higher is the flow area, and the smaller is the velocity of the tube-side fluid. If a high number of tubes is adopted, it may be necessary to increase the number of tube passes to get a reasonable velocity and good heat transfer coefficient.

Let's consider a countercurrent heat exchanger such as that shown in Fig. 7-10a. This heat exchanger can be transformed in a two-pass heat exchanger with about the same number of tubes by changing the heads as shown in Fig. 7-10b.

If both exchangers are put in operation with the same flow rates, the tube-side fluid velocity in case *b* will be twice as high as that in case *a*. The film heat transfer coefficient for turbulent flow within the tubes changes with the power 0.8 of the velocity. Then, with the change performed, the coefficient h_i increased by $2^{0.8} = 1.74$.

If we assume that the friction factor does not change too much with the velocity increase in turbulent flow, the pressure drop can be considered to be proportional to the square of the velocity and to the length

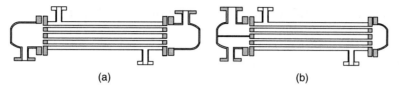

(a)   (b)

FIGURE 7-10   (*a*) Countercurrent and (*b*) double-pass heat exchangers.

of the fluid path. Since the path length in case *b* is twice that of case *a,* the pressure drop increases eight times with the modification performed.

The effect of this modification on the overall *U* depends on the relative importance of the tube-side coefficient with respect to other resistances. It must be kept in mind that the overall heat transfer coefficient is given by Eqs. (7-1-2) and (7-1-3), and it may happen, especially if the tube-side film is not the controlling resistance, that the improvement achieved in the overall heat transfer coefficient *U* may be much lower than a 1.74 factor, making the modification unworthy.

Obviously, the higher the number of tubes in the heat exchanger, the larger must be the shell diameter. Once the number of tubes, tube passes, and tube array have been selected, it is necessary to determine the necessary shell diameter. This can be done with the help of App. D and with the tube distribution diagrams in App. E.

The shell diameter is related to the separation of the baffles because an increase in the shell diameter must be compensated during design with a reduction in baffle spacing to keep a high shell-side fluid velocity.

Tube Length.   The selection of the tube length was analyzed earlier in connection with the other design parameters. We should add that sometimes it may be convenient to consult the tube suppliers to know what are the commercial lengths available on the market. If the heat exchanger is designed with a tube length that is not a submultiple of a commercial size, a waste of material may result.

Baffle Type and Spacing.   The most popular type of baffle is the segmental baffle, and the height of the circular segment corresponding to the baffle window is usually about 25 percent of the shell diameter. Baffle spacing determines the shell-side fluid velocity. The smaller the spacing, the higher is the fluid velocity, the higher is the film coefficient, and the higher is the pressure drop. Baffle spacing should be adopted to get the highest convection film coefficient within the pressure-drop limitations. The minimum baffle spacing allowed by the TEMA Standard is 1/5 of the shell diameter, but not less than 2 in.

The other function of baffles is to provide mechanical support to the tubes, mitigating bending and vibration. This is why TEMA Standards define maximum values for unsupported tube length. Since the tubes crossing the baffle window are supported every other baffle, the unsupported span is twice the baffle spacing, as shown in Fig. 7-11. Sometimes baffles with no tubes in the window area must be employed (baffle type NTIW).

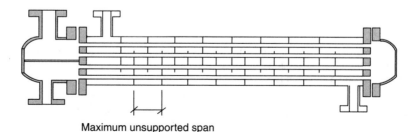

Maximum unsupported span

FIGURE 7-11   Maximum unsupported tube span.

7-4 CONVECTION FILM COEFFICIENT AND PRESSURE DROP IN THE TUBE SIDE

7-4-1 Convection Film Coefficient

The convection film coefficient for the fluid circulating inside the tubes of a heat exchanger can be calculated with the usual correlations for in-tube convection. A Reynolds number is defined as,

$$\text{Re} = \frac{D_i G_t}{\mu} \qquad (7\text{-}4\text{-}1)$$

where G_t = mass velocity = W/a_t
$\quad a_t$ = tube side flow area = $(N\pi D_i^2)/4n$ $\qquad (7\text{-}4\text{-}2)$
$\quad N$ = number of tubes
$\quad n$ = number of tube passes
$\quad D_i$ = tube internal diameter

And the results are correlated using this Reynolds number as follows:

1. For laminar flow (Re < 2,100), according to Sieder and Tate,[3]

$$\frac{h_i D_i}{k} = 1.86 \left(\text{Re} \cdot \text{Pr} \frac{D_i}{L} \right)^{0.33} \left(\frac{\mu}{\mu_w} \right)^{0.14} = 1.86 \left(\frac{4Wnc}{\pi k L N} \right)^{0.33} \left(\frac{\mu}{\mu_w} \right)^{0.14} \qquad (7\text{-}4\text{-}3)$$

In a multipass heat exchanger, L is still the length of one tube, and not the length of the total path $L \cdot n$ because in the exchanger heads, the temperature of the fluid becomes uniform, and the temperature profile must start developing again.

2. For the turbulent region (Re > 10,000), the correlation[4] is

$$\frac{h_i D_i}{k} = 0.023 \text{Re}^{0.8} \cdot \text{Pr}^{0.33} \left(\frac{\mu}{\mu_w} \right)^{0.14} \qquad (7\text{-}4\text{-}4)$$

The average deviation of this correlation was reported as +15 and −10 percent for Reynolds numbers higher than 10,000. For the particular case of water at moderates temperature and in turbulent regime, the following dimensional correlation is suggested:

$$h_i = 1{,}423(1 + 0.0146t) \frac{v^{0.8}}{D_i^{0.2}} \qquad (7\text{-}4\text{-}5)$$

where h_i = W/(m² · K)
$\quad t$ = °C (mean temperature of the water)
$\quad v$ = m/s
$\quad D_i$ = m

3. Transition zone (2,100 < Re < 10,000). This is a highly nonstable region, and it is not possible to find a suitable correlation representing the experimental results. The following correlation was suggested,[18] but the best recommendation is to avoid this region during design:

$$\frac{h_i}{cG_t} = 0.116 \left(\frac{\text{Re}^{0.66} - 125}{\text{Re}} \right) \left[1 + \left(\frac{D_i}{L} \right)^{0.66} \right] \text{Pr}^{-0.66} \left(\frac{\mu}{\mu_w} \right)^{0.14} \qquad (7\text{-}4\text{-}6)$$

7-4-2 Frictional Pressure Drop

The frictional pressure drop for fluids circulating in the tube side of a heat exchanger can be considered as the sum of two effects:

1. The pressure drop along the tubes
2. The pressure drop owing to the change in direction in the exchanger heads

The pressure drop along the tubes can be calculated with the Fanning equation:

$$\Delta p = 4 fn \frac{L}{D_i} \frac{G_t^2}{2\rho} \left(\frac{\mu}{\mu_w} \right)^a \tag{7-4-7}$$

where the exponent a is -0.14 for turbulent flow and -0.25 in laminar flow.

In this equation, nL is the total fluid path length corresponding to n passes. The friction factor in laminar flow is

$$f = \frac{16}{\text{Re}} \tag{7-4-8}$$

For heat exchanger tubes, the friction factor in the turbulent region can be determined with the following equation owing to Drew, Koo, and McAdams:[6]

$$f = 0.0014 + 0.125 \text{Re}^{-0.32} \tag{7-4-9}$$

It is normally accepted that Eq. (7-4-8) can be used for $2{,}100 < \text{Re}$, whereas Eq. (7-4-9) must be used for $\text{Re} > 2{,}100$.

The friction factors obtained with Eq. (7-4-9) are for smooth tubes. Some authors suggest to increase them by 20 percent for commercial steel heat exchanger tubes.

The pressure drop corresponding to the change in direction at the exchanger heads in multipass heat exchangers can be calculated as

$$\Delta p_r = 4n \frac{G_t^2}{2\rho} \tag{7-4-10}$$

where n is the number of tube passes. The total pressure drop thus will be

$$\Delta p_T = \Delta p_t + \Delta p_r \tag{7-4-11}$$

Example 7-4 A heat exchanger has 300 ¾-in BWG 16 tubes, 2 m length, arranged in two passes. It is used to heat up 58 kg/s of an oil whose properties are

$\rho = 790 \text{ kg/m}^3$

$c = 2{,}100 \text{ J/(kg} \cdot \text{K)}$

$k = 0.133 \text{ W/(m} \cdot \text{K)}$

Viscosity as a Function of Temperature [kg/(m · s)]					
T (°C)	57	67	77	147	177
μ [kg/(m · s)]	3.6×10^{-3}	3.02×10^{-3}	2.5×10^{-3}	7.4×10^{-4}	5.2×10^{-4}

The oil inlet temperature is 57°C, and the desired outlet temperature is 77°C. The heating medium is saturated steam at 927 kPa (condensation temperature = 177°C). Decide if the unit is appropriated for the service and, in that case, what fouling factor is allowed by the heat exchanger. Assume that the oil circulates into the tubes.

Solution We must analyze some particular characteristics of heat exchangers in which steam is the heating medium.

1. If we look at App. G, we notice that the heat transfer coefficients for condensing steam at pressures higher than atmospheric are extremely high. Then, in the expression

$$U = \left(\frac{1}{h_{io}} + \frac{1}{h_o} + R_f \right)^{-1}$$

the term $1/h_o$ is much smaller than the other two terms. Thus, even if we have a high error in the prediction of h_o, this error has little influence in the calculated U. It is therefore usual, in steam-heated exchangers, to assume for the condensing steam heat transfer film coefficient a value of 8,500 W/(m² · K), which is not necessary to verify. Thus, to calculate the overall heat transfer coefficient, it is only necessary to calculate the cold-side film coefficient, and it will be

$$U = \left(\frac{1}{h_{io}} + \frac{1}{8,500} + R_f \right)^{-1}$$

2. In these types of units, one fluid (the condensing steam) suffers a change of state. Since it is a pure-component fluid, the condensation is isothermal, and

$$Q = W_h \lambda = W_c c_c (t_2 - t_1)$$

where λ is the latent heat of condensation.

It can be seen in the F_t graphs that if one fluid is isothermal, F_t equals 1 independently of the number of shell passes. Thus

$$Q = UA \times \text{LMTD} = UA \frac{(T - t_1) - (T - t_2)}{\ln \frac{T - t_1}{T - t_2}} = UA \frac{t_2 - t_1}{\ln \frac{T - t_1}{T - t_2}}$$

The mean temperature of the cold fluid is $\frac{1}{2}(57 + 77) = 67$°C. At this temperature, the viscosity is 3.02×10^{-3} kg/ms. Then

$$\text{Pr} = \frac{c\mu}{k} = \frac{0.00302 \times 2,100}{0.133} = 47.68$$

$$a_t = \text{tube-side flow area} = \frac{1}{4}\pi D_i^2 \frac{N}{n} = \frac{\pi \times 0.0157^2 \times 300}{4 \times 2} = 0.029 \text{ m}^2$$

$$G_t = W/a_t = 58/0.029 = 1,997 \text{ kg/(m}^2 \cdot \text{s)}$$

$$\text{Re} = D_i G_t/\mu = 0.0157 \times 1,997/3.02 \times 10^{-3} = 10,381$$

$$h_i = 0.023 \text{Re}^{0.8} \cdot \text{Pr}^{0.33} \frac{k}{D_i} \left(\frac{\mu}{\mu_w}\right)^{0.14} = 0.023 \times 10,381^{0.8} \times 47.68^{0.333}$$

$$\times \frac{0.133}{0.0157}\left(\frac{\mu}{\mu w}\right)^{0.14} = 1,151\left(\frac{\mu}{\mu_w}\right)^{0.14}$$

As a first approximation, we assume that $\left(\dfrac{\mu}{\mu_w}\right)^{0.14} = 1,$ to be checked later. Then

$$h_i = 1,151 \,\text{W/(m}^2 \cdot \text{K)}$$

$$h_{io} = 1,151\frac{D_i}{D_o} = 1,151\frac{0.0157}{0.019} = 951 \,\text{W/(m}^2 \cdot \text{K)}$$

We shall calculate the mean tube wall temperature. The heat transferred from the steam to the tube walls per unit area can be expressed as

$$Q/A = h_o(T - T_w)$$

And the heat transferred from the tube walls to the oil is

$$Q/A = h_{io}(T_w - t)$$

If we take for t the mean temperature of the oil, which is 67°C, and we equate both expressions, we get

$$8,500 \times (177 - T_w) = 951 \times (T_w - 67)$$

from which we get

$$T_w = 165°C$$

We can obtain the viscosity at the wall temperature. Adopting $\mu_w = 5.2 \times 10^{-4}$ kg/(m · s), then

$$\left(\frac{\mu}{\mu_w}\right)^{0.14} = \left(\frac{0.00302}{0.00052}\right)^{0.14} = 1.27$$

By applying this correction to the h_{io} value, we get

$$h_{io} = 951 \times 1.27 = 1,207 \,\text{W/(m}^2 \cdot \text{K)}$$

It would be possible to perform a new iteration and recalculate T_w, but this is usually not necessary. The clean overall coefficient (not including fouling) thus will be

$$U_c = \left(\frac{1}{h_o} + \frac{1}{h_{io}}\right)^{-1} = \left(\frac{1}{8,500} + \frac{1}{1,207}\right)^{-1} = 1,056 \,\text{W/(m}^2 \cdot \text{K)}$$

The heat duty to be transferred in the unit is

$$Q = W_c c_c(t_2 - t_1) = 58 \times 2,100 \times (77 - 57) = 2.436 \times 10^6 \,\text{W}$$

The logarithmic mean temperature difference is

$$\frac{t_2 - t_1}{\ln\dfrac{T - t_1}{T - t_2}} = \frac{77 - 57}{\ln\dfrac{177 - 57}{177 - 77}} = 109.7°C$$

The heat exchanger area is

$$A = \pi D_o LN = \pi \times 0.019 \times 2 \times 300 = 35.8 \,\text{m}^2$$

With a temperature driving force of 109.7°C, the required heat transfer coefficient is

$$U = \frac{Q}{A \cdot \text{DMLT}} = \frac{2.436 \times 10^6}{35.8 \times 109.7} = 620 \text{ W/(m}^2 \cdot \text{K)}$$

Since this value is smaller than the calculated U_c, the unit is suitable for the required service. The fouling resistance allowed by this heat exchanger is

$$R_f = \left(\frac{1}{U} - \frac{1}{U_c} \right) = \frac{1}{620} - \frac{1}{1,056} = 6.6 \times 10^{-4} \text{ (m}^2 \cdot \text{K)/W}$$

Let's now calculate the pressure drop of the oil:

$$f = 0.0014 + 0.125 \text{ Re}^{-0.32} = 0.0014 + 0.125 \times 10,381^{-0.32} = 7.88 \times 10^{-3}$$

$$\Delta p_t = 4 fn \frac{LG_t^2}{D_i 2\rho} = \frac{4 \times 7.88 \times 10^{-3} \times 2 \times 2 \times 1,997^2}{0.0157 \times 2 \times 790} = 20,269 \text{ N/m}^2$$

$$\Delta p_r = \frac{4nG_t^2}{2\rho} = \frac{4 \times 2 \times 1,997^2}{2 \times 790} = 20,192$$

Then $\Delta p_T = \Delta p_t + \Delta p_r = 20,269 + 20,192 = 40,461 \text{ N/m}^2$

7-5 HEAT TRANSFER COEFFICIENT AND PRESSURE DROP AT THE SHELL SIDE

7-5-1 Kern Method

Calculation of the shell-side heat transfer coefficient and pressure drop is considerably more complex than that of the tube side. The first calculation methods were developed in the 1930s and 1940s. The method that has been the most popular was proposed by Kern.[1] This method was used extensively in heat exchanger design. However, the error can be high in some cases. At present, other methods are reported to be more accurate, and we can recommend the Kern method for preliminary estimation, but not as a final design tool.

Shell-Side Reynolds Number. In the region between two consecutive baffles of a heat exchanger, the main flow direction is normal to the tubes, as shown in Fig. 7-12. We see that the fluid velocity is subject to continuous fluctuations owing to the reduction in flow area when the fluid crosses a tube row in comparison with the flow area in the space between two consecutive rows.

Additionally, the width of the shell cross section changes from zero at the bottom and top of the shell to a maximum in the central plane. Hence it is not possible to define a single value for the cross-flow area. This means that in order to define a fluid velocity, the definition of the flow area must be arbitrary.

Kern considers a flow area that corresponds with the hypothetical tube row at the shell central plane. We called it *hypothetical* because it may happen that there is no tube row at the central plane, but rather two tube rows shifted a certain distance up and down the central plane. This hypothetical flow area is shown in Fig. 7-13.

We shall call P_t, or tube pitch, the separation between tube axes, and c is the clearance or free distance between two adjacent tubes in the same row (see Fig. 7-13). The number of clearances existing in the central row can be obtained approximately by dividing the shell diameter by P_t. If B is baffle spacing, the area of any of these clearances is cB, and it results that the shell flow area is

$$a_S = \frac{D_S cB}{P_t} \tag{7-5-1}$$

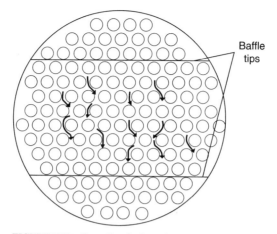

FIGURE 7-12   Cross-flow in the region between baffles.

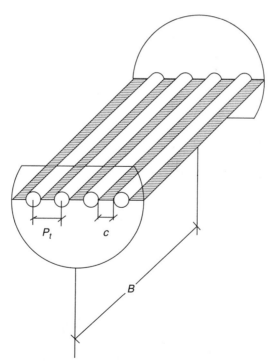

FIGURE 7-13   Shell-side flow area.

where D_s is the shell internal diameter. It is then possible to define a mass velocity for the shell fluid as

$$G_s = \frac{W}{a_s}$$

(7-5-2)

It must be noted that the free distance between tubes that correspond with the central plane of a tubes row coincides with the tube clearance c in triangular or square arrays. In the case of rotated triangular or square patterns (see Fig. 6-8), these magnitudes are different, and Kern does not indicate how to calculate as in these cases. However, square and triangular arrays are the most popular tubes patterns.

Equivalent Diameter. To complete the definition of the Reynolds number, Kern uses an equivalent diameter for the shell. The usual definition of an equivalent diameter is

$$D_e = 4 \times \text{hydraulic radius} = 4 \times \frac{\text{flow area}}{\text{wetted perimeter}}$$

(7-5-3)

Even though the main component of the fluid velocity is in the direction perpendicular to the tubes axes, Kern defines the hydraulic radius as if the fluid were flowing in a direction parallel to the tubes. The hydraulic radius in Eq. (7-5-3) thus is defined by the tube distribution pattern in the tubesheet. The ratio of flow area to wetted perimeter for square and triangular arrays can be obtained from Fig. 7-14, where the shaded area represents the flow area in Eq. (7-5-3).

The result for a square pattern is

$$D_e = \frac{4 \times (P_t^2 - \pi D_o^2/4)}{\pi D_o}$$

(7-5-4)

For the triangular array shown in Fig. 7-14, the wetted perimeter for the portion of the tubesheet represented corresponds to half of a tube. Then

$$D_e = \frac{4 \times (0.5P_t \times 0.86P_t - 0.5\pi D_o^2/4)}{0.5\pi D_o}$$

(7-5-5)

Table 7-1 shows the equivalent diameters calculated with Eqs. (7-5-4) and (7-5-5) for the most common arrays.

Then the shell Reynolds number is defined as

$$\text{Re}_s = \frac{D_e G_s}{\mu}$$

(7-5-6)

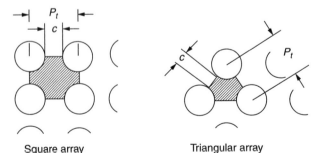

Square array   Triangular array

FIGURE 7-14   Tube patterns.

TABLE 7-1 Equivalent Diameters Calculated for the Most Common Arrays

Tube Diameter	Pattern	Pitch (see Fig. 6-8)	Equivalent Diameter
3/4 in (0.019 m)	Square	1 in (0.0254 m)	0.95 in (0.0241 m)
1 in (0.0254 m)	Square	1 1/4 in (0.0317 m)	0.99 in (0.0251 m)
1 1/4 in (0.0317 m)	Square	1 9/16 in (0.0397 m)	1.23 in (0.0312 m)
3/4 in (0.019 m)	Triangle	15/16 in (0.0238 m)	0.55 in (0.0139 m)
3/4 in (0.019 m)	Triangle	1 in (0.0254 m)	0.73 in (0.0185 m)
1 in (0.0254 m)	Triangle	1 1/4 in (0.0317 m)	0.72 in (0.0183 m)
1 1/4 in (0.0317 m)	Triangle	1 9/16 in (0.0397 m)	0.91 in (0.0231 m)

And the heat transfer and friction-factor data are correlated as a function of this Reynolds number.

Correlation for the Heat transfer Coefficient h_o. The correlation suggested by Kern for a shell with 25 percent cut segmental baffles is

$$\frac{h_o D_e}{k} = 0.36 \, \mathrm{Re}_s^{0.55} \, \mathrm{Pr}^{0.33} \left(\frac{\mu}{\mu_w}\right)^{0.14} \tag{7-5-7}$$

Graphical correlations for other baffle cuts can be found in ref. 7.

Pressure Drop. According to Kern, the pressure drop in the shell fluid is proportional to the number of times the fluid crosses the tube bundle. If N_B is the number of baffles, the shell fluid crosses the bundle $N_B + 1$ times. The pressure drop is also proportional to the length of the path at every bundle cross, which can be represented by the shell diameter. Thus

$$\Delta p_S = f \frac{(N_B + 1) D_S}{D_e} \frac{G_S^2}{2\rho} \left(\frac{\mu_w}{\mu}\right)^{0.14} \tag{7-5-8}$$

The friction factor can be correlated as a function of the shell Reynolds number. The correlations are presented graphically in the original source. A numerical regression of the curves leads to the following expressions:

1. For $\mathrm{Re}_5 < 500$,

$$f = \exp\left\{5.1858 - 1.7645 \ln(\mathrm{Re}_S) + 0.13357[\ln(\mathrm{Re}_S)]^2\right\} \tag{7-5-9}$$

2. For $\mathrm{Re}_5 > 500$,

$$f = 1.728 \, \mathrm{Re}_S^{-0.188} \tag{7-5-10}$$

Heat Exchangers with 2-4 Configuration. In the case of heat exchangers with two shell passes in a single shell with a longitudinal baffle (TEMA type F), such as that shown in Fig. 7-15, it must be considered that the flow area for the mass velocity calculation is half that of a heat exchanger with the same shell diameter without a longitudinal baffle. The number of bundle crosses to calculate the pressure drop will be twice the number in a heat exchanger with the same baffle spacing and no longitudinal baffle.

7-5-2 Factors Affecting the Performance Not Considered in the Kern Method

The Kern method has been criticized[8] for a number of reasons. Among them, we can mention the following:

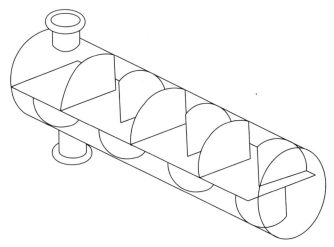

FIGURE 7-15 Shell with longitudinal baffle.

Bypass and Leakage Streams. The method assumes that the shell fluid flow is perpendicular to the tubes. This is not true for the following reasons:[14]

1. At the baffles window, the flow is parallel to the tubes.
2. In the region between baffle tips, part of the fluid diverts toward the periphery of the bundle where frictional resistance is lower, as shown in Fig. 7-16. This fraction of the flow is called the *bypass stream*. The magnitude of this stream is higher in heat exchangers having a big gap between the tube bundle and

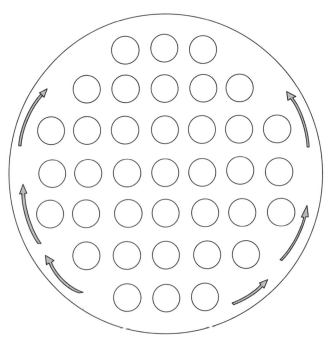

FIGURE 7-16 Bypass streams.

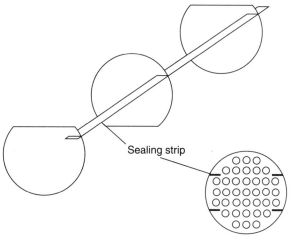

FIGURE 7-17   Location of sealing strips.

the shell diameter, as in TEMA type T floating-head heat exchangers. Obviously, this flow is ineffective for the heat transfer because it by passes the tube bundle.

One way to solve this problem is to install sealing strips, which are flat bars inserted in the baffles, as shown in Fig. 7-17. These sealing strips redirect the flow toward the center of the bundle.

A similar problem appears in multipass heat exchangers, where the separation between tubes must be increased to allocate the pass-partition plates into the exchanger headers. This creates additional bypass lanes, and part of the shell fluid flows through them without contacting the tubes. This fraction of the flow is called *pass-partition flow* (Fig. 7-18). Sometimes this effect is reduced by installing false tubes (closed at both ends) called *dummy tubes*[5] that act as deflectors (Fig. 7-19).

3. Since the baffle holes crossed by the tubes are drilled with a diameter slightly larger than the tube diameter, there is a clearance between the tube and the baffle, and part of the shell fluid leaks across this clearance. There is also leakage through the free area between the baffle and the shell. These baffle-shell

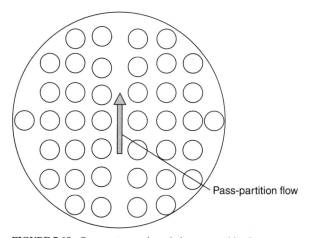

FIGURE 7-18   Bypass streams through the pass-partition lanes.

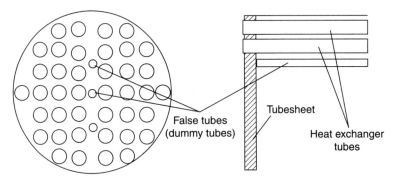

FIGURE 7-19   Installation of dummy tubes to block bypass lanes.

leaks are completely inefficient from the point of view of the heat exchange. This is not the case with the leaks between baffle and tube, which takes place with a certain amount of heat exchange with the tube fluid (see Fig. 7-20).

Thus the total fluid flow in the shell of a heat exchanger is the sum of the following fractions:

- A cross-flow fraction, which is the fraction that really circulates across the bundle
- A bypass fraction
- Leaks between the tubes and baffle
- Leaks between the baffle and shell

These flow fractions are represented in Figs. 7-20 and 7-21.

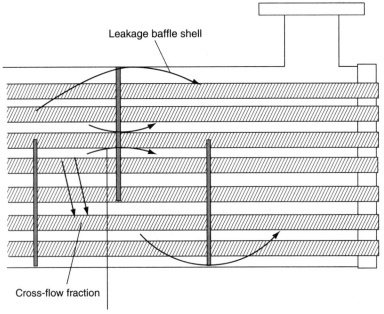

FIGURE 7-20   Leakage flows.

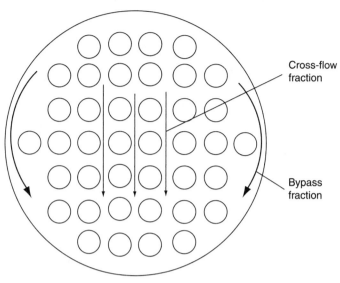

FIGURE 7-21   Bypass flows.

The relative magnitude of the several flow fractions is shown in Table 7-2 as a percentage of the total shell fluid flow. The experimental data were obtained in heat exchangers with the following range of parameters:

Shell diameter between 3 and 44 in

Tube diameters from $\frac{1}{4}$ to $\frac{3}{4}$ in

Baffle cut from 9.6 to 50 percent

Baffle-spacing:shell-diameter ratio from 0.1 to 0.8

Number of sealing strips from 0 to 3 for every 5 tube rows

Reynolds number between 2 and 100,000

Prandtl number between 0.7 and 9,000

It is easy to appreciate the serious deviation we may have when calculating a heat transfer coefficient with the assumption that 100 percent of the fluid circulates in cross-flow, as in Kern's method.

Tube Pattern.   As we explained, Kern defines an equivalent diameter for the shell-side flow. The definition of this equivalent diameter depends on the tube pattern. For square and triangular tubes, it is calculated with Eqs. (7-5-4) and (7-5-5). This approach was objected to by other authors for the following reasons:

1. The equivalent diameter is defined for a flow direction parallel to the tubes, whereas the shell-side flow is mainly normal to the tubes.

2. It is not possible to have a correlation that is valid for every geometry if geometric similitude does not exist.

TABLE 7-2   Relative Magnitude of the Several Flow Fractions

	Turbulent Flow	Laminar Flow
Cross-flow	10–45%	10–50%
Bypass	15–35%	30–80%
Baffle-shell leakage	6–21%	6–48%
Baffle-tube leakage	9–23%	9–10%

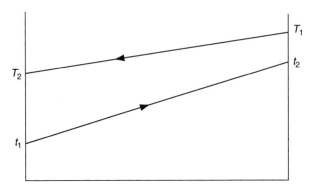

FIGURE 7-22   Theoretical temperature diagram.

This means that there should be a different correlation for each geometry. When geometric similitude exists, any characteristic length can be used to define the Reynolds number. Thus there is no need to define the equivalent diameter. This is why the concept of equivalent diameter was abandoned by most authors and the present approach is to use the tube diameter for the Reynolds number and obtain a different correlation for each tube pattern.

Effect of Temperature-Profile Distortion.   The effect of leakage and bypass flows is not only to reduce the heat transfer coefficients but also to produce a distortion in the temperature profiles that may seriously affect the behavior of the unit. For example, let's assume that it is desired to heat up a certain fluid from an inlet temperature t_1 to temperature t_2 using a hot stream available at T_1. If we know the mass flow rates of both streams, it is possible, with a heat balance, to calculate the outlet temperature T_2, resulting in a temperature diagram such as that in Fig. 7-22. Let's assume that the cold fluid is the shell-side stream.

Since the total flow decomposes in different fractions, and each one is heated up with different efficiency, their temperature evolutions also will be different. For example, let's consider the total flow divided in two fractions W_b and W_a that exchange heat with different efficiencies, as shown in Fig. 7-23. Curve b shows the evolution of the most efficient flow fraction.

The fraction corresponding to the less efficient streams (e.g., leakage between baffle and shell) can follow an evolution such as that of curve a. The resultant from both is curve c, which represents the heat balance.

It can be seen that the true temperature difference between the hot fluid and the stream W_b (which is mainly responsible of the heat transfer because W_a corresponds to the less efficient fractions) is lower than

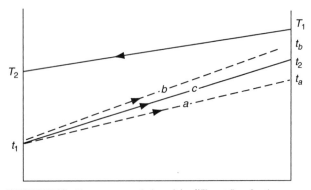

FIGURE 7-23   Temperature evolution of the different flow fractions.

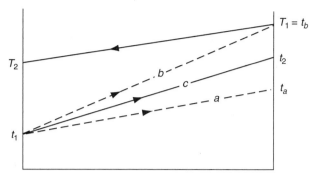

FIGURE 7-24 Limit case where flow of the "inefficient" fraction is large.

that resulting from a simple heat balance (curve c). Thus the required heat transfer area will be higher in comparison with an exchanger with no leaks.

The outlet temperature t_2 results from

$$t_2 = \frac{W_a t_a + W_b t_b}{W_a + W_b}$$

If the "inefficient" flow fraction is very high, to get the resulting outlet temperature t_2, it will be necessary that the "efficient" fraction b reaches a much higher temperature to compensate inefficiencies of stream a As a limit case, we may have a situation in which curve b reaches the hot fluid temperature at T_1, as shown in Fig. 7-24. In this case, an infinite heat transfer area would be necessary.

Thus special precautions must be taken when leakage flows are important and the temperature difference between hot and cold streams is small. In these cases, the use of global methods can lead to considerable errors because of the reduction in the true temperature difference with respect to the theoretical value calculated without considering the relative efficiency of the different flow fractions.

7-5-3 Bell Method

In 1950 at Delaware University, a research program in shell-and-tube heat exchangers was developed with the sponsorship of the Tubular Exchanger Manufacturers Association (TEMA) and the American Society of Mechanical Engineers (ASME). Many researchers worked in this program and published their conclusions in several reports over that decade.

The final synthesis of the study was published as a heat exchanger design method by Kenneth Bell[9] in 1963. This is now known as *Bell* or *Delaware method*.

Fundamentals of the Method.   The method is based on heat transfer and pressure-drop experimental data corresponding to an ideal tube bank (a tube bank with infinite width). The ideal tube bank is materialized with a rectangular heat exchanger in pure cross-flow. Half-tubes are installed at the borders to simulate continuity (Fig. 7-25). The data corresponding to the ideal bank[17] are later corrected with coefficients that take into account the characteristics of a particular exchanger (such as leakage areas, baffle cut, bypass streams, etc.).

Ideal-Tube-Bank Data.   Data are correlated as a function of a Reynolds number defined as

$$\mathrm{Re}_m = \frac{W D_o}{S_m \mu} \tag{7-5-11}$$

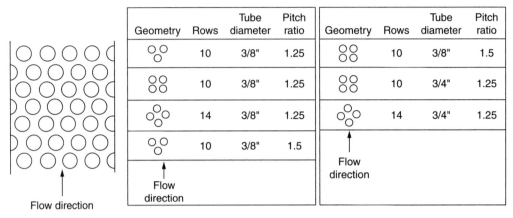

Geometry	Rows	Tube diameter	Pitch ratio	Geometry	Rows	Tube diameter	Pitch ratio
	10	3/8"	1.25		10	3/8"	1.5
	10	3/8"	1.25		10	3/4"	1.25
	14	3/8"	1.25		14	3/4"	1.25
	10	3/8"	1.5				

FIGURE 7-25 Experimental ideal tube banks in Delaware research program.

where W = shell-fluid mass flow
  D_o = external diameter of tubes
  μ = viscosity
  S_m = cross-flow area

S_m is the minimum cross-sectional flow area in the tube bank. In square or triangular arrays, this is the free area available for normal flow calculated at the central plane of a tube row (Fig. 7-26).

The mass velocity or mass flow density is defined as

$$G_m = \frac{W}{S_m} \tag{7-5-12}$$

And the experimental data are expressed as a function of a friction factor f and a Colburn coefficient j defined as

$$f = \frac{2\Delta p_b \rho}{4 G_m^2 N_c} \left(\frac{\mu}{\mu_w} \right)^{-0.14} \tag{7-5-13}$$

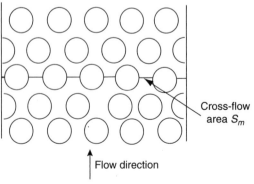

Cross-flow area S_m

Flow direction

FIGURE 7-26 Cross-flow area.

where Δp_b = tube-bank pressure drop

N_c = number of tube rows perpendicular to the flow

and

$$j = \frac{h_B}{cG_m} \mathrm{Pr}^{2/3} \left(\frac{\mu_w}{\mu} \right)^{0.14} \qquad (7\text{-}5\text{-}14)$$

where h_B is the heat transfer coefficient corresponding to the ideal tube bank. (In the definition of h_B, LMTD is used to express the temperature difference. In a heat exchanger, N_c will be the number of tube rows included between the edges of two consecutive baffles.)

The experimental data are plotted as a function of the Reynolds number for different tube patterns and different pitch:diameter ratio. This graph is shown in Fig. 7-27a. The data in the figure correspond to an ideal tube bank with 10 tube rows. A numerical regression of these data is presented in Fig. 7-27b.

It can be seen that the friction-factor curves corresponding to square arrays present a minimum, which corresponds to the regime transition from laminar to turbulent.[15] In the staggered arrays (i.e., triangular or rotated square), the transition is not clear. The difficulty in appreciating the regime transition is due to the fact that this phenomenon is evidenced by a sudden change in the pressure drop. Since the flow area is not constant, the fluid experiences accelerations and deaccelerations, then the inertial terms of the Navier-Stokes equation are not nil and the turbulence effects are masked by them. The effect is more pronounced in staggered arrays.

However, it is normally accepted that the laminar-flow region extends up to a Reynolds number on the order of 100. In this region, the inertial effects are small compared with the viscous forces, and f is approximately proportional to the reciprocal of the Reynolds number.

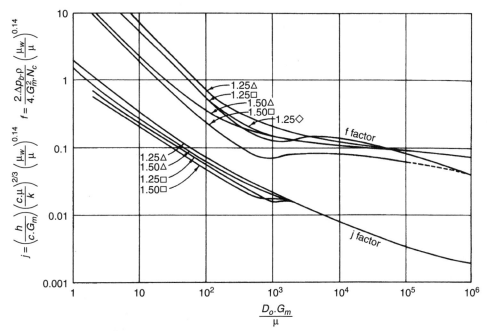

FIGURE 7-27 (a) j and f factors for the ideal tube bank. *(From O. P. Bergelin, G. A. Brown, and S. C. Doberstein,"Heat Transfer and Fluid Friction During Flow Across Banks of Tubes," Trans ASME 4:958, 1952. With permission from the American Society of Mechanical Engineers.)* (b) Numerical regression of the curves in part a.

CORRELATION FOR SHELL SIDE j FACTOR

Mathematical regression of curves in Figure 7-27

$$j = (h/c\,G_m)\,Pr^{2/3}\,(\mu_w/\mu)^{0.14} \quad \text{(See variables definition in the text)}$$

INTERVAL	$Re_m < 100$		$100 < Re_m < 3000$			$Re_m > 3000$	
	$j = a.Re_m{}^b$		$\ln j = a + b \ln Re_m + c (\ln Re_m)^2$			$j = a.Re_m{}^b$	
Geometry and PT/Do ratio	a	b	a	b	c	a	b
Triangle 1.25	1.81	−0.72	1.70	−1.25	0.065	0.275	−0.38
Square 1.25	0.97	−0.62	3.29	−1.90	0.121	0.275	−0.38
Rotated sq 1.25	1.81	−0.72	0.32	−0.76	0.025	0.275	−0.38
Triangle 1.5	1.34	−0.68	−0.4548	−0.68	0.00718	0.275	−0.38
Square 1.5	0.88	−0.64	1.3132	−1.2937	0.0743	0.275	−0.38

CORRELATION FOR SHELL SIDE f FACTOR

Mathematical regression of curves in Figure 7-27

$$f = (2.\rho.\Delta p/4\,N_c\,G_m{}^2)\,(\mu_w/\mu)^m$$

(See variables definition in the text)

$$\ln f = a + b \ln Re_m + c (\ln Re_m)^2 + d (\ln Re_m)^3$$

INTERVAL	$Re_m < 100$				$100 < Re_m < 1300$				$1300 < Re_m < 3000$				$Re_m > 3000$			
Geometry and PT/Do ratio	a	b	c	d	a	b	c	d	a	b	c	d	a	b	c	d
Triangle 1.25	4.376	−1.025	0	0	5.293	−1.864	0.1584	−0.00472	5.293	−1.864	0.1584	−0.00472	5.293	−1.864	0.1584	−0.00472
Square 1.25	3.923	−0.984-	0	0	3.871	−0.498	−0.2052	0.0221	3.871	−0.498	−0.2052	0.00221	−7.907	1.774	−0.16	0.00407
Rotated sq 1.25	3.923	−0.984	0	0	6.30	−2.403	0.226	−0.00721	6.3	−2.403	0.226	−0.0072	6.30	−2.403	0.226	−0.0072
Triangle 1.5	3.196	−1.176	0.0557	0	3.472	−1.547	0.1425	−0.00454	3.472	−1.547	0.1425	−0.00454	3.472	−1.547	0.1425	−0.00454
Square 1.5	3.04	−1.13	0.031	0	−8.015	5.152	−1.166	0.0771	−8.015	5.152	−1.166	0.0771	−6.666	1.152	−0.0981	0.00237

FIGURE 7-27 (*Continued*)

The region between Reynolds numbers 100 and 4,000 corresponds to a transition regime. The occasional appearance of turbulent eddies can be observed in this range. The friction factor is higher than could be expected if calculated with the laminar-flow correlation. The difference corresponds to the energy wasted in eddy formation.

In staggered arrays (i.e., triangular and rotated-square arrays), eddy appearance starts at the outermost downstream tube row, and the phenomenon propagates upstream as the Reynolds number is increased. In square arrays, eddies appear simultaneously in the entire tube bundle, provoking a sudden increase in the friction factor. The higher the distance between tube rows, the lower is the friction-factor increase.

Fully turbulent flow corresponds to $\mathrm{Re}_m > 4,000$. In this region, the friction factor is proportional to a power of Re_m ranging from -0.2 to -0.4.

Influence of the Number of Tube Rows. It is known that in laminar flow the heat transfer coefficient decreases with increasing distance from the start of heating. [For example, see Eq. (7-4-3) for in-tube convection.]This is due to the fact that with increasing distance to the tube inlet, the temperature gradient at the tube wall decreases and it also decreases the heat transfer coefficient. This phenomenon also exists during flow across tube banks.. For large heat exchangers in deep laminar flow, it can result in a decrease in the average heat transfer coefficient by a factor of 2 or more compared with what would have been predicted based on flow across a 10-row tube bank such as that used in the Delaware research work.

Bell proposed to introduce a correction factor that depends on the total number of tube rows in the fluid path across the exchanger, which is

$$N_c' = (N_c + N_w)(N_B + 1) \tag{7-5-15}$$

where N_c = number of tube rows in each cross-flow section (number of tube rows between baffle tips)
  N_w = effective number of tube rows in the baffle window [defined in Eq. (7-5-31)]

The correction factor is

1. For $\mathrm{Re}_m < 20$,

$$X = \left(\frac{N_c'}{10}\right)^{-0.18} \tag{7-5-16a}$$

2. For $\mathrm{Re}_m > 100$,

$$X = 1 \tag{7-5-16b}$$

3. For $20 \le \mathrm{Re}_m \le 100$,

$$X = 1 - \left\{(1.217 - 0.0121\,\mathrm{Re}_m)\left[1 - \left(\frac{N_c'}{10}\right)^{-0.18}\right]\right\} \tag{7-5-16c}$$

Thus, for a real exchanger, the shell-side heat transfer coefficient is related to the 10-row ideal tube bank by

$$h_{\text{exchanger}} = h_{\text{ideal bank}} \cdot X \tag{7-5-17}$$

The number of tube rows does not have any effect on the friction factor.

Other Correction Factors for Tubular Heat Exchangers. As we have mentioned, the data corresponding to an ideal tube bank must be corrected to take into account the effects of leakage, bypass streams, and baffle windows when used in real heat exchanger design.

TABLE 7-3 Influence of By-pass Area in Shellside Film Coefficient and Shellside Pressure Drop

Percent Bypass Area with Respect to Cross-Flow Area	Laminar Flow		Turbulent Flow	
	$\dfrac{\Delta p \text{ with Bypass}}{\Delta p \text{ without Bypass}}$	$\dfrac{h \text{ with Bypass}}{h \text{ without Bypass}}$	$\dfrac{\Delta p \text{ with Bypass}}{\Delta p \text{ without Bypass}}$	$\dfrac{h \text{ with Bypass}}{h \text{ without Bypass}}$
Ideal bank	100%	100%	100%	100%
17.8% bypass without sealing devices	42%	76–79%	43%	77–80%
27% bypass without sealing device	22%	60–65%	31%	69–73%
17.8% bypass with one sealing device	68%	90–94%	67%	90–94%
27% bypass with one sealing device			63%	85–88%

Bypass Effects. Some studies were performed in nonideal tube banks where part of the fluid had the potential to circulate through the peripheral lane between the tube bundle and the shell.[19] It was observed that even in cases where the bypass area was not very important, a big fraction of the total flow deviates toward this region.

The fraction of the total flow bypassing the bundle could be as high as 75 percent in laminar flow and 50 percent in turbulent flow, with a consequent reduction in heat transfer coefficient and pressure drop (the bypass lane has a lower frictional resistance than the tube bundle). The results are shown in Table 7-3 (these data correspond to a particular exchanger; they are not general). It can be observed that the influence of the sealing devices is very important The effect of the sealing devices is illustrated in Fig. 7-28. All the experiments reported in the table were performed comparing a tube bank with bypass with an ideal tube bank at the same total flow.

These sealing effects are introduced in Bell's method by means of a correction factor for h, namely,

$$\xi_h = \frac{h_{BP}}{h_B} \qquad (7\text{-}5\text{-}18)$$

where h_{BP} = coefficient of the bank with bypass area
h_B = coefficient of an ideal bank

And for pressure drop, it is defined as

$$\xi_{\Delta p} = \frac{\Delta p_{BP}}{\Delta p_B} \qquad (7\text{-}5\text{-}19)$$

where Δp_{BP} = pressure drop of the bank with bypass area
Δp_B = pressure drop of an ideal bank

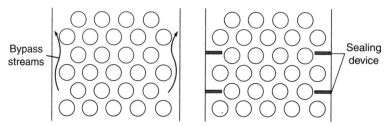

FIGURE 7-28 Effect of sealing strips.

TABLE 7-4 Values of α

	Laminar	Turbulent
For $\xi\Delta p$	5	4
For ξ_h	1.5	1.35

The correction factors are calculated as

$$\xi_h = \xi_{\Delta p} = \exp\left[-\alpha F_{BP}\left(1 - \sqrt[3]{\frac{2N_S}{N_C}}\right)\right] \qquad (7\text{-}5\text{-}20)$$

where F_{BP} = bypass fraction $= \dfrac{S_{BP}}{S_m} = \dfrac{\text{bypass area}}{\text{cross-flow area}}$ $\qquad (7\text{-}5\text{-}21)$

N_S = number of pairs of sealing devices in the bank
N_C = number of tube rows in a cross-flow section (number of tube rows included between edges of two consecutive baffles)
α = coefficient depending on the flow regime obtained from Table 7-4. As it can be seen, the values of α are also different for heat transfer and pressure drop calculations

If $N_S > \frac{1}{2}N_C$, take

$$\xi_h = \xi_{\Delta P} = 1 \qquad (7\text{-}5\text{-}22)$$

The cross-flow and bypass areas for a heat exchanger can be calculated as

$$S_m = \text{cross-flow area} = (D_S - N_{CL}D_o)B \qquad (7\text{-}5\text{-}23)$$

where D_S = shell diameter
N_{CL} = number of tubes in the central row
B = baffle spacing

$$S_{BP} = \left[D_S - (N_{CL} - 1)P_t - D_o\right]B \qquad (7\text{-}5\text{-}24)$$

where P_t is the separation between tube axes (tube pitch).

Note: Equation (7-5-24) can be used for triangular and square arrays. In the case of rotated arrays, P_t must be substituted by the separation between tube axes in a direction perpendicular to the flow in the central row. It is also assumed that pass-partition lanes in the direction parallel to the flow are adequately sealed with dummy tubes.

Effect of the Baffle Window.   The effect of the baffle window on heat transfer and pressure drop will be analyzed separately.

Effect of the baffle window on pressure drop.   Up to now, we have analyzed the pressure drop corresponding to a cross-flow section, which is the portion of the tube bank limited by the edges of two consecutive baffles, as shown in Fig. 7-29. We shall now consider the pressure drop in the baffle window. It was observed that in order to correlate the experimental data properly, it was necessary to use a velocity v_z defined as the geometric mean between the cross-flow velocity and the window velocity. Thus

$$v_m = \text{cross-flow fluid velocity} = \frac{G_m}{\rho} \qquad (7\text{-}5\text{-}25)$$

$$v_w = \text{fluid velocity through window} = \frac{G_w}{\rho} = \frac{W}{S_w\rho} \qquad (7\text{-}5\text{-}26)$$

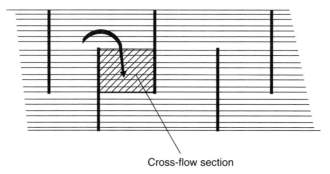

Cross-flow section

FIGURE 7-29 Cross-flow section.

where S_w is the free-flow area through the window, that is,

$$S_w = \pi\left(\frac{D_S}{2}\right)^2 \frac{A}{360} - \left(\sin\frac{A}{2}\right)\left(\frac{D_S}{2}\right)\left(\frac{D_S}{2} - BC\right) - N_{WT}\frac{\pi}{4}D_o^2 \qquad (7\text{-}5\text{-}27)$$

where N_{WT} is the number of tubes in the window, BC is the baffle cut, and A is the baffle central angle (Fig. 7-30). Then

$$v_Z = \sqrt{v_m v_W} \qquad (7\text{-}5\text{-}28)$$

(The need to define this velocity arises from the observation that by increasing the baffle spacing, the pressure drop through the window changes despite the fact that v_w remains constant.)

Pressure Drop Through a Baffle Window in Laminar Flow (Re < 100). The fluid velocity is not uniform throughout the window, and it has components in the directions parallel and normal to the tubes. The total pressure drop through the window will be the sum of three contributions. The first is due to the flow parallel to the tubes, another is due to the normal flow, and the third one is due to direction changes. Data are correlated as a function of the parameter

$$\frac{v_Z\mu}{c}$$

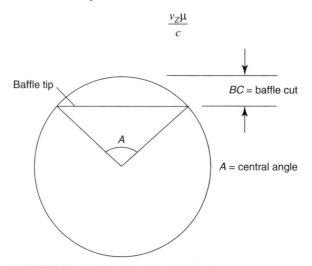

FIGURE 7-30 Baffle central angle and baffle cut.

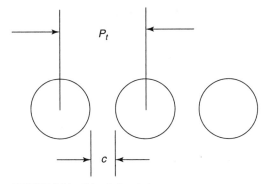

FIGURE 7-31 Tube pitch and clearance.

where c is the clearance between tubes which is $c = P_t - D_o$ (Fig. 7-31).

The pressure drop through the window in laminar flow thus is

$$\Delta p_{wl} = 28 \left(\frac{v_z \mu}{c} \right) N_W + 26 \left(\frac{v_z \mu}{D_v} \right) \left(\frac{B}{D_v} \right) + \left(\frac{2\rho v_z^2}{2} \right)$$ (7-5-29)

where D_v is the hydraulic diameter of the window which is

$$D_v = \frac{4 S_w}{\pi N_{WT} D_o + \pi D_S \dfrac{A}{360}}$$ (7-5-30)

N_W is a parameter called the *effective number of tubes in the window* and is defined as

$$N_W = \frac{0.8 BC}{\sigma}$$ (7-5-31)

where σ is the distance between two tube rows measured in the flow direction (Fig. 7-32).

On the right side of Eq. (7-5-29), the first term is the contribution to pressure drop owing to the flow normal to the tubes; the second is the parallel- flow contribution and the last one is due to direction change.

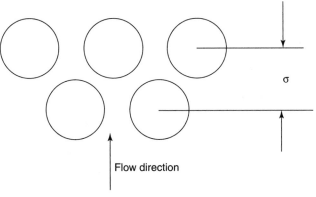

Flow direction

FIGURE 7-32 Definition of σ.

Pressure Drop Through the Baffle Window in Turbulent Flow (Re >100).   The correlation for pressure drop through the window in turbulent flow is

$$\Delta p_{wt} = (2 + 0.6N_w)\frac{\rho v_z^2}{2} \tag{7-5-32}$$

Effect of the baffle window on the heat transfer coefficient.

The influence that the tubes located at the window exert on the heat transfer was studied by plugging the cross-flow tubes so that the heat transfer was solely due to the window tubes. It was observed that the experimental data can be correlated adequately by multiplying the ideal-tube-bank coefficient by a correction factor ϕ that includes all the effects exerted by the window. Thus

$$h_{NL} = \phi h_B \tag{7-5-33}$$

where h_B = ideal-tube-bank coefficient
   h_{NL} = heat transfer coefficient for a heat exchanger with baffles but no leakages (since the effect of leakage has not been considered yet)

And it was found that ϕ can be correlated as

$$\phi = 1 - r + 0.524r^{0.32}\left(\frac{S_m}{S_w}\right)^{0.03} \tag{7-5-34}$$

where

$$r = \frac{\text{number of tubes in window}}{\text{total number of tubes}} = \frac{N_{wt}}{N}$$

Thus the shell-side heat transfer coefficient for a heat exchanger with no leakage can be calculated as

$$h_{NL} = \left[jcG_m\,\text{Pr}^{2/3}\left(\frac{\mu_w}{\mu}\right)^{0.14}\right]\phi\xi_h X \tag{7-5-35}$$

The term within the brackets corresponds to h_B, and the others are the correction factors.

Effect of Leakage.
Effect of leakage on pressure drop.

Owing to the leakage existing between baffles and tubes and between baffles and shell, both pressure drop and heat transfer coefficients differ from the values of an ideal bank. This subject was studied by Sullivan and Bergelin.[10,12] The mathematical treatment was simplified by Bell, who assumed that the ratio between leakage flow rate and cross-flow flow rate is independent of the flow regime and depends only on the ratio between leakage area and cross-flow area.

This is a simplification of the problem. Some other methods allow calculation of the individual flow rates of the different flow fractions as a function of the total flow and geometric parameters. Then, according to Bell, the ratio between Δp_L (pressure drop for a heat exchanger with leakage) and Δp_{NL} (pressure drop for a similar heat exchanger with the same total flow but without leakage) can be represented by a curve of the type shown in Fig. 7-33. The upper curve corresponds to a heat exchanger where the leakage occurs between tubes and baffles exclusively, and the lower curve corresponds to a heat exchanger with leakage only between the shell and baffles.

It can be appreciated that $[1 - (\Delta p_L/\Delta p_{NL})]$ for tube-baffle leakage is approximately half the value of $[1 - (\Delta p_L/\Delta p_{NL})]$ corresponding to shell-baffle leakage. Thus the mathematical treatment can be simplified using a single curve, which is that of tube-baffle leakage (Fig. 7-34).

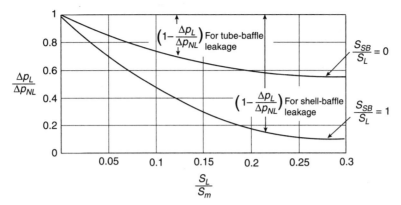

FIGURE 7-33 Baffle leakage effect on pressure drop. *(Courtesy of Professor K. Bell)*

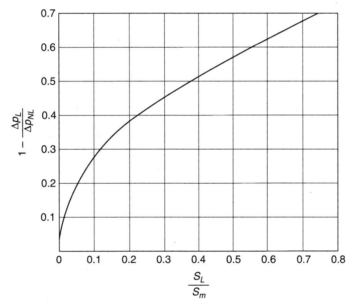

FIGURE 7-34 Correlation for the leakage effect.

The ordinate of the graph represents the value $(1 - \Delta p_L/\Delta p_{NL})$ for a heat exchanger with leakage only between tubes and baffles. The mathematical expression of the curve is[11]

$$\left(1 - \frac{\Delta p_L}{\Delta p_{NL}}\right)_o = 0.57\frac{S_L}{S_m} + 0.27\left[1 - \exp\left(-20\frac{S_L}{S_m}\right)\right] \qquad (7\text{-}5\text{-}36)$$

and it is

$$\left(1 - \frac{\Delta p_L}{\Delta p_{NL}}\right)_{exchanger} = \left(1 - \frac{\Delta p_L}{\Delta p_{NL}}\right)_o \frac{S_{TB} + 2S_{SB}}{S_L} \qquad (7\text{-}5\text{-}37)$$

where S_{TB} = leakage area between tubes and baffle
S_{SB} = leakage area between baffle and shell

and

$$S_{TB} + S_{SB} = S_L \tag{7-5-38}$$

S_{TB} and S_{SB} can be calculated as follows:

$$S_{TB} = N_{BT}\frac{\pi}{4}\left(D_{BT}^2 - D_o^2\right) \tag{7-5-39}$$

where N_{BT} = number of tubes passing across the baffle
D_{BT} = diameter of the baffle holes

and

$$S_{SB} = \left(\frac{360 - A}{360}\right)\frac{\pi}{4}\left(D_S^2 - D_B^2\right) \tag{7-5-40}$$

where D_S = shell diameter
D_B = baffle diameter

Effect of leakage on heat transfer.   With similar simplifications to those assumed in pressure drop calculations, it is possible to draw curves showing the ratio h_L/h_{NL} (where h_L = heat transfer coefficient for a heat exchanger with leakage, and h_{NL} = heat transfer coefficient for a heat exchangers with no leakage and operating with the same flow rate) as a function of S_L/S_m. Here again, it can be observed that the value of $(1 - h_L/h_{NL})$ for tube-baffle leakage is roughly half the value corresponding to shell-baffle leakage. Then it is also possible to use a single curve, which can be correlated as[11]

$$\left(1 - \frac{h_L}{h_{NL}}\right)_o = 0.45\frac{S_L}{S_m} + 0.1\left[1 - \exp\left(-30\frac{S_L}{S_m}\right)\right] \tag{7-5-41}$$

And then

$$\left(1 - \frac{h_L}{h_{NL}}\right)_{exchanger} = \left(1 - \frac{h_L}{h_{NL}}\right)_o \frac{S_{TB} + 2S_{SB}}{S_L} \tag{7-5-42}$$

With these equations, knowing $(1 - h_L/h_{NL})$ and $(1 - \Delta p_L/\Delta p_{NL})$, it is possible to calculate h_L/h_{NL} and $\Delta p_L/\Delta p_{NL}$, which are the correction factors to be applied to correct the values of h and Δp corresponding to an exchanger without leakage.

Conclusions.
   Heat Transfer.   The shell-side heat transfer coefficient of a heat exchanger can be obtained as

$$h = \left[jcG_m\mathrm{Pr}^{-2/3}\left(\frac{\mu}{\mu_w}\right)^{0.14}\right]\left(\phi\xi_h X\frac{h_L}{h_{NL}}\right) \tag{7-5-43}$$

which equals

Ideal-tube-bank coefficient × window correction × bypass correction × number-of-tube-rows correction × leakage correction

Pressure Drop. The total shell-side pressure drop must be calculated by adding the pressure drops corresponding to all cross-flow sections and all baffle windows. The pressure drop in a cross-flow section without leakage correction is

$$\Delta p_{BP} = \frac{4 f N_c G_m^2}{2\rho}\left(\frac{\mu_w}{\mu}\right)^{0.14}\xi_{\Delta p} \tag{7-5-44}$$

$$= \Delta p \text{ ideal tube bank} \times \text{bypass correction}$$

The pressure drop through each window can be obtained as follows:

1. Laminar flow:

$$\Delta p_w = 28\frac{v_z\mu}{(P_t - D_o)}N_c + 26\frac{v_z\mu BC}{D_v D_v} + \frac{2\rho v_z^2}{2} \tag{7-5-45}$$

2. Turbulent flow:

$$\Delta p_w = (2 + 0.6 N_w)\frac{\rho v_z^2}{2} \tag{7-5-46}$$

The total pressure drop is obtained by adding all cross-flow sections and all windows as

$$\Delta p \text{ total} = 2\Delta p_{BP}\left(1 + \frac{N_W}{N_C}\right) + \left[(N_B - 1)\Delta p_{BP} + N_B\Delta p_W\right]\left(\frac{\Delta p_L}{\Delta p_{NL}}\right) \tag{7-5-47}$$

In Eq. (7-5-47), the first term on the right side represents the pressure drop through the inlet and outlet sections, the first term within the square brackets represents the remaining cross-flow sections, and the second term in the square brackets corresponds to the pressure drops through the windows.

For example, let's consider the heat exchanger in Fig. 7-35, where N_B is the number of baffles (=4), Sections I and V are the inlet and outlet sections. These are treated as no-leakage sections. Since it can be observed that the number of tube rows crossed by the fluid in these two sections is $N_w + N_c$ instead of N_w, the correction factor $(N_W + N_C)/N_C$ is applied.

The pressure drop through these two sections is

$$2\Delta p_{BP}\left(1 + \frac{N_w}{N_c}\right)$$

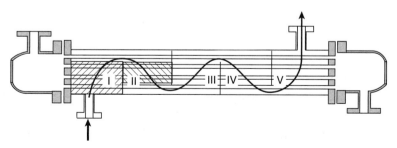

FIGURE 7-35 A heat exchanger with four baffles.

The pressure drop through the cross-flow sections identified as sections II, III, and IV is

$$(N_B - 1)\Delta p_{BP} \left(\frac{\Delta p_L}{\Delta p_{NL}} \right) = 3\Delta p_{BP} \left(\frac{\Delta p_L}{\Delta p_{NL}} \right)$$

And the pressure drop through the four windows is

$$N_B \Delta p_W \left(\frac{\Delta p_L}{\Delta p_{NL}} \right) = 4\Delta p_W \left(\frac{\Delta p_L}{\Delta p_{NL}} \right)$$

7-5-4   Computational Techniques

Nowadays, heat exchanger design is performed almost exclusively using commercial software, as are many other activities. The use of commercial software makes the designer's task faster and easier, and in addition, the designer responsibility is backed by the prestige of the software supplier.

These programs are developed by companies that have specialists in heat transfer and sponsor research programs at universities or have their own research facilities. Normally, their design programs are more elaborated than the simple methods presented in this chapter. For example, Heat Transfer Research, Inc. (HTRI) uses a calculation method that subdivides the heat exchanger into a series of three-dimensional cells (Fig. 7-36). First, the heat exchanger is divided into increments in the axial direction. Then it is divided into rows in the vertical direction and, finally, into sections in the horizontal direction. The number of rows and sections depends on the geometry of the heat exchanger (i.e., type of baffle, baffle-cut orientation, number of tube passes, etc.). The total number of cells necessary is such that in each one there is a predominant type of shell flow (axial in window or cross-flow) and only one pass in tubes.

In each cell, the program calculates the heat transfer and pressure drop using HTRI proprietary correlations and local values of temperature, pressure, and physical properties. The model is solved with finite-elements analysis, and the heat exchanged is calculated as

$$Q = \Sigma \Delta Q_i = \Sigma \Delta A_i U_i \left(\Delta T_{ml} \right)_i \qquad (7\text{-}5\text{-}48)$$

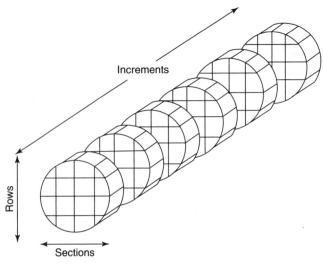

FIGURE 7-36   Heat exchanger cells.

This rigorous technique allows one to predict the behavior of heat exchangers with complex geometries. It must be noted that this model uses neither the concept of logarithmic mean temperature difference nor the F_t correction factor.

7-5-5 Accuracy of Correlations

As we have explained previously, the heat exchanger design task is presently performed almost exclusively with computational techniques. However, we must realize that for decades, all heat exchangers (most of them still in operation) were designed using Kern's and Bell's method. The information regarding accuracy of the correlations must be analyzed with care. As we have seen, there are certain geometries for which we know that these methods present problems. For example, we know that Kern's method cannot be applied to a TEMA type T floating-head heat exchanger without sealing strips or with unsealed pass-partition lanes.

Whitley[18] presents a study of the errors found in heat transfer coefficient and pressure-drop predictions obtained with the Kern and Bell methods, among others. In some cases, Kern's correlations are reported to present errors as high as 850 percent. However, it is not informed what are the types of heat exchangers presenting so high deviations. Probably if those "pathologic" cases were not considered, the errors may have been considerably lower.

According to Palen and Taborek,[14] the Bell-Delaware method allows one to predict shell-side film coefficients that range from 50 percent lower to 100 percent higher than the real values, whereas the error range for pressure-drop prediction varies from 50 percent lower to 200 percent higher than the real values. For the entire set of data analyzed, the mean error for heat transfer was 15 percent below the real values (safe side) for all Reynolds numbers, whereas the mean error for pressure-drop prediction was about 100 percent above the real values for Reynolds numbers lower than 10. Commercial program suppliers offer very little information about the correlations used, and comparative studies among their results and accuracy have not been not published.

Example 7-5 It is desired to heat up 38.88 kg/s of a liquid A from 40 to 61°C using process hot stream B, available at 104°C with a flow rate of 33.33 kg/s. The intention is to use an existing heat exchanger whose characteristics are indicated below. A 0.0009 (m² · K)/W fouling resistance is anticipated. Verify if the unit is suitable for the service, and calculate the fluid pressure drops.

Note: A downloadable spreadsheet with the calculations of this example is available at http://www.mhprofessional.com/product.php?isbn=0071624082

Fluids Data

	Fluid A (Tube Side)	Fluid B (Shell Side)
W (kg/s)	38.88	33.33
ρ (kg/m³)	716	578
c [J/(kg · K)]	2140	2640
μ [kg/(m · s)]	0.62×10^{-3}	0.16×10^{-3}
k [W/(m · K)]	0.129	0.0917

Heat Exchanger Data

N = number of tubes = 414 two-tube passes, type AES

Tubes ¾ in BWG 16 ($D_o = 0.019$ m, $D_i = 0.0157$ m)

Pattern Δ, $P_t = 1$ in (0.0254 m)

L = tube length = 4.267 m

Central baffle spacing $B = 0.234$ m

Shell diameter $D_s = 0.609$ m

Tubes layout and other geometric dimensions according to Fig. 7-37.

1. *Tube-side film coefficient calculation:*

$$a'_t = \text{tube flow area} = \frac{\pi D_i^2}{4} = \frac{\pi \times 0.0157^2}{4} = 1.936 \times 10^{-4}\,\text{m}^2$$

$$a_t = \text{total tube-side flow area} = \frac{a'_t N}{n} = \frac{1.935 \times 10^{-4} \times 414}{2} = 0.0401\,\text{m}^2$$

$$G_t = \text{mass velocity} = \frac{W_t}{a_t} = \frac{38.88}{0.0401} = 970\,\text{kg/(m}^2 \cdot \text{s)}$$

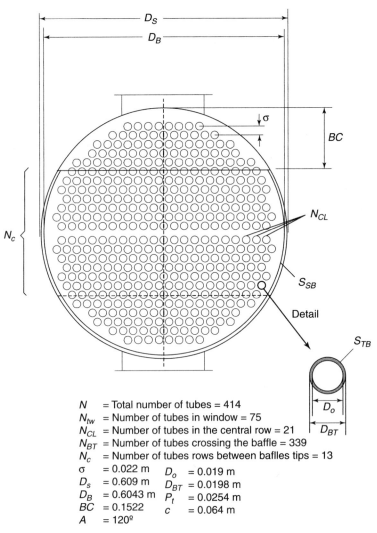

N = Total number of tubes = 414
N_{tw} = Number of tubes in window = 75
N_{CL} = Number of tubes in the central row = 21
N_{BT} = Number of tubes crossing the baffle = 339
N_c = Number of tubes rows between baffles tips = 13
σ = 0.022 m D_o = 0.019 m
D_s = 0.609 m D_{BT} = 0.0198 m
D_B = 0.6043 m P_t = 0.0254 m
BC = 0.1522 c = 0.064 m
A = 120º

FIGURE 7-37 Figure for Example 7-5.

$$\text{Re}_t = \text{Reynolds number} = \frac{DiG_t}{\mu} = \frac{0.0157 \times 970}{0.62 \times 10^{-3}} = 24,568$$

$$\text{Pr} = \text{Prandtl number} = \frac{c\mu}{k} = \frac{2,140 \times 0.62 \times 10^{-3}}{0.129} = 10.29$$

$$h_i = 0.023 \frac{k}{D_i} \times \text{Re}^{0.8} \times \text{Pr}^{0.33} \left(\frac{\mu}{\mu_w}\right)^{0.14} = 0.023 \times \frac{0.129}{0.0157} \times 24,568^{0.8} \times 10.29^{0.33} = 1,327 \text{ W/(m}^2 \cdot \text{K)}$$

All fluid properties have been considered constant, so the correction factor μ/μ_w is unity. The same criteria will be adopted for the shell-side fluid.

$$h_{io} = h_i \frac{D_i}{D_o} = 1,327 \frac{0.0157}{0.019} = 1,096 \text{ W/(m}^2 \cdot \text{K)}$$

2. *Tube-side pressure drop:*

$$f = \text{friction factor} = 1.2 \times (0.0014 + 0.125 \text{Re}^{-0.32}) = 7.58 \times 10^{-3}$$

$$\Delta p_t = \text{pressure drop in tube} = 4fn\frac{L}{D_i}\frac{G_t^2}{2\rho} = 4 \times 7.58 \times 10^{-3} \times 2 \times \frac{4.267 \times 970^2}{0.0157 \times 2 \times 716} = 10,843 \text{ N/m}^2$$

$$\Delta pr = \frac{4.n.G_t^2}{2\rho} = 5,256 \text{ N/m}^2$$

$$\Delta p_T = \Delta p_t + \Delta p_r = 16,099 \text{ N/m}^2$$

3. *Shell-side heat transfer coefficient:* Both Kern's and Bell's method will be used.

3-a. *Kern's method:*

$$a_S = \text{shell-side flow area} = \frac{D_S cB}{P_t} = \frac{0.609 \times (6.35 \times 10^{-3}) \times 0.234}{0.0254} = 0.0356 \text{ m}^2$$

$$G_S = \text{mass velocity} = \frac{W}{a_S} = \frac{33.33}{0.0356} = 936.2 \text{ kg/(m}^2 \cdot \text{s)}$$

$$D_e = \text{equivalent diameter} = 0.0182 \text{ m}$$

$$\text{Re}_S = \text{Reynolds number} = \frac{D_e G_S}{\mu} = \frac{0.0182 \times 936.2}{0.16 \times 10^{-3}} = 106,492$$

$$h_o = \text{shell coefficient} = 0.36\frac{k}{D_e}\text{Re}_S^{0.55}\text{Pr}^{1/3} = 0.36\frac{0.0917}{0.0182}106,492^{0.55}\left(\frac{2640 \times 0.16 \times 10^{-3}}{0.0917}\right)^{1/3}$$

$$= 1,756 \text{ W/(m}^2 \cdot \text{K)}$$

3-b. *Bell's method:*

$$S_m = \text{cross-flow area} = (D_S - N_{CL}D_o)B = (0.609 - 21 \times 0.019) \times 0.234 = 0.0491 \text{ m}^2$$

$$G_m = \text{mass velocity} = \frac{W}{S_m} = \frac{33.33}{0.0491} = 678.3 \text{ kg/(m}^2 \cdot \text{s)}$$

$$\mathrm{Re}_m = \text{Reynolds number} = \frac{D_o G_m}{\mu} = \frac{0.019 \times 678.3}{0.16 \times 10^{-3}} = 80{,}544$$

$$j = \text{Colburn coefficient} = 0.00376$$

(In this case, $P_t/D_o = 1.33$. The curves corresponding to $P_t/D_o = 1.25$ were used.)

3-b-1. *Bypass correction:*

$$F_{BP} = \text{bypass fraction} = \frac{\left[D_S - (N_{CL} - 1)P_t - D_o\right]B}{S_m} = \frac{[0.609 - (21-1)0.0254 - 0.019] \times 0.234}{0.0491} = 0.390$$

$$\therefore \alpha = 1.35$$

If two pairs of sealing strips are used,

$$\xi_h = exp\left\{-1.35 F_{BP}\left[1 - (2 \times N_s/N_c)^{0.333}\right]\right\} = exp\{-1.35 \times 0.39 \times [1 - (2 \times 2/13)^{0.333}]\} = 0.842$$

Please note that if no sealing strips were used ($N_s = 0$), the bypass correction factor would be 0.59 and this reduces the heat transfer coefficient to about 30 percent. Observe that Kern's method remains insensitive to these important effects

3-b-2. *Window correction:*

$$r = \frac{N_{wt}}{N} = \frac{75}{414} = 0.181$$

$$S_W = \text{flow area at window} = \frac{\pi D_S^2}{4}\frac{A}{360} - \left(\sin\frac{A}{2}\right)\left(\frac{D_S}{2}\right)(0.5D_S - BC) - N_{wt}\pi\frac{D_o^2}{4}$$

$$= \frac{\pi \times 0.609^2}{4} \times \frac{120}{360} - (\sin 60°) \times 0.3045 \times (0.3045 - 0.152) - \frac{\pi \times 75 \times 0.019^2}{4} = 0.0357 \text{ m}^2$$

$$\phi = \text{correction factor} = 1 - r + 0.524 r^{0.32}\left(\frac{S_m}{S_W}\right)^{0.03} = 1 - 0.181 + 0.524 \times 0.181^{0.32}\left(\frac{0.0491}{0.0357}\right)^{0.03} = 1.125$$

3-b-3. *Leakages correction:*

$$S_{TB} = \text{tube-baffle leakage area} = N_{BT}\frac{\pi}{4}\left(D_{BT}^2 - D_0^2\right) = 339\frac{\pi}{4}\left(0.0198^2 - 0.019^2\right) = 0.00826 \text{ m}^2$$

$$S_{SB} = \text{baffle-shell leakage area} = \frac{360 - A}{360}\frac{\pi}{4}\left(D_S^2 - D_B^2\right) = \frac{360 - 120}{360}\frac{\pi}{4}\left(0.609^2 - 0.6043^2\right) = 0.003 \text{ m}^2$$

$$S_L = \text{total leakage area} = S_{TB} + S_{SB} = 0.003 + 0.0082 = 0.0112 \text{ m}^2$$

$$\frac{S_L}{S_m} = \frac{0.0112}{0.0491} = 0.228$$

$$\left(1 - \frac{h_L}{h_{NL}}\right)_o = 0.45\frac{S_L}{S_m} + 0.1\left[1 - exp\left(-30\frac{S_L}{S_m}\right)\right] = 0.45 \times 0.228 + 0.1(1 - 1.07 \times 10^{-3}) = 0.203$$

$$1 - \frac{h_L}{h_{NL}} = \left(1 - \frac{h_L}{h_{NL}}\right)_o \frac{S_{TB} + 2S_{SB}}{S_L} = 0.203 \frac{0.00826 + 2 \times 0.003}{0.0112} = 0.257$$

$$\therefore \frac{h_L}{h_{NL}} = 1 - 0.257 = 0.743$$

3-b-4. The correction for number of tube rows is negligible in turbulent flow.

3-b-5. Shell-side heat transfer coefficient:

$$h = \left[jcG_m \Pr^{-2/3}\left(\frac{\mu_w}{\mu}\right)^{-0.14}\right]\phi\xi_h\left(\frac{h_L}{h_{NL}}\right)$$

$$= (0.00376 \times 2{,}640 \times 678.3 \times 4.606^{-0.66} \times 1) \times 1.1251 \times 0.8426 \times 0.743 = 1{,}714 \text{ W/(m}^2 \cdot \text{K)}$$

Both methods in this particular situation give similar results, but this is not normally the case.

4. Shell-side pressure drop

4-a. Kern's method:

$$f = \text{friction factor} = 1.728\,\text{Re}_S^{-0.188} = 1.728(106{,}492)^{-0.188} = 0.196$$

$$\Delta p = f\frac{G_S^2 D_S(N_B + 1)}{2\rho D_e} = \frac{0.196 \times 936.2^2 \times 0.609 \times 18}{2 \times 578 \times 0.0182} = 89{,}500 \text{ N/m}^2$$

4-b. Bell's method:

4-b-1. Friction factor:

$$f = 0.095 \quad \text{(with Re}_m = 80{,}544; \text{ see Fig. 7-27)}$$

4-b-2. Bypass correction:

$$\xi_{\Delta p} = \exp\left\{-\alpha F_{BP}\left[1 - \left(\frac{2N_s}{N_c}\right)^{1/3}\right]\right\} = \exp\left\{-4 \times 0.39 \times \left[1 - \left(\frac{2 \times 2}{13}\right)^{1/3}\right]\right\} = 0.602$$

4-b-3. Pressure drop in a cross-flow section without leakage:

$$\Delta p_{BP} = \frac{4fN_C G_m^2 (\mu_w/\mu)^{0.14}\xi_{\Delta p}}{2\rho} = \frac{4 \times 0.095 \times 13 \times 678.3^2 \times 0.602}{2 \times 578} = 1183.6 \text{ N/m}^2$$

4-b-4. Pressure drop through window:

$$V_Z = \text{mean velocity} = \frac{W}{\rho\sqrt{S_m S_W}} = \frac{33.33}{578\sqrt{0.0491 \times 0.0357}} = 1.38 \text{ m/s}$$

$$\Delta p_W = \text{pressure drop through window} = (2 + 0.6N_W)\frac{\rho v_Z^2}{2}$$

Thus, since

$$N_W = 0.8\frac{BC}{\sigma} = \frac{0.8 \times 0.15225}{P_t \sin 60°} = 5.54$$

$$\therefore \Delta p_W = (2 + 0.6 \times 5.54) \times 578 \times \frac{1.38^2}{2} = 2{,}930 \text{ N/m}^2$$

4-b-5 *Leakage correction:*

$$\left(1-\frac{\Delta p_L}{\Delta p_{NL}}\right)_o = 0.57\frac{S_L}{S_m}+0.27\left[1-\exp\left(\frac{-20S_L}{S_m}\right)\right] = 0.57 \times 0.228 + 0.27 \times (1-e^{-4.56}) = 0.397$$

$$1-\frac{\Delta p_L}{\Delta p_{NL}}=\left(1-\frac{\Delta p_L}{\Delta p_{NL}}\right)_o\frac{S_{TB}+2S_{SB}}{S_L}=0.397 \times 1.27 = 0.504$$

$$\therefore \frac{\Delta p_L}{\Delta p_{NL}}=1-0.504=0.496$$

4-b-6 *Pressure drop through the entire unit:*

$$\Delta p_S = 2\Delta p_{BP}\left(1+\frac{N_W}{N_c}\right)+\left[(N_B-1)\Delta p_{BP}+N_B\Delta p_W\right]\frac{\Delta p_L}{\Delta p_{NL}}$$

$$= 2 \times 1,183.6 \times \left(1+\frac{5.54}{13}\right)+(16 \times 1,183.6 + 17 \times 2930) \times 0.496 = 37,445 \text{ N/m}^2$$

Note the difference between the two methods.

5. *Verification of the heat transfer area:*

$$U=\left(\frac{1}{h_o}+\frac{1}{h_{io}}+R_f\right)^{-1}=\left(\frac{1}{1,714}+\frac{1}{1,096}+9\times 10^{-4}\right)^{-1}=417 \text{ W/(m}^2 \cdot \text{ K)}$$

$$Q=W_c c_c(t_2-t_1)=38.88 \times 2,140 \times (60-40)=1.664 \times 10^6 \text{ W}$$

$$T_2=T_1-\frac{Q}{W_h c_h}=104-\frac{1.664\times 10^6}{2,640 \times 33.33}=85°\text{C}$$

$$R=\frac{104-85}{60-40}=0.95 \quad \text{and} \quad S=\frac{60-40}{104-40}=0.312$$

For a 1-2 configuration, we get $F_t = 0.967$. Then

$$\text{DMLT}=\frac{(T_1-t_2)-(T_2-t_1)}{\ln\dfrac{T_1-t_2}{T_2-t_1}}=\frac{(104-60)-(85-40)}{\ln\dfrac{104-60}{85-40}}=44.5°\text{C}$$

$$A \text{ (calculated)}=\frac{Q}{U\Delta T}=\frac{1,664,064}{417 \times (44.5 \times 0.967)}=92.7 \text{ m}^2$$

$$A \text{ (real)} = \pi D_o NL = \pi \times 0.019 \times 414 \times 4.267 = 105.4 \text{ m}^2$$

$$\text{Excess area}=\frac{105.4-92.7}{92.7}\times 100=14\%$$

It is possible now to estimate the wall temperature by equating the heat transfer rate on both sides of the tube wall, as in Example 5-5. A more rigorous expression considering the individual fouling resistances would be

$$\left(\frac{1}{h_{io}}+R_{fi}\right)^{-1}(T_w-t)=\left(\frac{1}{h_o}+R_{fo}\right)^{-1}(T-T_w)$$

From which T_w and the viscosity correction factors can be calculated, feeding back the procedure if necessary.

7-6 DESIGN AND RATING OF HEAT EXCHANGERS

Two different types of problems can exist in relation with heat exchanger design or evaluation:

1. To verify adequacy of an existing or proposed unit (rating)
2. To design a new unit to perform a certain service

In what follows we shall see how to proceed in either of these cases.

7-6-1 Rating Existing Units

This task consists of deciding whether a unit will allow us to reach a proposed process objective. For example, let's assume that we want to cool down a certain process stream with known mass flow from a temperature T_1 to a final T_2. It was decided to use cooling water with a certain inlet temperature and with a proposed flow rate. There is a heat exchanger available whose geometric dimensions are known, and it is desired to know if the objective will be reached with this unit.

To do this, we assume that the desired outlet temperatures will be obtained, so it is possible to calculate the LMTD. Since the exchanger configuration is known, we can calculate the F_t correction factor and the effective temperature difference. Since the flow rates and flow areas are known, it is possible to calculate the heat transfer coefficients h_o and h_{io}. By adding the fouling resistance based on Eq. (7-1-2), we obtain the overall heat transfer coefficient U. It is then possible to calculate a heat transfer area by means of

$$A \text{ (calculated)} = \frac{Q}{U \Delta T} \qquad (7\text{-}6\text{-}1)$$

This is the surface area required by a heat exchanger whose overall heat transfer coefficient is U to exchange a heat-duty Q with a "driving force" ΔT. This value can be compared with the real area of the unit we are verifying. If $A_{calc} > A_{real}$, the heat exchanger will not be able to transfer the required heat.

If the exchanger is put into operation anyway, a heat-duty Q' will be transferred, with $Q' < Q$. The result is that the outlet temperature of the hot fluid will be higher than desired, and the outlet temperature of the cooling water will be lower than proposed. In case it happens that $A_{calc} < A_{real}$, the unit will transfer a higher amount of heat than originally proposed.

In any case, if calculation of the real outlet temperature of the fluids is desired, it would be necessary to solve the following set of equations:

$$Q = W_h c_h (T_1 - T_2) \qquad (7\text{-}6\text{-}2)$$

$$Q = W_c c_c (t_2 - t_1) \qquad (7\text{-}6\text{-}3)$$

$$Q = U A_{real} \cdot \text{DMLT} \cdot F_t \qquad (7\text{-}6\text{-}4)$$

$$F_t = f(T_1, T_2, t_1, t_2) \qquad (7\text{-}6\text{-}5)$$

In this system, the unknowns are Q, F_t, T_2, and t_2. The resolution of this equation system, as it is written, presents the difficulty that it cannot be solved algebraically, and iterative procedures are necessary. In Sec. 7-7 we shall explain a technique that allows a straightforward solution and avoids this iterative procedure.

However in many cases, it is not necessary to know the exact outlet temperatures, but it is enough to know if the unit is capable of transferring, at a minimum, the required duty. Continuing with our example, we should consider that if it were really necessary to establish and maintain an exact value for the process-fluid outlet temperature with an oversized heat exchanger, it is always possible to adjust it by reducing the cooling-water flow.

The other thing that always must be checked is that the fluid pressure drops are lower than the allowable values.

In what follows we shall develop a detailed procedure with all the steps for verification of an existing heat exchanger. The flow rates and inlet temperatures will be known, as well as the outlet temperature of one fluid. In this case, with a heat balance, we can calculate the outlet temperature of the other fluid.

It is also possible that the outlet temperatures of both fluids have been defined, in which case we shall use the heat balance to calculate one of the flow rates. Whatever the situation, the steps are as follows:

1. Write the heat balance, that is,

$$Q = W_h c_h (T_1 - T_2) = W_c c_c (t_2 - t_1) \tag{7-6-6}$$

and then calculate Q and the remaining unknown, flow rate or temperature.

2. With the four temperatures known, calculate LMTD as

$$\text{LMTD} = \frac{(T_1 - t_2) - (T_2 - t_1)}{\ln \dfrac{T_1 - t_2}{T_2 - t_1}} \tag{7-6-7}$$

3. Calculate R and S parameters:

$$R = \frac{T_1 - T_2}{t_2 - t_1} \tag{7-6-8}$$

$$S = \frac{t_2 - t_1}{T_1 - t_1} \tag{7-6-9}$$

4. For the exchanger configuration (i.e., 1-2, 2-4, etc.), obtain F_t from graphs or analytical expressions included in appendices (graphs in App. B).

5. Obtain the physical properties of the fluids at the mean temperatures. These will be $(T_1 + T_2)/2$ and $(t_1 + t_2)/2$ for the hot and cold fluids, respectively These mean temperatures will be called T and t.

In order to calculate the heat transfer coefficients, it will be necessary to calculate (μ/μ_w) for each fluid, where μ is the viscosity at the mean temperature, and μ_w is the viscosity at the tube wall temperature. The tube wall temperature is not known beforehand. Thus it is necessary to assume a tube wall temperature that will be verified later. To make a first guess, it must be considered that the tube wall temperature will have an intermediate value between T and t and will be closer to the temperature of the fluid with higher film coefficient h. Anyway, the exponent 0.14 means that the importance of this factor is not very high, and normally, the first guess is a "good enough" value.

6. The flow area for the fluid flowing into the tubes is calculated as

$$a_t = \frac{Na_t'}{n} \tag{7-6-10}$$

The mass velocity of the tube-side fluid can be calculated as

$$G_t = \rho v_t = W_t / a_t \tag{7-6-11}$$

7. Calculate the Reynolds number for the tube-side fluid:

$$\text{Re}_t = \frac{D_i G_t}{\mu} \tag{7-6-12}$$

8. Calculate h_i with Eqs. (7-4-3) through (7-4-5) based on the Reynold number.
9. Correct for the external diameter:

$$h_{io} = h_i \frac{D_i}{D_o}$$

$$(7\text{-}6\text{-}13)$$

10. Calculate the shell-side heat transfer coefficient based on the procedures explained in Sec. 7-5.
11. With the calculated values of h_{io} and h_o, the tube wall temperature assumed in step 5 can be checked. To do this, it is necesssary to equate the heat transfer rates at both sides of the wall:

$$h_{io}(T_w - t) = h_o(T - T_w)$$

$$(7\text{-}6\text{-}14)$$

if the shell-side fluid is the hot fluid, or

$$h_{io}(T - T_w) = h_o(T_w - t)$$

$$(7\text{-}6\text{-}15)$$

if the tube-side fluid is the hot fluid. And then we can solve Eq. (7-6-15) for T_w.

12. The clean heat transfer coefficient can be calculated as

$$U_c = \left(\frac{1}{h_{io}} + \frac{1}{h_o} \right)^{-1}$$

$$(7\text{-}6\text{-}16)$$

13. The total heat transfer coefficient can be obtained as

$$U = \left(\frac{1}{U_c} + R_f \right)^{-1}$$

$$(7\text{-}6\text{-}17)$$

14. The calculated area is

$$A_{calc} = \frac{Q}{U \Delta T}$$

$$(7\text{-}6\text{-}18)$$

15. The real area of the heat exchanger is calculated with its geometric dimensions:

$$A_{real} = N \pi D_o L$$

$$(7\text{-}6\text{-}19)$$

16. If $A_{real} > A_{calc}$, it will be possible to use the equipment from the heat transfer point of view. The excess area will be

$$\text{Percent excess} = \frac{A_{real} - A_{calc}}{A_{calc}} \times 100$$

$$(7\text{-}6\text{-}20)$$

Tube-side pressure-drop calculation: With Re_t calculated in step 7, obtain the friction factor with Eq. (7-4-8) or Eq. (7-4-9), and calculate the straight-tube pressure drop with Eq. (7-4-7). Then calculate the pressure drops in the headers with Eq. (7-4-10). The total pressure drop will be the sum of both effects, as indicated in Eq. (7-4-11).

17. Shell-side pressure-drop calculation: It can be calculated with the methods explained in Sec. 7-5.
18. The heat exchanger will be suitable for the required service if the two following conditions are satisfied:

a. $A_{real} > A_{calc}$, as explained in step 16
b. Both the tube-side and shell-side calculated Δp values must be lower than allowable values.

7-6-2   Design of a Heat Exchanger

If a totally new heat exchanger must be designed to perform a certain service, all its geometric characteristics must be defined by the designer. In general terms, the procedure is to propose the heat exchanger geometry and then apply the rating methodology detailed in Sec. 7-6-1.

In this case, the goal is to minimize the difference between the assumed and calculated areas because any excess area results in an unnecessary cost. Additionally, the heat transfer coefficients must be as high as possible because then the heat transfer area will be at a minimum. The obvious limitation to the increase in heat transfer coefficients is the allowable pressure drop for both fluids.

As a general rule, we can say that the design will be optimal when the Δp values of both fluids are close to the maximum allowable values (because then the heat transfer coefficients also will be close to the maximum) and the heat transfer area is enough, but with little excess, to transfer the required heat duty. If any of these conditions are not satisfied, it will be necessary to change the geometry of the unit.

We have explained in Sec. 7-3 how the several geometric parameters affect the performance of the unit, and these criteria must be considered to improve the design. To initially adopt the heat exchanger geometry, the following procedure is suggested:

1. With a heat balance, the unknown process variable (flow rate or temperature) of one of the streams can be found. Then calculate LMTD.

2. Select the number of shell passes or shells configuration with the criteria explained in Sec. 7-2 and with the procedure explained in Example 7-3. If there is no limitation owing to F_t considerations, a heat exchanger with one shell pass will be selected. It must be remembered that the countercurrent configuration allows the highest ΔT because $F_t = 1$, and this possibility must be investigated. Many times, however, a pure countercurrent configuration is avoided because it makes the removable-bundle construction difficult or because it makes a very high tube length or installation of shells in series necessary. So usually exploration of alternatives starts with one shell pass and $2n'$ tubes passes, with n' being any integer number, calculate F_t.

3. Considering the type of fluids to be handled, a first guess of the overall heat transfer coefficient can be obtained using the table in App. G.

4. With the assumed U, an approximate value of the heat transfer area can be calculated as

$$A' = \frac{Q}{U \cdot \text{DMLT} \cdot F_t} \tag{7-6-21}$$

5. With the criteria explained in Sec. 7-3, it is possible to choose the tube diameter, pattern, and pitch and decide the allocation of fluids on the shell and tube sides.

6. Select the number of tubes per pass. This is normally selected so as to have a reasonable tube-side fluid velocity. For example, if we adopt a fluid velocity of 1 m/s, the number of tubes in parallel will be

$$n_p = \frac{W_t}{\rho a_t' (1 \text{ m/s})}$$

7. Select the number of tubes, tube length, and number of tube passes. Here, it is necessary to find a combination of number of tubes and tube length to satisfy the value of A' calculated in step 3, which is

$$A' = N\pi D_o L \tag{7-6-22}$$

At the same time, the number of tube passes must be selected in such a way that the quotient between N and n_p is an integer. This means that one must adopt the number of tube passes n as

$$n = N/n_p \tag{7-6-23}$$

adjusting n_p to get an integer. At the same time, the heat exchanger length must be reasonable, and the unit must be proportionate (see step 8).

8. Selection of shell diameter. The tables in App. D show the number of tubes that can be allocated in a certain shell diameter for different exchanger types. Once the number of tubes has been selected with the help of these tables, it is possible to find the necessary shell diameter.

   Usually the number of tubes obtained from the tables will not match exactly with the number of tubes calculated in step 7. Anyway, at this stage we are only looking for a first approximation to the final design, to be verified later.

9. With the number of tubes determined in step 8, calculate the corrected heat transfer area as

$$A = N\pi D_o L \qquad (7\text{-}6\text{-}24)$$

   This is usually somewhat different from the area calculated in step 7 because we may have slightly changed the number of tubes.

10. Finally, the baffle separation must be selected. As a first guess. we shall try to obtain a Reynolds number that gives a reasonably high shell-side heat transfer coefficient. This may require a few trial calculations with Eq. (7-5-7).

Once the heat exchanger is completely defined, we can proceed to the rating procedure explained in Sec. 7-6-1.

Example 7-6 It is desired to cool down 33.33 kg/s of methanol from 65°C to 30°C using cooling water at 25°C. The cooling water outlet temperature must not be higher than 40°C because of cooling system restrictions. Pressure drops of 1.50×10^5 N/m² for the process stream and 1×10^5 N/m² for the cooling water are allowed. The total fouling resistance must be 5×10^{-4} (s · m² · K)/J. The physical properties of methanol are

$k = 0.21$ J/(m · s · K)

$\rho = 800$ kg/m³

$c = 2{,}508$ J/(kg · K)

Viscosity [kg/(m · s)]:

T (°C)	74	60	50	40	30
μ	0.00030	0.00036	0.00040	0.00047	0.00054

Note: A downloadable spreadsheet with the calculations for this example is available at http://www.mhprofessional.com/product.php?isbn=0071624082

Solution

a. *Propose a tentative geometry:*

1. Calculation of water-mass flow:

$$Q = W_h c_h (T_1 - T_2) = 33.33 \times 2{,}508 \times (65 - 30) = 2.925 \times 10^6 \text{ W}$$

$$\therefore W_c = \frac{Q}{c_c(t_2 - t_1)} = \frac{2.925 \times 10^6}{4{,}180(40 - 25)} = 46.66 \text{ kg/s}$$

2. Select a tentative overall heat transfer coefficient:

   With the help of the table in App. G, we choose as a first guess

$$U = 450 \text{ W/(m}^2 \cdot \text{K)}$$

3. Exchanger configuration:

We shall calculate the F_t factor for different configurations:

$$R = \frac{65-30}{40-25} = 2.33 \quad \text{and} \quad S = \frac{40-25}{65-25} = 0.375$$

With these values, for a 1-2 configuration (one shell pass), $F_t < 0.75$. For two shell passes, with these parameters, we get $F_t = 0.82$. It thus will be necessary to use a configuration with two shell passes and $2n$ tubes passes, where n is an integer. We can obtain this configuration by connecting two shells in series (one pass each), as show in Fig. 7-8. (except that in this case the cold fluid will be the tube-side fluid).The number of tube passes will be decided later.

4. Calculation of LMTD:

$$\text{LMTD} = \frac{(T_1 - t_2) - (T_2 - t_1)}{\ln \dfrac{T_1 - t_2}{T_2 - t_1}} = \frac{(65-40)-(30-25)}{\ln \dfrac{65-40}{30-25}} = 12.4°C$$

Then ΔT $\qquad = \text{LMTD} \cdot F_t = 12.4 \times 0.82 = 10.17.$

5. Estimation of the heat transfer area:

   With the assumed U, we can estimate an area

$$A' = \frac{Q}{U \Delta T} = \frac{2.925 \times 10^6}{450 \times 10.17} = 639.4 \text{ m}^2$$

   Since this area will be split into two shells, they will have approximately 320 m² each.

6. Selection of tube diameters and pattern.

   Let's adopt 3/4-in BWG 14 tubes. According to the table in App. H, the internal diameter is 0.0147 m. We shall adopt a triangular pattern with 0.0254-m pitch.

7. Selection of tube length, number of tubes, and number of tube passes.

   The cooling water will circulate inside the tubes. As a first guess, we shall adopt a water velocity of 2 m/s. The flow area of each tube is

$$a_t' = \frac{\pi D_i^2}{4} = \frac{\pi \times 0.0147^2}{4} = 1.69 \times 10^{-4} \text{ m}^2$$

   To have a velocity of 2 m/s, the number of tubes in parallel must be

$$n_p = \frac{W}{\rho a_t' v} = \frac{46.66}{1{,}000 \times 1.69 \times 10^{-4} \times 2} = 138 \text{ tubes}$$

   If we assume, for the first trial, a 6-m tube length, the total number of tubes must be

$$N = \frac{A}{\pi D_o L} = \frac{320}{\pi \times 0.019 \times 6} = 893 \text{ tubes in each shell}$$

   Thus the number of tube passes in each shell will be

$$n = \frac{N}{n_p} = \frac{893}{138} = 6.47$$

   We shall adopt six tube passes in each shell. Then the proposed configuration consists in two 1-6 shells connected in series, making an overall 2-12 configuration.

We can see in the tables of App. D that with a triangular pattern with ¾-in tubes separated by 1 in and six tube passes, the tube number most close to 892 corresponds to 834 tubes in a 33-in shell diameter (0.838 m).

The area of each shell is

$$A = \pi D_o LN = \pi \times 0.019 \times 6 \times 834 = 298.6 \text{ m}^2$$

This means that the total area of the unit is 597 m². Thus we shall propose an arrangement consisting in two shells in series with 834 tubes each and with a tube length of 6 m arranged in six passes per shell. As a first trial, we shall adopt a baffle spacing of 0.3 m. Thus each shell shall have 19 baffles.

b. *Verification of the proposed design:*

1. Calculation of the tube-side film coefficient:

$$\text{Flow area } a_t = a_t' \frac{N}{n} = 1.69 \times 10^{-4} \frac{834}{6} = 0.0236 \text{ m}^2$$

$$\text{Water velocity} = \frac{W_c}{\rho a_t} = \frac{46.66}{1,000 \times 0.0235} = 1.98 \text{ m/s}$$

The mean temperature of the cooling water is 32.5°C.

$$h_i = 1,423(1+0.0146t)\frac{v^{0.8}}{D_i^{0.2}} = 1,423(1+0.0146 \times 32.5)\frac{1.98^{0.8}}{0.0147^{0.2}} = 8,420 \text{ W/(m}^2 \cdot \text{K)}$$

Then

$$h_{io} = h_i \frac{D_i}{D_o} = 8,420 \frac{0.0147}{0.019} = 6,515 \text{ W/(m}^2 \cdot \text{K)}$$

2. Calculation of the shell-side film coefficient:

$$a_s = \text{flow area} = \frac{D_s cB}{P_t} = \frac{0.838 \times 6.35 \times 10^{-3} \times 0.3}{0.0254} = 0.0629 \text{ m}^2$$

$$G_S = \text{mass velocity} = \frac{W}{a_s} = \frac{33.33}{0.0629} = 530 \text{ kg/(m}^2 \cdot \text{s)}$$

For a 30-degree triangular tube pattern with 19-mm tubes separated by 25.4 mm, the equivalent shell-side diameter is 18.5 mm (0.0185 m).

$$\text{Re}_s = \frac{D_e G_s}{\mu} = \frac{0.0185 \times 530}{0.0004} = 24,526$$

$$h_o = 0.36 \text{Re}_S^{0.55} \text{Pr}^{0.33} \frac{k}{D_e} = 0.36 \times 24,526^{0.55} \times 4.77^{0.33} \times \frac{0.21}{0.0185} = 1,777 \text{ W/(m}^2 \cdot \text{K)}$$

We have assumed that the viscosity correction factor is unity. We can now verify this assumption. It will be

$$h_o(T - T_w) = h_{io}(T_w - t)$$

where T and t are the mean temperatures of methanol and water, respectively. Then

$$1,777(47 - T_w) = 6,514(T_w - 32.5)$$

From which we get $T_w = 35.5°C$. This temperature practically coincides with the mean temperature of water, so we shall not perform any correction to the water film coefficient. For methanol, the viscosity at 35.5°C is 0.0005 kg/ms; thus

$$\left(\frac{\mu}{\mu_w}\right)^{0.14} = \left(\frac{0.0004}{0.0005}\right)^{0.14} = 0.96$$

Then

$$h_o = 1,777 \times 0.96 = 1,706 \text{ W/(m}^2 \cdot \text{K)}$$

3. Calculation of the overall heat coefficient:

$$U = \left(\frac{1}{h_{io}} + \frac{1}{h_o} + R_f\right)^{-1} = \left(\frac{1}{1,706} + \frac{1}{6,514} + 5 \times 10^{-4}\right)^{-1} = 807 \text{ W/(m}^2 \cdot \text{K)}$$

4. Calculation of the heat transfer area.

$$A = \frac{Q}{U\Delta T} = \frac{2,925,700}{807 \times 10.16} = 357 \quad \text{(total for both shells)}$$

This value is sensibly smaller than the 597 m^2 we had proposed. The design thus must be modified because this excess area cannot be accepted. Nevertheless, we shall calculate the pressure drops for both fluids to know how the proposed design must be modified

5. Calculation of the cooling-water pressure drop.

For the water flowing inside tubes,

$$\text{Re}_t = \frac{D_i v \rho}{\mu} = \frac{0.0147 \times 1.98 \times 1,000}{8.2 \times 10^{-4}} = 35,495$$

The friction factor is

$$f = 0.0014 + 0.125\,\text{Re}^{-0.32} = 5.78 \times 10^{-3}$$

The pressure drop in the straight tubes in each shell is

$$\Delta p_t = 4f\frac{Ln}{D_i}\frac{\rho v^2}{2} = 4 \times 5.77 \times 10^{-3} \times \frac{6 \times 6}{0.0147} \times \frac{1,000 \times 1.98^2}{2} = 1.1079 \times 10^5 \text{ N/m}^2$$

And the pressure drop in the return headers is

$$\Delta p_r = 4n\frac{\rho v^2}{2} = 4 \times 6 \times 1,000 \times \frac{1.98^2}{2} = 4.704 \times 10^4 \text{ N/m}^2$$

So the pressure drop for each shell is

$$\Delta p_T = \Delta p_r + \Delta p_t = 1.578 \times 10^5 \text{ N/m}^2$$

And for the complete unit, it will be 3.156 × 10^5 N/m^2, which largely exceeds the maximum allowable.

6. Calculation of the methanol pressure drop:

$$f = \text{friction factor} = 1.728\,\text{Re}^{0.188} = 0.258$$

$$\Delta p_s = f \frac{G_S^2 D_S (N_B + 1)}{2 D_e \rho} \left(\frac{\mu_w}{\mu} \right)^{0.14} = \frac{0.258 \times 530^2 \times 0.838 \times 20}{2 \times 0.0185 \times 800} \frac{1}{0.96} = 4.280 \times 10^4 \ \text{N/m}^2$$

And for both shells, $\Delta p = 8.56 \times 10^4 \ \text{N/m}^2$.

c. *Modification of the proposed design:* The design we have rated has an excessive pressure drop in the water side and an excess in heat transfer surface. In a next design step, we can reduce the water velocity by decreasing the number of tube passes. This will affect the heat transfer coefficient, but since the controlling resistances are in the methanol side and fouling, the net effect on the overall U will not be important. We also must reduce the heat transfer area.

An obvious solution is to decrease the tube length, which also will help to reduce the pressure drop. However, to lower the cost of the unit, it is usually more effective to reduce the number of tubes.

Since with the help of a spreadsheet it is very easy to explore alternatives, we shall analyze the option of maintaining the 6-m tube length as well as the option of using 4-m tube length. In both cases we shall reduce the number of tubes and the number of tube passes with the objective of obtaining the smaller area requested for the heat transfer without exceeding the fluid pressure drops.

In the shell side, we see that the pressure drop calculated in the first trial is lower than acceptable. This means that we still have some margin to reduce the baffle spacing and increase the shell-side film coefficient. However, since we shall reduce the shell diameter, and this also will increase the velocity, we shall, for the moment, keep the 0.3-m spacing.

Second trial: We shall try the following parameters:

D_s	n	N	L	B	A
736 mm	4	646	6 m	0.3 m	462 m²

And the calculated result is

Re_t	Re_s	h_{io}	h_o	U	Δp_s	Δp_t	A Required
30,500	27,925	5,778	1,832	820	95,200	159,000	350 m²

We see that we still have a 24 percent excess area, and the water pressure drop is still excessive. We can move forward in the same direction with a smaller number of tubes and a smaller number of tube passes.

Third trial: We shall try the following parameters:

D_s	n	N	L	B	A
685 mm	2	578	6	0.3	414 m²

And we get the following results:

Re_t	Re_s	h_{io}	h_o	U	Δp_s	Δp_t	A Required
17,100	29,970	3,627	1,905	768	100,870	27718	374 m²

We still have a 9 percent excess area. We shall further reduce the number of tubes. It is interesting to note that the decrease in the water velocity does not influence the overall U because of the relatively low shell-side film coefficient.

Fourth trial: We shall try the following parameters:

D_s	n	N	L	B	A
635 mm	2	492	6	0.30	352 m²

And the results are

Re$_t$	Re$_s$	h_{io}	h_o	U	Δp_s	Δp_t	A Required
20,000	32,367	4,126	1,987	802	107,377	37,122	358 m^2

In this case, the required area is larger than the actual area. The defect is −1.6 percent. However, since the shell-side pressure drop is still lower than allowable, we can get some improvement in U reducing the baffle spacing.

Fifth trial: We shall try with the following parameters:

D_s	n	N	L	B	A
635	2	492	6	0.27	352 m^2

With the following results:

Re$_t$	Re$_s$	h_{io}	h_o	U	Δp_s	Δp_t	A Required
20,000	35,900	4,126	2,106	821	144,000	37,122	352 m^2

And this design can be considered acceptable.
     We shall now see what design results using 4-m tubes.

Sixth trial: The following geometry is proposed:

D_s	n	N	L	B	A
0.787 m	4	750	4	0.22	358 m^2

And the results are

Re$_t$	Re$_s$	h_{io}	h_o	U	Δp_s	Δp_t	A Required
12,700	35,600	5,127	1,766	932	143,800	92,575	308 m^2

This design has 14 percent excess area, so from the heat transfer point of view it is more attractive. However, even though the heat transfer areas of the last two designs are very similar, the cost of the unit in trial 5 probably will be considerably lower owing to the economies in the tubesheet and manufacturing labor. Note that the tube-side coefficient is considerably higher in trial 6 than in trial 5, but this difference has little importance in the overall coefficient.

7-6-3 Design Programs

Heat exchanger design is usually performed with the help of computer programs. The structure of these design programs is usually complex owing to the great number of parameters that must be adopted (e.g., TEMA type, number of shell and tubes passes, diameter and number of tubes, clearances, sealing strips, tube pattern and pitch, tube length, type of baffles and baffle cut, baffles spacing, and shell diameter). Many of these variables depend on user preference, project specifications, or layout restrictions and are defined by the user in the program input. Thus the optimization path followed by the program depends on which are the design parameters defined by the user. For example, we include the flow diagram in Fig. 7-38, which shows the design strategy of a program for a specific situation.

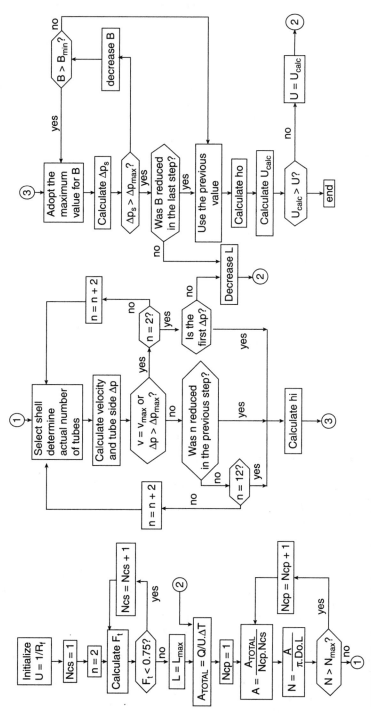

FIGURE 7-38 Structure of a heat exchanger design program.

The program starts by assuming an initial value of the overall heat transfer coefficient. We know that this coefficient is given by

$$\frac{1}{U} = \frac{1}{h_o} + \frac{1}{h_{io}} + R_f \tag{7-6-25}$$

The maximum possible value that U can have is $1/R_f$, a situation that is reached when the film coefficients are very high. Thus the program starts by adopting for U the value given by $U_{max} = 1/R_f$. The area calculated with this first approach therefore will be the minimum, and it will increase with the evolution of the calculations.

Then the heat exchanger configuration is adopted. In this case, the use of countercurrent heat exchangers was removed from consideration owing to cleaning difficulties. The simplest configuration thus will be a single shell with one shell pass and $2n'$ tube passes, with n' being any integer. More complex configurations will be obtained by adding shells in series.

For the simplest configuration, the F_t factor is calculated. If it is lower than 0.75, the program moves to a 2-4 configuration, and so on, finally obtaining the simplest structure that allows operating with a satisfactory F_t. At the beginning, it is assumed that each shell has two tube passes (remember that the number of tube passes may be changed without affecting F_t).

The program also assumes as an initial guess for the tube length the maximum allowable value that has been defined by design restrictions. Then the total heat transfer area is calculated as

$$A_{total} = Q/U\Delta T$$

where A_{total} represents the area of all the individual shells that are necessary.

If there is also a restriction on the maximum shell diameter or maximum number of tubes per shell, it may be necessary to use shells in parallel. At the beginning of the calculation, it is assumed that no parallel trains are necessary.

The area of each shell will be

$$A = \frac{A_{total}}{N_{cs} N_{cp}}$$

where N_{cs} is the number of shells in series and N_{cp} is the number of parallel branches.

Then the number of tubes per shell is determined. If this number is higher than the maximum allowable per shell, parallel branches will be added. The shell diameter then can be selected, and the real number of tubes per shell is determined by means of a computerized version of the tables in App. D.

Then the tube-side fluid velocity and pressure drop are calculated. If in the first calculation these values are higher than the maximum allowed, the program continues with the calculation of heat transfer coefficients because, since the initially assumed U is very high, in successive calculation steps the number of tubes will increase and Δp_T will decrease.

If Δp_T is lower than allowable, the program always will try to increase the velocity by increasing the number of tube passes (up to a maximum of 12) unless any recent trial has shown that this would result in an unacceptable pressure drop. Then the tube-side film coefficient is calculated, and the program moves to shell-side calculations.

The maximum baffle spacing allowed by the TEMA Standards is assumed (which is defined by tube support considerations), and in successive calculation steps it will be reduced if the shell-side pressure drop allows. If, even with the highest spacing, the allowable Δp is exceeded, the tube length is reduced, which results in an increase of the number of tubes in parallel. Finally, the shell-side heat transfer coefficient is calculated, which allows calculation of the overall coefficient U with Eq. (7-6-25).

This value is compared with the previously assumed value. If the calculated U is smaller than the previous U, the calculated value is adopted, restarting the calculations.

7-7 HEAT EXCHANGER EFFECTIVENESS

7-7-1 Calculation of Heat Exchanger Outlet Temperatures

In Sec. 7-6-1 we mentioned that it is sometimes necessary to calculate the heat exchanger outlet temperatures of the fluids when the inlet temperatures are known. This may be necessary in the following cases:

- When one wants to know the behavior of the heat exchanger in conditions different from those of the design
- In process-simulation programs (In this case, every piece of equipment is represented by a module or transfer function that calculates the outlet streams for any inlet condition.)
- When one wants to adapt an existing heat exchanger to a new process condition

In these types of problems, we know all the geometric characteristics of the heat exchanger and the inlet streams. If we assume that the physical properties are constant, we can calculate the film heat transfer coefficients h and then the overall U. As explained in Sec. 7-6-1, the equation system that must be solved is

$$Q = W_h c_h (T_1 - T_2) \qquad (7\text{-}7\text{-}1)$$

$$Q = W_c c_c (t_2 - t_1) \qquad (7\text{-}7\text{-}2)$$

$$Q = UA \frac{(T_1 - t_2) - (T_2 - t_1)}{\ln \frac{T_1 - t_2}{T_2 - t_1}} F_t \qquad (7\text{-}7\text{-}3)$$

$$F_t = f(T_1, T_2, t_1, t_2) \qquad (7\text{-}7\text{-}4)$$

In this system, the unknowns are Q, T_2, t_2, and F_t. The system can be solved mathematically by an iterative procedure (Fig. 7-39).

When using this algorithm in computer programs, it is necessary to restrict the size of the iteration steps to achieve convergence in the calculations and to avoid indeterminate solutions in the F_t calculations. A more direct method for solving this equation system was proposed by W.Kays and A.M.London.[16] This method is particularly useful for cases where the outlet temperatures are not known. They define a new parameter called *effectiveness of the heat exchanger*, which will be   designated with the symbol ε and is defined as

$$\varepsilon = \frac{Q}{(Wc)_{\min}(T_1 - t_1)} \qquad (7\text{-}7\text{-}5)$$

where

$$(Wc)_{\min} = W_c c_c \qquad \text{if } W_c c_c < W_h c_h$$
$$(Wc)_{\min} = W_h c_h \qquad \text{if } W_h c_h < W_c c_c$$

Another parameter called *number of transfer units* (NTU) is defined as

$$\text{NTU} = \frac{UA}{(Wc)_{\min}} \qquad (7\text{-}7\text{-}6)$$

And a parameter called *heat capacity ratio R'* is defined as

$$R' = \frac{(Wc)_{\min}}{(Wc)_{\max}} \qquad (7\text{-}7\text{-}7)$$

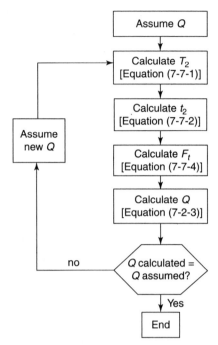

FIGURE 7-39 Flow diagram for solving the equation system (7-7-1) through (7-7-4).

where $(Wc)_{\max}$ is the Wc product that was not defined as $(Wc)_{\min}$. The general form of the correlations is

$$\varepsilon = f(R', \text{NTU})$$

And the function f depends on the heat exchanger configuration. The curves representing these functions are shown in Fig. 7-40.

As can be seen, there are effectiveness curves for any kind of configuration, such as countercurrent, co-current, 1-2, 2-4, 3-6, etc. Curves for other configurations may be found in ref. 16. These graphs can be constructed through mathematical solutions to Eqs.(7-7-1) through (7-7-4). Analytical expressions of the effectiveness functions also can be developed theoretically, and they are included in Table 7-5.

Use of the effectiveness concept will be better understood with completion of the following examples.

Example 7-7 Calculate the outlet temperature of the oil in the heat exchanger in Example 7-4 when the heat exchanger is initially put into service and the fouling factor is 0.

Note: A downloadable spreadsheet to perform these calculations is available at http://www. mhprofessional.com/product.php?isbn=0071624082

Solution This is a specific case in which one of the fluids exchanges latent heat isothermally.

A fluid that is capable of yielding or receiving heat without changing its temperature can be considered a fluid with an infinite heat capacity. (The heat capacity is the ratio between the heat received or yielded and the temperature change that the fluid undergoes.). Thus

$$(Wc)_{\min} = W_c c_c \quad \text{and} \quad R' = 0 \qquad \text{since } (Wc)_{\max} = \infty$$

The area of the heat exchanger is $A = 35.8$ m². When the unit is put into service, R_f will equal 0, and then $U = U_c = 1056$ W/(m² · K). Then

$$\text{NTU} = \frac{UA}{(Wc)_{\min}} = \frac{35.8 \times 1,056}{58 \times 2,100} = 0.310$$

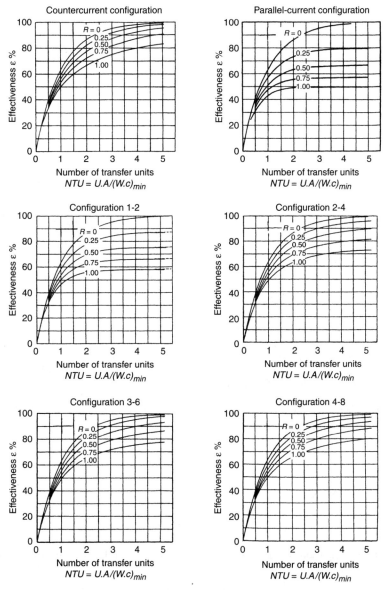

FIGURE 7-40 Heat exchanger effectiveness graphs. (*From W. Kays and A.M London, Compact Heat Exchangers. New York: McGraw-Hill, 1952- Courtesy of Professor W. Kays.*)

TABLE 7-5 Heat Exchanger Effectiveness

Configuration	Effectiveness Expression
Parallel flow	$\varepsilon = \dfrac{1 - \exp[(-\text{NTU}(1 + R')]}{1 + R'}$
Countercurrent flow	$\varepsilon = \dfrac{1 - \exp[-\text{NTU}(1 - R')]}{1 - R'\exp[-\text{NTU}(1 - R')]}$ for $R' \neq 1$
	$\varepsilon = \dfrac{\text{NTU}}{1 + \text{NTU}}$ for $R' = 1$
1-2 exchanger (one shell pass; even number of tube passes)	$\varepsilon = \dfrac{2}{(1 + R') + \sqrt{R'^2 + 1} \dfrac{1 + \exp(-\text{NTU}\sqrt{R'^2 + 1})}{1 - \exp(-\text{NTU}\sqrt{R'^2 + 1})}}$
n-2n exchanger (n shell passes; 2n, 2n + 2, 2n + 4, etc. tube passes)	$\varepsilon = \dfrac{(1 - \varepsilon_a R')^n - (1 - \varepsilon_a)^n}{(1 - \varepsilon_a R')^n - R'(1 - \varepsilon_a)^n}$ for $R' \neq 1$
	$\varepsilon = \dfrac{n\varepsilon_a}{1 + (n - 1)\varepsilon_a}$ for $R' = 1$
	where ε_a is calculated as in 1-2 exchanger with NTU = NTU/n

With NTU = 0.310 and $R' = 0$, from the graph corresponding to the 1-2 configuration we obtain $\varepsilon = 0.267$. Thus

$$\varepsilon = 0.267 = \frac{Q}{(Wc)_{\min}(T_1 - t_1)} = \frac{W_c c_c(t_2 - t_1)}{W_c c_c(T_1 - t_1)} = \frac{t_2 - 57}{177 - 57}$$

Isolating t_2, we get $t_2 = 89°C$.

Example 7-8 Four 1-2 heat exchangers are connected as shown in Fig. 7-41, thus making a 4-8 configuration. The cold-fluid mass flow is 18 kg/s, and its specific heat is 3,000 J/(kg · K). The hot-fluid mass flow is 25 kg/s, and its specific heat is 2,889 J/(kg · K). The inlet temperatures are $T_1 = 170°C$ and $t_1 = 87°C$. The heat transfer area of each shell is 40 m², and the overall heat transfer coefficient is 1,012 W/(m² · K). Calculate the outlet temperatures.

Solution We can consider the whole assembly as a single unit with a 4-8 configuration. The NTU will be

$$\text{NTU} = \frac{UA}{(Wc)_{\min}} = \frac{1,012 \times 160}{1.8 \times 3,000} = 3$$

Since $(Wc)_{\min}$ is in this case $W_c c_c$,

$$R' = \frac{(Wc)_{\min}}{(Wc)_{\max}} = \frac{18 \times 3,000}{25 \times 2,880} = 0.75$$

Entering the graph corresponding to a 4-8 configuration with NTU = 3 and $R' = 0.75$, we obtain $\varepsilon = 0.8$. Then

$$\varepsilon = \frac{W_c c_c(t_2 - t_1)}{W_c c_c(T_1 - t_1)} = \frac{t_2 - 87}{170 - 87} = 0.8$$

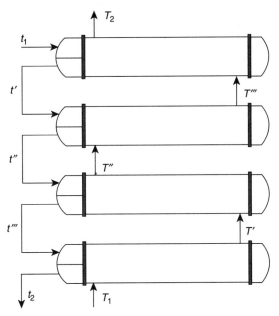

FIGURE 7-41   Example 7-8 heat exchangers.

Isolating t_2, we get $t_2 = 153.4$. The heat exchanged is then

$$Q = W_c c_c (t_2 - t_1) = 54{,}000(153.4 - 87) = 3.585 \times 10^6 \text{ W}$$

The outlet temperature of the hot fluid can be obtained from

$$T_2 = T_1 - \frac{Q}{W_h c_h} = 170 - \frac{3.585 \times 10^6}{25 \times 2{,}880} = 120.2°C$$

If you want to calculate the intermediate temperatures, each one of the four shells can be considered individually. The heat transfer area of each shell is 40 m², and the number of transfer units is

$$\text{NTU} = \frac{UA}{(Wc)_{\min}} = \frac{40 \times 1{,}012.5}{18 \times 3{,}000} = 0.75$$

With NTU = 0.75 and R' = 0.75, from the graph corresponding to a 1-2 configuration, we find $\varepsilon = 0.433$. If we consider the lowest shell,

$$\varepsilon = \frac{Q}{W_c c_c (T_1 - t''')} = \frac{W_c c_c (t_2 - t''')}{W_c c_c (T_1 - t''')} = \frac{t_2 - t'''}{T_1 - t'''}$$

$$t''' = \frac{t_2 - \varepsilon T_1}{1 - \varepsilon} = \frac{153.4 - 0.433 \times 170}{1 - 0.433} = 140.72°C$$

Temperature T' can be obtained from a heat balance:

$$W_c c_c (t_2 - t''') = W_h c_h (T_1 - T')$$

$$\therefore T' = T_1 - \frac{W_c c_c (t_2 - t''')}{W_h c_h} = T_1 - R'(t_2 - t''') = 160.49°C$$

Following the same procedure with the other units, we have

$$t'' = \frac{t''' - \varepsilon T'}{1 - \varepsilon} = \frac{140.72 - 0.433 \times 160.49}{1 - 0.433} = 125.62°C$$

$$T'' = T' - R'(t''' - t'') = 160.49 - 0.75(140.72 - 125.62) = 149.16°C$$

$$t' = \frac{t'' - \varepsilon T''}{1 - \varepsilon} = \frac{125.62 - 0.433 \times 149.16}{1 - 0.433} = 107.63°C$$

$$T''' = 149.16 - 0.75(125.62 - 107.63) = 135.66°C$$

7-7-2 Physical Interpretation of the Effectiveness

We have explained in Sec. 7-2 that the maximum heat exchange that is possible thermodynamically between two streams is achieved with a countercurrent heat exchanger with infinite area. If $W_c c_c = (Wc)_{min}$, the temperatures diagram is as shown in Fig. 7-42a. In this case, the outlet temperature of the cold fluid reaches the inlet temperature of the hot fluid. The heat exchanged is

$$Q = W_c c_c (t_2 - t_1) = (Wc)_{min} (T_1 - t_1) \qquad (7\text{-}7\text{-}8)$$

If, on the other hand, $W_h c_h = (Wc)_{min}$, the temperature diagram is that of Fig. 7-42b. In this case, the hot fluid leaves the exchanger at a temperature equal to the cold-fluid inlet temperature. The heat exchanged is

$$Q = W_h c_h (T_1 - T_2) = (Wc)_{min} (T_1 - t_1) \qquad (7\text{-}7\text{-}9)$$

We see that in both cases the heat exchanged can be expressed as

$$Q = (Wc)_{min} (T_1 - t_1) \qquad (7\text{-}7\text{-}10)$$

We can then assign a physical interpretation to the definition of the effectiveness given by Eq. (7-7-5), saying that the effectiveness of a heat exchanger is the ratio between the heat exchanged in the unit and the maximum amount of heat that is possible thermodynamically to exchange between the two streams. This

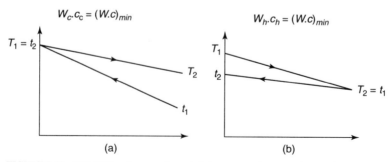

FIGURE 7-42   Temperature diagrams for an infinite-area countercurrent heat exchanger.

thermodynamic maximum could be achieved by putting the streams in contact in a countercurrent heat exchanger with infinite area.

It is important to understand that the effectiveness is not a concept related to a good or bad design of a heat exchanger. If it is necessary to design a heat exchanger to perform a certain process service, defined by the flow rates of the fluids and by the desired inlet and outlet temperatures, the effectiveness of the unit is already defined, and its value is

$$\varepsilon = \frac{t_2 - t_1}{T_1 - t_1} \qquad \text{if } W_c c_c = (Wc)_{min} \tag{7-7-11}$$

or

$$\varepsilon = \frac{T_1 - T_2}{T_1 - t_1} \qquad \text{if } W_h c_h = (Wc)_{min} \tag{7-7-12}$$

This value is independent on the exchanger design, so it must not be thought that a unit with a low effectiveness is related to poor design.

Physical Interpretation of the Number of Transfer Units
The number of transfer units has been defined as

$$NTU = \frac{UA}{(Wc)_{min}} \tag{7-7-13}$$

Despite the fact that the definition includes U and A, which are values obtained after the thermal design, the product of these two magnitudes is constant and independent of the exchanger design once the pass configuration has been adopted. This can be better understood if we write

$$UA = \frac{Q}{F_t \cdot DMLT} \tag{7-7-14}$$

We see that all the magnitudes on the right side of the equation are defined by the process conditions once the configuration is adopted. This means that NTU can be calculated before performing the thermal design as

$$NTU = \frac{Q}{(Wc)_{min} F_t \cdot DMLT} \tag{7-7-15}$$

The concept of transfer unit was used much more extensively in the theory of mass-transfer operations than in heat transfer. The basic idea is that the length of a mass transfer or heat transfer device can be expressed as the product of two magnitudes, that is,

$$L = HTU \cdot NTU \tag{7-7-16}$$

where HTU is the height of a transfer unit. This is a magnitude that does depend on the characteristics of the device; this means the area per unit length and the heat transfer coefficient. For example, in the case of a single-pass countercurrent heat exchanger, the height of the transfer unit can be calculated as

$$HTU = \frac{(Wc)_{min}}{U \pi D_o N} \tag{7-7-17}$$

Thus the definitions of NTU and HTU satisfy Eq. (7-7-16).

This means that once the process is defined, the number of transfer units is also defined. Then, depending on the characteristics of the heat exchanger that will be used (i.e., number and diameters of tubes, etc.), we shall have different values of U and then different heights of the transfer unit. The length of the heat exchanger then can be calculated as the product of HTU and NTU.

7-7-3   Analysis of the Effectiveness Graphs

An analysis of the effectiveness graphs, especially for countercurrent and parallel-current configurations, may help to clarify some concepts. Looking at the countercurrent-configuration graph, we see that effectiveness tends toward 1 when the number of transfer units tends toward infinity, and this is true for any R'. This conclusion is in agreement with the interpretation of the effectiveness definition explained previously.

In the case of a parallel-flow configuration, if the heat transfer area tends toward infinity, the outlet temperatures of both streams would be equal. The effectiveness is lower than 1 because the heat exchanged is lower than in the countercurrent case. For the particular case in which the heat capacities $W \times c$ of both fluids are identical, the limit value of the effectiveness is 0.5. A temperatures diagram for a parallel-flow infinite-area heat exchanger is shown in Fig. 7-43. When the Wc values of both streams are equal, the outlet temperature T_2 will be

$$T_2 = t_2 = \tfrac{1}{2}(T_1 + t_1)$$

and

$$Q = Wc[T_1 - \tfrac{1}{2}(T_1 + t_1)] = \tfrac{1}{2}Wc(T_1 - t_1)$$

Then, from the definition of effectiveness, we get $\varepsilon = \tfrac{1}{2}$.

We also can analyze the case when $R' = 0$. Since

$$R' = \frac{(Wc)_{min}}{(Wc)_{max}}$$

R' will be 0 when the heat capacity of one of the fluids is infinite. This situation, as explained previously, exists when there is a change of phase of a single-component fluid. In this case, that stream can receive or yield energy without changing its temperature. In this case. the thermal diagram is as shown in Fig. 7-44.

We can see that the diagram is the same for countercurrent or parallel-current configurations. In particular, if the area is infinite, it may happen that the outlet temperature of the cold fluid equals that of the hot fluid even for the parallel-current case.

This means that the effectiveness versus NTU curves coincide for any configuration in which $R' = 0$. (The same happens with the F_t curves. This means that when one of the fluids is isothermal, it does not matter what the configuration is.)

For all other cases different from $R' = 0$, at the same R' and NTU, the countercurrent exchanger has a higher effectiveness than the parallel-current exchanger, which means that it has a higher heat transfer capacity. This can be appreciated in the effectiveness graphs by comparing ordinates for the same abscissa.

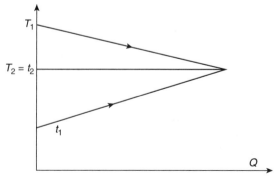

FIGURE 7-43   Thermal diagram in a parallel-flow heat exchanger with infinite area.

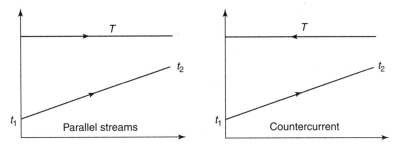

FIGURE 7-44 Thermal diagrams for parallel flow or countercurrent flow are identical when one fluid is isothermal.

We see that for low NTU values, the differences in effectiveness for both types are not very large. The parallel-current heat exchanger loses effectiveness at high NTU values because at the entrance end the temperature difference may be high (which means a high driving force), but it greatly reduces toward the outlet end, and the performance of the unit deteriorates. The difference between both types is important with high NTU values. On the other hand, for small NTU values, both units behave in a similar way.

GLOSSARY

A = heat transfer area (m^2)

c = specific heat (J/kg . K)

F_t = LMTD correction factor (dimensionless)

f = friction factor (dimensionless)

G = mass velocity = W/flow area = $\rho v (kg/s)$

h = film coefficient (W/m^2 . K)

h_i = internal h (W/m^2 . K)

h_o = external h (W/m^2 . K)

h_{io} = internal h referred to the external area = $h_i D_i/D_o$ (W/m^2 . K)

h_B = ideal tube-bank h (W/m^2 . K)

h_{BP} = film coefficient for a tube bank with bypass (W/m^2 . K)

h_{NL} = h of a tubes bank without leakage (W/m^2 . K)

h_L = h of a tube bank with leakage (W/m^2 . K)

j = Colburn coefficient (dimensionless)

k = thermal conductivity (W/m . K)

Q = heat exchanged per unit time (W)

R = parameter for F_t calculation (dimensionless)

R_f = fouling ressistance (m^2 . K/W)

S = parameter for F_t calculation (dimensionless)

T = hot-fluid temperature, mean or generic value (K or °C)

t = cold-fluid temperature, mean or generic value (K or °C)

U = overall heat transfer coefficient (W/m^2 . K)

U_c = clean overall heat transfer coefficient (W/m^2 . K)

v = velocity (m/s)

W = mass flow (kg/s)

X = correction factor for number of tube rows (dimensionless)

Δp = pressure drop (N/m^2)

$\Delta p_t = \Delta p$ in straight tube length (N/m^2)

$\Delta p_r = \Delta p$ in return headers (N/m^2)

Δp_T = total Δp for the tube-side fluid (N/m^2)

Δp_s = Shell side fluid Δp (N/m^2)

Δp_B = ideal tube bank Δp (N/m^2)

$\Delta p_{BP} = \Delta p$ of a tube bank with bypass (N/m^2)

$p_L = \Delta p$ for an exchanger section considering leakage (N/m^2)

$\Delta p_{NL} = \Delta p$ for an exchanger section with no leakage (N/m^2)

ξ_h = bypass correction coefficient for h calculation (dimensionless)

$\xi_{\Delta p}$ = bypass correction coefficient for Δp calculation (dimensionless)

α = constant in Eq. (7-5-20) (dimensionless)

ε = heat exchanger effectiveness (dimensionless)

ϕ = window correction factor (dimensionless)

μ = viscosity (kg/m.s) or (cP)

ρ = density (kg/m^3)

Subscripts

1 = inlet

2 = outlet

c = cold

h = hot

i = internal

m = cross-flow

o = external

s = shell

t = tubes

w = wall or window

Glossary of the Heat Exchanger Geometry

A = heat transfer area (m^2)

A = baffle central angle (°)

a_1 = external area per unit length of tube = πD_o (m^2/m)

a'_t = flow area of a tube = $\pi D_i^2/4$ (m^2)

a_t = total flow area for the tubeside fluid = $Na't/n$ (m^2)

B = baffle spacing (m)

BC = baffle cut (m)

c = clearance between tubes (m)

D_B = baffle diameter (m)

D_{BT} = diameter of the baffle hole (m)

D_e = shell-equivalent diameter according to Kern (m)

D_i = internal diameter of a tube (m)

D_o = external diameter of a tube (m)

D_S = internal shell diameter (m)

F_{BP} = bypass fraction (dimensionless)

L = tube length (m)

N = number of tubes (dimensionless)

n = number of tube passes (dimensionless)

N_B = Number of baffles (dimensionless)

N_{BT} = number of tubes crossing a baffle (dimensionless)

N_c = number of tube rows between baffles borders (dimensionless)

N_{CL} = number of tubes in the central row (dimensionless)

N_S = number of pairs of sealing strips (dimensionless)

N_W = effective number of tube rows in a window (dimensionless)

N_{wt} = number of tubes in window (dimensionless)

N_p = number of tubes in parallel (number of tubes per pass)

P_t = distance between tube axes (tube pitch) (m)

S_L = total leakage area (m^2)

S_m = cross-flow area (m^2)

S_{SB} = leakage area between shell and baffle (m^2)

S_{TB} = lakage area between tubes and baffle (m^2)

S_{BP} = bypass area (m^2)

S_W = flow area at window (m^2)

REFERENCES

1. Kern D.: *Process Heat Transfer.* New York: McGraw-Hill, 1954.

2. Underwood A. J.: *J Inst Petroleum Technol* 20:145–158, 1934.

3. Sieder E. N., Tate G. E.: "Heat Transfer and Pressure Drop of Liquids in Tubes," *Ind Eng Chem* 28:1429–1436, 1936.

4. Dittus F. W., Boelter L. M.: *University of California Publ Eng* 2:443, 1930.

5. Lord R. C., Minton P. E., Slusser R. P.: "Design of Heat Exchangers," *Chem Eng* January 26:98–115, 1970.

6. Drew T. B., Koo E. C., McAdams W. H.: *Trans AICHE* 28:56–72, 1932.

7. Ludwig E.: *Applied Process Design for Chemical and Petrochemical Plants.* Gulf Publishing Company, Houston, 1964.

8. Taborek J.: "Survey of Shell Side Flow Correlations," in *Heat Exchangers Design Handbook:* Hemisphere Publishing Co., 1981.

9. Bell K.: Bulletin No. 5 "Final Report of the Cooperative Research Program on Shell andTube Heat Exchangers," University of Delaware, Newark, Delaware, 1963.

10. Bergelin O. P., Bell K., Leighton M. D.: "Heat Transfer and Fluid Friction during Flow across Banks of Tubes VI—The Effect of Internal Leakages within Segmentally Baffled Exchangers," *Trans ASME* 80:53–60, January 1958.

11. Rohsenow, H.: *Handbook of Heat Transfer.* New York: McGraw-Hill, 1973.

12. Bergelin O. P., Brown G. A., Colburn A. P.: "Heat Transfer and Fluid Friction during Flow across Banks of Tubes V. A. Study of a Cylindrical Baffled Heat Exchanger without Internal Leakages," Trans ASME 74:841–850, 1952.

13. Bowman R. A., Mudller A. C., Tagle W. M.: *Trans ASME* 62:283–294, 1940.

14. Palen J., Taborek J.: "Solution of Shell Side Pressure Drop and Heat Transfer by Stream Analysis Method," *Chem Eng Progr Symp Series* 65(92):53–63, 1969.

15. Bergelin O. P., Brown G. A., Doberstein S. C.: "Heat Transfer and Fluid Friction during Flow across Banks of Tubes-IV—A Study of the Transition Zone Between Viscous and Turbulent Flow," *Trans ASME* 74:953–960, 1952.

16. Kays W. M., London A. L.: *Compact Heat Exchangers.* New York: McGraw-Hill, 1951.

17. Sullivan F. W., Bergelin O. P.: "Heat Transfer and Fluid Friction in a Shell and Tube Heat Exchanger with a Single Baffle," *Chem Eng Progr Symp. Series* 18(52):85–94, 1956.

18. Whitley D. L.: "Calculating Heat Exchangers Shell Side Pressure Drop," *Chem Eng Progr* (9):57–65 1961.

19. Bergelin O. P., Bell K. J., Leighton M. D.: "Heat Transfer and Fluid Friction during Flow across Banks of Tubes VI: By Passing between Tube Bundle and Shell," *Chem Eng Progr* 29(55):48–58, 1959.

CHAPTER 8
FINNED TUBES

One way to increase the heat transfer capacity of heat exchanger tubes consists of attaching to them metal pieces, called *fins,* that extend the heat transfer surface. This makes it possible to have a larger external surface for the same internal surface. In cases where the heat transfer controlling coefficient is the external film, this may result in a significant increase in the heat-flux density per unit internal area.

Fins can be transverse to the tube axis, as in Fig. 8-5, or longitudinal, as in Fig. 8-1. This chapter will examine two applications of finned tubes in the process industry: hairpin exchangers with longitudinal fins (either double-tube or multitube) and air coolers, usually provided with transverse fins.

8-1 DOUBLE-TUBE HEAT EXCHANGERS WITH LONGITUDINAL FINS

8-1-1 Use of Fins in Double-Tube Heat Exchangers

Let's consider a double-tube heat exchanger such as those studied in Chap. 5 with longitudinal fins welded to the internal tube, as shown in Fig. 8-1. We shall assume that the hot fluid circulates through the annulus and the cold fluid through the internal tube. The fins are in contact with the hot fluid, receiving heat by convection, and they transfer this heat by conduction to the base of the fin and to the internal tube.

We shall analyze the situation at a generic cross section of the heat exchanger. In the considered section, the annulus fluid temperature is T, and the cold fluid temperature is t. Since the fins are welded to the tube, the temperature at the base of the fins T_b is practically equal to the tube-wall temperature. In the rest of the fin surface, however, the temperature will be higher because it is in contact with the hot fluid. Figure 8-2 is a plot of the temperature at the fin surface T_f as a function of the inverse radial coordinate x (the origin of the radial coordinate is at the fin tip).

Thus

$$T_f = T_f(x)$$

It can be seen that there is a temperature gradient dT_f/dx such that there is a conduction heat flow toward the base of the fin.

With relation to Fig. 8-3, if we consider a slice of fin with height dx and length dz, the hot fluid delivers heat to the fin through a heat transfer area

$$dA = 2\,dz\,dx \tag{8-1-1}$$

The heat flow through this surface, assuming that the surface is clean, will be

$$dQ_f = (2\,dz\,dx)h_f[T - T_f(x)] \tag{8-1-2}$$

where h_f is the convection heat transfer coefficient between the hot fluid and the metal. In the presence of a fouling resistance on the fin surface, it is convenient to incorporate it into the convection coefficient, defining a coefficient corrected by fouling as

$$h'_f = \left(\frac{1}{h_f} + R_{fo}\right)^{-1} \tag{8-1-3}$$

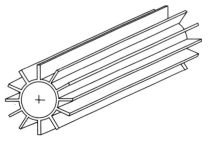

Section of a double-tube
heat exchanger

Tube with longitudinal fins

FIGURE 8-1   Longitudinal fins.

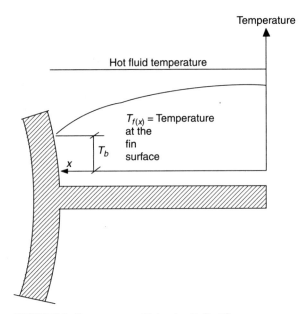

FIGURE 8-2   Temperature profile in a longitudinal fin.

in which case

$$dQ_f = (2dz\,dx)h'_f\,[T - T_f(x)] \tag{8-1-4}$$

In what follows, we shall express everything per unit tube length. We shall call

N_f = number of fins per tube

A_o = area of tube in correspondence with the external diameter per unit length (plain tube area), that is,

$$A_o = \pi D_o \text{ (m}^2\text{/m of tube)} \tag{8-1-5}$$

A_f = fin surface per unit tube length (consider both sides of the fins). If H is the fins height, it will be

$$A_f = 2HN_f \quad \text{(m}^2\text{/m of tube) (neglecting the area of the fin shoulder)} \tag{8-1-6}$$

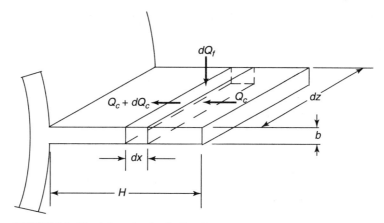

FIGURE 8-3   Heat balance in a longitudinal fin.

A_D = exposed external area of tube (not covered by fins), that is,

$$A_D = A_o - N_f b \tag{8-1-7}$$

where b is the fin thickness.

All the heat penetrating into the fin is transmitted by conduction toward the base of the fin owing to the temperature gradient. In the control volume considered in Fig. 8-3, it can be seen that the conduction heat flow Q_c increases toward the base of the fin as a consequence of the heat incorporated through the lateral surface, so in steady state, it must be $dQ_c = dQ_f$.

The conduction heat flow, which enters the internal tube through the base of the fin, equals the convection heat transferred to the entire surface of the fin. Then, if Q_b (Watt) is the heat flow entering the internal tube through the base of the fin (which corresponds to the value of Q_c at $x = H$), and if dQ_b/dz is the flow per unit tube length (W/m), we can write

$$\frac{dQ_b}{dz} = \int_{Af} h_f' \, [T - T_f(x)] dA = h_f' \, (T - T_f)_m A_f \tag{8-1-8}$$

where the term in parentheses is a mean temperature difference between the fluid and the fin surface.

If T_b is the temperature at the base of the fin, we can write

$$(T - T_f)_m = \Omega(T - T_b) \tag{8-1-9}$$

where Ω is a coefficient whose value is between 0 and 1 because at any point on the fin surface, the temperature difference between the annulus fluid and the metal is smaller than that existing at the base of the fin.

Then

$$\frac{dQ_b}{dz} \, \text{(W/m)} = A_f \Omega h_f' \, (T - T_b) \tag{8-1-10}$$

The heat flow penetrating into the internal tube is the sum of that penetrating through the base of the fin (Q_b) plus the convection heat transferred from the annulus fluid to the bare portion of the tube (area A_D).

Calling Q_D the heat flow through A_D, the heat flow per unit length through the bare area is

$$\frac{dQ_D}{dz} \ (\text{W/m}) = h_f' \, A_D (T - T_b) \tag{8-1-11}$$

And the total heat flow per unit length through the internal tube wall is

$$\frac{dQ}{dz} = A_D h_f' \, (T - T_b) + A_f \, h_f' \, \Omega (T - T_b) = h_f' \, (T - T_b)(\Omega A_f + A_D) \tag{8-1-12}$$

So the calculation can be performed as if the entire metal were at a uniform temperature (that of the tube wall) but affecting the area of the fin by an efficiency coefficient that indicates that this area is not as effective as the bare tube for heat transfer. This is why Ω is called *fin efficiency.*

The efficiency of a fin depends on its geometry, on the thermal conductivity of its material, and also on the convection heat transfer coefficient of the external side. In certain cases, it is possible to deduce analytical expressions to calculate the fin efficiency as a function of these parameters. For example, in the next section we shall deduce the expressions to calculate the efficiency of a fin with a relatively simple geometry such as that shown in Fig. 8-1.

Since the usual practice is to refer the heat transfer coefficients to the external plain area of the tube A_o, we shall define a heat transfer coefficient to be used with this area as

$$h_{fo}' = \frac{h_f' \, (A_D + \Omega A_f)}{A_o} \tag{8-1-13}$$

And the heat transfer between the external fluid and the tube can be expressed as

$$\frac{dQ}{dz} = h_{fo}' \, A_o (T - T_b) \tag{8-1-14}$$

From now on, the procedure is the same as for the case of plain tubes.

From the tube at T_b, heat is transmitted to the internal fluid through the resistances corresponding to the internal film coefficient and internal fouling; that is,

$$\frac{dQ}{dz} = \left(\frac{1}{h_{io}} + R_{fi} \right)^{-1} A_o (T_b - t) \tag{8-1-15}$$

Isolating the temperatures differences in Eqs. (8-1-14) and (8-1-15) and summing both equations we get

$$\frac{dQ}{dz} = \left(\frac{1}{h_{io}} + \frac{1}{h_{fi}'} + R_{fi} \right)^{-1} A_o (T - t) \tag{8-1-16}$$

Defining

$$U = \left(\frac{1}{h_{io}} + \frac{1}{h_{fi}'} + R_{fi} \right)^{-1} \tag{8-1-17}$$

it is

$$\frac{dQ}{dz} = U \pi D_o (T - t) \tag{8-1-18}$$

Equations (8-1-14) and (8-1-15) allows calculation of the heat-flux density (watts per unit tube length) in a section of a heat exchanger where the fluid temperatures are T and t.

As was demonstrated in Chap. 5, this expression can be integrated into all the exchanger length, and for a countercurrent or co-current configuration, we get

$$Q = UA_o(\text{LMTD}) \tag{8-1-19}$$

where A_o is now the total area of the exchanger (not per unit length).

8-1-2 Derivation of the Fin Efficiency for the Case of Longitudinal Fins with Constant Section

The efficiencies of different types of fins can be calculated theoretically with certain assumptions. We shall see how the analytical expression for fin efficiency in the case of rectangular fins can be obtained. The reader not interested in this type of mathematical derivation can move to the next section.

We shall assume that the fin is surrounded by a hot fluid at temperature T, so the transfer of heat takes place from the exterior into the tube. However, the final expressions are valid if the heat transfer is in the opposite direction.

We shall assume that the hot fluid temperature does not vary with the radial coordinate x. It is also assumed that the fin thickness b is small enough to neglect the temperature gradients across its width. We also shall assume that the heat transferred through the top of the fin (fin shoulder) is negligible owing to the small thickness.

If we consider the volume element shown in Fig. 8-3, with thickness b, length dz, and height dx, we can see that a conduction heat flow Q_c is entering the control volume through the right face (bdz) and that a conduction heat flow $Q_c + dQ_c$ is leaving through the left face. Additionally, a convection heat flow dQ_f enters through the lateral area $2dxdz$. In steady state, the sum of these terms is zero, so it must be

$$dQ_c = dQ_f \tag{8-1-20}$$

Q_c is a conduction heat flow and can be expressed as

$$Q_c = -k\frac{dT_f(x)}{dx}(bdz) \tag{8-1-21}$$

where $T_f(x)$ is the metal temperature at a certain x, which is assumed constant through the fin thickness.

The convection heat entering the fin can be expressed as a function of the temperature difference between the external fluid and the metal. This is

$$dQ_f = h'_f[T - T_f(x)](2dx\,dz) \tag{8-1-22}$$

Then

$$\frac{dQ_f}{dx} = h'_f\,2dz[T - T_f(x)] \tag{8-1-23}$$

Differentiating Eq. (8-1-21), we get

$$\frac{dQ_c}{dx} = -k\frac{d^2T_f(x)}{dx^2}(bdz) \tag{8-1-24}$$

Combining Eqs. (8-1-20), (8-1-23), and (8-1-24), we get

$$-kb\frac{d^2T_f(x)}{dx^2} - 2h'_f[T - T_f(x)] = 0 \tag{8-1-25}$$

Since T is a constant for the integration in x, we can also write

$$kb\frac{d^2[T-T_f(x)]}{dx^2} - 2h'_f[T-T_f(x)] = 0 \qquad (8\text{-}1\text{-}26)$$

And defining a variable $\theta = [T - T_f(x)]$,

$$kb\frac{d^2\theta}{dx^2} - 2h'_f\,\theta = 0 \qquad (8\text{-}1\text{-}27)$$

Rearranging, we get

$$\frac{d^2\theta}{dx^2} - \frac{2h'_f\,\theta}{kb} = 0 \qquad (8\text{-}1\text{-}28)$$

The general solution to this differential equation is

$$\theta = C_1 e^{mx} + C_2 e^{-mx} \qquad (8\text{-}1\text{-}29)$$

where

$$m = \left(\frac{2h'_f}{kb}\right)^{1/2} \qquad (8\text{-}1\text{-}30)$$

Now we must evaluate the integration constants. We shall call θ_0 the value of θ at $x = 0$ (at the fin shoulder), so from Eq 8-1-29

$$\theta_0 = C_1 + C_2 \qquad (8\text{-}1\text{-}31)$$

Since we assumed that there is no heat flow through the fin shoulder, for $x = 0$, the temperature gradient must also be zero

$$\left(\frac{d\theta}{dx}\right)_{x=0} = 0 \qquad (8\text{-}1\text{-}32)$$

Since $d\theta/dx = m(C_1 e^{mx} - C_2 e^{-mx})$, this condition requires that $C_1 - C_2 = 0$. Then

$$C_1 = C_2 = \frac{\theta_0}{2} \qquad (8\text{-}1\text{-}33)$$

Thus Eq. (8-1-29) results in

$$\theta = \theta_0 \frac{e^{mx} + e^{-mx}}{2} = \theta_0 \cosh\, mx \qquad (8\text{-}1\text{-}34)$$

If we call θ_b the value of θ for $x = H$ (at the base of the fin), it results,

$$\theta_b = \theta_0 \cosh\, mH \qquad (8\text{-}1\text{-}35)$$

The conduction heat flow for different values of x can be obtained as follows: Combining Eqs. (8-1-24) and (8-1-25), we get

$$\frac{dQ_c}{dx} = 2h'_f\, dz[T - T_f(x)] = 2h_f dz\theta \qquad (8\text{-}1\text{-}36)$$

Differentiating with respect to x, we get

$$\frac{d^2 Q_c}{dx^2} = 2h'_f \, dz \frac{d\theta}{dx} \qquad (8\text{-}1\text{-}37)$$

Combining with Eq. (8-1-21) gives

$$\frac{d^2 Q_c}{dx^2} - \frac{2h'_f}{kb} Q_c = 0 \qquad (8\text{-}1\text{-}38)$$

As before, the solution to this differential equation is

$$Q_c = C'_1 e^{mx} + C'_2 e^{-mx} \qquad (8\text{-}1\text{-}39)$$

Since for $x = 0$, $Q_c = 0$, the result is

$$C'_1 = -C'_2 \qquad (8\text{-}1\text{-}40)$$

Differentiating Eq. (8-1-39) and calculating the value of the derivative at $x = 0$, we get

$$\left. \frac{dQ_c}{dx} \right|_{x=0} = mC'_1 - mC'_2 \qquad (8\text{-}1\text{-}41)$$

And according to Eq. (8-1-36),

$$\frac{dQ_c}{dx} = 2h'_f \, dz\theta_0 = mC'_1 - mC'_2 \qquad (8\text{-}1\text{-}42)$$

And combining with Eq. (8-1-40), we get

$$C_1 = \frac{h'_f \, dz\theta_0}{m} \qquad \text{and} \qquad C_2 = -\frac{h'_f \, dz\theta_0}{m} \qquad (8\text{-}1\text{-}43)$$

Thus

$$Q_c = \frac{h'_f \, dz\theta_0}{m} \left(e^{mx} - e^{-mx} \right) = \frac{2h'_f \, dz\theta_0}{m} \sinh mx \qquad (8\text{-}1\text{-}44)$$

$$Q_b = Q_c|_{x=H} = \frac{h'_f \, dz\theta_0}{m} \left(e^{mH} - e^{-mH} \right) = \frac{2h'_f \, dz\theta_0}{m} \sinh mH \qquad (8\text{-}1\text{-}45)$$

Dividing Eqs. (8-1-45) and (8-1-35), we get

$$\frac{Q_b}{\theta_b} = \frac{2h'_f \, dz}{m} \tanh mH \qquad (8\text{-}1\text{-}46)$$

which can be written as

$$Q_b = h'_f (2H dz) \frac{\tanh mH}{mH} (T - T_b) \qquad (8\text{-}1\text{-}47)$$

Comparing with Eq. (8-1-10), the result is

$$\Omega = \frac{\tanh mH}{mH} \tag{8-1-48}$$

Interpretation of the Results. The film heat transfer coefficient for the fluid external to the tube is given by Eq (8-1-13):

$$h'_{fo} = \frac{h'_f(A_D + \Omega A_f)}{A_o}$$

It is evident from this expression that by increasing the fin area, we improve the value of h'_{fo}, which also improves the overall heat transfer coefficient.

However, the effect of increasing A_f is weighted by the fin efficiency. If the efficiency Ω is small, an increase in the fin area will not improve the overall heat transfer coefficient significantly. The following table shows the values of Ω calculated with Eq. (8-1-48) for different values of *mH*:

mH	0.0	0.1	0.2	0.3	0.4	0.5	0.6	0.7	0.8
Ω	1	0.997	0.987	0.971	0.949	0.924	0.895	0.863	0.830
mH	0.9	1.0	1.1	1.2	1.3	1.4	2.0	3.0	4.0
Ω	0.795	0.762	0.728	0.695	0.663	0.632	0.482	0.332	0.25

It can be observed in this table that for high values of *mH,* the efficiency of the fin is small. Thus, for example, if $mH = 4$, $\Omega = 0.25$, which means that every 4 m² of fin area is equivalent to 1 m² of bare tube surface. In these cases, the installation of fins may not be cost-effective, and it is more convenient to increase the number or length of tubes.

Since

$$mH = \left(\frac{2h'_f}{kb}\right)^{1/2} H$$

we can analyze the different factors affecting the value of the efficiency as follows:

1. *Heat transfer coefficient h_f.* The higher h_f is, the higher is the product *mH,* and the smaller is the fin efficiency. This means that the installation of fins is advantageous when the convection film coefficients are small, such as in cases where the fluids external to the tubes are gases or viscous fluids. On the other hand, with fluids that have intrinsic high values of h_f, the percentage improvement obtained with fins is small.

2. *Fin height.* The greater the fin height, the smaller is the efficiency. Obviously, if the fin height is increased, A_f also increases, so h'_{fo} always will increase regardless the efficiency reduction. But it must be interpreted that every square meter of additional surface obtained by increasing the fin height is less effective every time. A situation is reached where the marginal benefit is so low that the results are uneconomical.

3. *Fin thickness.* The greater the thickness, the higher is the fin efficiency, but the lower is the number of fins that can be installed in the tube. Usually it is better to use small-thickness fins. BWG 20-gauge (0.9-mm-thick) plate is typical.

4. *Thermal conductivity of the fin material.* The higher the thermal conductivity, the higher is the fin efficiency.

8-1-3 Film Coefficients and Frictional Pressure Drop for Longitudinal Fins

To use the preceding expressions, it is necessary to calculate the heat transfer coefficients between the annulus fluid and the metal (fins and bare tube surface). As in the case of double-tube heat exchangers with plain tubes, the correlations are based on Reynolds and Nusselt numbers defined with an equivalent diameter.

The equivalent diameter is four times the hydraulic radius. The hydraulic radius is the quotient between the flow area and the transference perimeter. For heat transfer, this perimeter is that of the fins and the bare portion of the internal tube. To calculate the friction factor, the perimeter of the external tube also must be considered.

The expressions are

$$D_e(\text{heat}) = \frac{4(\pi D_{io}^2/4 - \pi D_o^2/4 - N_f bH)}{\pi D_o - N_f b + 2HN_f} \tag{8-1-49}$$

and

$$D'_e(\text{friction}) = \frac{4(\pi D_{io}^2/4 - \pi D_o^2/4 - N_f bH)}{\pi D_o - N_f b + 2HN + \pi D_{io}} \tag{8-1-50}$$

In these expressions, D_{io} is the internal diameter of the external tube, and N_f is the number of fins per tube. With these equivalent diameters, Reynolds numbers for heat transfer and friction can be defined, considering that the flow area for velocity calculation is

$$a_s = \pi D_{io}^2/4 - \pi D_o^2/4 - N_f bH \tag{8-1-51}$$

According to Kern,[1] the correlations to calculate the heat transfer coefficients in the case of finned tubes are different from those for plain tubes, and he proposes a graph whose values are correlated adequately with the following expressions:

Friction factor. For Re < 2,100,

$$f = \frac{16}{\text{Re}} \tag{8-1-52}$$

For Re > 2,100,

$$f = 0.109 \, \text{Re}^{-0.255} \tag{8-1-53}$$

Heat transfer coefficient. For Re < 2,100,

$$\text{Nu} = 0.359 \, \text{Re}^{0.324} \, \text{Pr}^{0.33} \left(\frac{\mu}{\mu_w}\right)^{0.14} \tag{8-1-54}$$

For Re > 10,000,

$$\text{Nu} = 0.01783 \, \text{Re}^{0.835} \, \text{Pr}^{0.33} \left(\frac{\mu}{\mu_w}\right)^{0.14} \tag{8-1-55}$$

The main difference is that the dimensionless group L/D is not included in the laminar-flow correlation. The author does not mention if the correlation is valid in a particular range of L/D.

The expression for turbulent flow is quite similar to that used for plain tubes, but the coefficients are about 30 percent lower. This is due to the fact that longitudinal fins reduce turbulence.

As usual, there is a transition regime for Reynolds number values between 2,100 and 10,000, where the correlations are not accurate. A linear interpolation can be done in this region, and the results are as follows:

For $2,100 < \text{Re} < 6,000$,

$$\text{Nu} = \left[4.28 + 4.8 \times 10^{-3} (\text{Re} - 2,100) \right] \text{Pr}^{0.33} \left(\frac{\mu}{\mu_w} \right)^{0.14} \tag{8-1-56}$$

For $6,000 < \text{Re} < 10,000$,

$$\text{Nu} = \left[23 + 3.75 \times 10^{-3} (\text{Re} - 6,000) \right] \text{Pr}^{0.33} \left(\frac{\mu}{\mu_w} \right)^{0.14} \tag{8-1-57}$$

Calculation of Wall Temperatures. To evaluate the quotients (μ/μ_w), it is necessary to evaluate the wall temperatures at the annulus side and at the internal-tube side. Since an important temperature drop takes place along the fin, both temperatures will be different. Calling $\overline{T}$ and $\overline{t}$ the mean temperatures of the hot and cold fluids, the mean wall temperature at the annulus side is

$$\overline{t} + \frac{Q}{h_f (A_D + A_f) LN}$$

if the annulus fluid is the cold fluid or

$$\overline{T} - \frac{Q}{h_f (A_D + A_f) LN}$$

if the annulus fluid is the hot fluid. And the wall temperature corresponding to the internal side is

$$\overline{t} + \frac{Q}{h_i A_i LN}$$

if the cold fluid is in the internal tube or

$$\overline{T} - \frac{Q}{h_i A_i LN}$$

if the tube-side fluid is the hot fluid.

8-1-4 Multipass Heat Exchangers with Longitudinal Flow

In Chap. 5 we explained some constructive aspects of hairpin heat exchangers (see Fig. 5-15). When the fluid circulating through the annular space has a low heat transfer coefficient, these units can be built using tubes with longitudinal fins. In these cases, the calculation methods presented earlier can be used, but it is necessary to have the precaution of modifying the equations according to the geometry. That is,

$$\text{Flow area } a_s = \pi D_{io}^2/4 - N\pi D_o^2/4 - NN_f bH \tag{8-1-58}$$

$$D_e(\text{heat}) = \frac{4(\pi D_{io}^2/4 - N\pi D_o^2/4 - NN_f bH)}{N\pi D_o - NN_f b + 2HNN_f} \tag{8-1-59}$$

$$D'_e(\text{friction}) = \frac{4(\pi D_{io}^2/4 - N\pi D_o^2/4 - NN_f bH)}{N\pi D_o - NN_f b + 2HNN_f + \pi D_{io}} \tag{8-1-60}$$

In this case, N is the number of tubes.

It is also possible to build shell-and-tube heat exchangers with longitudinal fins. In this case, transverse baffles will not be installed, and the shell-side fluid circulates with longitudinal flow, so we can use the equations presented earlier to calculate the heat transfer coefficients. This construction allows us to have two or more tube passes with one shell pass.

Example 8-1 When a natural gas stream suffers an isoenthalpic expansion, reducing its pressure, a temperature decrease takes place. This is due to the Joule-Thompson effect (because in real gases enthalpy is not only a function of temperature but also of pressure). If the gas contains some water vapor, this temperature decrease can lead to the formation of solid components called *hydrates*. These are icelike substances, but they form at temperatures about 15°C and can cause obstructions in piping systems. It is then necessary to heat the gas before the expansion.

A stream consisting in 75,330 kg/h of natural gas, molecular weight 16.88, at 20°C and 4,000 kPa, must be heated up to 40°C before its expansion. A hot-oil stream, available at 200°C, will be used for this service. The hot-oil return temperature should be 150°C.

The physical properties of the streams are as follows:

	Hot Oil	Gas
Mean temperature,°C	175	30
Specific heat, J/(kg · K)	2800	2510
Viscosity, cP	0.46	0.0119
Density, kg/m³	790	30.71 (= Mp/zRT)
Thermal conductivity, W/(m · K)	0.112	0.0358

Available pressure drops are 0.5 bar for the hot oil and 0.1 bar for the gas. Fouling resistances should be 0.00053 (m² · K)/W for each fluid. A suitable heat exchanger must be designed. The selected tubes are 25.4 mm OD, 19.86 mm ID, with 20 fins per tube. Fins will be 12.5 mm high and 0.889 mm (0.035 in) thick. Fin thermal conductivity is 43.25 W/(m · K).

Note: Downloadable interactive spreadsheet with the calculations of this example is available at http://www.mhprofessional.com/product.php?isbn=0071624082

Solution

$$Q = W_c c_c (t_2 - t_1) = (75,330/3,600) \times 2,510 \times 20 = 1.05 \times 10^6 \text{ W}$$

$$W_h = Q/c_h(T_1 - T_2) = 1.05 \times 10^6/2,800 \times 50 = 7.5 \text{ kg/s} = 27,000 \text{ kg/h}$$

After some preliminary trials, a shell-and-tube heat exchanger with the following configuration is proposed:

Number of tubes: 70 in four tube passes. From a geometric analysis, we find that a shell diameter of 489 mm is necessary for this number of tubes. We shall calculate the required tube length.

Hot fluid inside tubes:

$$\text{Flow area} = a_t = \frac{N}{n}\frac{\pi D_i^2}{4} = \frac{70}{4}\frac{\pi \times 0.01986^2}{4} = 0.00542 \text{ m}^2$$

$$G_t = \text{mass flow density} = \frac{W_h}{a_t} = \frac{7.5}{0.00542} = 1{,}383 \text{ kg/(m}^2 \cdot \text{s)}$$

$$\text{Re}_t = \frac{D_i \times G_t}{\mu} = \frac{0.01986 \times 1{,}383}{0.46 \times 10^{-3}} = 59{,}730$$

$$\text{Pr} = \frac{2{,}800 \times 0.46 \times 10^{-3}}{0.112} = 11.7$$

$$h_i = 0.023\frac{k}{D_i}\text{Re}^{0.8}\text{Pr}^{0.33} = 0.023 \times \frac{0.112}{0.01986} \times 59{,}730^{0.8} \times 11.7^{0.33} = 1{,}900 \text{ W/(m}^2 \cdot \text{K)}$$

$$h_{io} = h_i\frac{D_i}{D_o} = 1{,}900\frac{19.86}{25.4} = 1{,}485 \text{ W/(m}^2 \cdot \text{K)}$$

External coefficient
Flow area

$$as = \left(\frac{\pi D_i^2}{4} - \frac{N\pi D_o^2}{4} - NN_f bH\right) = \frac{\pi \times 0.489^2}{4} - \frac{70 \times \pi \times 0.0254^2}{4}$$
$$- 70 \times 20 \times 0.889 \times 10^{-3} \times 0.0125 = 0.1367 \text{ m}^2$$

$$\text{Velocity} = \frac{W}{a_s \rho} = \frac{75{,}330}{3{,}600 \times 0.1367 \times 30.7} = 4.98 \text{ m/s}$$

Equivalent diameter (friction)

$$= \frac{4(\pi D_{io}^2/4 - N\pi D_o^2/4 - NN_f bH)}{N\pi D_o - NN_f b + 2HNN_f + \pi D_{io}}$$
$$= \frac{4(\pi 0.489^2/4 - 70\pi 0.0254^2/4 - 70 \times 20 \times 0.889 \times 10^{-3} \times 0.0125)}{70 \times \pi \times 0.0254 - 70 \times 20 \times 0.889 \times 10^{-3} + 2 \times 70 \times 20 \times 0.0125 + \pi \times 0.489} = 0.01338 \text{ m}$$

Equivalent diameter (heat)

$$= \frac{4(\pi D_{io}^2/4 - N\pi D_o^2/4 - NN_f bH)}{N\pi D_o - NN_f b + 2HNN_f}$$
$$= \frac{4(\pi 0.489^2/4 - 70\pi 0.0254^2/4 - 70 \times 20 \times 0.889 \times 10^{-3} \times 0.0125)}{70 \times \pi \times 0.0254 - 70 \times 20 \times 0.889 \times 10^{-3} + 2 \times 70 \times 20 \times 0.0125} = 0.01391 \text{ m}$$

$$\text{Re (friction)} = \frac{D'ev\rho}{\mu} = \frac{0.01338 \times 4.98 \times 30.7}{0.0119 \times 10^{-3}} = 171{,}900$$

$$\text{Re (heat)} = \frac{Dev\rho}{\mu} = \frac{0.01391 \times 4.98 \times 30.7}{0.0119 \times 10^{-3}} = 178,700$$

For the first trial, we shall neglect the viscosity correction factor. Thus

$$\text{Pr} = \frac{2,510 \times 0.0119 \times 10^{-3}}{3.58 \times 10^{-2}} = 0.834$$

$$\text{Nu} = 0.01783\text{Re}^{0.835}\,\text{Pr}^{0.33} = 0.01783 \times 178,700^{0.835} \times 0.834^{0.33} = 407$$

$$h_f = 407\frac{k}{D_e} = 407 \times \frac{3.58 \times 10^{-2}}{0.01391} = 1,050 \text{ W/(m}^2 \cdot \text{K)}$$

Fouling corrected coefficient $h'_f = \left(\dfrac{1}{h_f} + R_{fo}\right)^{-1} = \left(\dfrac{1}{1,050} + 5.3 \times 10^{-4}\right)^{-1} = 674$

Calculation of fin efficiency:

$$m = \left(\frac{2h'_f}{kb}\right)^{1/2} = \left(\frac{2 \times 674}{43.25 \times 0.889 \times 10^{-3}}\right)^{1/2} = 187.3$$

$$mH = 187.3 \times 0.0125 = 2.34$$

$$\Omega = \frac{\tan\text{h}(mH)}{mH} = 0.419$$

$$A_D = \pi D_o - N_f b = 0.0254\pi - 20 \times 0.889 \times 10^{-3} = 0.0620 \text{ m}^2/\text{m}$$

$$A_f = N_f 2H = 20 \times 2 \times 0.0125 = 0.5 \text{ m}^2/\text{m}$$

$$A_o = \pi D_o = \pi \times 0.0254 = 0.0798 \text{ m}^2/\text{m}$$

$$h'_{fo} = \frac{h'_f(A_D + \Omega A_f)}{A_o} = \frac{674(0.0620 + 0.419 \times 0.5)}{0.0798} = 2,297 \text{ W/(m}^2 \cdot \text{K)}$$

Internal fouling resistance referred to the external diameter is

$$R_{fio} = R_{fi} \times \frac{D_o}{D_i} = 5.3 \times 10^{-4} \times \frac{0.0254}{0.0198} = 6.8 \times 10^{-4} \text{ (m}^2 \cdot \text{K)/W}$$

$$U = \left(R_{fio} + \frac{1}{h_{io}} + \frac{1}{h'_{fo}}\right)^{-1} = \left(6.8 \times 10^{-4} + \frac{1}{1,485} + \frac{1}{2,297}\right)^{-1} = 560 \text{ W/(m}^2 \cdot \text{K)}$$

$$\text{LMTD} = \frac{(200 - 40) - (150 - 20)}{\ln\dfrac{200 - 40}{150 - 20}} = 144.48°\text{C}$$

LMTD correction factor:

$$R = \frac{20 - 40}{150 - 200} = 0.4 \quad \text{and} \quad S = \frac{200 - 150}{200 - 20} = 0.277 \quad F_t \cong 1$$

$$A = \frac{Q}{U(\text{DMLT})} = \frac{1{,}050{,}400}{560 \times 144.48} = 12.98 \text{ m}^2$$

$$L = \frac{A}{\pi D_o N} = \frac{12.98}{\pi \times 0.0254 \times 70} = 2.31 \text{ m}$$

Viscosity correction factor: The hot and cold fluid mean temperatures are

$$\overline{T} = 175°C \quad \text{and} \quad \overline{t} = 30°C$$

Wall temperature at the external side $= \overline{t} + \dfrac{Q}{h_f (A_D + A_f) LN} = 30 + \dfrac{1{,}050{,}400}{1{,}050(0.0617 + 0.5) \times 2.31 \times 70} = 41°C$

Wall temperature at the internal side $= \overline{T} - \dfrac{Q}{h_i A_i NL} = 175 - \dfrac{1{,}050{,}400}{70 \times \pi \times 0.0199 \times 2.31 \times 1{,}930} = 120.8°C$

The gas viscosity at 41°C is sensibly equal to the viscosity at 30°C, so the factor $(\mu/\mu_w)^{0.14}$ is unity. We also shall neglect this correction for the hot oil.

Pressure-drop calculations:

Shell side

$$f = 0.109 \, \text{Re}^{-0.255} = 0.109 \times 171{,}900 - 0.255 = 5.04 \times 10^{-3}$$

$$\Delta p = 4f \frac{L}{D'e} \rho \frac{v^2}{2} = 4 \times 5.04 \times 10^{-3} \times \frac{2.31}{0.0133} \times 30.07 \times \frac{4.98^2}{2} = 1{,}332 \text{ N/m}^2 = 0.013 \text{ bar}$$

Tube side

$$f = 1.2(0.0014 + 0.125/\text{Re}\,0.32) = 6.11 \times 10^{-3}$$

$$\Delta p_t = 4f \frac{Ln}{D_i} \rho \frac{v^2}{2} = 4 \times 6.11 \times 10^{-3} \times \frac{2.3 \times 4}{0.0199} \times \frac{790 \times 1.76^2}{2} = 13{,}800 \text{ N/m}^2$$

$$\Delta p_r = 4n\rho \frac{v^2}{2} = 4 \times 4 \times \frac{790 \times 1.76^2}{2} = 19{,}382 \text{ N/m}^2$$

$$\Delta p = \Delta p_t + \Delta p_r = 33{,}187 \text{ N/m}^2$$

It can be observed that the fins transform the shell-side coefficient from 674 to 2,297 W/(m² · K).

However, the relatively high value of the fouling resistance, in addition to a hot-oil coefficient of 1,539 W/(m² · K), dampens the effect of the fins on the overall heat transfer coefficient.

Since the hot oil is close to its allowable Δp, it is not possible to improve h_i. So it may be necessary to evaluate whether the higher cost from the use of finned tubes still can be an attractive option against the use of a conventional heat exchanger with a larger area.

8-2 AIR COOLERS

Air coolers are heat exchangers in which heat is removed from a process stream in a cooling or condensing operation using air as a refrigerant medium. The obvious advantage over a water-cooled exchanger is that a cooling water circuit is not necessary, thus avoiding the cooling tower, circulation pumps, water-treating systems, and piping.

As a counterpart, air coolers are bigger units and more difficult to allocate in a plant. Air coolers usually are installed on piperacks or at the highest level of process structures to reduce the required ground area.

The cost of these units is usually higher than that of water-cooled shell-and-tube heat exchangers. They are the only option in locations where cooling water is not available, as is usually the situation in oil and gas fields, or in process plants where the cooling-water systems are operating at their maximum capacity and it is necessary to install additional process coolers.

A typical application is as distillation column overhead condensers as an alternative to shell-and-tubes condenser, which will be studied in Chap. 10. They are also used to cool down final products before sending them to storage tanks, as refrigeration-cycle condensers, as interstage coolers in compressor circuits, etc.

Even though they can be constructed with plain tubes, it is usual to employ finned tubes with transverse fins, and this is the reason why they are included in this chapter. The minimum temperature at which the process stream can be cooled down is about 5°C above the ambient air temperature. The air temperature varies during the day and during different seasons of the year, so it is necessary to have meteorologic information to decide which air temperature will be used for design. The usual practice is to choose an air temperature that is not exceeded during a certain percentage of time over the year (e.g., 95 percent of the time). This value is obtained from statistics covering a certain number of years. In very critical applications, it is possible to choose the absolute historical maximum of the site, but it is important to evaluate this subject to avoid unnecessary oversizing.

8-2-1 Components of an Air Cooler

An air cooler is composed of[3,7]

1. One or more tubes bundles
2. Fans to circulate air through the external side of the bundle (air circulation is usually in cross-flow.)
3. A plenum between the fans and the bundle
4. A supporting structure that must be high enough to allow the entrance of air below the bundle with a reasonably low velocity
5. Platforms and ladders for maintenance
6. Optionally, louvers to regulate the airflow or systems to modify the angle of the fan blades to control the process temperature, thus saving electric power (see Fig. 8-4)

Tube Bundle. The tube bundle includes a set of tubes, headers, lateral frame, and tube supports. The tube length is usually defined by layout considerations. As a general rule, the longer the tubes, the lower will be the exchanger cost per unit area, but in very long exchangers with few tubes, it may be difficult to locate the fans to cover the entire surface of the bundle.

Since the film heat transfer coefficients on the air side are usually low, finned tubes are used to increase the heat transfer area on the external side. There are several types of fins. The most common are[6]

1. *Tension-wrapped.* This is the most common type because of economics. It is a rectangular-section aluminum ribbon tension-wrapped around the tube. The contact between the fin and the tube is through the ribbon edge. Since this contact is not perfect, fin efficiency is affected.
2. *Embedded.* These fins are built by helically wrapping a strip of aluminum to embed it in a precut helical groove and then peening back the edges of the groove against the base of the fin to tightly secure it.
3. *Extruded.* These fins are extruded from the wall of an aluminum tube that is integrally bonded to the base tube for the full length.

232   CHAPTER EIGHT

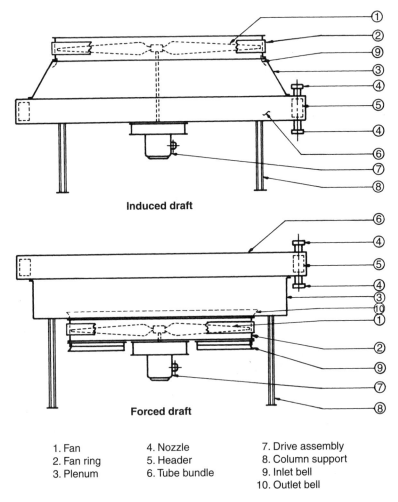

Induced draft

Forced draft

1. Fan
2. Fan ring
3. Plenum
4. Nozzle
5. Header
6. Tube bundle
7. Drive assembly
8. Column support
9. Inlet bell
10. Outlet bell

FIGURE 8-4   Air cooler configurations.
Courtesy of Hudson Products Corporation.

4. *Footed.* These fins are built by wrapping an aluminum strip that is footed at the base as it is wrapped on the tube. The tube is completely covered by the fin feet.

5. *Others.* These are tubes at whose exterior surface fins are fixed by other means, such as brazing, welding, or hot immersion (see Fig. 8-5).

Finned tubes can be built in diameters ranging from ⅝ to 6 in. For air coolers, the most popular diameter is 1 in. Fin heights ranges from ½ to 1 in. (The most popular sizes are ½ and ⅝ in.) Finned tubes are manufactured with 275–433 fins per meter (7–11 fins per inch). The ratio between the fin area and the plain-tube external area (without fins) varies from 7 to 25. The larger the tubes, the lower is the unit cost for the same heat transfer area.

The tube bundle section is rectangular and typically consists of 2 to 10 tube rows in a triangular pattern. The separation between tubes axes is typically 2½ tube diameters. The free area for airflow allowed by the tubes is roughly 50 percent of the projected area.

Temperature

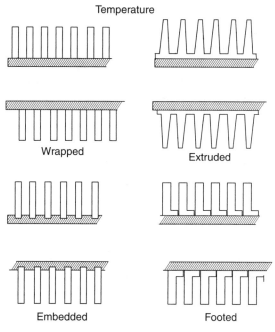

Wrapped

Extruded

Embedded

Footed

FIGURE 8-5   Different fin types.

The tubes are roller-expanded or welded to tubesheets that are part of the headers. The headers are rectangular boxes. One of their faces is the tubesheet, and the opposite face is the cover. The cover can be bolted to allow removal or welded to the header. When the cover is welded, threaded holes are drilled just opposite each tube. The holes are provided with screwed plugs that can be removed for internal tube cleaning (Fig. 8-6).

The headers can have pass-partition plates, as in the case of shell-and-tube heat exchangers. This allows one to increase the process fluid velocity and/or obtain a more countercurrent circulation configuration. The tube bundles are usually installed in a horizontal position, with the air always entering at the lower side and discharging vertically upward. The pass partition can be done horizontally or vertically. Horizontal partitions allow approximating the configuration of a countercurrent pattern, with the process stream entering at the upper pass.

Fans.   Fans are of the axial-flow type. The air cooler is a *forced-draft cooler* when the fan pushes air through the bundle. If the bundle is located at the suction side of the fans, it is called an *induced-draft cooler.*

Fans normally have 2–20 paddles. The fan diameter is related to the bundle width. Usually, the fan diameter is limited to 4 or 5 m, but fans with up to 20-m diameters have been manufactured. Fan paddles can be manufactured in aluminum, fiberglass, plastic, or steel and can be straight or contoured.

The paddles pitch can be adjusted either manually or automatically to regulate the airflow. The automatic control is done with pneumatic devices installed in the fan central hub (that rotate with the fan) and whose air supply is through a rotating coupling. Each bundle is usually air fed by at least two fans. This is a safety measure to avoid complete air loss in case of a fan failure, as well as a means to achieve certain control of the airflow (e.g., switching off one fan in winter).

Fan coverage is defined as the ratio between the projected area of the fans and the projected area of the bundles they serve. This ratio must be 0.4 at a minimum to achieve good air distribution.

Fans are driven by electric motors, and velocity reduction is achieved with reducing gears or V belts. The tangential tip speed always must be lower than 60 m/s for mechanical reasons.

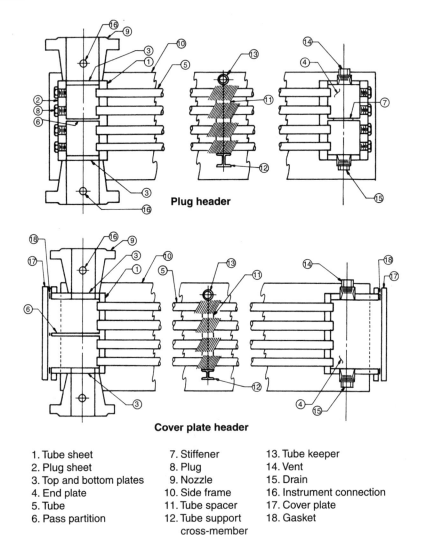

Plug header

Cover plate header

1. Tube sheet	7. Stiffener	13. Tube keeper
2. Plug sheet	8. Plug	14. Vent
3. Top and bottom plates	9. Nozzle	15. Drain
4. End plate	10. Side frame	16. Instrument connection
5. Tube	11. Tube spacer	17. Cover plate
6. Pass partition	12. Tube support cross-member	18. Gasket

Typical construction of tube bundles with plug and cover plate headers

FIGURE 8-6   Header types.
(*Courtesy of Hudson Products Corporation.*)

The fan power requirements depend on the airflow rate and air pressure drop, which is related to the number of tubes in the flow direction. Sometimes it is preferred to split the service into a higher number of fans to avoid high-power motors. Most air cooler fan power is lower than 35 kW.

Plenums.   The air plenum is a totally enclosed space that allows accommodation of the airflow between the fan and the tube bundles to produce a uniform air distribution. Plenums can be designed as pyramidal or straight-section.

Structure.   The structure is formed by columns and beams to support the unit at enough height over ground level to allow the air to enter below the bundle without excessive velocity. In process plants, it is

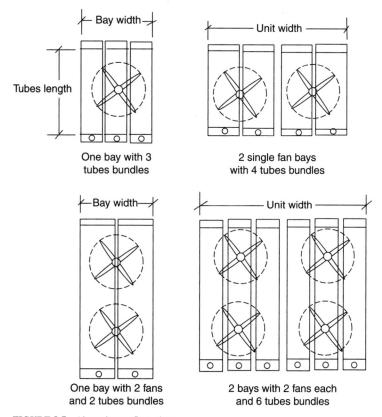

Bay width

Tubes length

One bay with 3
tubes bundles

Unit width

2 single fan bays
with 4 tubes bundles

Bay width

One bay with 2 fans
and 2 tubes bundles

Unit width

2 bays with 2 fans each
and 6 tubes bundles

FIGURE 8-7 Air cooler configurations.

common for air coolers to be installed on the piperacks to avoid occupying ground area. In these cases, the piperack structure supports the air cooler.

Configuration.   The assembly of one or more tube bundles, served by one ore more fans, complete with a plenum, structure, and auxiliary devices, is called a *bay*. When large flow rates are handled, it is common for the unit to be divided in several bays to facilitate transport and construction. Figure 8-7 shows some air cooler configurations.

Forced Draft versus Induced Draft.   Advantages of induced-draft coolers are

1. A better distribution of air on the entire section of the bundle

2. Less possibility of hot-air recirculation to the intake section. (This is so because in these units the hot air is discharged upward at a velocity that is about two or three times the intake velocity, owing to the ratio between the fan area and the cross section of the bundle.)

3. Higher capacity in case of fan failure owing to a natural draft effect. (This effect is much greater in induced-draft units than in forced-draft units.)

4. Less affected by rain, snow, or climatic effects because most of the top face of the bundle is covered

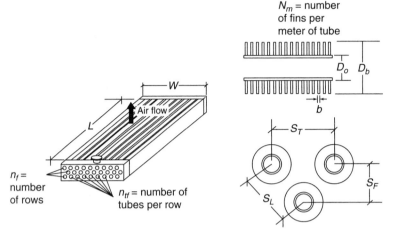

FIGURE 8-8   Air cooler nomenclature.

Disadvantages are

1. Higher power consumption, especially if the air temperature increase is important
2. Air outlet temperature limited to about 90°C to avoid potential damage to the fan components owing to high temperature
3. Fan maintenance more difficult and must be done in the hot air generated by the natural convection effect

8-2-2   Heat Transfer in Air Coolers

Heat transfer in air coolers can be described by the general equation

$$Q = UA\Delta T \tag{8-2-1}$$

But it is necessary to clarify the meaning of the different terms of the equation owing to the particular issues resulting from the presence of fins and a cross-flow configuration. The nomenclature we shall use is explained in Fig. 8-8.

Heat Transfer Area.   We shall call

A_o = plain tube external area per unit length of tube = πD_o.

A_D = exposed area of tube (not covered by fins) per unit tube length = $A_o(1 - bN_m)$.

A_f = fin area per unit tube length = $\pi 2 N_m (D_b^2 - D_o^2)/4$.

Usually the heat transfer area in Eq. (8-2-1) is taken as the plain-tube area. This is

$$A = A_o L n_f n_{tf} \tag{8-2-2}$$

Overall Heat Transfer Coefficient U.   It was already explained in Sec. 8-1 that the fin surface behaves differently from the exposed tube surface, and in order to define a film heat transfer coefficient for the external side, a fin efficiency Ω must be considered, so

$$h'_{fo} = h'_f (A_D + \Omega A_f)/A_o \tag{8-2-3}$$

The meaning of the fin efficiency was explained in Sec. 8-1. It depends on the geometry and on the parameter $(bh'_f/k)^{1/2}$, where h'_f is the film coefficient for the external surface of the fins, and k is the thermal conductivity of the fin. The higher the thermal conductivity of the material, the higher is the fin efficiency.

Additionally, the lower the external film coefficient h'_f, the higher is the fin efficiency. This is so because if h'_f is intrinsically high; a significant increase in the heat transfer rate is not achieved with the use of fins. On the other hand, for low heat transfer coefficients, as is the case with air coolers, the improvement is considerable.

The value of h'_f includes the effect of fouling resistance on the external side. Usually this resistance is neglected because it may be considered that air is a clean fluid, but if it is desired to include this effect, it must be calculated as

$$\frac{1}{h'_f} = \frac{1}{h_f} + R_{fo} \tag{8-2-4}$$

where h_f is the clean coefficient.

To calculate the overall heat transfer resistance, the internal film coefficient and internal fouling resistance are included, and we get

$$\frac{1}{U} = \frac{1}{h'_{fo}} + \frac{1}{h_{io}} + R_{fio} \tag{8-2-5}$$

where h_{io} is the internal coefficient referred to the external tube area, $h_{io} = h_i \times D_i/D_o$, and R_{fio} is the internal fouling resistance, also referred to the *external diameter* ($R_{fio} = R_{fi} \times D_o/D_i$). This correction is usually neglected.

Fin Efficiency.   In Sec. 8-1 we performed the analytical derivation to calculate the efficiency of a longitudinal fin. Air coolers have transverse fins, and the mathematical deduction of fin efficiency is more complex. The analytical expressions for different fin geometries can be found in ref. 4.

The graph in Fig. 8-9 allows one to obtain the efficiency of constant-thickness circular fins, which are the most common type. The nomenclature for the geometric parameters can be found in Fig. 8-8. Figure 8-10 includes the mathematical expressions of these curves.[4]

To obtain the fin efficiency, it is necessary first to calculate the film coefficient h_f. This calculation will be explained later.

Temperatures Difference.   In an air cooler, air circulates in a perpendicular direction to the tube-side fluid. The tube-side fluid, in turn, flows trough several tube rows (usually between two and eight) and can have several tube passes. Let's consider an air cooler such as that shown in Fig. 8-11, where the tube-side fluid performs only one pass, so all the tubes are in parallel.

We also can assume that the air velocity only has a component in the direction perpendicular to the tubes, so there is no effect of longitudinal mixing. This can be interpreted as if fictitious baffles (as indicated by the vertical lines in the figure) are present. The inlet temperatures of the hot fluid and air are T_1 and t_1, respectively.

At point A, at the first horizontal tube row, and in correspondence with plane 1–1', the temperature difference between the two streams is $T_1 - t_1$. If we move to plane 2–2' , in position A', there will be a smaller temperature difference between the hot fluid and the inlet air because the process fluid temperature decreases. Similarly, at plane 3–3', in position A'', the temperature difference will be even smaller, and it continues to decrease as we move along the tubes in the first row.

Let's now analyze the second tube row. The air temperature leaving the first row varies along the tube because the heat transferred from the first row varies from point to point as the ΔT for heat transfer decreases. At the second row, the temperature differences also vary, and the temperature changes experienced by the air are different from those in the preceding row. This means that in every tube segment, defined by its longitudinal position and by the tube row to which it belongs, the rate of heat transfer is different, and the temperature change of each fluid is also different. The mean outlet temperature of the process stream will be the mixing temperature of the streams coming from all tube rows.

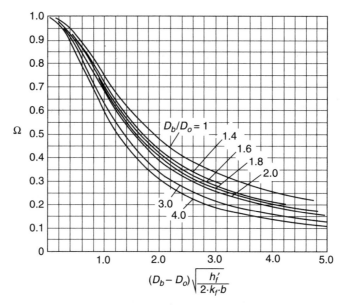

FIGURE 8-9   Efficiency of annular fins of constant section.

$$m = \sqrt{\frac{2 \cdot h_f'}{k_f \cdot b}} \qquad H = \frac{D_b - D_o}{2}$$

$$Y = \left(H + \frac{b}{2} \right)\left(1 + 0.35 \ln\frac{D_b}{D_o} \right)$$

$$\Omega = \frac{\tan h(mY)}{mY}$$

FIGURE 8-10   Mathematical expressions for the curves in Fig. 8-9.

To calculate the total heat transferred in the unit, it is possible to develop mathematical models that predict the temperatures evolution with finite-increments techniques, analytical solutions, or iterative methods. The most common approach is to adopt the same methodology used in shell-and-tube heat exchangers and calculate the heat transfer rate by defining a factor F_t as

$$Q = UA(\text{LMTD})F_t \tag{8-2-6}$$

where LMTD is the logarithmic mean temperature difference assuming a countercurrent configuration. The correction factor F_t can be calculated with the methods mentioned earlier.

Another possible model we may have used is to consider that the air suffers axial mixing after each tube row and adopts a uniform temperature before passing to the next row. In this case, the air temperature is a function of the position of the row under consideration but not of the longitudinal coordinate. With this assumption, the mathematical derivations of the F_t relations are simpler and can be obtained analytically. However, in long-tube air coolers, this hypothesis is far from reality, and the nonmixed model must be used.

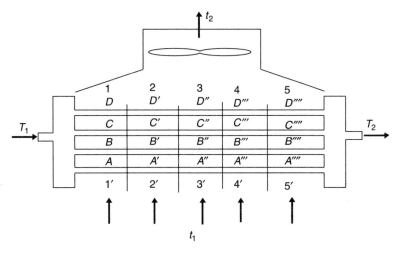

FIGURE 8-11 Single-pass air cooler with no longitudinal mixing of the air stream.

In the configuration in Fig. 8-11, the tube-side fluid performs only one pass through the tubes. But usually, multipass units such as that represented in Fig. 8-12 are employed. In this case, the mathematical model is even more complicated, and the F_t value is different from that of a single-pass unit.

A mathematical model to calculate the true temperature difference in this type of equipment, with different numbers of tube passes, for mixed and unmixed airflow was proposed by Cordero and Pignotti.[5] The graphs in App. C allow calculation of the F_t factors using the same parameters as those employed for shell-and-tube heat exchangers:

$$R = \frac{T_1 - T_2}{t_2 - t_1} \qquad (8\text{-}2\text{-}7)$$

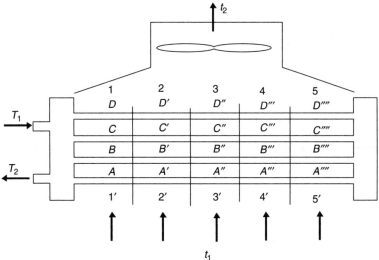

FIGURE 8-12 Double-pass air cooler with no longitudinal mixing of the air stream.

$$S = \frac{t_2 - t_1}{T_1 - t_1} \tag{8-2-8}$$

These graphs are presented for configurations with one, two, or three tube passes. If the unit has more than three passes, F_t is practically 1.

The passes performed by the hot fluid always must have opposite direction to the airflow. This means that the hot-fluid inlet is always in the upper pass, as shown in Fig. 8-12.

Calculation of the Heat Transfer Coefficient. A method to calculate the film heat transfer coefficients and the pressure drop for a finned tube bundle was proposed by Kern[1] and shall be described below. The method is valid for a tube bank with transverse fins of constant thickness in which the tubes are arranged in a triangular pattern, as shown in Fig. 8-8. This is the geometry usually employed in air coolers.

The following nomenclature for the heat transfer coefficients will be used:

h_f = film coefficient for the air side, for the tube and fin surface, without including either fouling effects or the different efficiency of the fin surface with respect to the bare-tube portion.

h'_f = coefficient h_f corrected for external fouling. It is related to h_f by the expression

$$\frac{1}{h'_f} = \frac{1}{h_f} + R_{fo} \tag{8-2-9}$$

where R_{fo} is the external fouling resistance. It is frequently assumed that air is a clean fluid, and then, $h_f = h'_f$.

h'_{fo} = film coefficient corrected for fin efficiency, and considering as reference area the external surface of the plain tube,

$$h'_{fo} = (\Omega A_f + A_D)\frac{h'_f}{A_o} \tag{8-2-10}$$

h_i = internal heat transfer coefficient (process side). This is calculated with the correlations included in Chaps. 4 and 10 for forced convection and condensation depending on the case.

h'_i = internal coefficient corrected for fouling

R_{fi} = the internal fouling resistance. Then

$$\frac{1}{h'_i} = \frac{1}{h_i} + R_{fi} \tag{8-2-11}$$

h'_{io} = internal heat transfer coefficient corrected for fouling, referred to the exterior area of the tube, $h'_i(D_i/D_o)$.

U = overall heat transfer coefficient based on the external area of the tubes. Thus

$$\frac{1}{U} = \frac{1}{h'_{io}} + \frac{1}{h'_{fo}} \tag{8-2-12}$$

The film coefficient h_f is obtained with the following correlation:

$$j = \frac{h_f D_e}{k}\mathrm{Pr}^{-1/3} = 0.0959(\mathrm{Re}_S)^{0.718} \tag{8-2-13}$$

In this expression, D_e is an equivalent diameter for heat transfer defined as

$$D_e = \frac{2(A_f + A_D)}{\pi(\text{projected perimeter})} \tag{8-2-14}$$

Projected perimeter is the sum of all external dimensions in the plant view of a finned tube per unit length. That is,

$$\text{Projected perimeter} = 2(D_b - D_o)N_m + 2(1 - bN_m) \tag{8-2-15}$$

The Reynolds number for heat transfer is defined as

$$\text{Re}_S = \frac{D_e G_S}{\mu} = \frac{D_e W_c}{a_S \mu} \tag{8-2-16}$$

where W_c is the airflow (air is always the cold stream) and a_S is the flow area, which is calculated by subtracting the projected area obstructed by the tubes from the cross section of the bay a_F ($a_F = WL$).

$$a_S = WL - n_{tf} L[D_o + N_m(D_b - D_o)b] \tag{8-2-17}$$

n_{tf} is the number of tubes in a row. If even and odd tubes rows have different number of tubes, a mean value is used.

Fin Efficiency. The fin efficiency is calculated with the graph of Fig. 8-9 or the analytical expressions of Fig. 8-10. For other geometries, ref. 1 can be used.

Calculation of Air Pressure Drop. The pressure drop of the air through the bundle can be calculated as

$$\Delta p = \frac{f G_S^2 L_p}{2\rho D'_e} \left(\frac{D'_e}{S_T}\right)^{0.4} \left(\frac{S_L}{S_T}\right)^{0.6} \tag{8-2-18}$$

where f = friction factor
$\quad G_S$ = mass flow density, the same as defined for heat transfer coefficient calculation ($G_S = W_c/a_S$)
$\quad L_p$ = length of path = $n_f S_F$
$\quad S_F$, S_L, S_T see Fig. 8-8
$\quad D'_e$ = equivalent diameter for friction

D'_e is different from the equivalent diameter used for heat transfer coefficient calculations and is defined as

$$D'_e = \frac{4 \times \text{net free volume}}{L n_{tf}(A_f + A_D)} \tag{8-2-19}$$

The net free volume is the volume between the central planes of two consecutive rows minus the portion of that volume occupied by the tubes and fins. That is,

$$\text{Net free volume} = WLS_F - n_{tf}\pi\frac{D_o^2}{4}L - N_m n_{tf} L\pi\left(\frac{D_b^2 - D_o^2}{4}\right)b \tag{8-2-20}$$

The friction factor f is obtained as a function of a Reynolds number Re'_S, defined as

$$\text{Re}'_S = \frac{D'_e G_S}{\mu} = \frac{D'_e W_c}{a_S \mu} \tag{8-2-21}$$

And the correlation is

$$f = 1.276 \, \text{Re}'_S{}^{-0.14} \tag{8-2-22}$$

Fan Power Consumption.   The power consumption of each fan can be calculated as

$$\text{Power} = \frac{\text{volumetric flow of each fan} \times \text{pressure developed}}{\text{fan efficiency} \times \text{driver efficiency}} \tag{8-2-23}$$

The volumetric flow of the fan is the air-mass flow (kg/s) divided by the air density at the fan ρ_{fan} (kg/m^3):

$$\text{Fan volumetric flow} = \frac{(W_c/\text{number of fans})}{\rho_{\text{fan}}} = \frac{(W_c/\text{number of fans}) R T_{\text{fan}}}{M_A p} \tag{8-2-24}$$

In induced-draft fans, the air temperature at the fan is higher than the ambient air temperature; thus its density is lower, and the volumetric flow is higher than in the case of a forced draft. This means that the power ratio between an induced-draft fan and a forced-draft fan is approximately equal to the ratio of the outlet and inlet absolute air temperatures.

The pressure developed by the fan is the sum of the velocity head ($\rho_{\text{fan}} v_{\text{fan}}{}^2/2$) and the frictional pressure drop through the bundle, calculated as indicated earlier. The air velocity at the fan depends on the adopted fan diameter. The smaller this diameter, the higher will be the velocity for the same volumetric flow.

The efficiency of the fans employed in air coolers is usually about 75 percent, whereas the efficiency of the electric motor and transmission system is about 95 percent.

8-2-3 Detailed Design Procedure

Preliminary Design.   As usual in heat transfer equipment, the design procedure of an air cooler consists of proposing a preliminary design and verifying its heat transfer capacity by calculating the heat transfer coefficient and the corrected mean temperature different, as was explained in the preceding section. Depending on the results of the verification, the proposed geometry or the airflow will be modified in an attempt to obtain a solution with an acceptable combination of heat transfer area, fan power, and process-side pressure drop.

It thus is necessary to have guidelines to set up a preliminary design that can be used as starting point. The design input data usually consist of the following:

1. Hot-fluid process conditions (W_h, T_1, and T_2)

2. Inlet air temperature t_1

3. Some restriction or preference regarding the footprint. It is possible that the width or length of the unit will be subject to some restriction or be defined for layout considerations. When the size of the unit is large, the more economical solution is obtained with the longest acceptable tubes. If the unit will be mounted on a pipe rack, the length is fixed and must coincide with the width of the pipe rack.

4. The characteristics of the finned tubes (i.e., tube diameter, fin diameter and thickness, number of fins per unit tube length, and fin material) usually are adopted from the beginning and are seldom modified during the design process. They are usually defined by the project specifications or by the manufacturing workshop facilities. The most popular finned tube is 1 in with aluminum fins ⅝-in height with 11 fins per inch (433 fins per meter), and the tubes are arranged in a triangular equilateral pattern separated by 2½ in between axes.

5. Sometimes there are also preferences regarding the maximum power per fan, which in some cases can affect unit design, requiring a higher number of fans.

To come up with a preliminary design, a first guess of the heat transfer coefficient is assumed, and an air face velocity and number of tube rows are adopted based on typical values for different applications.

TABLE 8-1 Typical Values of the Overall Heat Transfer Coefficients in Air Coolers [W/(m² · K)], Using 5/8-in Aluminum Fins in 1-in OD Tubes with 393 Fins per Meter

Condensation		Gas Cooling		Liquid Cooling	
Amine regenerator	570–670	Air or combustion gas at 50 psig (ΔP = 1 psi)	60	Machine cooling water	740–900
Ammonia	600–700	Air or combustion gas at 100 psig (Δp = 2 psi)	112	Fuel oil Re-forming or	115–170 480
Refrigerant 12	420–500	Air or combustion gas at 100 psig (Δp = 5 psi)	170	platforming liquids	
Heavy naphtha	400–500	Hydrocarbon gas at 15–50 psig (Δp = 1 psi)	170–220	Light gas oil	450–550
Light gasoline	540	Ammonia reactor stream	500–600	Light hydrocarbons	510–680
Light hydrocarbons	540–600	Gas hydrocarbon at 15–50 psig (Δp = 1 psi)	170–220	Light naphtha	510
Light naphtha	450–550	Gas hydrocarbon at 50–250 psig (Δp = 3 psi)	280–340	Process water	680–820
Reactor effluent, re-forming, platforming	450–550	Gas hydrocarbon at 250–1500 psig (Δp = 5 psi)	400–500	Distillation residue Tar	60–120 30–60
Steam	800–1200				

(Courtesy of Hudson Products Corporation.)

Table 8-1 shows some typical values of the overall heat transfer coefficient for different process services that can be used as an initial estimation. The number of tube rows is related to the temperatures approximation between the air and the process flow. A parameter Z is defined as

$$Z = \frac{T_1 - T_2}{T_1 - t_1} \tag{8-2-25}$$

This parameter can be calculated from the process conditions. The higher the value of Z, the deeper must be the tube bundle within practical and economic considerations. Since an isothermal condenser has $Z = 0$, it should have a low depth, typically no more than four rows. A process cooler with a high temperature range (e.g., $Z > 0.8$), on the other hand, should be as deep as possible, say, 8–10 tube rows.

The air face velocity V_F, defined as

$$V_F = \frac{W_c}{\rho_A W L} \tag{8-2-26}$$

is also related to the number of rows because if the number of rows is high, the velocity should be reduced to avoid excessive air pressure drop. Table 8-2 can be used for the preliminary design.

The procedure is

1. Calculate the heat duty Q.
2. Adopt a heat transfer coefficient with the help of Table 8-1.
3. Calculate Z.
4. From Table 8-2, obtain the number of tube rows n_f and air face velocity v_F.
5. Assume an air outlet temperature. This temperature must be adopted considering the process temperatures. In multipass heat exchangers, some temperature cross is possible. In single-pass units, the air outlet temperature must be lower than the process outlet temperature.
6. Calculate the air mass flow as

$$W_c = \frac{Q}{c_c(t_2 - t_1)} \tag{8-2-27}$$

TABLE 8-2   Design Recommendations as a Function of Z Parameter

Z	n_f = Number of Rows	V_F (m/s)
0.4	4	3.3
0.5	5	3
0.6	6	2.8
0.8–1	8–10	2–2.4

(Courtesy of Hudson Products Corporation.)

7. Calculate the face area ($a_F = WL$) as

$$a_F = \frac{Wc}{\rho_A v_F} \qquad (8\text{-}2\text{-}28)$$

8. Select W or L according to layout convenience, and calculate the other one. The resulting unit should be proportionate. If the unit is excessively long and narrow, it will be necessary to install more fans to adequately cover the surface of the bundle. The more usual solution is to install two fans per bay, so it may be convenient to have an L/W ratio approximately equal to 2.5.

9. Calculate the LMTD. As a first approximation, the correction factor can be neglected, unless an air outlet temperature considerably higher than the process outlet temperature was adopted.

10. Calculate the plain tube area required for heat transfer as

$$A = \frac{Q}{U(\text{MLDT})} \qquad (8\text{-}2\text{-}29)$$

11. Calculate the required number of tubes as

$$N = \frac{A}{\pi D_o L} \qquad (8\text{-}2\text{-}30)$$

12. Calculate the number of tubes per row n_{tf} as

$$n_{tf} = \frac{N}{n_f} \qquad (8\text{-}2\text{-}31)$$

13. With the separation between tubes axes S_T (see Fig. 8-8), verify that

$$W = S_T n_{tf} \qquad (8\text{-}2\text{-}32)$$

In cases of agreement, proceed with the detailed design. Otherwise, modify the geometry within the design restrictions.

Detailed Design.   The following steps should be followed. We assume that the tube length was adopted, which is the case when the unit will be mounted on a pipe rack.

1. With the finally proposed geometry, recalculate the number of tubes $N = A/\pi D_o$.

2. n_{tf} = number of tubes per row = N/n_f.

3. W = unit width = $S_T \times n_{tf}$. If W is larger than the maximum bay width, which is usually limited for transport and erection considerations, it may be necessary to divide the unit in two or more bays. Each bay will handle part of the job. The reader can adapt the procedure for this case.

4. Airflow = $W_c = v_F \rho_A WL$.

5. Air outlet temperature = $t_2 = t_1 + Q/W_c c_c$.

6. Adopt the number of hot-fluid passes n.
7. Calculate F_t and LMTD.
8. Calculate the hot-fluid velocity into the tubes.
9. With suitable correlations, calculate the heat transfer coefficient h_{io} and pressure drop Δp_t.
10. In case $\Delta p_t > \Delta p$ allowable, if the number of tube passes can be reduced without seriously affecting h_{io}, or it is considered that h_{io} will not be controlling, reduce n and go back to step 7. If it is not possible to reduce the number of passes, it will be necessary to increase the number of tubes (thus increasing the area) and go back to step 1.
11. Calculate the external-side coefficient h_o and the overall coefficient U.
12. Verify $A = Q/U\Delta T$.
13. If the calculated area is larger than the proposed area, investigate if it is possible to improve h_{io} and U by increasing the number of tube passes without compromising the tube-side pressure drop. If it is not possible to increase the number of passes, it will be necessary to increase the heat transfer area A and go back to step 1. If the calculated area is considerably lower than the proposed area, the design must be changed by reducing the number of tubes.
14. Calculate the air-side pressure drop and fan power. Distribute the total service in a suitable number of fans with diameters that allow covering at least 40 percent of the bay face section.

In this procedure we have assumed the tube length, number of rows, and air face velocity are constant. We obviously assume that these values have been selected with good criteria. If the final design results are unsatisfactory (e.g., an excessively long and narrow unit), the designer should think about modifying these parameters as well.

The effect of airflow on the design also must be investigated. If the temperature approach between air and process is small, an increase in the airflow (or v_F) may have a considerable impact on the LMTD, and considerable savings in area can be obtained.

The graphs shown in Figs. 8-13 and 8-14 offer a tool to perform a fast design. They represent the value of h'_{fo} and pressure drop per tube row as a function of the face velocity v_F, calculated with the correlations presented earlier, for two of the most popular tube pattern (1-in tubes with aluminum fins 15.87 mm high and 0.381 mm thick with 394 or 433 fins per meter in a triangular pattern with 60.32 mm of pitch).

Example 8-2 Design an air cooler to cool down 180,000 kg/h (50 kg/s) of heavy-gas oil from 76°C to 66°C. The maximum allowable tube length is 7 m. The tubes will be 25.4 mm OD and 19.3 mm ID. They will be provided with aluminum fins 15.875 mm in height (5/8 in). The fin thickness will be 0.381 mm, and the tubes will have 394 fins per meter. The tubes will be arranged in a triangular pattern with 60.32 mm of pitch.

The air design temperature is 30°C. A pressure drop of 50 kPa is allowed for the process side. An internal fouling resistance of 0.0003 (m² · K)/W must be used. Fouling is considered negligible on the air side.

The following physical properties will be used:

	Product	Air
Density, kg/m³	770	
Viscosity, cP	5.6	0.0209
Thermal conductivity, W/(m · K)	0.14	0.0267
Specific heat, J/(kg · K)	2452	1011

The unit will be installed at sea level (atmospheric pressure 101.3 kPa).

Note: A downloadable interactive spreadsheet with the calculations of this example is available at http://www.mhprofessional.com/product.php?isbn=0071624082

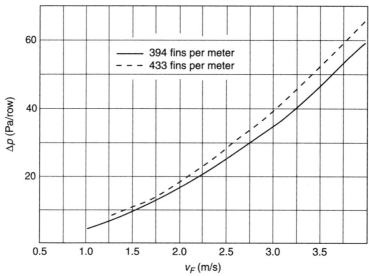

FIGURE 8-13   Pressure drop per tube row.

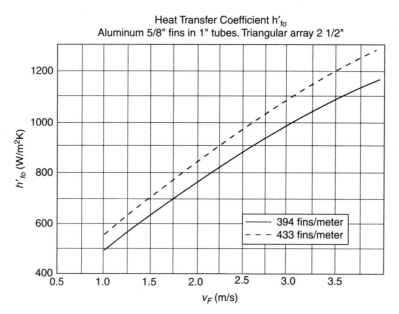

FIGURE 8-14   Heat transfer coefficient.

Solution   *Preliminary design:* We shall follow the steps indicated earlier in the text.

1. $Q = W_h c_h (T_1 - T_2) = 50 \times 2452 \times (76 - 66) = 1.226 \times 10^6$ W
2. This particular application is not specifically mentioned in Table 8-1, but we can assume that the value of the overall heat transfer coefficient will be higher than 170 W/(m² · K) (corresponding to fuel oil) and lower than 450 W/(m² · K) (corresponding to light-gas oil). We shall adopt 250 W/(m² · K).
3. $Z = (T_1 - T_2)/(T_1 - t_1) = 10/(76 - 30) = 0.22$
4. We shall adopt four tube rows and a face velocity of 3.3 m/s (see Table 8-2).
5. We shall tentatively assume an air outlet temperature of 45°C.
6. The airflow will be

$$W_c = Q/c_c (t_2 - t_1) = 1.226 \times 10^6/(1{,}011 \times 15) = 80.84 \text{ kg/s}$$

7. The inlet air density is $\rho = Mp/RT = 29 \times 1/0.082 \times 303 = 1.16$ kg/m³. The face area to get a face velocity of 3.3 m/s is

$$a_F = W_c / v_F \rho = 80.84/(3.3 \times 1.16) = 21.1 \text{ m}^2$$

8. Using 7-m tubes, the width of the unit is $W = a_F/L = 21.1/7 = 3.02$ m. The geometry looks reasonable.
9. LMTD $= [(T_1 - t_2) - (T_2 - t_1)]/\ln(T_1 - t_2)/(T_2 - t_1) = [(76 - 45) - (66 - 30)]/\ln(31/36) = 33.4°C$
10. $A = Q/(U \cdot \text{LMTD}) = 1.226 \times 10^6/(250 \times 33.4) = 146.8$ m²
11. $N = A/(\pi D_o L) = 146.8/(3.14 \times 0.0254 \times 7) = 262$
12. $n_{tf} = 262/4 = 65.5$
13. For the proposed geometry, $S_L = S_T = 0.0603$ m. Then $W = S_T \times n_{tf} = 65.5 \times 0.0603 = 3.94$ m.

This is larger than required for a face velocity of 3.3 m/s. For a first trial, we shall keep this geometry, and we shall correct the airflow, that is,

$$W_c = v_F \times \rho \times W \times L = 3.3 \times 1.16 \times 3.94 \times 7 = 105.5 \text{ kg/s}$$

The air outlet temperature then will be

$$t_2 = t_1 + Q/W_c c_c = 30 + 1.226 \times 10^6/(105.5 \times 1011) = 41.5$$

$$\text{LMTD} = [(76 - 41.5) - (66 - 30)]/\ln(34.5/36) = 35.2$$

Detailed design: The values adopted for the first trial are

1. $N = 262$
2. $n_{tf} = 65.5$ (two rows with 65 and two rows with 66)
3. $W = 3.94$ m
4. $Wc = 105.5$ kg/s
5. $t_2 = 41.5$
6. We shall assume two passes for the hot fluid.
7. LMTD = 35.2
$$R = (T_1 - T_2)/(t_2 - t_1) = 10/11.5 = 0.87$$
$$S = (t_2 - t_1)/(T_1 - t_1) = 11.5/46 = 0.25$$
$$F_t = {\sim}1. \text{ Then } \Delta T = 35.2°C.$$

Calculations for the hot fluid: The flow area for the tube-side fluid is

$$a_t = N(\pi D_i^2/4)/n = 262(3.14 \times 0.0193^2/4)/2 = 0.0380 \text{ m}^2$$

The velocity of the fluid into the tubes is

$$v_t = W_h/(\rho \times a_t) = 50/(770 \times 0.038) = 1.7 \text{ m/s}$$

9. $Re_t = D_i v_t \rho/\mu = 0.0193 \times 1.7 \times 770/5.6 \times 10^{-3} = 4{,}511$

$$f = 1.2(0.0014 + 0.125/Re^{0.32}) = 0.0118$$

$$\Delta p_t = 4fn(L/D_i)\rho v_t^2/2 = 4 \times 0.0118 \times 2 \times (7/0.0193) \times 770 \times 1.7^2/2 = 38{,}095 \text{ Pa}$$

$$\Delta p_r = 4n\rho v_t^2/2 = 4 \times 2 \times 770 \times 1.7^2/2 = 8{,}900$$

$$\Delta p = 46{,}996 \text{ Pa}$$

10. We see that the flow regime corresponds to the transition zone, which is not recommended. However, to achieve a Reynolds number of 10,000, it would be necessary to double the velocity, which would make the pressure drop higher than allowable. In this case, the pressure drop is close to the allowable, so it is not possible to improve the internal film coefficient.

For the transition regime, the correlation is

$$h_i = 0.116c\rho v\left(\frac{Re^{0.66}-125}{Re}\right)[1 + D_i/L)^{0.66}]Pr^{-0.66}$$

$$= 0.116 \times 2{,}452 \times 770 \times 1.7\left(\frac{4{,}511^{0.66}-125}{4{,}511}\right)\left[1+(0.0193/7)^{0.66}\right]\left(\frac{2{,}452 \times 5.6e^{-3}}{0.14}\right)^{-0.66}$$

$$= 534 \text{ W/(m}^2\cdot\text{K)}$$

$$h_{io} = h_i \times D_i/D_o = 534 \times 0.0193/0.0254 = 405 \text{ W/(m}^2 \cdot \text{K)}$$

Calculations for the air side: The coefficient h'_{fo} can be obtained directly from Fig. 8-14 (because the tube pattern coincides with the parameters of the graph) with a face velocity of 3.3 m/s. However, in order to illustrate how to proceed in a general case, we shall use the complete procedure.

A_o = plain tube area per meter of tube = $\pi D_o = 3.14 \times 0.0254 = 0.0798 \text{ m}^2/\text{m}$

A_D = bare tube surface per meter of tube = $A_o(1 - bN_m) = 0.0798(1 - 0.000381 \times 394) = 0.0678 \text{ m}^2/\text{m}$

A_f = fin surface = $\pi.2N_m(D_b^2 - D_o^2)/4 = 3.14 \times 2 \times 394 \times (0.05715^2 - 0.0254^2)/4 = 1.621 \text{ m}^2/\text{m}$

Projected perimeter = $2(D_b - D_o)N_m + 2(1 - bN_m) = 2(0.05715 - 0.0254) \times 394 + 2(1 - 0.000381 \times 394) = 26.718 \text{ m}$

Equivalent diameter = $2(A_f + A_D)/\pi(\text{projected perimeter}) = 2 \times (1.621 + 0.0678)/(\pi \times 26.718) = 0.0403 \text{ m}$

$a_S = WL - n_{tf}L[D_o + N_m(D_b - D_o)b] = 3.94 \times 7 - 65.5 \times 7[0.0254 + 394 \times (0.05715 - 0.0254) \times 0.000381] = 13.748 \text{ m}^2$

Re_S (heat) = $D_e W_c/a_S\mu = 0.0403 \times 105.5/(13.748 \times 0.0209 \times 10^{-3}) = 14{,}784$

$j = 0.0959Re_S^{0.718} = 0.0959 \times 14784^{0.718} = 94.57$

$Pr - c_c\mu/k = 1{,}011 \times 0.0209 \times 10^{-3}/0.0267 - 0.79$

Since a fouling factor on the air side is not considered,

$$h_f = h'_f = j(k/D_e)Pr^{1/3} = 94.57 \times (0.0267/0.0403) \times 0.79^{1/3} = 58.1 \text{ W/(m}^2 \cdot \text{K)}$$

Calculation of the fin efficiency:

$$m = (D_b - D_o)[h'_f/(2k_fb)]^{1/2}$$

$$k_f = \text{aluminum thermal conductivity} = 200 \text{ W/(m} \cdot \text{K)}$$

$$m = (0.05715 - 0.0254)[58.1/(2 \times 200 \times 0.000381)]^{1/2} = 0.62$$

The fin efficiency can be calculated with Fig. 8-10 or with the analytical expressions of Fig. 8-11. Thus

$$m = \sqrt{\frac{2 \times 58.1}{200 \times 0.000381}} = 39.05$$

$$H = \frac{0.05715 - 0.0254}{2} = 0.01587$$

$$Y = \left(0.01587 + \frac{0.000381}{2}\right)\left(1 + 0.35\ln\frac{0.05715}{0.0254}\right) = 0.0206$$

$$\Omega = \frac{\tanh mY}{mY} = 0.83$$

External coefficient:

$$h'_{fo} = h'_f(A_D + \Omega Af)/A_o = 58.1(0.0678 + 0.83 \times 1.621)/0.0798 = 1{,}028 \text{ W/(m}^2 \cdot \text{K)}$$

Overall coefficient:

$$U = (1/h'_{fo} + 1/h_{io} + R_{fi})^{-1}$$

The internal fouling resistance is corrected for the internal/external diameters ratio. Then

$$R_{fi} = 0.0003 \times (0.0254/0.0193) = 0.00039 \text{ (m}^2 \cdot \text{K)/W}$$

Then

$$U = (1/405 + 1/1{,}028 + 0.00039)^{-1} = 261 \text{ W/(m}^2 \cdot \text{K)}$$

This reasonably agrees with the assumed 250 W/(m^2 · K), indicating a slight area excess.

It is important to note that the controlling resistance is that of the internal fluid, and it was calculated using the correlation for transition regime. Considering a safety margin in the heat transfer area may be a good idea in these cases.

Calculation of the fan power: We must calculate the pressure drop of the air when flowing past the tubes. To calculate the equivalent diameter for friction, we must first calculate the net free volume of the bundle with Eq. (8-2-20). For this geometry, $S_F = 52.23$ mm. Thus

Net free volume = $WLS_F - n_{tf}\pi LD_o^2/4 - N_m n_{tf} L\pi b(D_b^2 - D_o^2)/4 = 3.93 \times 7 \times 0.05223 - 65.5 \times 3.14 \times 7 \times 0.0254^2/4 - 394 \times 65.5 \times 7 \times 3.14 \times 0.000381(0.05715^2 - 0.0254^2)/4 = 1.066 \text{ m}^3$

Equivalent diameter $= D'_e = 4 \times$ net free volume$/[Ln_{tf}(A_f + A_o)] = 4 \times 1.066/[7 \times 65.5(1.621 + 0.0798)] = 0.0055$ m

$$Re'_S = D'_e Wc/a_S\mu = 0.0055 \times 105.5/(13.748 \times 0.0209 \times 10^{-3}) = 2022$$

$$f = 1.276 Re'^{0.14}_S = 0.440$$

Pressure drop through the bundle: Mean air density in the tube bundle (at an average temperature of 35.7°C) $= 1.146$ kg/m^3. Thus

$$\Delta p = \frac{fG_S^2 n_f S_F}{2\rho D'_e}\left(\frac{D'_e}{S_T}\right)^{0.4}\left(\frac{S_L}{S_T}\right)^{0.6} =$$

$$= 0.44 \times \frac{(105.5/13.748)^2 \times 4 \times 0.05223}{2 \times 1.146 \times 0.0055}\left(\frac{0.0055}{0.06032}\right)^{0.4}\left(\frac{0.06032}{0.06032}\right)^{0.6} = 164.7 \text{ Pa}$$

This is $164.7/4 = 41$ Pa per row. We could have obtained this value from Fig. 8-13 for a face velocity of 3.3 m/s.

The pressure that must be developed by the fans is the sum of frictional pressure drop and velocity pressure:

$$\text{Pressure developed} = \Delta p + \tfrac{1}{2}\rho_{fan} v_{fan}^2$$

If forced draft is used, the density of the air at the fan corresponds to the inlet temperature:

$$\rho_{fan} = Mp/RT = 29 \times 101{,}300/8{,}306 \times 303 = 1.167 \text{ kg/m}^3$$

If two fans of 2.5 m diameter are used, the air velocity at the fan is

$$v_{fan} = W_c/2\rho_{fan}(\tfrac{1}{4}\pi \times 2.5^2) = 105.5/(2 \times 1.167 \times 4.906) = 9.21 \text{ m/s}$$

The pressure developed by the fans thus will be

$$164.7 + \tfrac{1}{2} \times 1.167 \times 9.21^2 = 214.2 \text{ Pa}$$

The volumetric flow in the fans will be $W_c/(2\rho_{fan}) = 105.5/(2 \times 1.167) = 45.2$ m^3/s.
The driver power will be

Volumetric flow $\times$ pressure developed/(fan efficiency $\times$ driver efficiency) $= 45.2 \times 214.2/(0.75 \times 0.95) = 13{,}600$ W $= 13.6$ kW

The ratio between the projected area of the fan and the face area of the bay should be at a minimum 40 percent to have a uniform distribution.

Projected area of the fans: $2 \times 4.906 = 9.81$ m^2.

Face area $= 21.1$ m^2.

$$9.81/21.1 = 0.46 > 0.40 \qquad \text{OK}$$

Temperature Control of the Process Side.   If the temperature difference between the process side and air is small, the changes in air temperature may have a pronounced effect on the unit capacity. Usually, the air design temperature is taken as the maximum possible temperature for the meteorologic conditions of the site. This means that on cold days, the capacity of the unit will be higher than design. In most cases, this is not an issue because the higher the heat removed, the better it is for the process. However, in air coolers operating as distillation columns over condensers, temperature fluctuations may lead to undesirable pressure fluctuations. In other cases it may be desired to keep the temperature within a narrow range for other process reasons.

The more common systems to control the process temperatures include[2]

1. *Adjust the blade pitch.* This changes the characteristic curve of the fan and makes it possible to change the airflow without changing the fan speed. The required blade angle decreases as the air temperature drops, and this conserves fan power. Different systems can be used to modify the blade angle. The simpler systems require stopping the fan and making the change manually. This system is used for seasonal changes (winter/summer) There exist hydraulic systems that make it possible to change the blade angle with the fan in operation, and these can be used for automatic temperature control.

2. *Variable-speed drivers.* Usually electronic speed controllers that allow changing the motor rpms are used.

3. *Adjustable louvers.* Placing these at the air inlet permits control of airflow. These again can be automatic or manual. Even though they can control the airflow, louvers do not reduce fan power requirements.

4. A simple control in units with more than one fan can be obtained by switching off one or more of the fans in winter.

GLOSSARY

A = heat transfer area (m^2)

A_o = plain tube area per unit length ($= \pi D_o$) (m^2/m)

A_D = bare area of the tube per unit length (not covered by fins) (m^2/m)

A_f = area of fins per unit length of tube (m^2/m)

a_S = airflow area (m^2)

a_F = face area (m^2)

A_i = internal area of the tube per unit length (m^2/m)

b = fin thickness (m or mm)

D_i = internal diameter of the tube (m)

D_o = external diameter of the tube (m)

D_b = fin diameter (m)

D_e = equivalent diameter for heat transfer (m)

D'_e = equivalent diameter for friction (m)

D_{io} = internal diameter of the external tube

f = friction factor

F_t = LMTD correction factor

G = mass flow density [$kg/(s \cdot m^2)$]

G_S = mass flow density of air based on area a_S [$kg/(s \cdot m^2)$]

h = convection film coefficient [$W/(m^2 \cdot K)$]

h_f = convection film coefficient for tube and fins [$W/(K \cdot m^2)$ of fin and tube area]

h'_f = convection film coefficient for tube and fins corrected by fouling [$W/(K \cdot m^2)$ of fin and tube area]

h'_{fo} = convection film coefficient for tubes and fins corrected by fouling and by fin efficiency, referred to the plain area of tube [$W/(K \cdot m^2)$ of plain tube surface]

h_i = internal film coefficient [$W/(m^2 \cdot K)$]

h_{io} = internal coefficient referred to the external area [W/(m$^2 \cdot$ K)]

H = height of longitudinal fin

j = dimensionless number defined by Eq. (8-2-13)

k = thermal conductivity [W/(m $\cdot$ K)]

k_f = thermal conductivity of the fin [W/(m $\cdot$ K)]

L = tube length (m)

$L_p = n_f \times S_F$ (m)

M = molecular weight

M = parameter defined in Fig. 8-10

m = parameter defined in Eq. (8-1-29)

N = number of tubes or number of tubes per shell

n = number of tube passes

N_m = number of transverse fins per meter of tube

n_f = number of tube rows

n_{tf} = number of tubes per row

N_f = number of longitudinal fins of the tube

Q = heat flow (W)

Q_f = heat flow on the fin surface (W)

Q_c = conduction heat flow through the section of the fin (W)

R = universal gas constant [8,306 (Pa $\cdot$ m^3)/(K $\cdot$ kmol)]

R = dimensionless number to obtain F_t

Re = Reynolds number

Re$_S$ = air Reynolds number for heat transfer

Re$'_S$ = air Reynolds number for friction

R_{fi} = iternal fouling resistance [(m$^2 \cdot$ K)/W]

R_{fo} = external fouling resistance [(m$^2 \cdot$ K)/W]

S = dimensionless number to obtain F_t

S_L = geometric parameter (see Fig. 8-8)

S_T = geometric parameter (see Fig. 8-8)

S_F = geometric parameter (see Fig. 8-8)

T = hot-fluid temperature

t = cold-fluid temperature

T_b = temperature at the fin base (K)

T_f = fin temperature

U = overall heat transfer coefficient [W/(m$^2 \cdot$ K)]

v = velocity (m/s)

v_F = face velocity (m/s)

v_{fan} = air velocity at the fan (m/s)

W = mass flow (kg/s)

Z = dimensionless parameter ($Z = RS$)

ρ = density (kg/m^3)

ρ_{fan} = air density at fan (kg/m^3)

θ = temperature difference between the external fluid and the fin surface (K)

μ = viscosity [kg/(m · s)]

Ω = fin efficiency

Subscripts

A = air

1 = inlet

2 = outlet

c = cold

h = hot

REFERENCES

1. Kern D.: *Process Heat Transfer.* New York: McGraw-Hill, 1954.

2. Ludwig E.: *Applied Process Design for Chemical and Petrochemical Plants.* Gulf Publishing Company, Houston, Texas.

3. *Engineering Data Book,* 11th ed. Gas Processors Suppliers Association, Tulsa Oklahoma, 1998.

4. Weierman C.: "Correlations Ease the Selection of Finned Tubes". *Oil and Gas Journal* 74, 36, pp 97–100, 1976.

5. Pignotti A., Cordero G.: "Mean Temperature Difference in Multipass Crossflow" and "Mean Temperature Difference Charts for Air Coolers" Journal of Heat Transfer 105: pp 584–597, 1983.

6. Brown R., Ganapathy V.: "Design of Air Cooled Exchangers" Chem. Eng March 27, 1978, pp 107–124.

7. A.P.I Standard 661- Air Cooled Heat Exchangers for General Refinery Service, 6th ed. Feb 2006.

CHAPTER 9
PLATE HEAT EXCHANGERS

9-1 OPERATING PRINCIPLES AND GENERAL DESCRIPTION

A thin planar metallic plate separating two circulating fluids at different temperatures can act as a heat transfer surface, as indicated in Fig. 9-1. To complete the device, two more plates must be added at both sides to form the channels where the fluids circulate.

The plates are kept separated by means of elastomeric gaskets, as shown in Fig. 9-2. The assembly is clamped together in a frame by clamping bolts that must exert the necessary pressure to maintain the plates in position. A schematic drawing of a plate heat exchanger is shown in Fig. 9-3.

The gaskets are such that the separation between plates is only a few millimeters to achieve high fluid velocities and hence high heat transfer coefficients. This assembly is the basis of an elemental plate heat exchanger.

The main objection to this type of device is that it has important limitations regarding the maximum flow rates that it can handle, and it would not be practical to indefinitely increase the size of the plates when higher flow rates and heat duty are required. In this case it is necessary to increase the number of plates, making both fluids circulate through alternate channels separated by the heat transfer plates, as shown in Fig. 9-4. This is basically the arrangement of a plate heat exchanger.

Basically, a plate heat exchanger consists of a pack of corrugated metal plates with portholes for the passage of the two fluids. Heat transfer takes place through the plates. The plate pack is assembled between a fixed frame plate and a movable pressure plate and is compressed by tightening bolts (Fig. 9-5). The plates are fitted with gaskets that seal the interplate channels. The number of plates is defined by the heat transfer requirements. The plates and pressure plate are suspended by an upper carrying bar and located by a lower guiding bar, both of which are fixed to a support column.

The plates are corrugated. The corrugations provide reinforcement as well as a large number of plate-to-plate contact points. These contact points provide full plate support and make it possible to have high operating pressures with very thin-gauge plates. The usual gap between consecutive plates is about 5 mm.

The plates form the channels where the fluids circulate. The fluids enter and exit these channels through portholes located at the corners of the plates. These portholes may be perforated or not, as required by the desired circulation scheme. The portholes form the distribution headers, which distribute the fluids into the circulation channels.

The fluids enter the unit through inlet nozzles perforated in the frames and pass into the distribution headers. From the headers, the fluids distribute into the circulation channels following a specific flow pattern defined during the design stage. This is achieved by the positioning of elastomeric gaskets to blank off certain flowpaths and by blinding specific portholes of individual heat exchange plates. A suitable design of gaskets and portholes allows different flow configurations. Some models of heat exchange plates are shown in Fig. 9-14.

Coming back to Fig. 9-4, we see that this unit has nine plates. Figure 9-6 is an expanded view showing the configuration of plates. Plates 1 and 9 are closure plates (they do not act as heat transfer surfaces), and plates 2 to 8 are the heat transfer plates. It can be seen that each branch in which the cold stream is divided exchanges heat through the plates forming its channel with two branches of the hot stream, and vice versa, except for the outermost branches, which only exchange through one plate.

In this case, the plates are arranged for diagonal flow because the inlet and outlet ports to each channel are located at diagonally opposed corners. In some models, the flow pattern can be parallel, as shown in Fig. 9-13a, c, and d and in Fig. 9-15.

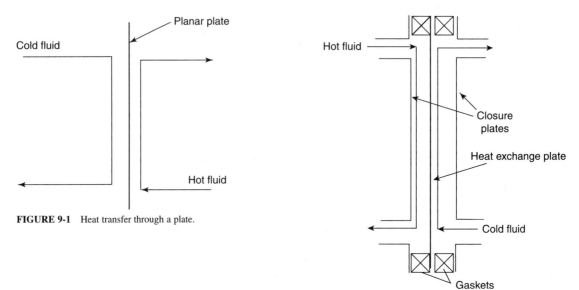

FIGURE 9-1 Heat transfer through a plate.

FIGURE 9-2 Principle of the plate heat exchanger.

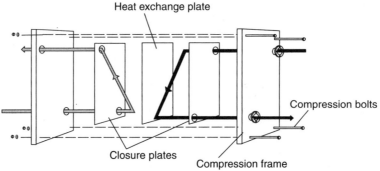

FIGURE 9-3 Elemental plate heat exchanger.

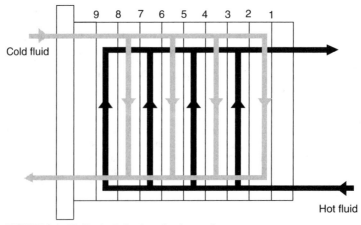

FIGURE 9-4 Fluids circulation in a plate heat exchanger.

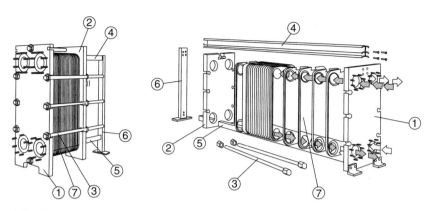

1-Fixed frame plate
3-Tightening bolts
5-Lower guiding bar
7-Plates

2-Movable pressure plate
4-Upper carrying bar
6-Support column

FIGURE 9-5   Component parts of a plate heat exchanger.

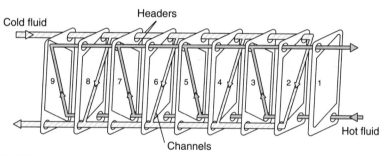

FIGURE 9-6   Expanded view of the heat exchanger of Fig. 9-4.

The gasket design of the different plates makes the fluid circulate according the desired flow configuration. For example, the hot fluid enters the exchanger through the lower-right porthole of plate 1. The gasket design of plate 2 prevents the fluid from entering the channel formed by plates 1 and 2, so it must cross plate 2.

The cold fluid, which enters the exchanger through the upper-right porthole of plate 9 (always looking at the exchanger from plate 1), when it reaches and crosses plate 2, can get into the channel formed by plates 1 and 2 and circulates through this channel, exiting through the lower-left porthole of plate 2 (Fig. 9-7).

The complete diagram of plates for this exchanger is shown in Fig. 9-8. Figure 9-9 is a schematic representation of the exchanger configuration.

The exchanger must always be looked at from plate 1. When the fluid streams cross a plate, it is because the porthole at the corresponding position is perforated. For example, in the heat exchanger in Fig. 9-9, we see that plate 1 is only crossed by the hot fluid at the lower-right and upper-left positions, so the plate is perforated in those corners.

This type of plate heat exchanger is referred to by some vendors as a *gasketed plate heat exchanger* because there are other types of plate heat exchanger that are all-welded (without gaskets) as we shall see later on in the text.

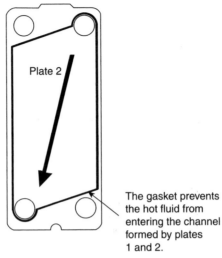

The gasket prevents
the hot fluid from
entering the channel
formed by plates
1 and 2.

FIGURE 9-7 The gasket design defines the flow circulation.

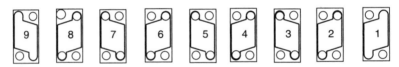

FIGURE 9-8 Gasket configuration.

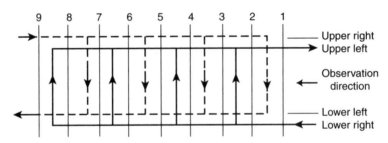

FIGURE 9-9 Schematic representation.

9-2 SERIES-PARALLEL COMBINATION

In the exchanger we have just described, all the channels where fluid circulates are in parallel. However, sometimes it is necessary, in order to achieve higher velocities, to arrange plate heat exchangers with more than one pass, the same as in shell-and-tube heat exchangers. For example, let's assume that the hot fluid must perform two passes, whereas the cold fluid requires only one. The configuration of plates can be as indicated in Fig. 9-10.

The hot fluid performs two passes, and each pass is made up of two channels in parallel. The cold fluid path, in turn, consists of one pass with four channels in parallel. The plates of this exchanger can be built easily

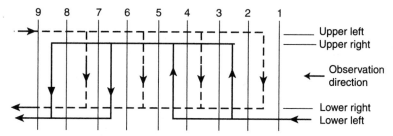

FIGURE 9-10 Schematic representation of a heat exchanger with two passes in one of the streams.

from the circulation diagram in Fig. 9-10,where you can see in which position each plate is crossed by a fluid stream. Then, looking at Fig. 9-10, we can deduce the perforation diagram of the plates. It is as follows:

Plate 1: Only one perforation at the lower left

Plate 2: Both lower perforations and one perforation at the upper left

Plate 3: Perforated at the four positions

Plate 4: Same as plate 3

Plate 5: One perforation at the lower right and both upper perforations

Plate 6: Same as plate 5

Plate 7: Must be perforated at all four positions

Plate 8: Same as plate 7

Plate 9: Both lower perforations and one perforation at the upper left

The gaskets, diagram can be deduced as follows: The channel formed by plates 1 and 2 must allow the communication of the lower-right and upper-left headers. This effect is achieved by the design of the gasket of plate 2, which must have these perforations communicated. The channel formed by plates 2 and 3 must allow communication of the lower-left and upper-right headers, and at the same time, it must prevent the fluid from the other headers from entering into this channel. This effect is achieved with the design of the gasket of plate 3. The complete diagram of plates and gaskets is shown in Fig. 9-11.

Another representation of the fluid circulation in this exchanger can be seen in Fig. 9-12. In this way, it is possible to arrange the plates so as to achieve any desired circulation pattern and modify the fluid

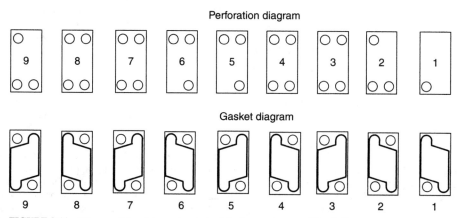

FIGURE 9-11 Diagram of the plates and gaskets of the heat exchanger of Fig. 9-10.

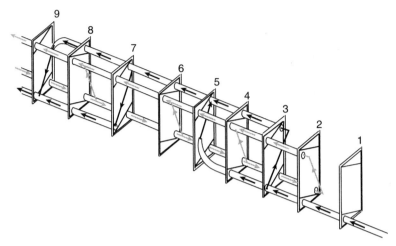

FIGURE 9-12   Expanded representation of the heat exchanger of Fig. 9-10.

velocities as dictated by the heat transfer requirements. Obviously, the lower the number of channels in parallel for each fluid, the higher is its velocity, and the higher is the heat transfer coefficient, a counterpart being an increase in the pressure drop.

9-3   COMPONENTS OF A GASKETED PLATE HEAT EXCHANGER

9-3-1   Heat Exchanger Plates

Heat exchanger plates are corrugated. The shape of the corrugations is a characteristic of each plate model and is carefully studied by the manufacturers. The purpose of the corrugations is to provide turbulence in order to increase the heat transfer coefficients and, at the same time, to increase the structural strength of the assembly.

Each plate must be supported mechanically to resist the pressure difference of the fluids on both sides of the plate. The most severe case is when one of the fluids is at its maximum pressure and the other one is at zero pressure. This situation may occur during startup or when the unit is being emptied for cleaning. It is necessary to avoid plate deformations owing to these pressure effects. The plate's corrugations provide many contact points between plates, thus preventing deformations.

There are many models of corrugated plates. In the "washboard" type, corrugations are parallel among themselves and perpendicular to the flow direction in the channel. This type of corrugation is shown in Figs. 9-13a and 9-14a.

The most common corrugation pattern is the herringbone pattern, which is shown in Figs. 9-13b and 9-14b. In this case, the corrugations of consecutive plates are rotated 180 degrees, so between two plates there are a large number of contact points where the corrugations cross. This allows very rigid plate support and a high degree of turbulence in the fluids.

Other models, not used very often, are shown in Fig. 9-13c and d. Each manufacturer has several types of plate for different applications. Selection of the plates for any application depends on the process requirements in terms of heat transfer coefficients and allowable heat exchanger pressure drop. The models that allow higher heat transfer coefficients for a given flow rate also produce higher pressure drops. Thus, for any application, the designer must choose the plate type offering the best balance of both effects.

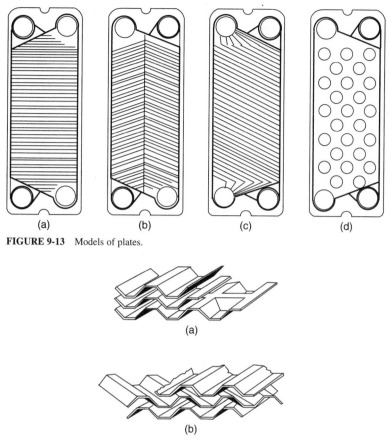

FIGURE 9-13 Models of plates.

FIGURE 9-14 Different corrugations:
(a) plates with parallel corrugations,
(b) plates with transversal corrugations.

The plates are manufactured by cold forming in hydraulic presses. Each manufacturer owns the matrixes corresponding to its standard models and usually has a certain stock of nonperforated plates. When a heat exchanger is purchased, the manufacturer designs the fluids circulation pattern, and then the plates are perforated in the positions required by the design. Thus plate heat exchangers must be considered to be semistandard units, and since the perforation and assembly are a rather fast operation, the delivery time is shorter than with other types of heat exchangers.

The effective heat transfer area of a plate can be as small as 0.033 m^2 in laboratory models up to 3 m^2 in larger models. The maximum number of plates that can be assembled in a unit is limited by the compression strength that can be achieved with the bolts and maldistribution of flow among the flow channels, but a maximum of 500 plates is possible in certain cases.

Construction Materials. Plates can be made from different materials, such as stainless steel, copper and its alloys, aluminum, titanium, nickel or nickel-molybdenum alloys, and others. Plate thickness is between 0.5 and 1.2 mm. Since the plates are so thin, it is necessary to use corrosion-resistant materials. This is why carbon steel is seldom employed in plate manufacturing, and stainless steel is used as a minimum quality. The plate material must be suitable for cold forming. Certain chrome, zirconium, and titanium alloys are difficult to cold form and are not used in plate heat exchanger manufacture. Table 9-1 indicates how easily the plates can be cold formed using different materials.

TABLE 9-1   Cold-Forming Characteristics of Various Materials

Material	Type
Common steel	a
Stainless steel (Mo < 4%)	a
Stainless Steel (Mo > 4%)	b
Titanium	b
Copper	b
Copper-nickel (70/30 or 90/10)	a o b
Bronze	a
Monel	b
Incolloy 825	b
Nickel	a
Inconel	b
Hastelloy B	c
Hastelloy C	b
Aluminum	b
Zirconium	d
Tantalum	a

Key: a = all types can be cold formed; b = some types can be cold formed;
c = very difficult to cold form; d = impossible to cold form.

9-3-2 Gaskets

For plate gaskets, elastomeric materials are employed, such as natural rubber, styrene butadiene rubber (SBR), nitrile rubber (acrylonitrile butadiene), butyl rubber (a copolymer of isobutylene and a small amount of isoprene), silicones, or other elastomers such as neoprene, hypalon, or viton.[7] Sometimes, compressed fibers are used for high-temperature applications. Plastic materials of the polytetrafluoroethylene (PTFE) type (e.g., Teflon or Fluon) are not suitable for gasket manufacture because they have poor elastic recovery behavior.

The selection of a suitable gasket material is very important in plate heat exchanger design because the gaskets are the limiting factor for the maximum operating temperature of the exchanger. Table 9-2 is a guide showing the maximum temperatures at which the different materials can be used. For higher temperatures the vendors must be consulted.

TABLE 9-2   Maximum Service Temperature for Gaskets

Material	Max. Temperature
Natural rubber, styrene, neoprene	70°C
Nitrile rubber, viton	100°C
Butyl rubber	120°C
Silicone	140°C

Regarding chemical resistance, SBR is suitable for general applications in aqueous systems at low temperatures. Nitrile rubber combines the general-purpose characteristics of SBR with a good resistance to fats and aliphatic hydrocarbons. Butyl rubber offers excellent chemical resistance to many chemical agents, such as acids, alkalys, and some ketones and amines, but is not very resistant to fats. Silicone is used in a limited number of applications but is the most suitable material for sodium hypochlorite and some low-temperature applications. Fluoroelastomers such as viton are more expensive but are the most suitable for high-temperature oils (140°C). With a proper formulation, they can be used for 98% sulfuric acid at temperatures up to 100°C.

The cross sections of gaskets depend on the plate design. They can be rectangular, trapezoidal, or oval. The width normally ranges from 5–15 mm. Gasket height before compression is 15–50 percent higher than the compressed height.

Gaskets are installed into grooves formed in each plate and glued with special cements. The cement must provide a good union during service and opening of the heat exchanger but must be removed easily when the gasket has to be replaced.

9-3-3 Frame

The frame is formed by two robust plates (frame plates) at both ends of the unit. These plates must absorb the force resulting from the fluid pressure. For example, a 5-bar pressure acting on a 1-m^2 surface exerts a force of 500,000 N (~50 tons). The force that must be applied on the frame plates to keep the whole assembly tight is also important. This force is absorbed by the tightening bolts.

Usually, one of the frame plates contains the inlet and outlet nozzles and is fixed to the system piping. The other frame plate, also called the *movable pressure plate,* can be slid back for inspection and cleaning when the unit is unbolted (see Fig. 9-17). With this configuration, it is not necessary to remove the piping components for maintenance. However, this construction is possible only when both fluids perform a single pass through the exchanger, as in Fig. 9-15. In a multipass exchanger, it is unavoidable to have piping connections in both frame plates.

Frames are usually free-standing or, for smaller units, attached to structural steel work. Figure 9-16 shows some frame models.

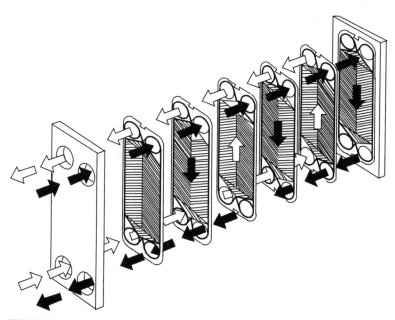

FIGURE 9-15 When both fluids make a single pass, all inlet and outlet nozzles are in the fixed frame plate.

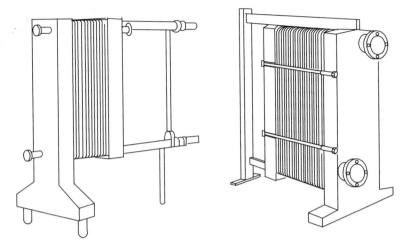

FIGURE 9-16 Types of frames.

9-4 USE AND LIMITATIONS OF GASKETED PLATE HEAT EXCHANGERS

9-4-1 General Characteristics

The gasketed plate heat exchanger is suitable for services where the process conditions are moderate. It was conceived initially in the 1930s to meet the hygienic demands of the dairy industry. The exchanger can be completely disassembled for cleaning, as shown in Fig. 9-17, which is a primary requirement in the food industry.

Another characteristic is that the heat transfer area can be enlarged, if necessary, by adding more plates. The only requirement is to oversize the tightening bolts in length and diameter initially so that they can contain a higher number of plates and supply the required closure force.

Maintenance is very simple, and cleaning is effected by brushing or with a pressure water hose. Chemical cleaning is also very effective owing to the high turbulence that can be achieved in the circulating medium.

Gasket replacement is also very simple, and if a plate is corroded, it can be changed easily. Another advantage, mentioned previously, is that delivery time is shorter than for other types of heat exchangers because they are semistandard units.

The heat transfer coefficients of plate heat exchangers are usually much higher than those of shell-and-tube heat exchangers, and the heat transfer area is smaller. Additionally, since they are compact in nature, they require a smaller footprint than shell-and-tube units, especially considering the additional space required for bundle removal.

Owing to their compactness, these heat exchangers found an important application in the marine industry in the refrigeration circuits of ship diesel motors, where seawater is used to cool down the primary water, which, in turn, cools the motor cylinders.

9-4-2 Pressure and Temperature Limitations

The maximum operating temperatures are related to the resistance of the gaskets. We have already treated this subject. The maximum operating pressures are limited by the ability to achieve a tight closure without gasket leakage. This depends on the size of the plates, the frame type, and the number of plates, so it is not possible to give general rules. In some cases, it is possible to operate at pressures as high as 30 bar. However, this type of equipment is seldom used for pressures higher than 10 bar.

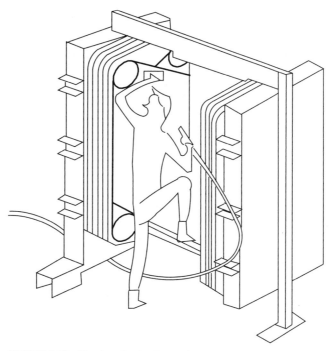

FIGURE 9-17   Cleaning of a plate heat exchanger.

Regarding capacity, some vendors offer plates capable of operating with flow rates up to 3,600 m^3/h. The maximum flow rates are limited by the size of the plate portholes that form the distribution headers.

9-5   WELDED AND SEMIWELDED PLATE HEAT EXCHANGERS

Plate heat exchangers also can be supplied in a semiwelded construction. The plates are welded in pairs, and each pair is supplied with gaskets that allow coupling with other similar pairs. The resulting assembly allows maintenance and cleaning of the channels of one fluid, whereas the channels of the other fluid are hermetically welded. This type of construction is employed in cases of clean fluids where leakage cannot be tolerated (e.g., ammonia or caustic soda), whereas the other fluid is innocuous and requires cleaning (e.g., cooling water).

It is also possible to construct plate heat exchangers that are 100 percent welded. In this case, cleaning is not possible in any of the sides, but a very economical gasketless design results. Welding is performed by brazing, and this type of heat exchangers is called a *brazed plate heat exchanger.*

9-6   PLATE HEAT EXCHANGERS VERSUS SHELL-AND-TUBE HEAT EXCHANGERS

The main objection to the gasketed plate heat exchanger is the possibility of leakage. This usually disqualifies this type of equipment in critical process services and is the reason why it is limited in acceptance in the refinery and petrochemical industry.

Plate heat exchangers are used for moderate pressures and temperatures and innocuous fluids. These are the conditions normally found in the food and beverage industry, and the additional cleaning facility explains the wide acceptance of the plate heat exchanger in such applications.

In the food industry, it is also frequent that heating operations must be carried out without submitting the products to high temperatures or excessive heating time to avoid decomposition or alterations of flavor. The plate heat exchanger is very suitable in these cases because the high heat transfer coefficients allow working with lower wall temperatures and lower residence time.

When the decision is a matter of cost, there are several factors that may favor one or the other type of exchanger. One the most important factors are the materials of construction. Plate heat exchangers are manufactured with stainless steel as the minimum-quality material. Therefore, when corrosion is not an issue, the shell-and-tube heat exchanger may have some advantage because it can be built with cheaper materials. On the other hand, if high-cost corrosion-resistant materials are necessary, the plate heat exchanger is usually favored because it is a lighter unit.[5]

9-7 TYPICAL HEAT TRANSFER COEFFICIENTS

Some typical values of heat transfer coefficients for plate heat exchangers are listed in Table 9-3. Values of specific pressure drop are included as well.[4] This concept will be explained later.

TABLE 9-3   Typical Overall Heat Transfer Coefficients

Service	Physical Properties of Product η = Kinematic Viscosity (m^2/s) ρc = Heat Capacity per Unit Volume $[J/(m^3 \cdot K)]$ K = Thermal Conductivity $[J/(s \cdot m \cdot K)]$	Overall Heat Transfer Coefficient		Specific Pressure Drop
		Transversel Corrugation Plate $W/(m \cdot K)$	Herringbone Corrugation Plate $W/(m \cdot K)$	m of Liquid Column per Transfer Unit
Water to water or steam	$\eta = 0.6 \times 10^{-6}$ $\rho c = 4.18 \times 10^6$ $K = 0.62$	3,100–3,900	3,000–3,700	1.5–3
Viscous aqueous solution to water or steam	$\eta = 50 \times 10^{-6}$ $\rho c = 3.97 \times 10^6$ $K = 0.39$	1,000–1,200	700–800	11–20
Mineral oil to water or steam	$\eta = 50 \times 10^{-6}$ $\rho c = 2.09 \times 10^6$ $K = 0.11$	450–580	300–350	15–30
Mineral oil to mineral oil	$\eta = 50 \times 10^{-6}$ $\rho c = 2.09 \times 10^6$ $K = 0.11$	210–270	120–190	18–40
Mineral oil to water or steam	$\eta = 100 \times 10^{-6}$ $\rho c = 2.09 \times 10^6$ $K = 0.11$	330–400	190–290	22–51
Organic solvent to water or steam	$\eta = 1 \times 10^{-6}$ $\rho c = 2.09 \times 10^6$ $K = 0.19$	1,850–2,100	1,500–1,950	2.5–3
Vegetable oil to water or steam	$\eta = 100 \times 10^{-6}$ $\rho c = 2.09 \times 10^6$ $K = 0.15$	870–1,000	810–930	7–10

9-8 HEAT TRANSFER AND PRESSURE-DROP CORRELATIONS FOR PLATE HEAT EXCHANGERS

Both heat transfer rate and pressure drop are highly influenced by the plate geometry and the corrugation design. Since each plate model is exclusive from a specific manufacturer, the thermal design of this type of equipment is always performed by the vendor, which has the correlations applicable to his plates.

So these heat exchangers are usually purchased on the basis of a process specification, where flow rates, inlet and outlet temperatures, fouling, and allowable pressure drops are specified by the purchaser, and the vendor selects a unit capable to satisfy these operating conditions. Nevertheless, we shall present some general correlations that may be useful to analyze changes in the operating conditions and to perform preliminary sizing.

Reynolds Number.   An equivalent diameter can be defined for the channel as

$$D_e = \frac{4 \times \text{flow area}}{\text{wetted perimeter}} = \frac{4be}{2(e + b\phi_a)} \cong \frac{2e}{\phi_a} \tag{9-8-1}$$

where b = channel width
$\quad\quad\;\; e$ = channel depth

The channel depth e equals the thickness of the corrugated plate minus the thickness of the metal sheet. Since the plates are in touch with each other, the thickness of the corrugated plates s can be obtained by dividing the length of the plate pack L (which is always shown in the vendor drawings) by the number of plates (see Figure 9-18). ϕ_a is a multiplication factor representing the enhancement of the heat transfer area owing to the corrugations. Is a quotient between the actual area of a plate and its planar projected area. It is usually between 1.1 and 1.25. This factor should be supplied by the vendor (see Example 9-1).

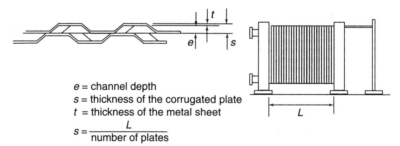

e = channel depth
s = thickness of the corrugated plate
t = thickness of the metal sheet
$$s = \frac{L}{\text{number of plates}}$$

FIGURE 9-18   Plate nomenclature.

A Reynolds number then can be defined as

$$\text{Re} = \frac{D_e v \rho}{\mu} = \frac{2ev\rho}{\mu\phi_a} \tag{9-8-2}$$

The velocity v is defined as

$$v = \frac{W}{\rho n_p be} \tag{9-8-3}$$

where W = mass flow (kg/s)
$\quad\quad\;\; \rho$ = density (kg/m^3)
$\quad\quad\;\; n_p$ = number of channels in parallel

Typical velocities in plate heat exchangers for aqueous fluids range from 0.3–0.9 m/s, but local velocities can be considerably higher owing to the effect of corrugations.

Film Coefficients.   For a plate heat exchanger, the usual dimensionless numbers are defined:

$$\text{Nusselt number} = \text{Nu} = \frac{hD_e}{k} \qquad (9\text{-}8\text{-}4)$$

$$\text{Colburn factor} = J_H = \frac{\text{Nu}}{\text{Re} \cdot \text{Pr}^{0.33}} \left(\frac{\mu}{\mu w}\right)^{0.14} \qquad (9\text{-}8\text{-}5)$$

The experimental results are usually expressed with correlations of the type $J_H = f(\text{Re})$, where the correlation parameters depend on the plate geometry.

An important factor in the correlations is the corrugation angle. Figure 9-19 shows a graph of J_H versus Re for herringbone corrugations at given parameters of the corrugation angle. This is a general graph that may change depending on the plate manufacturer.

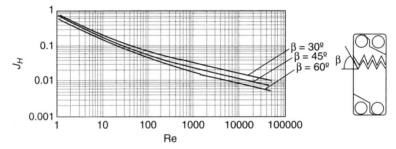

FIGURE 9-19   Colburn J_H factor for plate heat exchangers with herringbone corrugations.

Pressure Drop.   The pressure drop through the channels of a plate heat exchanger can be expressed as a function of a friction factor in the usual way:

$$\Delta p_{\text{plates}} = \frac{4 f L_e n_s \rho v^2}{2 D_e} \qquad (9\text{-}8\text{-}6)$$

where n_s is the number of channels in series and L_e is the effective length of the channel, which is the diagonal distance between inlet and outlet portholes. The friction factor is correlated as a function of the Reynolds number. Figure 9-20 is a graph of f versus Re for herringbone corrugations at given parameters of the corrugations angle. This is a general graph, and specific plate models may present deviations with respect to it.

The pressure drop in the unit is the sum of the pressure drop in the channels and the pressure drop in the distribution headers. The pressure drop in the distribution headers depends on the flow configuration.[3] By knowing the pass configuration, it is possible to estimate this additional pressure drop. As an estimation, we can consider 1.5 velocity heads for each pass, that is,

$$\Delta p_{\text{headers}} = \frac{1.5 n_s G_c^2}{2\rho} \qquad (9\text{-}8\text{-}7)$$

where

$$G_c = \frac{4W}{\pi D_c^2} \qquad (9\text{-}8\text{-}8)$$

Here, D_c is the diameter of the distribution header. Then

$$\Delta p_{\text{exchanger}} = \Delta p_{\text{plates}} + \Delta p_{\text{headers}} \qquad (9\text{-}8\text{-}9)$$

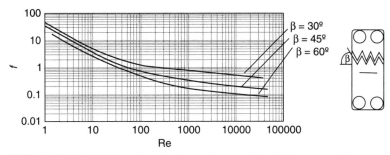

FIGURE 9-20   Friction factor for a plate heat exchanger with herringbone corrugations.

If the pressure drop in the distribution header is significant, it may result in an uneven flow distribution among the channels of the same pass. This uneven distribution may affect the thermal behavior. The most advanced computation programs take this effect in consideration and calculate the flow distribution among the channels.

9-9   LMTD CORRECTION FACTOR

The heat transfer area is calculated as in other types of exchangers by means of

$$Q = UA\Delta T \tag{9-9-1}$$

where

$$\frac{1}{U} = \frac{1}{h_c} + \frac{1}{h_h} + Rf \tag{9-9-2}$$

and

$$\Delta T = \text{MLDT} \cdot F_t \tag{9-9-3}$$

The LMTD correction factor F_t takes into consideration the efficiency reduction that results as a consequence of the partial co-current flow when the flow configurations of both streams are different. Experimental results of F_t factors for some flow configurations can be found in ref. 2. Marriot[1] presents a graph of F_t as a function of the number of transfer units for different flow configurations (Fig. 9-21).

It should be noted that even configurations with the same number of passes in both streams present an F_t factor smaller than 1, which could be explained as the result of border effects and stagnant zones in the channel.

9-10   FOULING RESISTANCES

Fouling resistances in plate heat exchangers are usually much smaller than in shell-and-tube heat exchangers. The reasons are pointed out by Marriot:[1]

1. The high degree of turbulence keeps solids in suspension.
2. Heat transfer surfaces are smooth. For some types, a mirror finish may be available.
3. There are no dead spaces where fluids can stagnate and solids deposit, such as, for example, in the vicinity of baffles in shell-and-tube heat exchangers.

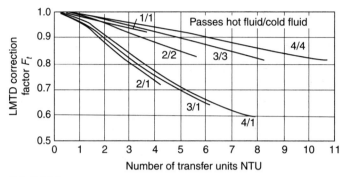

FIGURE 9-21 LMTD correction factor for plate heat exchangers. From J. Marriot, *Chem. Eng.*, April 5 1971. Reproduced with permission.)

4. Since the plates are built with higher-quality materials, there are no corrosion products to which fouling may adhere.

5. High film coefficients tend to lead to lower surface temperatures for the cold fluid (the cold fluid is usually the culprit as far as fouling is concerned).

6. Since cleaning is a very simple operation, the interval between cleanings is usually smaller.

A good estimation of the fouling resistance is very important in plate heat exchangers. This is due to the usually high heat transfer film coefficients in the circulating fluids, which frequently make fouling the controlling resistance with a direct incidence in the cost of the unit.[8]

It can be an expensive mistake to specify for a plate heat exchanger the same fouling factors used for shell-and-tube heat exchangers. Table 9-4 indicates some typical fouling resistances for plate heat

TABLE 9-4 Fouling Resistances for Plate Heat Exchangers[7]

Fresh pure water, condensate, etc.	0.2–0.5×10^{-4}
Lake, river, or sea water	0.5–1×10^{-4}
Sewer or brackish water	1–2×10^{-4}
Acetic acid	0.5–1×10^{-4}
Acetone	0.5–1×10^{-4}
Benzene	0.5–1×10^{-4}
Carbon tetrachloride	0.5–1×10^{-4}
Difenile (Dowtherm, etc.)	0.5–1×10^{-4}
Ethyl acetate	0.5–1×10^{-4}
Ethyl eter	0.5–1×10^{-4}
Ethanol	0.5–1×10^{-4}
Etylene glycol, 25% weight	0.5–1×10^{-4}
n-Heptane	0.5–1×10^{-4}
Methanol	0.5–1×10^{-4}
Methyl ethyl ketone	0.5–1×10^{-4}
n-Octane	0.5–1×10^{-4}
Toluene	0.5–1×10^{-4}
ClH, 96% weight	0.5–1×10^{-4}
Sulfuric acid, 30% weight	Depends on material
Magnesium chloride brine	Depends on material
Beer, milk	0.1–0.3×10^{-4}
Wine	0.5–1×10^{-4}
Sugar solution, 17°Bx	1.5–2×10^{-4}

Note: Values are given in $(h \cdot m^2)/kcal$. To convert to SI, they must be multiplied by 0.86.

exchangers. The table assumes operation at an economical pressure drop (~30 kPa/NTU). The total resistance must be obtained adding the individual resistances corresponding to both sides. Some plate heat exchanger manufacturers prefer to specify a percentage of excess area above that required for clean conditions instead of using fouling resistances.

Example 9-1 To cool down 45,000 kg/h of alcohol from 55.4°C to 43°C, a heat exchanger with the configuration indicated in Fig. 9-22 is available. Cooling water at 25°C will be used. The water outlet temperature should not be higher than 40°C. A fouling resistance of 0.0001 $(m^2 \cdot K)/W$ will be adopted for each fluid. Verify if the unit is suitable, and calculate the pressure drops for each fluid.

Physical properties:

	Alcohol	Water
Thermal conductivity, W/(m · K)	0.148	0.62
Viscosity, kg/ms	0.65×10^{-3}	0.75×10^{-3}
Specific heat, J/(kg · K)	3400	4,200
Density, kg/m³	765	1,000
Prandtl Number (Pr)	15.4	5.16

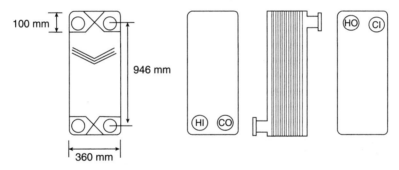

Characteristics of the plate
Effective area per plate: 0.319 m²
Channel depth: $e = 0.005$ m
Area increment factor $\phi = 1.14$
Channel width $b = 0.36$ m
Corrugations angle = 45 degrees

Number of plates: 61
(59 heat transfer plates and 2
closure plates)
Passes in series per fluid: 3
Number of channels per pass: 10
Area of the unit: 18.87 m²

FIGURE 9-22 Illustration for Example 9-1.

Solution

$$\text{Heat exchanged} = W_h c_h (T_1 - T_2) = 45,000/3,600 \times 3,400 \times (55.4 - 43) = 5.27 \times 10^5 \text{ W}$$

$$\text{Water flow rate} = Q/[cc(t_2 - t_1)] = 5.27 \times 10^5/[4,200(40 - 25)] = 8.36 \text{ kg/s} = 30,096 \text{ kg/h}$$

$$a_s = \text{flow area per channel} = eb = 0.005 \times 036 = 0.0018 \text{ m}^2$$

$$\text{Equivalent diameter } D_e = 2e/\phi_a = 2 \times 0.005/1.14 = 0.00877 \text{ m}$$

	Alcohol	Water
W_1 = mass flow per channel = $W/10$ (kg/h)	4,500	3,009
Velocity = $W_1/(\rho a_s \times 3,600)$	0.9	0.46
Re = $D_e v \rho/\mu$	9,370	5,400
J_H (graph)	0.012	0.0145
f	0.2	0.22
$h = J_H(k/D_e)\text{Re} \cdot \text{Pr}^{0.33}$	4,640	9,600
$\Delta p_{\text{plates}} = \dfrac{4fns\,\rho v^2}{2}\left(\dfrac{L_e}{D_e}\right)$ (For simplicity, we took L_e as 946 mm)	80,208 N/m^2	30,128 N/m^2
Flow area of the distribution headers = $\pi Dc^2/4$ m^2	0.00785	0.00785
Mass velocity in the distribution headers G_c [kg/(s · m^2)]	1,592	1,061
$\Delta p_{\text{headers}} = \dfrac{1.5nsG_c^2}{2\rho}$	7,454	2,532
Δp_{total}	87,662	32,660

The nozzle arrangement of the figure corresponds to a countercurrent configuration.

$$\text{MLDT} = \frac{(T_1 - t_2) - (T_2 - t_1)}{\ln\dfrac{(T_1 - t_2)}{(T_2 - t_1)}} = \frac{(55.4 - 40) - (43 - 25)}{\ln\dfrac{15.4}{18}} = 16.7°\text{C}$$

$$U = \left(\frac{1}{h_c} + \frac{1}{h_h} + Rf_1 + Rf_2\right)^{-1} = \left(\frac{1}{4,640} + \frac{1}{9,600} + 2\times 0.0001\right)^{-1} = 1,900$$

$$Q = UA \cdot \text{MLDT} = 1,900 \times 18.87 \times 16.7 = 599,000 \text{ W}$$

The unit is approximately 13 percent oversized.

9-11 NUMBER OF TRANSFER UNITS AND SPECIFIC PRESSURE DROP

Both the friction factor and the heat transfer coefficient are functions of the Reynolds number. When different plate types are compared for the same process service, it may happen that one of them has better J_H values for the same Reynolds number, which may make it look more attractive. But it may present too high a pressure drop, so it may be necessary to work with lower velocities. This means that the comparison cannot be effected at the same Reynolds number.

It is more useful to use other parameters, which are the number of transfer units (NTUs) and the specific pressure drop ε. The number of transfer units was already defined in relation to shell-and-tube heat exchangers. For a plate heat exchanger, the same definition applies:[4,6]

$$\text{NTU} = \frac{UA}{(Wc)_{\min}} \tag{9-11-1}$$

The specific pressure drop for each stream is defined as the quotient between pressure drop and the number of transfer units. This is

$$\varepsilon_h = \frac{\Delta p_h}{\text{NTU}} \qquad \varepsilon_c = \frac{\Delta p_c}{\text{NTU}} \tag{9-11-2}$$

where Δp_h and Δp_c are the pressure drops for the hot and cold streams, respectively.

To compare the performance of different plates, the analysis always must be performed with the same fluids. Then, if the physical properties of the fluids are constant, the heat transfer coefficients will be exclusively a function of the velocities and geometry.

If the heat exchanger configuration is adopted such that the velocities of both fluids are approximately the same (which is usual in plate heat exchanger design), then, for every geometry, the overall U will be a function of this velocity. This means that

$$U = U_{(v)}$$

where the subscript only denotes the functional dependence.

Since the area of the unit can be expressed as

$$A = A_p n_p n_s \qquad (9\text{-}11\text{-}3)$$

where A_p is the area of the plate, and

$$W = \rho v n_p b e \qquad (9\text{-}11\text{-}4)$$

NTU will be

$$\mathrm{NTU} = \frac{U_{(v)} A_p n_s}{(\rho v c)_{\min} be} \qquad (9\text{-}11\text{-}5)$$

And then

$$\frac{\mathrm{NTU}}{n_s} = \varphi_{(v)} \qquad (9\text{-}11\text{-}6)$$

for each model of plate. And

$$\varepsilon = \frac{\Delta p}{\mathrm{NTU}} = \frac{4 f (1/2) \rho v^2 L n_s}{\varphi_{(v)} n_s} = \varepsilon_{(v)} \qquad (9\text{-}11\text{-}7)$$

This means that ε is also a function of the velocity for each model of plate. So it is possible to use ε as a valid parameter to correlate the data instead of using the velocity.

Plate heat exchanger vendors can have curves of ε versus v for the different types of plates. These curves can be used to make a preliminary selection of the plate type for a certain service.

Since the specific pressure drop can be calculated for each service with only the process conditions [because $\mathrm{NTU} = Q/(F_t \cdot \mathrm{LMTD})$], with these curves, the manufacturer can select, for each type of plate, the fluid velocity necessary to achieve the desired specific pressure drop. This allows defining the number of plates in parallel in each pass by calculating U and selecting the number of passes in series with Eq. (9-11-6).

GLOSSARY

A = total heat transfer area (m^2)

A_p = heat transfer area of one plate (m^2)

b = channel width (m)

c = specific heat [J/(kg · K)]

D_c = distribution header diameter (m)

D_e = channel equivalent diameter (m)

e = channel depth (m) (see Fig. 9-18)

f = friction factor (dimensionless)

F_t = LMTD correction factor (dimensionless)

h = film coefficient [J/(s · m^2 · K)]

J_H = Colburn factor (dimensionless)

k = thermal conductivity [J/(s · m · K)]

L_e = effective plate length (m)

n_p = number of parallel channels per pass

n_s = number of passes

s = thickness of the corrugated plate (m) (see Fig. 9-18)

t = metal sheet thickness (m)

v = velocity (m/s)

W = mass flow (kg/s)

ρ = density (kg/m^3)

μ = viscosity (kg/ms)

β = corrugations angle (degrees)

ε = specific pressure drop (N/m^2)

ϕ_a = area correction factor owing to corrugations (dimensionless)

φ = number of transfer units per pass

REFERENCES

1. Marriot J.: "Where and How to Use Plate Heat Exchangers," *Chem Eng,* May 4, 1971, p. 127.

2. Buonopane R., Troupe R. A., Morgan J. C.: "Heat Transfer Design Method for Plate Heat Exchangers," *Chem Eng Progr* 59(7):57–61, 1963.

3. Watson E. L., McKillop A. A., Dunkley W. L., Perry R. L.: "Plate Heat Exchangers: Flow Characteristics," *Ind Eng Chem* 52(9):733–739, 1960.

4. *Thermal Handbook.* Sweden: Alfa Laval.

5. *Heat Transfer Handbook*: *Design and Application of Paraflow Heat Exchangers.* New York: APV Company.

6. Usher J.: "Evaluating Plate Heat Exchangers," in *Process Heat Exchange.* New York: McGraw-Hill, 1979.

7. Cowan C. T.: "Choosing Materials of Construction for Plate Heat Exchangers," in *Process Heat Exchange.* New York: McGraw-Hill, 1979.

8. Cross P. H.: "Preventing Fouling in Plate Heat Exchangers," in *Process Heat Exchange.* New York: McGraw-Hill, 1979.

CHAPTER 10
CONDENSATION OF VAPORS

10-1 *MECHANISMS OF CONDENSATION*

Condensation is the process by which a vapor transforms into a liquid. For condensation to take place, it is necessary to remove heat from the condensing fluid by means of a cooling medium.

If the vapor is a pure substance, as long as the pressure remains constant, condensation takes place isothermally. The temperature of the process is the saturation temperature of the vapor at the prevailing pressure. In this case, the amount of heat that must be removed per unit mass to achieve condensation of a saturated vapor is called the *latent heat of condensation* λ.

The process can be performed in a continuous fashion in a heat exchanger, which receives the specific name of *condenser*, as represented in Fig. 10-1. The heat balance can be expressed as

$$Q = W_h(i_2 - i_1) \tag{10-1-1}$$

For a pure vapor, the enthalpy difference is the latent heat of condensation λ; then

$$Q = W_h \lambda_h = W_c c_c (t_2 - t_1) \tag{10-1-2}$$

We have assumed that the cooling medium is a fluid that exchanges sensible heat. It is also possible to use as refrigerant a fluid that performs the opposite change of phase, removing heat from the condensing fluid and suffering a vaporization process (this is the case with ammonia or propane vaporization in refrigeration cycles).

In these cases, the enthalpy balance can be written as

$$Q = W_h \lambda_h = W_R \lambda_R \tag{10-1-3}$$

where the subscript R denotes the refrigerant.

In the foregoing discussion we have assumed that the condensing fluid is a pure substance that comes into the condenser at its saturation temperature. If, instead, the vapor is superheated, it will be necessary to extract from it an additional amount of heat to bring it to the saturation temperature, and the enthalpy balance will be

$$Q = W_h(i_2 - i_1) = W_h c_{hv}(T_1 - T_s) + W_h \lambda_h \tag{10-1-4}$$

where c_{hv} is the specific heat of the hot fluid at vapor state, T_s is the saturation temperature, and λ_h is the latent heat of condensation at the saturation temperature. The first term on the right side of Eq. (10-1-4) is called *sensible heat* (it is associated with a temperature change), whereas the second term is called *latent heat*.

If the vapor to be condensed is not a pure substance but a mixture of different components, the condensation will not be isothermal because there is a temperature range in which vapor and liquid may coexist. This temperature range extends between the dew-point temperature (where the first drop of condensate forms) and the bubble-point temperature (where complete condensation of the mixture is reached).

Bubble-point and dew-point temperatures of a mixture depend on its composition and pressure. Figure 10-2 is a condensation temperature versus composition diagram for a binary mixture at a certain pressure.

In this diagram, the evolution of a vapor that is cooled down from an initial temperature T_1 is represented. When the dew-point temperature T_R is reached, condensation starts. From then on, the composition

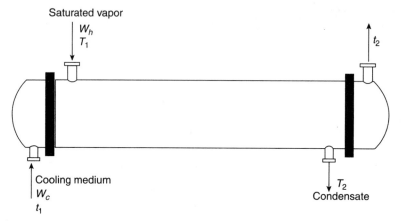

FIGURE 10-1 Shell-and-tube condenser.

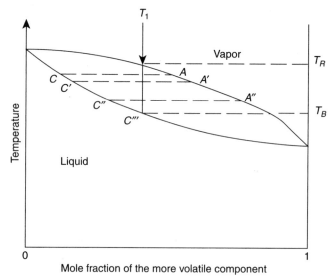

FIGURE 10-2 Condensation of a binary mixture.

of the vapor phase, while the temperature decreases, passes through points A, A', and A'', whereas the composition of the condensate liquid evolves through points C, C', C'', etc.

When the bubble-point temperature T_B is reached, the condensation is complete, and the liquid composition C''' coincides with the original composition of the vapor. We then can see that there is a temperatures range for the condensation of the mixture that extends between the dew-point and bubble-point temperatures.

The heat balance of the process is always expressed with Eq. (10-1-1), where i_1 is the inlet enthalpy and i_2 is the outlet enthalpy (which may correspond to the bubble-point or even lower temperature if the condensate is subcooled). Both enthalpies are evaluated at the temperatures corresponding to the fluid evolution.

Dropwise and Filmwise Condensation. To condense a vapor, it is necessary to extract heat from it. This heat extraction is performed with a refrigerant fluid that must be at a lower temperature than the fluid to be condensed.

Both streams, vapor and refrigerant, are separated by a solid wall that will be at an intermediate temperature between the fluid temperatures. For condensation to occur, it is necessary for the wall temperature to be below the dew-point temperature of the vapor at the prevailing pressure.

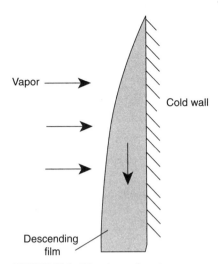

The condensation of a vapor over a cold surface may take place by two different mechanisms: dropwise or filmwise condensation. The first one takes place when the condensate exhibits little surface affinity for the wall. The vapor condenses in the form of small droplets that grow on the surface. These droplets act as nucleation centers for the condensation of additional vapor, increasing their size. When the weight of the droplets overcomes the surface attraction forces, the droplet falls down, leaving bare metal on which successive droplets of condensate may form.

The second mechanism is filmwise condensation. Let's consider a vertical-plane surface in contact with a vapor (Fig. 10-3). In film condensation, the vapor condenses on the cold surface, forming a continuous film. This film descends owing to the action of gravity, and the liquid flows downward.

FIGURE 10-3 Filmwise condensation on a vertical surface.

While the film drains, more liquid is incorporated into the film owing to vapor condensation. The film thickness increases downward because the amount of liquid in the film also increases.

All the condensation heat must be removed by the refrigerant fluid, which is on the other side of the solid wall. So all the condensation heat, transferred by the condensing vapor to the vapor-liquid interface, must go through the condensate film to pass into the refrigerant. This means that the condensate film constitutes an additional resistance to heat transfer. The values of the heat transfer coefficients for filmwise condensation thus are lower than those of dropwise condensation.

However, most fluids form condensate films. Additionally, dropwise condensation is an unstable process that turns into film condensation in a rather unpredictable way. Thus, since it is not possible to guarantee dropwise condensation, the usual assumption is to design on the hypothesis of film condensation, which implies a conservative approach.

10-2 CONDENSATION OF SINGLE-COMPONENT VAPORS

10-2-1 Filmwise Condensation: Nusselt Theory

Let's analyze the case of a vertical plate at temperature T_w in contact with a pure-component vapor saturated at temperature T_v. The vapor will condense over the solid surface, forming a condensate film, as shown in Fig. 10-4. The condensate flows downward owing to gravity.

Since more fluid incorporates into the film owing to vapor condensation as the condensate flows down, the liquid flow rate in the film, at a distance $x + dx$ from the upper edge of the plate is higher than the flow rate at distance x. This difference corresponds to the amount of vapor condensed on the surface Bdx per unit time.

If the mass of vapor condensing on the surface Bdx per unit time is designated as dW', we can define a vapor mass velocity toward the interface as

$$J = \frac{dW'}{Bdx} \quad [\text{kg/(s} \cdot \text{m}^2)]$$

(10-2-1)

If W is the mass flow of condensate in the film (kg/s), a mass balance gives

$$W|_{x+dx} - W|_x = dW' = JBdx$$

(10-2-2)

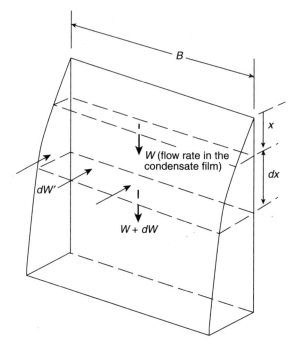

FIGURE 10-4   Liquid-film layer on a vertical surface.

So

$$dW = dW' = JBdx \qquad (10\text{-}2\text{-}3)$$

The model we shall develop in what follows is due to Nusselt[1] and is based on the following hypotheses:

1. The vapor is saturated, and it only delivers latent heat.

2. The drainage of the condensate film from the surface is by laminar flow only. It is assumed that the vapor velocity is low, so the vapor does not exert any drag force on the film. In other words, the shear stress at the liquid-vapor interface is nil.

3. A thermodynamic equilibrium exists at the interface. This means that the liquid temperature at the interface is also T_v, so there is no heat transfer resistance in the vapor phase.

4. The vapor condenses on the interface and delivers its latent heat of condensation to the liquid film. This amount of heat must be transmitted to the cold wall (and to the refrigerant that is on the other side) through the liquid film. It is assumed that the heat is transferred through the film by conduction.

Since the temperatures at both sides of the film are T_v (at the vapor-liquid interface) and T_w (at the solid surface), the temperature gradient for the heat transmission is $(T_v - T_w)/\delta_{(x)}$, where $\delta_{(x)}$ is the film thickness, which is a function of position x.

Condensation Heat Transfer Film Coefficient.   When a mass flow dW' (kg/s) condenses on the surface of the film, it delivers to the film an amount of heat per unit time given by

$$dQ = dW'\lambda \qquad (10\text{-}2\text{-}4)$$

This heat is delivered to a film surface area

$$dA = Bdx \qquad (10\text{-}2\text{-}5)$$

and must be transmitted by conduction through the liquid film. Applying Fourier's conduction law, we have

$$dQ = k_L \frac{(T_v - T_w)}{\delta_{(x)}} dA \qquad (10\text{-}2\text{-}6)$$

And the heat-flux density will be

$$q = \frac{dQ}{dA} = \lambda \frac{dW'}{dA} = \lambda J = \frac{k_L}{\delta_{(x)}}(T_v - T_w) \qquad (10\text{-}2\text{-}7)$$

We see that since the film thickness is a function of the position, the heat-flux density also will be.
    We can define the local heat transfer coefficient as

$$q_{(x)} = h_{(x)}(T_v - T_w) \qquad (10\text{-}2\text{-}8)$$

And comparing Eqs. (10-2-7) and (10-2-8), we get

$$h_{(x)} = \frac{k_L}{\delta_{(x)}} \qquad (10\text{-}2\text{-}9)$$

This means that if we know the film thickness at a certain position, we can calculate the local heat transfer coefficient by the simple Eq. (10-2-9). In what follows, we shall try to obtain a mathematical expression to calculate the film thickness at any location in the film.

Velocity Profile in a Descending Film.   The condensate layer drains owing to the force of gravity. At each point into the film, there is a fluid velocity that depends on the position. Figure 10-5 shows the velocity profile at a certain section of the condensate layer.
    In contact with the solid wall, the velocity is zero. According to the Nusselt's hypothesis, the velocity of the film at the vapor-liquid interface is not affected by the vapor. This means that there is no momentum transfer between the vapor and the liquid. Since there is no shear stress, the slope of the velocity profile at the interface must be 0.
    Let's consider a volume element belonging to the liquid layer, as shown in Fig. 10-5. Owing to the existence of a velocity profile, there are shear forces acting at the faces of the volume element, as indicated in the figure.
    The volume element must be in an equilibrium, resulting from the action of these forces and its own weight. If we analyze the forces acting on the volume element, we have

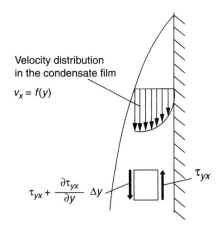

Velocity distribution in the condensate film

$v_x = f(y)$

FIGURE 10-5   Velocity distribution in the liquid film.

1. On the right side, near the cold vertical surface, there is a tangential force acting upward and tending to support the volume. This force is

$$-\tau_{yx} \Delta_x \Delta_z$$

2. On the opposite side there is a tangential force acting downward owing to the more rapid movement of the liquid downward as the distance from the surface is increased. This force is

$$\left(\tau_{yx} + \frac{\partial \tau_{yx}}{\partial y} \Delta y \right) \Delta x \Delta z$$

3. The weight of the volume element acting down, that is,

$$\rho g \Delta x \Delta y \Delta z$$

Equating these forces,

$$\rho g \Delta x \Delta y \Delta z + \frac{\partial \tau_{yx}}{\partial y} \Delta x \Delta y \Delta z = 0 \tag{10-2-10}$$

$$\therefore \frac{\partial \tau_{yx}}{\partial y} = -\rho g \tag{10-2-11}$$

Integrating,

$$\tau_{yx}' = -\rho g y + C_1 \tag{10-2-12}$$

Since for $y = \delta_{(x)}$, it is $\tau_{yx} = 0$

$$\therefore \tau_{yx} = \rho g (\delta_{(x)} - y) \tag{10-2-13}$$

(Note that even though $\delta_{(x)}$ is a function of x, it is a constant for the integration over y.) Since

$$\tau_{yx} = \mu \frac{dv_x}{dy} \tag{10-2-14}$$

we get

$$dv_x = \frac{\rho g (\delta_{(x)} - y)}{\mu} dy \tag{10-2-15}$$

Integrating again, we get

$$v_x = \frac{\rho g y \delta_{(x)}}{\mu} - \frac{\rho g y^2}{2\mu} + C_2 \tag{10-2-16}$$

For $y = 0$, $v = 0$ $\therefore C_2 = 0$, then

$$v_x = \frac{\rho g y \delta_{(x)}}{\mu} - \frac{\rho g y^2}{2\mu} \tag{10-2-17}$$

This is the expression of the velocity profile in a cross section of the film in which the film thickness is $\delta_{(x)}$.

The mean velocity in that section is

$$\bar{v} = \frac{1}{\delta_{(x)}} \int_0^{\delta(x)} v_x \, dy = \frac{1}{\delta_{(x)}} \int_0^{\delta(x)} \left(\frac{\rho g \delta_{(x)} y}{\mu} - \frac{\rho g y^2}{2\mu} \right) dy$$

$$= \frac{1}{\delta_{(x)}} \left(\frac{\rho g \delta^3_{(x)}}{2\mu} - \frac{\rho g \delta^3_{(x)}}{6\mu} \right) = \frac{1}{3} \frac{\rho g \delta^2_{(x)}}{\mu}$$

(10-2-18)

Condensate-Film Thickness. The condensate mass flow at any film section is

$$W = \rho \bar{v} B \delta_{(x)}$$

(10-2-19)

The change in the mass flow between a section at x and a section at $x + dx$ is

$$dW = d(\rho \bar{v} \delta_{(x)} B) = d\left(\frac{\rho^2 g \delta^3_{(x)} B}{3\mu} \right) = \frac{\rho^2 g \delta^2_{(x)} B}{\mu} d\delta_{(x)}$$

(10-2-20)

But, according to Eq. (10-2-3), $dW = dW' = JBdx$; then

$$Jdx = \frac{\rho^2 g \delta^2_{(x)}}{\mu} d\delta_{(x)}$$

(10-2-21)

but

$$J = \frac{q}{\lambda} = \frac{k(T_v - T_w)}{\delta_{(x)} \lambda}$$

(10-2-22)

$$\therefore \quad \frac{k}{\delta_{(x)}} \frac{(T_v - T_w)}{\lambda} dx = \frac{\rho^2 g \delta^2_{(x)} d\delta_{(x)}}{\mu}$$

(10-2-23)

$$\therefore \quad \frac{k(T_v - T_w)}{\lambda} dx = \frac{\rho^2 g \delta^3_{(x)} d\delta_{(x)}}{\mu}$$

(10-2-24)

Equation (10-2-24) is a differential equation that can be integrated to find $\delta(x)$ as a function of the position x. The integration is done with the following boundary conditions:

At $x = 0$, $\delta(x) = 0$.
At $x = x$ (generic), $\delta(x) = \delta(x)$.

And we get

$$\delta_{(x)} = \left[\frac{4k\mu(T_v - T_w)x}{\rho^2 g \lambda} \right]^{1/4}$$

(10-2-25)

This equation gives the film thickness at position x.

Condensation Heat Transfer Coefficient.   According to Eq. (10-2-9), the local heat transfer coefficient is

$$h_{(x)} = \frac{k}{\delta_{(x)}} = \left(\frac{k^3 \rho^2 g \lambda}{4 \mu \Delta T x} \right)^{1/4} \tag{10-2-26}$$

In practice, what interests us most is not the local heat transfer coefficient, but the mean value for the whole surface. Thus, if we consider a vertical plate of height L, we can define the mean coefficient as

$$Q = hA(T_v - T_w) \tag{10-2-27}$$

where Q is the heat flow transferred to the whole plate.
    For a surface element $dA = Bdx$, it will be

$$dQ = h_{(x)}(T_v - T_w)Bdx \tag{10-2-28}$$

Then

$$Q = \int_A dQ = \int_A h_{(x)}(T_v - T_w)B \, dx$$
$$= \left(\frac{k^3 \rho^2 g \lambda \Delta T^3}{4\mu} \right)^{1/4} \int_A \frac{dx}{x^{1/4}} B \tag{10-2-29}$$

Remembering that

$$\int \frac{dx}{x^{1/4}} = \frac{4}{3} x^{3/4} \tag{10-2-30}$$

the heat flow transferred will be

$$Q = \frac{4^{3/4}}{3} \left(\frac{k^3 \rho^2 \lambda \Delta T^3 L^3}{\mu} \right)^{1/4} B \tag{10-2-31}$$

Then, according to the definition given in Eq. (10-2-27),

$$h = \frac{Q}{A\Delta T} = \frac{Q}{LB\Delta T} = \frac{4^{3/4}}{3} \left(\frac{k^3 \rho^2 \lambda g}{\mu \Delta T L} \right)^{1/4} \tag{10-2-32}$$

This is

$$h = 0.943 \left(\frac{k^3 \rho^2 \lambda g}{\mu L \Delta T} \right)^{1/4} \tag{10-2-33}$$

Equation (10-2-33) is the final expresion that allows calculating the mean heat transfer coefficient for the condensation of a saturated vapor at temperature T_v over a vertical surface at T_w (where $\Delta T = T_v - T_w$) as a function of the physical properties of the condensate and the height of the plate. The expression also can be applied to condensation over vertical tubes.

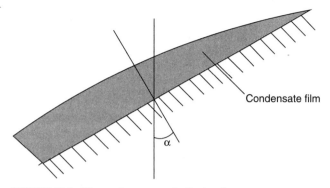

FIGURE 10-6   Film condensate on an inclined surface.

It can be demonstrated that if condensation occurs over an inclined surface whose normal forms a certain angle α with the vertical (see Fig. 10-6), the heat transfer coefficient is

$$h = 0.943 \left(\frac{k^3 \rho^2 \lambda g \sin \alpha}{\mu L \Delta T} \right)^{1/4} \qquad (10\text{-}2\text{-}34)$$

This expression can be obtained easily if in the deduction of the velocity profile it is considered that the component of the gravity acceleration in the direction of movement is $g \sin \alpha$ instead of g.

Condensation Over Horizontal Tubes.   When a vapor condenses over the exterior surface of a horizontal tube, as in Fig. 10-7, it also forms a condensate film on the tube, and it is possible to develop a similar model to that for vertical surfaces. The deduction is more complex since the system geometry also is, and will not be developed here (see ref. 2).

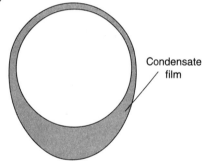

FIGURE 10-7   Film condensate on a horizontal tube.

The final expression is

$$h = 0.725 \left(\frac{k^3 \rho^2 \lambda g}{\mu D_o \Delta T} \right)^{1/4} \qquad (10\text{-}2\text{-}35)$$

Colburn-Type Expressions.   Equations (10-2-34) and (10-2-35) are called *Nusselt equations*. They can be written in terms of dimensionless numbers to obtain the so-called Colburn equations. These equations are expressed as a function of a condensate-film Reynolds number. We shall analyze what the expressions of the film Reynolds number are for the most usual geometries.

Condensation Over a Vertical Plate.   Let's consider a cross section at the lower end of the liquid layer, as shown in Fig. 10-8. The flow area for the liquid flow at this section is $B\delta_f$, where δ_f is the film thickness at the bottom end. The total amount of vapor condensing per unit time in the whole film is W_f (kg/s).

The amount of heat transferred per unit time then will be

$$Q = W_f \lambda \qquad (10\text{-}2\text{-}36)$$

(Note that W_f is the mass flow at the bottom end of the film.)

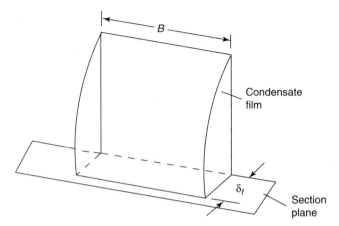

FIGURE 10-8 Cross section at the bottom end of the liquid film.

It is possible to define an equivalent diameter for the bottom section of the condensate film as

$$D_{\text{eq}} = \frac{4 \times \text{flow area}}{\text{wetted perimeter}} = \frac{4\delta_f B}{B} = 4\delta_f \qquad (10\text{-}2\text{-}37)$$

And we can define a mass velocity G as

$$G = \frac{\text{mass flow}}{\text{flow area}} = \frac{W_f}{B\delta_f} \qquad (10\text{-}2\text{-}38)$$

We also can define a Reynolds number as

$$\text{Re}_f = \frac{D_{\text{eq}} G}{\mu} = \frac{4\delta_f W}{\delta_f \mu B} = \frac{4W_f}{B\mu} \qquad (10\text{-}2\text{-}39)$$

Calling

$$G' = W_f / B \qquad (10\text{-}2\text{-}40)$$

The Reynolds number thus can be expressed as

$$\text{Re}_f = \frac{4G'}{\mu} \qquad (10\text{-}2\text{-}41)$$

where G' must be interpreted as a mass flow per unit film width $[(\text{kg})/(\text{m} \cdot \text{s})]$.

Condensation Over a Vertical Tube. It is also possible to define a Reynolds number for condensation over a vertical tube. A cross section at the base of the tube showing the film layer is shown in Fig. 10-9. If p is the tube perimeter ($= \pi D$), the equivalent diameter will be

$$D_{\text{eq}} = \frac{4p\delta_f}{p} = 4\delta_f \qquad (10\text{-}2\text{-}42)$$

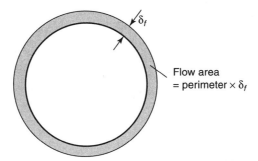

FIGURE 10-9 Condensate layer cross section at the base of a vertical tube.

and

$$G = \text{mass velocity} = \frac{W_f}{p\delta_f} \qquad (10\text{-}2\text{-}43)$$

then

$$\text{Re}_f = \frac{W_f 4\delta_f}{p\delta_f \mu} = \frac{4W_f}{p\mu} \qquad (10\text{-}2\text{-}44)$$

It is possible to define a mass flow per unit tube perimeter as

$$G' = \frac{W_f}{p} \qquad (10\text{-}2\text{-}45)$$

And then again,

$$\text{Re}_f = \frac{4G'}{\mu} \qquad (10\text{-}2\text{-}46)$$

Condensation Over a Vertical Tube Bundle. When the condensation takes place over a bundle of N vertical tubes, if W is the total condensate mass flow, the flow over each tube is

$$\text{Mass flow} = W/N \qquad (10\text{-}2\text{-}47)$$

And the Reynolds number is

$$\text{Re}_f = \frac{4G'}{\mu}$$

where

$$G' = \frac{W}{pN} \qquad (10\text{-}2\text{-}48)$$

The heat transfer coefficients can be expressed as a function of the film Reynolds number, as we shall explain in what follows.

According to Eq. (10-2-36),

$$Q = W\lambda \tag{10-2-49}$$

(From now on, we shall use W to express the total mass flow rate, dropping the subscript f, which was used to emphasize that it is the condensate mass flow at the bottom of the film.) At the same time,

$$Q = hA\Delta T \tag{10-2-50}$$

where A = condensation area
$= pL$ for condensation over a tube
$= pLN$ for condensation on a tube bundle
$= BL$ for a plane surface

Combining Eqs. (10-2-49) and (10-2-50), we get

$$h = \frac{\lambda W}{A\Delta T} \tag{10-2-51}$$

which can be expressed as

$$h = \frac{\lambda W}{pL\Delta T} = \frac{\lambda G'}{L\Delta T} \qquad \text{for condensation on a tube} \tag{10-2-52}$$

$$h = \frac{\lambda W}{pLN\Delta T} = \frac{\lambda G'}{L\Delta T} \qquad \text{for condensation on a tube bundle} \tag{10-2-53}$$

$$h = \frac{\lambda W}{BL\Delta T} = \frac{\lambda G'}{L\Delta T} \qquad \text{for condensation on a vertical plane} \tag{10-2-54}$$

Thus, in all cases, the result is

$$\Delta T = \frac{\lambda G'}{hL} \tag{10-2-55}$$

Substituting Eq. (10-2-55) in Eq. (10-2-33), we get

$$h\left(\frac{\mu^2}{k^3\rho^2 g}\right)^{1/3} = 1.47\left(\frac{4G'}{\mu}\right)^{-1/3} \tag{10-2-56}$$

Equation (10-2-56) is called a *Colburn-type expression* for condensation on vertical surfaces.

Horizontal Tubes.   When a vapor condenses over the surface of a horizontal tube, it also may form a condensate film. The film covers the tube surface and drains by gravity action, increasing its thickness from top to bottom, as shown in Fig. 10-10.

It is also possible to define a mass flow per unit film length. We shall define

$$G'' = W/LN$$

where N is the number of tubes and L is the tube length. Dimensions of G'' are also kg/(m · s).

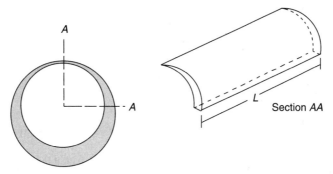

FIGURE 10-10   Condensate film on a horizontal tube.

Operating similarly to the case of vertical surfaces, substituting in Eq. (10-2-35), we come to

$$h\left(\frac{\mu^2}{k^3\rho^2 g}\right)^{1/3} = 1.51\left(\frac{4G''}{\mu}\right)^{-1/3} \tag{10-2-57}$$

Example 10-1   We must condense 1,400 kg/h of methanol vapors coming from a distillation column at 120 kPa. The condensation temperature at this pressure is 68.8°C. An existing condenser with 100 tubes, 19 mm external diameter, and 3 m in length will be used. Calculate which should be the wall temperature of the tubes if the condenser is installed in vertical and horizontal positions.

Solution   Physical properties of liquid methanol at 68.8°C are

$\rho = 738$ kg/m^3 $k = 0.161$ W/(m · K)

$\mu = 0.32$ cP $c = 3,773$ J/(kg · K)

$\lambda = 1,090$ kJ/kg

$W = 1,400$ kg/h $= 0.389$ kg/s

$$\text{Condenser area} = \pi D_o LN = 3.14 \times 0.019 \times 3 \times 100 = 17.9 \text{ m}^2$$

The heat flow yielded during condensation will be

$$Q = W\lambda = 0.389 \times 1,090 = 423.8 \text{ kW}$$

For a vertical condenser,

$$G' = W/\pi D_o N = 0.389/(3.14 \times 0.019 \times 100) = 6.52 \times 10^{-2} \text{ kg/(m · s)}$$

$$h\left(\frac{\mu^2}{k^3\rho^2 g}\right)^{1/3} = 1.47\left(\frac{4G'}{\mu}\right)^{-1/3} = 1.47\left(\frac{4 \times 6.52 \times 10^{-2}}{0.32 \times 10^{-3}}\right)^{-0.33} = 0.161$$

$$h = 0.161\left(\frac{\mu^2}{k^3\rho^2 g}\right)^{-1/3} = 0.161\left[\frac{(0.32 \times 10^{-3})^2}{0.161^3 \times 738^2 \times 9.8}\right]^{-0.33} = 887$$

Thus heat flow must be

$$Q = h(T - T_w)A$$

$$\therefore T_w = T - \frac{Q}{Ah} = 68.8 - \frac{423,800}{887 \times 17.9} = 42.1°C$$

In the case of a horizontal condenser, according to the definition of G'', we get

$$G'' = W/NL = 0.389/3 \times 100 = 1.29 \times 10^{-3} \text{ kg/(m} \cdot \text{s)}$$

But it must be considered that in a horizontal tube bundle, part of the condensate formed on a tube drains over the tubes below, thus increasing their condensate flow rate. Nusselt equations, then, would not be valid. Kern[2] proposes to correct the correlations using in this case a G'' value calculated as

$$G'' = \frac{W}{N^{2/3}L}$$

where N is the total number of tubes in the bundle. Then

$$G'' = 0.389/(3 \times 100^{0.66}) = 6.2 \times 10^{-3} \text{ kg/(m} \cdot \text{s)}$$

$$h\left(\frac{\mu^2}{k^3\rho^2 g}\right)^{1/3} = 1.51\left(\frac{4G''}{\mu}\right)^{-1/3} = 1.51\left(\frac{4 \times 6.2 \times 10^{-3}}{0.32 \times 10^{-3}}\right)^{-0.33} = 0.35$$

$$h = 0.35\left(\frac{\mu^2}{k^3\rho^2 g}\right)^{-1/3} = 0.35\left[\frac{(0.32 \times 10^{-3})^2}{0.161^3 \times 738^2 \times 9.8}\right]^{-0.33} = 1,927 \text{ W/(m}^2 \cdot \text{K)}$$

And the necessary wall temperature is

$$T_w = T - \frac{Q}{Ah} = 68.8 - \frac{423,800}{1,967 \times 17.9} = 56.7°C$$

We can see that a lower wall temperature (and hence a lower refrigerant temperature) is required for the vertical condenser because of the smaller heat transfer coefficient.

10-2-2 Film Condensation in Turbulent Flow

One of the hypotheses assumed by Nusselt was that of laminar flow in the condensate film. This is true for film Reynolds numbers [Eqs. (10-2-41) and (10-2-46)] lower than 1,600 (some authors extend the laminar Reynolds number range up to 2,100).

Dukler[3] developed a semitheoretical model for turbulent-flow condensate films. In those cases where the vapor velocity is still low (Nusselt hypothesis about nonexistence of shear stress at the vapor liquid interface continues being valid), the Dukler solution can be represented by the graph in Fig. 10-11.

In this graph, the abscissa corresponds to the film Reynolds number, and the parameter is the Prandtl number of the condensate. For low Reynolds numbers, all curves are close to Nusselt's prediction (dashed line in Fig. 10-11).

However, for high Reynolds numbers, the coefficients calculated with Dukler's model are considerably higher than those obtained with Nusselt's equations (except for very low Prandtl numbers, which correspond to liquid metals). This is explained by the presence of important convective effects in the condensate film when flow becomes turbulent.

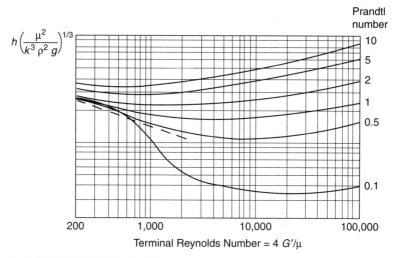

FIGURE 10-11 Dukler's plot of the condensate film coefficient. (*From* Chem. Eng. Progr. *55(10): 62–67, 1959. Reproduced with permission from the American Institute of Chemical Engineers.*)

10-2-3 Effect of Vapor Velocity

Up to now, we have assumed that the shear stress is nil at the vapor-liquid interface. This hypothesis may be far from reality in cases where the vapor velocity is considerable, situations found more frequently when condensation takes place inside tubes. We shall examine some models that take into account this effect.

Condensation Inside Horizontal Tubes.   In the case of condensation into horizontal tubes, two different flow regimes may exist: stratified and annular. These are represented in Fig. 10-12. (There are other two-phase intermediate flow patterns, but these are the two most distincts ones.) The former of these regimes corresponds to the case where the vapor velocity is low. In this case, gravitational forces prevail, and film

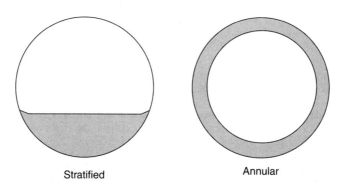

Stratified   Annular

FIGURE 10-12 Flow patterns in condensation within horizontal tubes.

coefficients can be obtained with a model similar to Nusselt's (the film regime is always laminar in horizontal tubes).

Kern[2] proposed a correction to take into account the flooded portion of the tube and suggested the following correlation for the stratified flow pattern:

$$h = 0.761 \left[\frac{k_L{}^3 \rho_L (\rho_L - \rho_V) g}{G'' \mu_L} \right]^{1/3} = 0.815 \left[\frac{k_L{}^3 \rho_L (\rho_L - \rho_V) g \lambda}{\mu_L D_i \Delta T} \right]^{1/4} \qquad (10\text{-}2\text{-}58)$$

When the vapor velocity is high, vapor displaces the liquid against the tube wall, and an annular flow patterns results. Several correlations were proposed for this annular flow regime.

The Boyko-Krushilin correlation[5] allows calculation of a mean condensation coefficient for a stream whose inlet liquid mass fraction is x_i and whose outlet mass fraction is x_o (x = mass of liquid present/total mass). The expresion is

$$\frac{hD_i}{k_L} = 0.024 \left(\frac{D_i G_T}{\mu_L} \right)^{0.8} \text{Pr}_L{}^{0.43} \left[\frac{(\rho_m/\rho_v)_i{}^{0.5} + (\rho_m/\rho_v)_o{}^{0.5}}{2} \right] \qquad (10\text{-}2\text{-}59)$$

where G_T is the mass flow density calculated with the total flow (vapor + liquid). Then

$$(\rho_m/\rho_v)_i = 1 + \frac{\rho_L - \rho_V}{\rho_V} x_i \qquad (10\text{-}2\text{-}60)$$

$$(\rho_m/\rho_v)_o = 1 + \frac{\rho_L - \rho_V}{\rho_V} x_o \qquad (10\text{-}2\text{-}61)$$

Another correlation is that of Carpenter and Colburn:[6]

$$\frac{h\mu_L}{k_L \rho_L{}^{1/2}} = 0.065 . \text{Pr}_L{}^{1/2} \left(\frac{f G_{V_m}^2}{2\rho_V} \right)^{1/2} \qquad (10\text{-}2\text{-}62)$$

where

$$G_{V_m} = \left(\frac{G_{v_i}^2 + G_{V_i} G_{V_o} + G_{v_o}^2}{3} \right)^{1/2} \qquad (10\text{-}2\text{-}63)$$

In Eq 10-2-62 f is the friction factor evaluated for a single-phase system at a Reynolds number of

$$\text{Re}_{V_m} = \frac{D_i G_{V_m}}{\mu_V} \qquad (10\text{-}2\text{-}64)$$

What we still must decide is when we should use the stratified or annular model. A simple recipe consists in calculating the coefficient with both models and adopting the higher.

Condensation Inside Vertical Tubes.   The effect of the circulating vapor will be different based on whether it circulates upward or downward. In the first case the effect of vapor is to slow down the liquid drainage, whereas in the second case the effect is to accelerate it. Let's consider first the case of vapor circulating downward. If the vapor velocity is low, it is possible to use the same model as for condensation outside tubes and employ Dukler's graph. If, on the other hand, vapor velocities are high, the annular model described for horizontal tubes may be used. Again, the recommendation is to try both models and adopt the higher coefficient or an average of both.[7]

For vapor flowing upward, the design must be done keeping the vapor velocity low enough to avoid flooding, which may happen when vapor prevents film drainage. One criterion suggested by Wallis[8] establishes that flooding will be avoided if it is

$$v_G^{*1/2} + v_L^{*1/2} < 0.6$$

where

$$v_G^* = \left[\rho_G v_G^2 / (\rho_L - \rho_G) g D_i \right]^{1/2} \tag{10-2-65}$$

and

$$v_L^* = \left[\rho_L v_G^2 / (\rho_L - \rho_G) g D_i \right]^{1/2} \tag{10-2-66}$$

If it is checked that flooding will not be an issue, the shear stress at the vapor-liquid interface will be low enough to have no effect on the liquid-film drainage, and h can be calculated with the methods employed for low vapor velocity.

Condensation Outside Tubes with High Vapor Velocities. In distillation-column condensers, where water is the cooling medium, condensation is usually in the shell side, so the interiors of the tubes can be cleaned easily (the condensing vapor is a clean fluid). In this case, baffled exchangers are used, and baffle spacing determines the velocity of the vapor.

There is little information in the open literature about design methods for these cases. A model proposed by Mueller[7] consists of adding a convective effect depending on vapor velocity to the coefficient calculated with Dukler's correlation, The model is based on the following equations:

$$\text{Nu} = \frac{h_f' D}{k_L} = \left(X^4 \cdot \text{Re}_{LG}^2 + \text{Nu}_f^4 \right)^{1/4} \tag{10-2-67}$$

where

$$\text{Re}_{LG} = \frac{DG\rho_L}{\mu_L \rho_G} \tag{10-2-68}$$

$$G = \frac{\text{vapor mass flow}}{\text{shell flow area}} \tag{10-2-69}$$

and

$$X = 0.9 \left(1 + \frac{1}{RH} \right)^{1/3} \tag{10-2-70}$$

h_f' is the coefficient corrected by vapor velocity. The parameters included in the correlations are defined as

$$R = \frac{\rho_L \mu_L}{\rho_G \mu_G} \tag{10-2-71}$$

$$H = \frac{c_L (T_{\text{sat}} - T_w)}{\text{Pr}_L \, \lambda} \tag{10-2-72}$$

$$\text{Nu}_f = \frac{h_f D}{k_L} \qquad (10\text{-}2\text{-}73)$$

h_f is the coefficient obtained with Dukler's model without considering the effect of vapor velocity.

It must be noted that if vapor velocity tends to be 0, in Eq. (10-2-67), Nu = Nu$_f$, and the coefficient coincides with Dukler's model. This means that this model is more general, and it is convenient to employ it in all cases. In the following section we shall see some application examples of these models for condensers design, but before that, we shall explain the use of this type of equipment in distillation installations.

For readers who are not familiar with the distillation operation, we include App. A, which illustrates some basic concepts about this subject.

10-3 SINGLE-COMPONENT VAPORS CONDENSERS

A condenser is a piece if equipment in which a vapor is condensed by heat extraction with a cooling medium. In process industry, the most usual applications are the condensers of distillation columns (see Fig. 10-13). The vapors exiting the top of the distillation column come into the condenser. Heat is removed by the cooling medium, and vapor condenses.

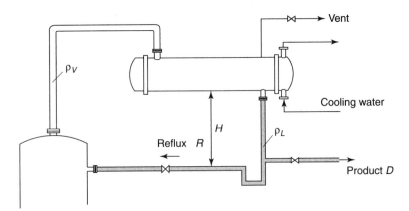

FIGURE 10-13 Horizontal gravity-flow condenser.

The condensate stream is split in two other streams. One of them is returned to the column as reflux R to wash the vapors ascending through the column, thus extracting the heavy components, and the other is withdrawn as distillate product D.

The cooling medium is usually water. If no cooling water is available, air-cooled condensers may be employed. Sometimes a process stream that must be heated is used as coolant.

The more common type is the shell-and-tube condenser. This is a heat exchanger with some modifications with respect to those used in single-phase applications.

Figure 10-14 shows two possible ways to install the condenser. The figure at the left corresponds to an installation with forced reflux return and is the more usual. The condensate discharges from the condenser into a reflux accumulator. The reflux pumps take the condensate from the accumulator, pumping it back to the column and extracting the distillate. The condenser is usually installed above the accumulator, and the accumulator must be at enough height above the pumps to provide the required net positive suction head (the condensate is a liquid at its bubble-point temperature, so the levels difference must be higher than the required NPSH). If it is possible to install the condenser at a level above the top of the column, a gravity-return reflux system can be used. This system is shown in the figure at the right. The obvious advantage is

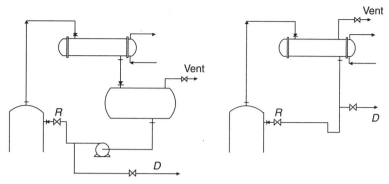

FIGURE 10-14   Forced reflux and gravity reflux condensers.

simplicity because pumps are not required. The liquid seal in the return line, shown in the right side figure, avoids vapor reverse flow through this line.

The vapors coming from the column suffer a pressure drop when they circulate through the condenser. In the case of gravity reflux, for the fluid to circulate, it is necessary that the hydrostatic pressure difference between the vapor and liquid legs be high enough to overcome the frictional pressure drop. Referring to Fig. 10-13, this can be expressed as

$$Hg(\rho_L - \rho_V) \geq \Delta p_{\text{friction}} \tag{10-3-1}$$

where

$$\Delta p_{\text{friction}} = \Delta p_{\text{condenser}} + \Delta p_{\text{piping}} + \Delta p_{\text{valve}} \tag{10-3-2}$$

During steady-state operation, the liquid level in the return pipe will stabilize in a position H that satisfies the equality in Eq. (10-3-1).

Usually the vapor density is much lower than that of liquid, and it can be written approximately as

$$H = \frac{\Delta p_{\text{friction}}}{\rho_L g} \tag{10-3-3}$$

Equation (10-3-3) allows calculating the height difference between the condenser bottom and the column that is required for the operation. This condition must be checked during the design of the system because if the height is not enough to allow the condensate discharge, the liquid level will reach the condenser, and as the condenser floods, the heat transfer surface available for condensation will be reduced.

The value of $\Delta p_{\text{friction}}$ depends on the vapor mass flow into the condenser. If the column is operated above its design capacity, the increase in $\Delta p_{\text{friction}}$ may cause condenser flooding, and this becomes a system bottleneck.

The gravity-return reflux system, even it presents the advantage of elimination of the reflux pump, makes it necessary to install the condenser at a high level, which in most cases is uneconomical because it requires expensive structures. Forced-return reflux systems are more stable than gravity-return systems because the hydraulics of the system are independent of the operating conditions of the column.

Condensers are usually provided with a vent line, as shown in Fig. 10-14. This is used to purge the condenser, eliminating the inert noncondensable gases that may be present in the system. These gases may consist of air or light components that enter the system dissolved in the feed stream and cannot be condensed. Air also can get into the system during plant shutdowns. If these gases are not purged, they reach the condenser, and because they cannot condense, they accumulate there, so their partial pressures increase with time.

This has two harmful effects on the condenser operation. Since the total pressure at which the condenser operates is the sum of the partial pressures of vapor and noncondensable gas, the higher the partial pressure

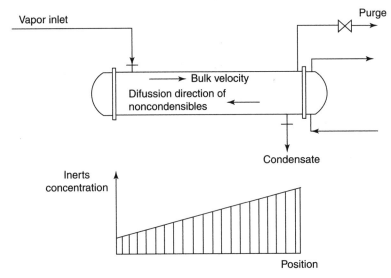

FIGURE 10-15 Inerts concentration profile in a condenser.

of the last one, the lower will be the partial pressure of the condensing vapors to maintain the total pressure. But the partial pressure of the condensing vapors is related to the condensing temperature. To condense the vapors at a lower partial pressure, a lower temperature is required. This would make a higher refrigerant flow rate or a lower refrigerant temperature necessary. If this is not possible, the capacity of the system is affected.

The second effect of the accumulation of inert components is a very pronounced reduction in the condensing heat transfer coefficient. Othmer[9] analyzed the case of condensation of water vapor at low pressure. With a temperature difference between the vapor and cooling media of 11°C, the condensing coefficient is reduced from 11,000 to 6,000 W/(m$^2 \cdot$ K) when the inert concentration is 1 percent and to 4,000 W/(m$^2 \cdot$ K) for a 2 percent concentration. This is due to the fact that inert gases accumulate over the heat transfer surface, thus forming a cold film through which the condensing vapor must diffuse to reach the cold surface. This creates an additional heat transfer resistance in the process.

To avoid inert gas buildup, part of the vapor-inert mixture is vented, either in a continuous or discontinuous way. This maintains a low inert concentration in the condenser. Purging is usually performed at the opposite end from the vapor inlet nozzle because this is the zone of the condenser where inert concentrations are higher.

To favor the inert purging, it is convenient to increase the vapor velocity into the condenser. This effect is explained as follows: As long as vapors condense, the mole fraction of inert gas in the mixture increases. This means that there is an inert concentration gradient across the condenser, as shown in Fig. 10-15. As a consequence of this gradient, inerts would tend to diffuse toward the inlet end. Since it is desired to keep a high inert concentration at the outlet end to minimize the loss of vapors with the purge, it is necessary to increase the vapor velocity near the outlet end to counteract diffusion. This is attained with a proper baffle spacing.

10-3-1 Different Types of Condensers

Condensers can be horizontal or vertical. Also, condensation can take place inside or outside the tubes. We shall briefly analyze the characteristics of each type.

Vertical Condenser with Condensation on the Shell Side. The installation of this type is represented in Fig. 10-16. The condensate formed on the tubes surface drains downward, forming a film. This film is interrupted by the presence of baffles (Fig. 10-17). Since the baffle holes are drilled with diameters only

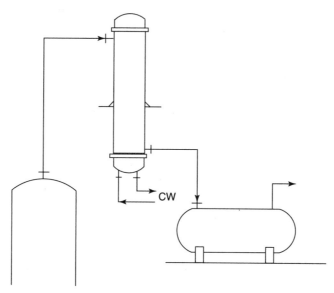

FIGURE 10-16 Vertical condenser installation.

0.8 mm higher than the tube diameter, the film thickness cannot be greater than this, so part of the condensate is retained by the baffle and is reentrained by the vapor.

Usually, in this type of unit, the condensation coefficient depends more on the vapor velocity than on the film thickness, and a conservative assumption is to calculate the resistance of the liquid film as if the baffles were not present.

Horizontal Condenser with Condensation on the Shell Side. This is a horizontal heat exchanger. The cooling medium, usually water, circulates inside the tubes. The vapors condense on the exterior surface of the tubes. A condensate layer is formed over the tubes. The condensate flows downward around the tube perimeter and drops to the bottom of the unit.

If vapor velocities are low, the condensation coefficient could be predicted with the Nusselt model. It was explained on Example 10-1 that a correction to the Nusselt model is necessary to take into account that the liquid draining from a tube may drop on the tubes below, thus increasing the condensate load on the lower tubes.

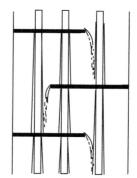

FIGURE 10-17 Effect of baffles in vertical shell-side condensation.

When baffles are installed in a horizontal condenser, they should have drain areas at the bottom so that condensate can flow toward the outlet. Otherwise, condensate would be trapped between baffles, obstructing the vapor circulation, as is the case in Fig. 10-18. Another solution, shown in Fig. 10-19, is to rotate the baffles 90 degrees.

Vertical Condenser with In-Tube Condensation. Figure 10-20 shows different ways to install a vertical condenser with in-tube condensation. A condensate film forms on the interior surface of the tubes. The film drains downward into the lower head. Vapor and liquid flow in the same direction. This favors the condensation heat transfer coefficient because it decreases the film thickness.

Horizontal Condenser with In-Tube Condensation. When cooling water is used as refrigerant, it is usually preferred to allocate the water in the tube side for cleaning facility. However, in air-cooled condensers,

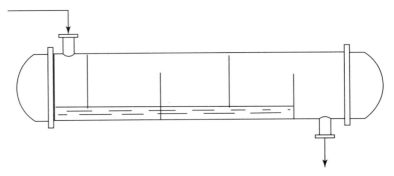

FIGURE 10-18 Wrong installation of baffles.

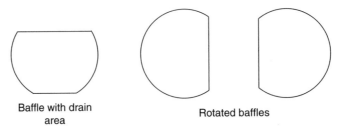

Baffle with drain
area

Rotated baffles

FIGURE 10-19 Baffles for horizontal condensers.

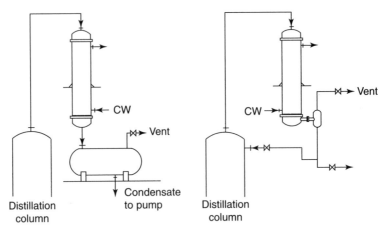

FIGURE 10-20 Vertical condenser installation.

or when a refrigeration cycle using boiling propane, ammonia, or other fluids is employed, condensation is usually on the tube side

If the condenser has only one tube pass, or in a U-tube condenser, each tube receives the same amount of vapor and produces the same amount of condensate. If a multipass heat exchanger with return heads is employed, however, vapor-liquid separation may take place into the heads. Then, in the following pass, the lower tubes will receive proportionately more liquid than the upper tubes. Since a mathematical model taking into account this phase separation would be complicated, the effect is usually not considered, and it is assumed that there is a uniform behavior within all tubes. Figure 10-21 shows a condenser refrigerated by a propane cycle.

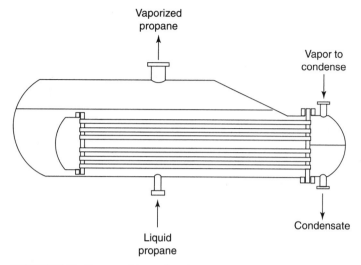

FIGURE 10-21   Propane-refrigerated condenser.

Partial Condensers.   Very frequently, the top product of a distillation column is not obtained as a liquid but as a vapor, for example, in the case of hydrocarbon stabilizer units. In these cases, the condenser only condenses the required reflux mass flow.

All the installation diagrams previously considered can be easily adapted for partial condensation. In this case, we shall not have the liquid product line, and the vapor outlet line (which in total condensers is used for inerts purging) must have the necessary diameter to evacuate the noncondensed vapors (see Fig. 10-22).

10-3-2   Design of Pure-Fluid Condensers

Process condenser design is performed with a similar methodology to that used for single-phase heat exchangers. The main difference lies in the fact that in order to calculate the film coefficients corresponding to the condensing vapor side, the correlations studied in previous sections of this chapter must be employed.

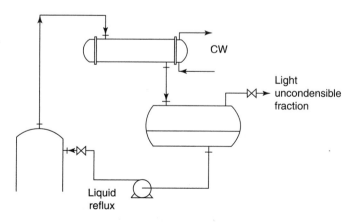

FIGURE 10-22   Partial condenser.

As usual, the types of problems that can exist are as follows:

1. To verify whether a unit with known geometry will accomplish a certain process service (rating)

2. To find the proper geometry of a unit to accomplish a certain process service (design)

Problems of the second type are solved by proposing a preliminary geometry based on typical heat transfer coefficients obtained from tables and rating the unit for the desired process conditions, so the task is transformed in the first type of problem.

We shall study first the case of single-component total condensers with no condensate subcooling and receiving vapors at the dew-point condition. These hypotheses, even though they look very restrictive, correspond to conditions usually found in binary distillation units in which the top product of the distillation is almost a single-component fluid.

Heat Balance of the Condenser and Mean Temperature Difference.   In the simple case of a pure-fluid condenser with cooling water as the refrigerant medium, the heat balance is

$$Q = W_h \lambda = W_c c_c (t_2 - t_1) \tag{10-3-4}$$

To calculate the heat transfer area of any heat exchanger device, the general expression is

$$Q = \int_0^A U(T - t) \, dA \tag{10-3-5}$$

where $T - t$ is the local temperature difference existing at any location in the unit, which can be obtained from a diagram such as that in Fig. 10-23 (which represents the particular case of a countercurrent condenser).

Usually, a mean temperature difference is defined as

$$Q = UA\Delta T_{\text{mean}} \tag{10-3-6}$$

If U can be considered constant, then

$$\Delta T_{\text{mean}} = \frac{\int_0^A (T - t) \, dA}{A} \tag{10-3-7}$$

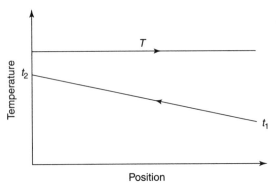

FIGURE 10-23 Temperatures diagram in the condenser of a pure substance.

When for each fluid there exists a proportionality constant between the amount of heat that the fluid yields or receives and the temperature variation it experiences, it can be demonstrated that for parallel or countercurrent configurations, ΔT_{mean} is the logarithmic mean temperature difference. Thus, for this case,

$$\Delta T_{mean} = \frac{(T - t_2) - (T - t_1)}{\ln \dfrac{(T - t_2)}{(T - t_1)}} \qquad (10\text{-}3\text{-}8)$$

The case of pure-component condensers can be analyzed as a particular case of a heat exchanger where the proportionality constant between the heat yielded by the condensing steam and its temperature variation is infinite. (A fluid with an infinite heat capacity can exchange heat without changing its temperature.) For this particular case, it is then valid to use the LMTD.

It can easily be seen that for an isothermal condenser, the LMTD for a countercurrent configuration coincides with that of a parallel-current configuration. If the condenser has two or more tube or shell passes, since these geometries correspond to different combinations, parallel-countercurrent, the value of ΔT_{mean} also will be the same. This means that the F_t factor, which corrects the temperature difference in multipass units, is always unity when one of the fluids is isothermal.

In a design problem, to transfer a certain amount of heat Q, the higher the temperature difference, the smaller will be the heat transfer area required. Usually the inlet temperature of the cooling medium t_1 is an input for the designer, for example, the available cooling water temperature. Thus, in order to have a high value of ΔT_{mean}, it is necessary to have a low refrigerant outlet temperature. According to Eq. (10-3-4), though, to transfer a certain amount of heat, the lower the refrigerant outlet temperature, the higher mass flow W_c will be necessary. Thus the equipment designer must adopt t_2 as a compromise between a bigger exchanger and a higher refrigerant consumption.

Condensation Heat transfer Coefficient.   The overall heat transfer coefficient can be expressed as

$$\frac{1}{U} = \frac{1}{h_c} + \frac{1}{h_R} + R_f \qquad (10\text{-}3\text{-}9)$$

In Eq. (10-3-9), h_R is the film coefficient corresponding to the cooling medium. This is calculated with the correlations usually employed for single-phase heat exchangers. The fouling resistance R_f is obtained from tables or from the design basis of the project. To calculate the condensation heat transfer coefficient, it is necessary to choose the proper correlation depending on the type of condenser. The following situations are possible:

Shell-Side Condensation: Vertical Condenser.   Use the model corresponding to Eqs. (10-2-67) through (10-2-73). The value of h_f is calculated with Dukler's graph.

Shell-Side Condensation: Horizontal Condenser.   Again, the model represented by Eqs. (10-2-67) through (10-2-73) is used. In this case, to calculate h_f, it must be considered that the condensate formed on the tube surface drops on the tubes below, increasing the condensate load and film thickness in the lower tubes. As explained previously, to take into account this effect, Kern [2] proposed to calculate h_f with Eq. (10-2-57) but using for G'' the following expression:

$$G'' = \frac{W_h}{L N^{2/3}} \qquad (10\text{-}3\text{-}10)$$

where N is the total number of tubes of the condenser and W_h is the total condensate mass flow.

Condensation Inside Horizontal Tubes.   Use Eq. (10-2-58) for stratified flow and Eq. (10-2-59) or Eq. (10-2-62) for annular flow, and choose the higher calculated coefficient.

Condensation Inside Vertical Tubes.   For downward vapor flow, use Dukler's model (low vapor velocity) and annular flow [Eqs. (10-2-59) to (10-2-62), high vapor velocities]. Choose the higher calculated coefficient.

Example 10-2   It is desired to condense 8,000 kg/h of acetone at a pressure of 120 kPa using cooling water at 35°C. At this pressure, the condensation temperature of acetone is 61°C. A cooling water flow rate of 100 m³/h will be used. After some preliminary trials, the following design was proposed:

N = number of tubes: 192

L = tube length: 4 m

Tube diameters: D_o: 19.05 mm; D_i: 14.4 mm; triangular array, 30 degrees; tube pitch: 23.81 mm

D_s = shell diameter = 438 mm

n = number of tube passes: 2

B = baffles spacing: 450 mm

Physical properties are as follows:

	Vapor	Liquid
Density, kg/m³	2.54	738
Viscosity, cP	0.0072	0.215
Thermal conductivity, W/(m · K)	0.0147	0.142
Specific heat, J/(kg · K)	1,381	2,255
Heat of condensation, kJ/kg	497.5	

Verify whether this unit is suitable to be installed in horizontal and vertical positions, and calculate the pressure drops for acetone and the cooling water. Use a total fouling factor of 0.0005 (m² · K)/W.

Solution   Water will flow inside the tubes.

$$\text{Heat balance } Q = W_h \lambda = (8,000/3,600) \times 497.5 = 1,105 \text{ kW}$$

$$\text{Water mass flow } W_c = 100,000/3,600 = 27.77 \text{ kg/s}$$

$$\text{Water flow area } a_t = (\pi D_i^2/4)N/n = (\pi \times 0.0144^2/4) \times 192/2 = 0.0156 \text{ m}^2$$

$$\text{Water velocity } = W_c/\rho a_t = 27.7/(1,004 \times 0.0156) = 1.76 \text{ m/s}$$

$$\text{Water outlet temperature} = 35 + Q/W_c c_c = 35 + 1.105 \times 10^6/(27.7 \times 4,180) = 44.5°C$$

Cooling-water film coefficient:

$$h_i = 1,423(1 + 0.0146t)V^{0.8}/D_i^{0.2} = 1,423(1 + 0.0146 \times 39.8) \times 1.76^{0.8}/0.0144^{0.2} = 8,250$$

$$h_{io} = h_i D_i/D_o = 8,250 \times 0.0144/0.019 = 6,250 \text{ W/(m}^2 \cdot \text{K)}$$

Shell-side heat transfer coefficient:

$$\text{Free distance between tubes } c = P_t - D_o = 23.8 - 19.05 = 4.8 \text{ mm}$$

$$\text{Shell-side flow area } a_s = (D_s cB)/P_t = (0.438 \times 0.0048 \times 0.45)/0.0238 = 0.0398 \text{ m}^2$$

The inlet vapor flow is 8,000 kg/h. At the outlet, the vapor flow is zero because all the vapor condenses. We shall consider a mean vapor flow $W_s = 8,000/2 = 4,000$ kg/h.

$$\text{Mass velocity } G_s = W_s/a_s = 4,000/(3,600 \times 0.0398) = 27.89 \text{ kg/(s} \cdot \text{m}^2)$$

According to Eq. (10-2-68),

$$\text{Re}_{LG} = (D_o G_s \rho_L)/(\rho_G \mu_L) = (0.019 \times 27.89 \times 738)/(0.215 \times 10^{-3} \times 2.54) = 715,800$$

According to Eq. (10-2-71),

$$R = \rho_L \mu_L/\rho_G \mu_G = (738 \times 0.215 \times 10^{-3})/(2.54 \times 0.0072 \times 10^{-3}) = 8,676$$

Liquid Prandtl number $\text{Pr}_L = c_L \mu_L/k_L = 2,255 \times 0.215 \times 10^{-3}/0.142 = 3.414$

According to Eq. (10-2-72),

$$H = c_L(T - T_w)/\text{Pr}_L \lambda$$

Let's assume a wall temperature of 50°C (to be checked later). Then

$$H = 2,255(61 - 50)/(3.414 \times 497,500) = 0.0146$$

According to Eq. (10-2-70),

$$X = 0.9(1 + 1/RH)^{0.33} = 0.9[1 + 1/(8,676 \times 0.0146)]^{0.33} = 0.902$$

To calculate h_f, we must consider horizontal and vertical installation separately.
Vertical condenser:

$$G' = W/(N\pi D_o) = [8,000/(3,600 \times 3.14 \times 192 \times 0.019)] = 0.194 \text{ kg/(m} \cdot \text{s)}$$

$$\text{Re}_f = 4G'/\mu_L = 4 \times 0.194/0.215 \times 10^{-3} = 3,609$$

From Dukler's graph, with $\text{Re}_f = 3,609$ and $\text{Pr}_L = 3.414$, we get an ordinate value = 0.28. This means that

$$h_f(\mu_L^2/\rho_L^2 g k_L^3)^{0.33} = 0.28$$

Then

$$h_f = 0.28(\rho_L^2 g k_L^3/\mu^2)^{0.33} = 0.28[738^2 \times 9.8 \times 0.142^3/(0.215 \times 10^{-3})^2]^{0.33} = 1,772 \text{ W/(m}^2 \cdot \text{K)}$$

$$\text{Nu}_f = h_f D_o/k_L = 1,772 \times 0.019/0.142 = 237$$

From Eq. (10-2-67),

$$\text{Nu} = hD_o/k_L = (X^4 \cdot \text{Re}_{LG}^2 + \text{Nu}_f^4)^{1/4} = (0.9024 \times 715,800^2 + 237^4)^{1/4} = 764 \text{ W/(m}^2 \cdot \text{K)}$$

Then

$$h = \text{Nu} \cdot k_L/D_o = 764 \times 0.142/0.019 = 5,711 \text{ W/(m}^2 \cdot \text{K)}$$

Horizontal condenser: The value of h_f is calculated from

$$h_f(\mu^2/k^3\rho^2 g)^{1/3} = 1.5(4G''/\mu)^{-1/3}$$

where

$$G'' = W/LN^{2/3} = 8,000/(3,600 \times 4 \times 192^{2/3}) = 0.0172 \text{ kg/ms}$$

Then

$$h_f[(0.215 \times 10^{-3})^2/(0.1423 \times 7{,}382 \times 9.8)]^{0.33} = 1.5(4 \times 0.0172/0.215 \times 10^{-3})^{-0.33}$$

from which $h_f = 1{,}413$ W/(m$^2 \cdot$ K). Then

$$\mathrm{Nu}_f = h_f D_o/k_L = 1{,}413 \times 0.019/0.142 = 189$$

X and Re_{LG} are the same as for the vertical condenser. Then

$$\mathrm{Nu} = hD_o/k_L = (X^4 \cdot \mathrm{Re}_{LG}^2 + \mathrm{Nu}_f^4)^{1/4} = (0.9024 \times 7{,}158{,}002 + 189^4)^{1/4} = 764$$

$$h = 764k_L/D_o = 568 \times 0.142/0.019 = 5{,}711 \text{ W/(m}^2 \cdot \text{ K)}$$

It produces the same coefficient from both horizontal and vertical condensers. *Verification of the wall temperature:*

$$h_{io}(T_w - t) = h_o(T - T_w)$$

$$6{,}250(T_w - 39.75) = 5{,}711(61 - T_w)$$

Solving for T_w gives $T_w = 49.9°C$, almost equal to the assumed 50°C. *Overall heat transfer coefficient:*

$$U = \left(\frac{1}{h_{io}} + \frac{1}{h_o} + R_f\right)^{-1} = \left(\frac{1}{6{,}250} + \frac{1}{5{,}711} + 0.0005\right)^{-1} = 1{,}197 \text{ W/(m}^2 \cdot \text{ K)}$$

$$\mathrm{DMLT} = \frac{44.5 - 35}{\ln\dfrac{61 - 35}{61 - 44.5}} = 20.9$$

$$\text{Required area} = Q/U \cdot \mathrm{DMLT} = 1{,}105 \times 10^6/1{,}197 \times 20.9 = 44.16 \text{ m}^2$$

$$\text{Condenser area} = \pi LND_o = \pi \times 4 \times 192 \times 0.019 = 45.8 \text{ m}^2 > 44.16 \text{ (okay)}$$

Shell-side pressure drop: For simplicity, we shall use Kern's method.

$$\text{Shell side equivalent diameter} = 0.0139 \text{ m}$$

Vapor Reynolds number at inlet:

$$\mathrm{Re} = \frac{WD_{eq}}{a_s\mu} = \frac{2.222 \times 0.0139}{0.0072 \times 10^{-3} \times 0.0398} = 107{,}699$$

The friction factor is 0.195

$$\Delta p_S \text{ for the inlet mass flow} = f\frac{(N_B + 1)D_S}{D_{eq}}\frac{G_S^2}{2\rho}\left(\frac{\mu_w}{\mu}\right)^{0.14} = \frac{0.195 \times 9 \times 0.438}{0.0139}\frac{55.82^2}{2 \times 2.5} = 34{,}000 \text{ N/m}^2$$

Since at the outlet the vapor velocity is 0, Δp_s calculated with the outlet mass flow also will be 0. The mean value then is 17,000 N/m^2.

10-4 *DESUPERHEATER CONDENSERS*

The vapor coming from the top of a distillation column is always in thermodynamic equilibrium with the liquid leaving the first tray of the column, so it is always at its dew-point temperature and comes into the condenser in this condition. However, there are other types of processes where it may be necessary to condense a vapor that comes into the condenser at a temperature higher than its dew-point temperature—this means in a superheated condition. Let's consider the case of a propane refrigeration cycle such as that represented in Fig. 10-24.

The chiller is a heat exchanger used to cool down a process stream by means of a liquid-propane stream that absorbs heat and vaporizes. The liquid-propane stream is indicated as (3) in the figure, and the vaporized propane is the stream at (4).

The vaporized propane is compressed by the compressor up to a pressure high enough to allow propane condensation using cooling water or atmospheric air as the cooling medium. So this pressure is basically defined by the available air or cooling-water temperature.

The condensed propane is later expanded in a valve down to the chiller pressure. This is an isoenthalpic evolution that produces a partial vaporization at the expense of a temperature drop, thus achieving a cold stream that closes the cycle. Figure 10-24 includes a representation of the cycle in a *T-S* diagram.

It must be noted that in evolution 4–1, which takes place in the compressor, the temperature of the fluid increases up to a value T_1 that is higher than the condensation temperature T_2, which corresponds to pressure p_1.

The condenser then will have two zones in series. In the first zone, there is heat transfer in single phase. Vapors cool down from the inlet temperature to the dew-point temperature corresponding to the condenser operating pressure. The second zone corresponds to the vapor condensation, approximately isothermal. In a countercurrent unit, the temperature diagram is that shown in Fig. 10-25.

The heat balance can be expressed by the following equations:

$$Q = Q_s + Q_\lambda \qquad (10\text{-}4\text{-}1)$$

$$Q_s = \text{sensible heat} = W_h c_{hv}(T_1 - T_2) \qquad (10\text{-}4\text{-}2)$$

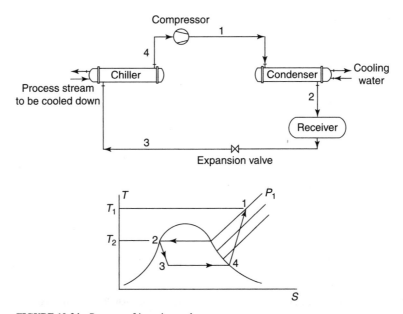

FIGURE 10-24 Propane refrigeration cycle.

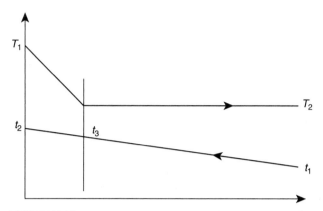

FIGURE 10-25   Temperature diagram of the compressor postcondenser.

$$Q_\lambda = \text{latent heat} = W_h \lambda \tag{10-4-3}$$

$$Q = W_c c_c (t_2 - t_1) \tag{10-4-4}$$

The temperature of the cooling water at the section corresponding to the transition between the two zones is

$$t_3 = t_2 - \frac{Q_s}{W_c c_c} \tag{10-4-5}$$

This configuration is equivalent to having two heat exchangers in series, and the heat transfer area of each zone can be calculated using the heat transfer coefficient and LMTD corresponding to that zone.
    For the desuperheating zone (sensible heat), the result will be

$$\Delta T_S = \text{MLDT}_S = \frac{(T_2 - t_3) - (T_1 - t_2)}{\ln \dfrac{(T_2 - t_3)}{(T_1 - t_2)}} \tag{10-4-6}$$

For the condensing zone (latent heat), the result will be

$$\Delta T_\lambda = \text{MLDT}_\lambda = \frac{(T_2 - t_3) - (T_2 - t_1)}{\ln \dfrac{(T_2 - t_3)}{(T_2 - t_1)}} \tag{10-4-7}$$

And the area of each zone can be calculated as

$$A_s = \frac{Q_s}{U_s \Delta T_s} \tag{10-4-8}$$

$$A_\lambda = \frac{Q_\lambda}{U_\lambda \Delta T_\lambda} \tag{10-4-9}$$

The overall heat transfer coefficients for both zones are usually very different because the heat transfer mechanisms also are very different.

In the desuperheating zone, heat transfer takes place by a mechanism called *reflashing*. If the temperature of the tube wall is lower than the vapor dew-point temperature, condensate will form on the tube wall despite the fact that the global fluid temperature corresponds to a superheated condition. When this condensate drains from the tube into the bulk of the hot vapor phase, it revaporizes (reflashes), extracting heat from the vapor. This is a new and very effective heat transfer mechanism that contributes to vapor cooling and improves the heat transfer coefficient.

However, most design methods do not take this effect into account and calculate U_s for the desuperheating zone with a vapor film coefficient obtained via the single-phase methods of Chap. 7. This is a conservative approach. In many cases, though, since the sensible-heat zone is considerably smaller than the latent-heat zone, the effect of this simplification on the total area of the condenser is not significant. The overall heat transfer coefficient for the condensing zone U_λ is calculated with the methods of this chapter.

The cooling-water flow rate must be selected high enough to avoid temperature crosses between both streams. For example, let's consider the situation shown in Fig. 10-26. Despite the fact that the water outlet temperature is lower than the hot-vapor inlet temperature, the temperature evolutions shown on this diagram are thermodynamically impossible because somewhere in the condenser, the cooling water should be heated above the temperature of the fluid from which it receives heat. The recommendation to avoid such situations is to choose the water flow rate so that its outlet temperature is always below the vapor condensation temperature T_2.

In the foregoing discussion, we assumed that the condenser configuration was countercurrent. If the condenser has two or more tube passes, the situation will be that of Fig. 10-27, where in each zone the hot

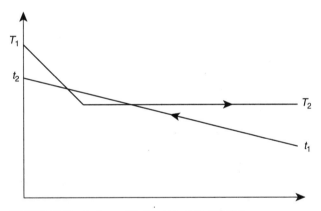

FIGURE 10-26 An impossible thermodynamic evolution.

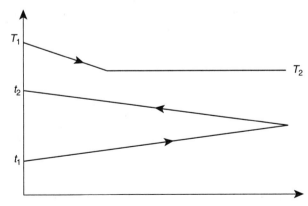

FIGURE 10-27 Temperature diagram in a condenser with two passes in the cooling side.

fluid is in contact with two (or more) passes of the cold fluid. In this case, it would not be valid to use the LMTDs defined by Eqs. (10-4-6) and (10-4-7).

According to Kern, provided that the outlet temperature of the cold fluid is below the vapor dew-point temperature, which means that the heat transfer mechanism in the desuperheating zone is that of reflashing, it is acceptable to use the temperatures differences defined by Eqs. (10-4-6) and (10-4-7) as if the configuration were pure countercurrent. The justification is that the error resulting from this simplification is somewhat compensated for by the error of not considering the reflashing mechanism in the calculation of the heat transfer coefficient.

This means that, independent of the pass configuration, a countercurrent diagram can be assumed, and a temperature t_3 (which in this case does not have any physical interpretation) is calculated with Eq. (10-4-5), and then the LMTDs for each zone are calculated.

Certain design programs calculate an F_t correction factor using the four extreme temperatures and then apply this correction factor to both LMTDs. This procedure does not look to have any other justification than to introduce an additional safety factor in the design.

Example 10-3   It is desired to condense 5,144 kg/h of propane coming from the discharge of a refrigeration compressor at 1,300 kPa and 60°C. Cooling water at 30°C will be used. At 1,300 kPa, the condensing temperature of propane is 37.2°C. A temperature rise of 4°C for the cooling water is adopted. After some trials, the following design is proposed:

N = number of tubes: 672

L = tube length: 4 m

Tube diameters: D_o: 19.05 mm; D_i: 14.4 mm; square array; tube separation: 23.81 mm

D_s = shell diameter: 762 mm

n = number of tube passes: 6

B = baffle spacing: 150 mm

Physical properties:

	Vapor	Liquid
Density, kg/m³ (at T_{sat})	28.1	476
Viscosity, cP	0.009	0.09
Thermal conductivity, W/(m · K)	0.020	0.088
Specific heat, J/(kg · K)	2,030	3,020
Heat of condensation, kJ/kg	314	

Verify if this unit is suitable if installed in a horizontal position. Calculate the pressure drops for the propane and for the cooling water. Assume fouling factors of 0.0003 (m² · K)/W for cooling water and 0.0001 (m² · K)/W for propane.

Solution   Water will flow in the tube-side
Heat balance:

$$Q_\lambda = W_h \lambda = (5{,}144/3{,}600) \times 314 = 449 \text{ kW}$$

$$Q_s = W_h c_h (T_1 - T_2) = (5{,}144/3{,}600) \times 2{,}030 \times (60 - 37.2) = 66.14 \text{ kW}$$

$$Q = Q_s + Q\lambda = 515.2 \text{ kW}$$

Water mass flow:

$$W_c = Q/c_c(t_2 - t_1) = 5.152 \times 10^6/(4{,}180 \times 4) = 30.81 \text{ kg/s} = 111{,}000 \text{ kg/h}$$

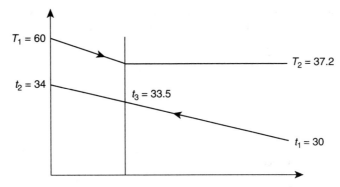

FIGURE 10-28   Illustration for Example 10-3.

The unit has six tube passes. We shall use the simplification of assuming a countercurrent configuration, and we shall correct the LMTD for both zones with a correction factor F_t calculated with the four extremes temperatures. See Fig. 10-28.

Temperature t_3, corresponding to the cooling water at the fictitious transition point, will be

$$t_3 = t_2 - Q_s/W_c c_c = 34 - 66,140/(30.81 \times 4,180) = 33.5°C$$

The logarithmic mean temperature differences for each zone will be

$$MLDT_s = [(60 - 34) - (37.2 - 33.5)]/\ln(26/3.7) = 11.4$$

$$MLDT_\lambda = [(37.2 - 33.5) - (37.2 - 30)]/\ln(3.7/7.2) = 5.2$$

With the four extreme temperatures, we calculate F_t. The parameters required for the calculation are

$$R = (T_1 - T_2)/(t_2 - t_1) = 5.7 \qquad \text{and} \qquad S = (t_2 - t_1)/(T_1 - t_1) = 0.133$$

Thus we get $F_t = 0.917$.

The corrected temperature differences are

$$\Delta T_s = MLDT_s \cdot F_t = 11.4 \times 0.917 = 10.4°C$$

$$\Delta T_\lambda = MLDT_\lambda \cdot F_t = 5.2 \times 0.917 = 4.8°C$$

Heat transfer coefficient for the cooling water:

$$\text{Flow area } a_t = (\pi D_i^2/4)N/n = (\pi \times 0.0144^2/4) \times 672/6 = 0.0182 \text{ m}^2$$

$$\text{Water velocity} = W_c/\rho a_t = 30.89/(1,000 \times 0.0182) = 1.7 \text{ m/s}$$

Film coefficient, tube side:

$$h_i = 1,423(1 + 0.0146t)v^{0.8}/D_i^{0.2} = 1,423(1 + 0.0146 \times 32)1.7^{0.8}/0.0144^{0.2} = 7,453 \text{ W/(m}^2 \cdot \text{K)}$$

$$h_{io} = h_i/(D_i/D_o) = 7,453 \times (14.4/19.05) = 5,633 \text{ W/(m}^2 \cdot \text{K)}$$

Sensible-heat zone: We shall first calculate the heat transfer coefficient for the desuperheating zone. To simplify calculations, we shall use Kern's method.

$$\text{Clearance between tubes } c = P_t - D_o = 23.8 - 19.05 = 4.8 \text{ mm}$$

$$\text{Shell-side flow area } a_s = (D_s cB)/P_t = (0.762 \times 0.0048 \times 0.15)/0.0238 = 0.0231 \text{ m}^2$$

The equivalent diameter for this array is

$$D_{eq} = 4 \times (P_t^2 - \pi D_o^2/4)/\pi D_o = 4 \times (0.0238^2 - 3.14 \times 0.019^2/4)/(3.14 \times 0.019) = 0.019 \text{ m}$$

The Reynolds number is

$$\text{Re}_K = D_{eq}W/(a_s\mu_G) = 0.019 \times 1.429/(0.0231 \times 0.009 \times 10^{-3}) = 130,600$$

(We use the subscript K to differentiate it from other Reynolds numbers that will be used in the calculations.)

$$h_o = 0.36(k_G/D_{eq})\text{Re}_K^{0.55} \cdot \text{Pr}_G^{0.33} = 0.36 \times (0.02/0.019) \times 130,600^{0.55} \times 0.928^{0.33} = 240 \text{ W/(m}^2 \cdot \text{K)}$$

Then

$$U_s = (1/h_{io} + 1/h_o + R_f) - 1 = (1/5,633 + 1/240 + 0.0004)^{-1} = 210 \text{ W/(m}^2 \cdot \text{K)}$$

The required area for the desuperheating zone is

$$A_s = Q_s/U_s\Delta T_s = 66,140/(210 \times 10.4) = 30.3 \text{ m}^2$$

Latent-heat zone: The vapor flow at the inlet is 5,144 kg/h. At the condenser outlet, the vapor flow is zero because it condenses totally. We consider an average mass flow $W_s = 5,144/2 = 2,572$ kg/h.

$$\text{Mass velocity } G_s = W_s/a_s = 2,572/(3,600 \times 0.0231) = 30.92 \text{ kg/(s} \cdot \text{m}^2)$$

By Eq. (10-2-68),

$$\text{Re}_{LG} = (D_o G_s\rho_L)/(\rho_G\mu_L) = (0.019 \times 30.92 \times 476)/(0.09 \times 10^{-3} \times 28.1) = 110,600$$

By Eq. (10-2-71),

$$R = \rho_L\mu_L/\rho_G\mu_G = (476 \times 0.09 \times 10^{-3})/(28.1 \times 0.009 \times 10^{-3}) = 169$$

$$\text{Liquid Prandtl number Pr}_L = c_L\mu_L/k_L = 3,020 \times 0.09 \times 10^{-3}/0.088 = 3.08$$

By Eq. (10-2-72),

$$H = c_L(T - T_w)/\text{Pr}_L\lambda$$

We shall assume a wall temperature of 33°C, which can be verified later. Then

$$H = 3,020(37.2 - 33)/(3.08 \times 314,000) = 0.013$$

According to Eq. (10-3-70),

$$X = 0.9(1 + 1/RH)^{0.33} = 0.9[1 + 1/(169 \times 0.013)]^{0.33} = 1.01$$

For a horizontal condenser, h_f is calculated from the expression

$$h_f(\mu^2/k^3\rho^2 g)^{1/3} = 1.5(4G''/\mu)^{-1/3}$$

where

$$G'' = W/LN^{2/3} = 5{,}144/(3{,}600 \times 4 \times 672^{\,2/3}) = 4.6 \times 10^{-3} \text{ kg/ms}$$

Then

$$h_f[(0.09 \times 10^{-3})^2/(0.088^3 \times 4{,}762 \times 9.8)]^{0.33} = 1.5(4 \times 4.6 \times 10^{-3}/0.09 \times 10^{-3})^{-0.33}$$

from which we get

$$h_f = 1{,}339 \text{ W/(m}^2 \cdot \text{K)}$$

$$\text{Nu}_f = h_f D_o/k_L = 1{,}339 \times 0.019/0.088 = 289$$

$$\text{Nu} = h'_f D_o/k_L = (X^4 \cdot \text{Re}_{LG}^2 + \text{Nu}_f^4)^{1/4} = (1.01^4 \times 110{,}600^2 + 289^4)^{1/4} = 374$$

$$h'_f = h_o = 374k_L/D_o = 547 \times 0.088/0.019 = 1{,}732 \text{ W/(m}^2 \cdot \text{K)}$$

Then, in the condensation zone, the overall heat coefficient will be

$$U_\lambda = (1/h_{io} + 1/h_o + R_f)^{-1} = (1/1{,}732 + 1/5{,}633 + 0.0004)^{-1} = 862 \text{ W/(m}^2 \cdot \text{K)}$$

Verification of the wall temperature:

$$T_w = T_2 - U_\lambda(T_2 - t)/h_o = 37.2 - 862 \times (37.2 - 31.7)/1{,}732 = 34.4$$

It is not considered necessary to correct H. Thus the area required for condensation is

$$A_\lambda = Q_\lambda/\Delta T_\lambda U_\lambda = 449{,}000/(4.8 \times 862) = 108 \text{ m}^2$$

Total area: The total area required then will be $108 + 30.3 = 138.3$ m². The condenser area is

$$A = \pi D_o LN = 3.14 \times 0.019 \times 4 \times 672 = 160 \text{ m}^2$$

Therefore, the unit has about 15 percent excess area.

10-5 CONDENSATION OF VAPOR MIXTURES

When the vapor to be condensed is not a pure substance, condensation is not isothermal. This is what happens in the condenser of a distillation column separating hydrocarbon mixtures. The vapors coming from the top of the distillation column are a mixture of components with different condensation temperatures.

When the mixture condenses, its temperature decreases while condensation progresses.

Initially, those components with higher condensation temperatures (called *heavier components*) condense in a higher proportion. As long as the condensation progresses, the temperature decreases, and the lighter components condense.

If the mixture contains an important proportion of heavy components and a few lights, most of the condensation heat corresponds to the components with higher condensation temperatures. Representing the temperature evolution as a function of the heat removed, a diagram such as that in Fig. 10-29a results. These curves are called *condensation curves*.

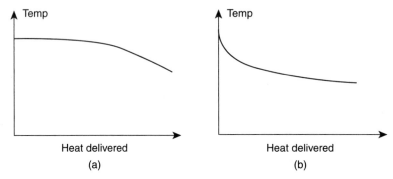

FIGURE 10-29   Different types of condensation curves.

On the other hand, for a mixture containing a higher proportion of light components, the condensation curve will be of the type shown in Fig. 10-29b. In this case, with the initial removal of a small amount of heat, the heavy components condense, and the condensation temperature decreases rapidly, whereas at the end of the condensation the temperature remains more constant.

Let's consider a countercurrent condenser and represent the T-Q diagram for the condensing fluid and the cooling water in both cases. The temperature difference between both streams changes along the condenser; then, to calculate the heat transfer area, we should integrate

$$A = \int \frac{dQ}{U \Delta T} \tag{10-5-1}$$

We have demonstrated that in cases where it a linear relationship exists between the heat received or yielded by each fluid and the temperature change that the fluid experiences, the preceding expression may be integrated analytically, resulting in

$$A = \frac{Q}{U(\text{LMTD})} \tag{10-5-2}$$

This situation is represented by the dashed lines in Fig. 10-30, which show an ideal evolution of the condensing fluid, assuming the existence of a linear relationship between heat and temperature, in which case it would be valid the use of the LMTD.

We see that in the case of Fig. 10-30a, the actual temperature difference ΔT is higher than that corresponding to the ideal situation, whereas in the case of Fig. 10-30b, it is lower. This means that in the first case, the use of LMTD would result in an error by excess in the calculation of the area, whereas in the second case we should have an error by deficit.

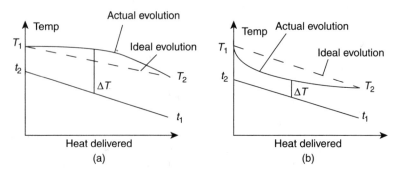

FIGURE 10-30   True temperature difference can be higher or smaller than the LMTD.

It is thus necessary to know the condensation curve in order to design the condenser. To obtain this curve, it is necessary to solve, for each temperature within the condensation interval, the equations representing the thermodynamic equilibrium between vapor and liquid, which requires us to know the liquid-vapor equilibrium relationship.

In very ideal systems, it is possible to use simplified models that assumed that the equilibrium constants are not a function of the phase composition and only depend on the temperature (such as Raoult's law). This is explained in more detail in App. A. But nowadays, these calculations invariably are made with simulation programs that use complex thermodynamic packages, making possible an exact prediction of phase equilibria. The condensation curve then is usually supplied to the exchanger designer as part of the process information.

10-5-1 Diffusional Processes in the Condensation of Multicomponent Mixtures

When we studied Nusselt's condensation theory, we described a model that considered that the vapor is in contact with a condensate film and the temperature in the bulk of the vapor phase is the same as the vapor-liquid interface temperature. The liquid film then was the only resistance to heat transfer.

In the case of multicomponent mixtures, the situation is different. When the vapor condenses at the interface, because the molecules that preferably condense are those of the heavier components, the mixture near the interface becomes richer in light components. This results in a concentration gradient in the direction normal to the interface.

In the case of a binary mixture, a graphic representation such as that in Fig. 10-31 is possible. The graph shows the mole fraction of the heavier component y_A as a function of the distance to the interface.

We see that at the interface, the concentration is y_i, whereas in the bulk of the vapor phase it is y, where $y_i < y$. This means that the proportion of A molecules that can reach the interface and condense is lower than that which would result if the concentration at the interface were equal to the bulk concentration.

The A molecules must diffuse through a layer with a higher proportion of B molecules to reach the interface. This adds an additional resistance to heat transfer.

Let's consider a condenser such as that represented in Fig. 10-32, in which the vapor temperature varies from inlet to outlet. The diagram also shows the cooling-water temperature evolution in a countercurrent arrangement.

Through a differential heat transfer area dA, the amount of heat exchanged per unit time is dQ. Considering the condensing fluid side, dQ must include the sensible heat required to reduce in dT the temperature of the vapor-liquid stream circulating through the condenser, plus the contribution of the latent heat that corresponds to the condensation of dW (kg/s) of vapor.

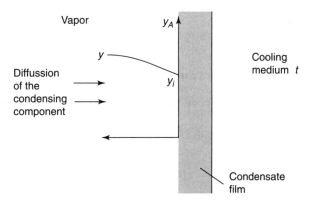

FIGURE 10-31 Concentration profile of the heavy component of a binary mixture in the vicinity of the condensation surface.

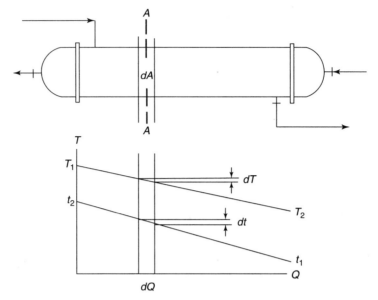

FIGURE 10-32 Temperature evolutions in a condenser.

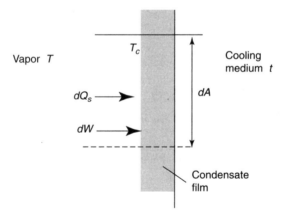

FIGURE 10-33 Condensate film in multicomponent condensation.

Schematically, the process can be represented by Fig. 10-33. Here, T, T_c, and t are the temperatures of the vapor, the interface, and the cooling media at a cross section of the condenser such as AA in Fig. 10-32.

If we consider a heat transfer differential area dA in the vicinity of this section, we can explain the condensation mechanism as follows:

1. The vapor at temperature T delivers heat to the interface at T_c. This transfer is by convection, and it can be expresed as

$$dQ_S = h_G (T - T_c) dA \qquad (10\text{-}5\text{-}3)$$

where dQ_s is the convection heat flow transferred from the gas to the liquid film per unit time, and h_G is the gas phase convection heat transfer coefficient.

2. In addition to the aforementioned heat flow, there exists a condensation of a mass vapor dW (kg/s) on dA. When this vapor condenses, it delivers to the condensate film its latent heat of condensation. Thus the total amount of heat that incorporates to the film on dA will be the sum of both terms

$$dQ = dQ_S + \lambda dW \tag{10-5-4}$$

This heat must be transmitted through the film to be removed by the cooling medium. The resistance to heat transfer consists of the liquid film, fouling, and the cooling-medium boundary layer. These resistances are in series, and it is possible to define a combined heat transfer coefficient as

$$U' = \left(\frac{1}{h'_f} + \frac{1}{h_R} + R_f \right)^{-1} \tag{10-5-5}$$

The first term within the parentheses represents the resistance of the condensate film, which can be calculated with the models we studied for pure fluids; the second term is the resistance of the refrigerant; and the third term is the fouling resistance.

The result thus will be

$$dQ = U'(Tc - t)dA \tag{10-5-6}$$

In order to obtain an expression for the heat transfer rate, let's consider a control volume, as indicated by the dashed lines in Fig. 10-34. This control volume includes the vapor phase contained in the portion of condenser corresponding to dA and extends up to the vapor-liquid interface without including it.

A vapor flow W_v at temperature T enters the control volume. A fraction dW of this vapor condenses over the surface dA. This mass leaves the control volume at the vapor state (because the interface is outside the control volume) and at a temperature T_c. The enthalpy change of this mass of vapor between its inlet and outlet conditions to the control volume is

$$dW c_v (T - T_c)$$

where c_V is the vapor specific heat.

Additionally, there is a convection heat transfer from the bulk of the vapor phase to the interface given by

$$dQs = h_G dA(T - T_c) \tag{10-5-7}$$

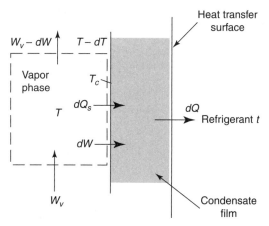

FIGURE 10-34 Control volume for energy balances in a multicomponent condenser.

The vapor flow W_V that enters the control volume at temperature T leaves at temperature $T - dT$. Then its enthalpy change is $W_V c_V\, dT$ (neglecting the mass flow reduction dW).

The enthalpy balance for the control volume then can be expressed as

$$W_V c_V\, dT + dW c_V\,(T - T_c) - dQs = 0 \qquad (10\text{-}5\text{-}8)$$

If it is accept that the second term is small in comparison with the first one (which is usually the situation), then

$$W_V c_V\, dT = h_G\,(T - Tc)\,dA \qquad (10\text{-}5\text{-}9)$$

Combining Eqs. (10-5-6) and (10-5-7), we get

$$\frac{dQ}{U'dA} = (T_c - t) \qquad (10\text{-}5\text{-}10)$$

$$\frac{dQ_S}{h_G dA} = (T - T_c) \qquad (10\text{-}5\text{-}11)$$

Adding both equations and defining $\gamma = dQ_S/dQ = W_V.c_V.dT/dQ$, we obtain

$$dQ\left(\frac{1}{U'dA} + \frac{\gamma}{h_G dA}\right) = T - t \qquad (10\text{-}5\text{-}12)$$

The overall heat transfer coefficient corresponds to the sum of all the resistances, so

$$\frac{1}{U} = \frac{1}{U'} + \frac{\gamma}{h_G} = \frac{1}{h_R} + \frac{1}{h'_f} + \frac{\gamma}{h_G} + R_f \qquad (10\text{-}5\text{-}13)$$

To be consistent with the methodology we always use, we can separate the resistances at both sides of the tube wall, defining a heat transfer coefficient for the hot fluid as

$$\frac{1}{h_h} = \frac{1}{h'_f} + \frac{\gamma}{h_G} \qquad (10\text{-}5\text{-}14)$$

So we get the usual expression

$$\frac{1}{U} = \frac{1}{h_h} + \frac{1}{h_R} + R_f \qquad (10\text{-}5\text{-}15)$$

The heat transfer area then can be calculated as

$$dA = \frac{dQ}{U(T - t)} \qquad (10\text{-}5\text{-}16)$$

and

$$A = \int \frac{dQ}{U(T - t)} \qquad (10\text{-}5\text{-}17)$$

To use this expression, it is necessary to know the condensation curve, which is sectioned in intervals of dQ. Superimposing this curve with the T-Q curve for the refrigerant, we can obtain for each interval the value of $(T - t)$ and the ratio of sensible heat to total heat γ. Since the film coefficients h_G, h_R, and h'_f can be calculated for the conditions of each interval, it is possible to calculate the integral numerically.

10-5-2 General Method for Condenser Design

The design method for a multicomponent condenser with condensation range includes the case of condensation of a pure vapor (making $\gamma = 0$). We thus shall develop a calculation method for the rating of a condenser with a condensation range whose results are completely general and applicable to any condensation case.

This procedure only covers the two-phase region. If the condenser also includes a desuperheating zone, this must be rated separately.

To simplify the treatment, we shall use Kern's correlations for calculation of single-phase heat transfer coefficients. As was explained in Chap. 7, this method can result in important errors, specially in floating-heads type condensers. However, the procedure can be easily adapted to be used with more elaborated methods for the prediction of shell-side heat transfer coefficients.

The input data to use this model include the condensation curve. This curve is obtained from the process simulation using programs such as Hysis, Aspen, ProII, etc. This curve represents the thermodynamic evolution of the hot stream while its temperature is reduced from T_1 to T_2. For each intermediate temperature, it is necessary to know the enthalpy of the stream, the vapor mass fraction (kilograms of vapor per total kilograms), and the molecular weight of the vapor. This subject is explained with a little more detail in App. A.

The number of intermediate points depends on the number of intervals in which it is desired to divide the calculations. Usually, five or six intervals are enough.

Two different types of condensation curves can be calculated: differential and integral.

Integral condensation assumes that vapor and condensate remain in contact and in thermodynamic equilibrium with each other along their path through the condenser. Vertical condensers with tube-side condensation are the best example of integral condensation. Other cases approximating integral condensation correspond to condensation within horizontal tubes, condensation outside vertical tubes, and condensation outside horizontal tubes in cross-flow (TEMA type X).

In differential condensation, the liquid condensate is separated from the vapor stream. This modifies the equilibrium and reduces the dew point of the remaining vapor. A typical example is the reflux counter-current condenser, in which vapor flows upward inside tubes, while the condensate film formed on the tube surfaces drains in countercurrent.

Horizontal condensers with shell types E or J are an intermediate case. To be conservative, in these cases one should use the differential condensation curve. However, the usual practice in industrial design has been to use the integral condensation curves in all cases.

The number of points in the condensation curve define the number of heat transfer area calculation intervals: n points of the condensation curve define $n - 1$ calculation intervals. Then, for the interval enclosed by points i and $i + 1$, it is possible to calculate

ΔQ = heat to be removed in the interval = $Wh(i_{i+1} - i_i)$, where i = specific enthalpy

ΔQ_s = vapor sensible heat = $W_v c_v (T_{i+1} - T_i)$

$\gamma = \Delta Q_s / \Delta Q$

Knowing the mass flow of the cold stream, its physical properties, and the inlet temperature, it is possible to calculate the temperature change that this stream will suffer in each interval. This is done by assuming a countercurrent configuration, and calculating $t_i - t_{i+1} = \Delta Q / W_c c_c$. The $T - Q$ curve for the refrigerant then is superimposed on the hot-stream temperature curve.

With the four extreme temperatures for the whole process, an LMTD correction factor F_t is calculated. Then, for each interval, it is possible to calculate the LMTD as

$$\text{LMTD} = [(T_{i+1} - t_{i+1}) - (T_i - t_i)]/\ln[(T_{i+1} - t_{i+1})/(T_i - t_i)]$$

And then apply the correction factor F_t to each interval. For a better understanding of the procedure, the following example is included.

Example 10-4 It is desired to condense 0.277 kg/s of a hydrocarbon mixture coming from the top of a distillation column at 49.2°C and 4.5 at(a). Condensation takes place in the range of 49.2–27.4°C. The refrigerant is cooling water, available at 20°C, with a mass flow of 3.031 kg/s. Vapor specific heat is 1,800 J/(kg · K). The condensation curve corresponds to the first five rows of the following table, and the calculations for each interval are shown on the next rows.

The LMTD correction factor was calculated with the four extreme temperatures ($T_1 = 49.2$, $T_2 = 27.4$, $t_1 = 20$, $t_2 = 28.2$) for a configuration with one shell pass and n tube passes resulting in $F_t = 0.783$.

Vapor temperature, °C	49.2		44.3		39.0		33.4		27.4
Vapor mass fraction, kg vap/kg total	1		0.728		0.476		0.233		0
Enthalpy of mixture, kJ/kg	375.4		281.6		187.2		93.8		0
Vapor mole weight, kg/kmol	58		56		54		52.5		51
Vapor mass flow, kg/s	0.277		0.201		0.132		0.064		0
Interval		I		II		III		IV	
ΔQ, kW		26		26		26		26	
ΔQ_s, kW		2.42		1.91		1.33		0.69	
γ		0.093		0.073		0.051		0.026	
Water temperature, °C	28.2		26.1		24.1		22.0		20
LMTD, °C		19.5		16.5		13.0		9.2	
LMTD · F_t		15.29		12.89		10.21		7.23	

The following properties for vapor and condensate may be considered constant in all the condensation range:

Liquid	Vapor
Density = 561 kg/m³	Compressibility factor = 0.9
Viscosity = 0.15 cP = 0.15×10^{-3} kg/(m · s)	Viscosity = 0.81×10^{-5} kg/ms
Thermal conductivity = 0.09 W/(m · K)	Thermal conductivity = 0.018 W/(m · K)
Specific heat = 2,500 J/(kg · K)	Specific heat = 1,800 J/(kg · K)
Prandtl number = 4.16	Prandtl number = 0.81

Note: A downloadable spreadsheet with the calculations of this example is available at http://www.mhprofessional. com/product.php?isbn=0071624082

A horizontal heat exchanger with the following geometry was proposed:

N = number of tubes: 42

n = number of tube passes: 2

L = tube length: 6.09 m

Tube diameters $D_o = 0.019$; $D_i = 0.0144$

D_s = shell diameter = 0.254 m

P_t = tube pitch = 0.0238 m

Triangular array 30 degrees

c = tube clearance = 0.0048 m

Segmental baffles with spacing $B = 0.063$ m

Fouling resistances of 0.0001 (m² · K)/W for the condensing vapor and 0.0003 (m² · K)/W for the cooling water must be considered.

Solution We shall calculate first the heat transfer coefficient for the cooling water.

$$a_t = \text{tube-side flow area} = (\pi D_i^2/4)N/n = 0.00341 \text{ m}^2$$
$$v = \text{velocity} = W_c/(1{,}000a_t) = 3.031/1{,}000 \times 0.00341 = 0.886 \text{ m/s}$$
$$t = \text{mean temperature of water} = 24°C$$
$$h_{io} = 1{,}423(1 + 0.0146t)v^{0.8}/D_i^{0.2} \times (D_i/D_o) = 3{,}092 \text{ (m}^2 \cdot \text{K)/W}$$

To calculate the vapor-side coefficient, we must first calculate the coefficient corresponding to the condensate film resistance h'_f. This requires initially calculating h_f (without considering the vapor drag effect over the condensate film) and correcting it later for vapor velocity.

We shall calculate a single h_f for the whole unit using Nusselt's condensation model:

$$h\left(\frac{\mu_L^2}{k_L^3 \rho_L^2 g}\right)^{1/3} = 1.51\left(\frac{4G''}{\mu_L}\right)^{-1/3}$$

where G'' is calculated as $G'' = W \text{ condensate}/LN^{0.66} = 0.277/(42^{0.66} \times 6.09) = 0.00386 \text{ kg/(m} \cdot \text{s)}$. Substituting the liquid physical properties, the result is $h_f = 1{,}386 \text{ W/(m}^2 \cdot \text{K)}$. Thus

$$\text{Nu}_f = h_f D_o/k_L = 1{,}386 \times 0.019/0.09 = 292.7$$

For each calculation interval, this coefficient must be corrected to include the effect of the vapor drag forces. The corrected condensate film coefficient is obtained with the following equations:

$$\text{Nu'}_f = \frac{h'_f D}{k_L} = \left(X^4 \cdot \text{Re}_{LG}^2 + \text{Nu}_f^4\right)^{1/4} \qquad \text{where } \text{Re}_{LG} = \frac{DG\rho_L}{\mu_L \rho_G}$$

$$G = \frac{\text{vapor mass flow}}{\text{shell-side flow area}} \qquad X = 0.9\left(1 + \frac{1}{RH}\right)^{1/3}$$

$$R = \frac{\rho_L \mu_L}{\rho_G \mu_G} \qquad H = \frac{c_L(T_{\text{sat}} - T_w)}{\text{Pr}_L \cdot \lambda}$$

(The heat of condensatio λ can be estimated as the quotient between "latent heat" and condensate mass flow.) Then

$$\lambda = \frac{Q - W_v c_v(T_1 - T_2)}{W_v} = \frac{104{,}000 - 0.277 \times 1{,}800 \times (49.2 - 7.4)}{0.277} = 336{,}000 \text{ J/kg}$$

This model makes it necessary to perform an iterative calculation to find T_w. For each interval, we shall assume a value of T_w that will be verified by means of

$$T_w = \frac{h_o T - U(T - t)}{h_o}$$

Once h'_f is calculated, we can obtain h_o (total heat coefficient of the vapor side) as

$$h_o = \left(\frac{1}{h_f} + \frac{\gamma}{h_G}\right)^{-1}$$

This requires calculation of h_G (single-phase vapor heat transfer coefficient). To simplify its calculation, we shall use Kern's method. A vapor mass velocity G_v is calculated as the quotient between the vapor mass flow of each interval and the shell-side flow area. this flow area is obtained as

$$a_s = D_s cB/P_t = 0.254 \times 0.0048 \times 0.063/0.023 = 0.00333 \text{ m}^2$$

The Kern Reynolds number for this geometry is calculated as $\text{Re}_K = D_{eq}G_v/\mu_v$. The equivalent diameter for this geometry is 0.0137 m, and then

$$h_G = 0.36 \times \text{Re}_K^{0.55} \times \text{Pr}_v^{0.33} \times kv/D_{eq}$$

For each interval, the overall U is calculated as

$$U = (1/h_o + 1/h_{io} + R_{fi} + R_{fo})^{-1}$$

And finally, the required area for the interval will be $A = \Delta Q/U\Delta T$.
    The results are shown in the following table:

ρ vapor	10.97	10.76	10.55	10.44	10.35
W vapor	0.274	0.201	0.132	.0646	0
Re	555,658	412,595	275,092	136,018	0
R	946	965	984	994	1,003
T_w	40	36	32	28	24
H	0.0187	0.0161	0.0133	0.0101	0.00662
X	0.919	0.920	0.924	0.930	0.947
Nu'_f	690	599	499	381	293
h'_f	3,271	2,841	2,367	1,807	1,386
Re_K	144,756	105,383	68,904	33,728	0
h_G	305	256	203	137	0
h_o	1,629	1,560	1,483	1,335	1,386
U	748	733	715	679	692
$A_{interval}$	2.3	2.78	3.65	5.25	

The total required area will be the summation of the intervals areas, which is 13.98 m². The area of the proposed unit is 15.2 m². This means an 8 percent excess area.

10-6 THE USE OF STEAM AS PROCESS HEATING MEDIUM

In preceding sections we considered the condensation of a process vapor using a refrigerant medium, which normally consists on a utility stream. However, the most common case of vapor condensation is when steam is used as a process heating medium. In this case, the condensing steam is the utility stream, whereas the process stream is the cold fluid receiving the heat. Because of the wide application of this process, we shall study some particular characteristics of steam installations.

10-6-1 Process Heating Systems

To deliver heat to a process using an external energy source, the following possibilities can be considered:

• Direct heating with a flame
• Heating with a fluid that delivers sensible heat
• Heating with a fluid that delivers latent heat

Direct heating with open flames in all process equipment would require the installation of expensive safety systems and better-quality materials to resist higher temperatures. The usual solution thus is to use some fluid as a vehicle to transport thermal energy where it is required.

This means that heat is delivered to a circulating fluid via burning a solid, liquid, or gas fuel in a central piece of equipment, which may be a furnace or steam generator, and this fluid transports the energy to all consumers, usually in a closed cycle. So there is only one piece of equipment working with open flame, and it is built under strict safety specifications by specialized companies, and its design is optimized to reach high thermal efficiencies, whereas in the rest of the equipment heating is performed safely and economically by the circulating fluid.

The circulating fluid used as heating medium may be a fluid that undergoes a change of phase, delivering its latent heat, or may be a single-phase fluid that performs its function by receiving and delivering sensible heat. For example, 1 kg of water in the vapor state condenses delivering an amount of energy equal to 2,000 J. If it is desired to deliver the same amount of heat using liquid water, which cools down from 100 to 50°C, 10 kg of water should be employed. This means that by using a fluid that exchanges latent heat, it is necessary to employ a lower mass flow to transport the same amount of energy.

The fluid most commonly used as a heating fluid is steam. Its main advantages are

• It has a high value of latent heat of condensation and vaporization.

• The raw material to produce it (softened or demineralized water) is relatively inexpensive.

• The film coefficients for steam condensation are extremely high, which means that the size of the heat transfer equipment will be relatively small.

In certain cases, it is necessary to deliver heat to a process at high temperature (200°C or more). In these cases, if steam were used, the vapor pressure of this steam would be too high. Then the cost of process equipment increases because a higher pressure makes it necessary to use higher plates thicknesses, and safety specifications are stricter.

In such situations, it is sometimes preferred to use as the heating medium any of the so-called thermal fluids. They are mineral oils or synthetic oils that have low vapor pressures so that they continue being liquids at high temperatures. The process equipment working with them then have low design pressures even for high operating temperatures.

Steam System Installations. Figure 10-35 shows schematically a steam system installation. We can distinguish the following elements:

Water-Treatment Unit. Dissolved solids may precipitate and produce scaling in the boiler and heat transfer equipment. The required quality of the water in the circuit depends on the boiler's operating pressure, and the dissolved solids level is maintained with periodic and continuous purges. To reduce purges, the makeup water is treated to eliminate hardness and other dissolved salts. The treatment unit may include clarifiers, filters, softeners, and ion-exchange or reverse-osmosis equipment depending on the quality of both the raw and treated water.

Deaereator and Condensate Tank. Since the steam condensate is free from ions, it is recovered to be recycled to the boiler. Thus the treating unit only needs to treat the makeup water that replaces purges and system losses. The deaereator eliminates dissolved air that can cause corrosion.

Boiler or Steam Generator. Here, the steam is generated by the combustion of any fuel. The steam pressure is controlled by regulating the fuel supply. Sometimes it is necessary to have two or more steam qualities, for example, a high-pressure steam for turbines operation or high-temperature process heating and a low-pressure steam for lower thermal level applications. In these cases, low-pressure steam is obtained by reducing the pressure in a control valve.

Industrial Process. The steam consumption in the different process equipment is represented schematically in Fig. 10-36 as a stirred jacketed vessel. As can be seen in the figure, the steam flow is regulated with an inlet valve, and the condensate is removed through a trap that avoids loss of uncondensed steam. We shall analyze in more detail the operation of these two elements for a better understanding of the steam condensation mechanism.

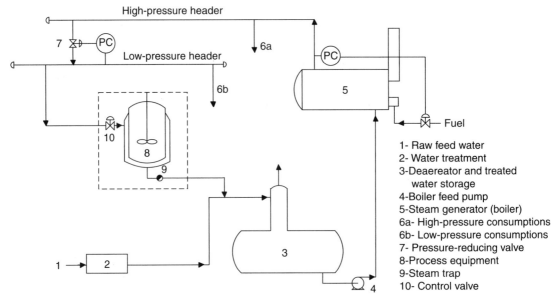

FIGURE 10-35   Components of a steam system.

1- Raw feed water
2- Water treatment
3-Deaereator and treated
   water storage
4-Boiler feed pump
5-Steam generator (boiler)
6a- High-pressure consumptions
6b- Low-pressure consumptions
7- Pressure-reducing valve
8-Process equipment
9-Steam trap
10- Control valve

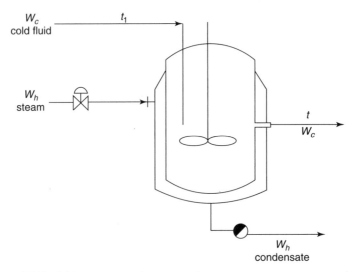

FIGURE 10-36   Steam-heated jacketed vessel.

Let's consider the steam-heated jacketed vessel shown in this figure. Since the liquid is agitated, we may consider that its temperature is uniform. Let t be the temperature of the process liquid in the vessel. The steam into the jacket is at its saturation temperature and condenses, delivering heat to the process fluid. The process fluid enters the vessel at a temperature t_1, and it immediately mixes with the vessel contents, increasing its temperature to t.

If the process fluid mass flow is W, the heat that must be delivered is

$$Q = W_c c_c (t - t_1) \tag{10-6-1}$$

This heat is delivered by the steam. Assuming that the inlet steam is saturated, the heat is

$$Q = W_h \lambda \tag{10-6-2}$$

because the steam condenses, delivering its heat of condensation.

Now, how can we know that only condensate exits through the jacket-bottom nozzle? If this connection were an open pipe or a manual valve at a fixed opening position, it is probable that an important fraction of the inlet steam would escape through the outlet nozzle without yielding its heat of condensation. Thus the steam consumption would be much higher than that resulting from Eq. (10-6-2).

Then it is necessary to install some device that only allows the passage of the condensate that has already delivered its latent heat, retaining the uncondensed steam. This function is performed by devices called *steam traps*. There are many different kinds of steam traps, and they will be described below. First, however, we must explain another problem that can exist in steam installations, which is the presence of air.

Effect of Air in Steam Systems.   Air can get into steam systems during shutdowns, when the installation cools down. When steam is again introduced into the system, a steam-air mixture will result. The presence of air in a steam system is harmful for the following reasons:

1. Steam will condense at a temperature that depends on its partial pressure. Since air is present in the mixture, the partial pressure of steam is reduced, and the condensation temperature will be lower, thus decreasing the temperature-driving force for heat transfer.

2. When steam condenses over cold surfaces, because air does not condense, the air-steam proportion in the vicinity of the heat transfer surfaces can be considerably higher than in the bulk of the mixture. This creates diffusional resistances because steam molecules must diffuse through an air-enriched layer to reach the heat transfer surfaces. This was already explained in Sec. 10-4. It was reported that only 1 percent of air in a steam system can result in a 50 percent decrease of the heat transfer coefficients.

3. Additionally, the presence of air increases corrosion rates.

The usual practice to deal with this problem is to purge the installation during startup. Manual purge valves are installed in cold and dead-end points, and they are left open for a certain amount of time, purging a steam-air mixture so that most of the air is eliminated. This obviously results in a loss of steam. But after startup, air still can get into the system, dissolved in the makeup water, so this purging should be done periodically. To reduce this procedure, steam traps also must perform the air purging, so they should allow the discharge of air and condensate but retain the steam.

Different Types of Steam Traps.   There are many types of steam traps, each one with its advantages and limitations. We shall describe the most common types.

Thermostatic Bellows Trap (Fig. 10-37). The principal element is a thermostatic bellows A filled with a fluid (usually an alcohol mixture) that can vaporize or condense depending on the temperature of the trap. The boiling point of this mixture is selected based on the temperatures range in which the trap has to operate.

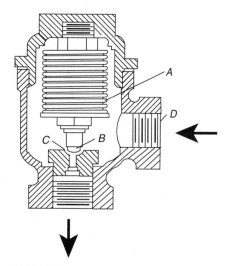

FIGURE 10-37   Thermostatic steam trap.

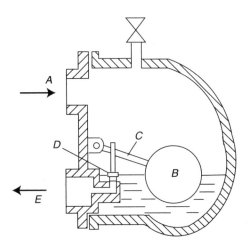

FIGURE 10-38   Float-type steam trap.

Thus these traps operate on the difference in temperature between steam and cooled condensate and air. Steam increases the temperature inside the thermostatic element, which causes the fluid to vaporize and expand, thus closing the discharge at *C*. As condensate and noncondensable gases back up in upstream piping (called the *cooling leg*), the temperature begins to drop, the thermostatic element contracts, and the valve opens. The amount of condensate backed up ahead of the trap depends on the load conditions, steam pressure, and size of the piping. The discharge of the trap is intermittent.

Thermostatic traps also can be used for venting air from a steam system. When air collects, the temperature drops, and the thermostatic air vent automatically discharges the air at slightly below-steam temperature throughout the entire operating pressure.

These traps are small in size and have high discharge capacity, but they offer little resistance to water hammer and may be affected by corrosion because the bellows is made of a very thin metal sheet.

Float-Type Trap (Fig. 10-38).   The float-type trap is a mechanical trap that operates on the density principle. A lever *C* connects the valve float *B* to the valve and seat *D*. Once condensate reaches a certain level in the trap, the float rises, opening the orifice and draining condensate. A water seal formed by the condensate prevents live-steam loss.

Since the discharge valve is under water, it is not capable of venting air and noncondensables. This is why these traps are usually provided with an auxiliary thermostatic vent, as shown in Fig. 10-39. When the accumulation of air and noncondensables causes a significant temperature drop, a thermostatic air vent *H* in the top of the trap discharges it.

This type of trap has a continuous condensate discharge because the float adopts an equilibrium position that allows evacuation of all the incoming condensate. These traps easily adapt to variable condensate flow and allow a stable operation without the process fluctuations caused by discontinuous-discharge traps.

Inverted-Bucket Steam Trap (Fig. 10-40).   The principal element is an inverted bucket *B* that can move vertically inside a casing. The bucket acts through a lever opening and closing the discharge valve. When the bucket is at the lower position, the valve opens, and when the bucket rises, the valve closes. The inlet port to the trap is located below the bucket. When the trap is full of condensate, the weight of the bucket makes it sink, opening the valve, and the condensate can discharge (left side of Fig. 10-40).

When steam comes into the trap, it displaces the condensate and gets trapped below the bucket. This makes the bucket rise and closes the valve (right side of Fig. 10-40). When the trap cools down, the steam condenses, and condensate gets below the bucket, which sinks again, reopening the discharge.

A small orifice is placed at the top of the bucket to vent air. Otherwise, air would accumulate below the bucket, and because it does not condense, the trap would remain closed. The air escapes through this orifice and accumulates at the top of the trap. When the trap opens, the air is eliminated by the condensate

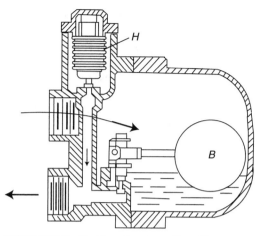

FIGURE 10-39 Float-type steam trap with auxiliary vent.

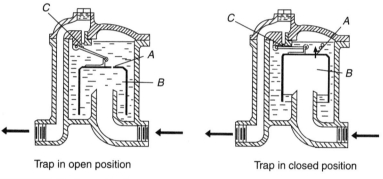

Trap in open position   Trap in closed position

FIGURE 10-40 Inverted-bucket steam trap.

pressure. Obviously, some steam can escape through the orifice too, so it must be of a small diameter to reduce such steam loss. This limits the air-purging capacity of this type of trap. However, the trap is robust and has good mechanical resistance.

Thermodynamic Trap (Fig. 10-41). This type of trap has a body A with inlet and outlet ports, a cover B, and a control disk C that is free to move vertically. The body has two concentric ring-shaped seats. The inner seat D surrounds the inlet orifice E and the exterior seat F close to the cover. Between these round seats is the discharge orifice G. Both seats are carefully ground, as well as the disk C, so that the disk closes against the two concentric faces at the same time, separating inlet from outlet.

The interior part of the cover has a central stop H that limits the lift of the disk, so when the disk is in the upper position, there is still some room between the upper face of the disk and the lower face of the cover. This space is called the *control chamber*. When the disk closes against the seats, the control chamber remains isolated from the inlet and outlet ports.

We shall see how this trap works (see Fig. 10-42). Let's assume that the trap is cold initially. When air and cold condensate come into the trap, they pass through the inlet orifice E, raise the disk, and flow radially from the center to the space between both seats D and F and discharge through the orifice G (Fig. 10-42*a*).

When steam and hot condensate come into the trap, since their specific volume is higher, the velocity below the disk increases. According to the Bernoulli principle, this velocity increase provokes a pressure reduction at the lower face of the disk, and it seats, closing the trap (Fig. 10-42*b*).

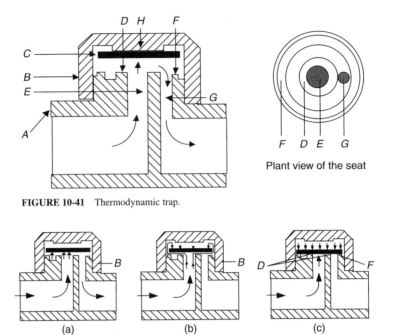

FIGURE 10-41 Thermodynamic trap.

Plant view of the seat

(a)   (b)   (c)

FIGURA 10-42 Operation of a thermodynamic trap.

But part of the steam could escape into the control chamber and be trapped there, where its velocity is converted completely into pressure. The steam trapped in the control chamber exerts a downward force on the disk equal to the product of the pressure times the area of the upper face of the disk, whereas the steam below exerts an upward force equal to the pressure times the area of the orifice E, which is sensibly smaller (Fig. 10-42c).

Then, while the trap is hot and there is steam into the control chamber, the trap remains closed. The control chamber losses heat to the ambient but maintains the temperature because it receives heat from the steam below the disk. When cold condensate and air come into the trap, it cools down. This makes the steam in the control chamber condense, so its pressure decreases. The condensate pressure then can raise the disk, and the cycle restarts.

This type of trap is robust, small, and economic. It has only one moving part, the disk, which is made from hardened stainless steel, so its service time is high. These traps cannot be used with pressures lower than 1 bar(g), and their capacity is small. Some steam is always lost when the trap opens, so their efficiency is not very high.

10-6-2 Regulation of the Steam Flow

Going back to Fig. 10-36, we shall now analyze the action of the inlet valve that regulates the steam flow to the jacket. We mentioned that the heat flow delivered to the unit is that necessary to heat the process stream from the inlet temperature t_1 up to t.

If the process flow W_c can fluctuate, it is necessary to adjust the steam flow to maintain the temperature of the process. Let's suppose that somebody is operating the valve so as to keep constant the temperature t. The heat flow is

$$Q = W_c c_c (t - t_1) \qquad (10\text{-}6\text{-}3)$$

The steam mass flow required to deliver this heat is

$$W_h = \frac{Q}{\lambda} \qquad (10\text{-}6\text{-}4)$$

(In what follows, as a simplification, we shall assume that λ is independent of pressure, and we shall neglect the sensible-heat terms in the steam enthalpy.) Combining the last two equations, we get

$$W_h = \frac{W_c c_c (t - t_1)}{\lambda} \qquad (10\text{-}6\text{-}5)$$

It is obvious that the higher the process stream flow W_c, the higher is the amount of steam that must be admitted into the jacket. It's intuitive, and we know from daily experience that in order to increase the steam flow, we must open the steam inlet valve to the jacket.

However, we also know that once the steady state is reached, it is not possible to admit more steam into the jacket than it can condense. So the inlet steam flow, which equals the amount of condensate exiting through the trap, depends on the heat transfer capacity of the jacket.

In other words, if U is the overall heat transfer coefficient and ΔT is the temperature difference between the steam in the jacket and the process liquid, the heat transferred will be

$$Q = UA\Delta T \qquad (10\text{-}6\text{-}6)$$

And then

$$W_h = \frac{UA\Delta T}{\lambda} \qquad (10\text{-}6\text{-}7)$$

This expression tells us that the steam flow entering the jacket is determined by the heat transfer capacity of the vessel (UA) and by the temperature difference. We can consider that U is independent of the process flow rate W_c within a certain range. But then we see that the valve opening, which regulates the steam flow, is not included directly in Eq. (10-6-7). Therefore, why can the action of the valve modify the steam flow?

The explanation lies in the term ΔT of Eq. (10-6-7). ΔT is the temperature difference between the fluids at both sides of the jacket wall. This is

$$\Delta T = T - t \qquad (10\text{-}6\text{-}8)$$

where T is the temperature at which steam condenses and t is the temperature of the liquid which is in the vessel.

T is the condensation temperature at the pressure existing into the jacket and depends on this pressure. But this pressure equals the pressure in the steam header minus the pressure drop in the valve. When the valve is opened, its Δp decreases, and the steam pressure into the jacket increases. This makes the steam condensation temperature increase and also increases ΔT in Eq. (10-6-7). This will be better understood with the following example.

Example 10-5 A jacketed vessel such as that in Fig. 10-36 has a heat transfer area of 10 m^2, and the overall heat transfer coefficient U is 350 W/(m^2 · K). The pressure in the steam header is 3 bar(abs). The specific heat of the process fluid is 2,500 J/(kg · K), and the inlet temperature is 20°C. The fluid must be heated in the vessel up to 70°C. Calculate the temperature in the steam jacket when the mass flow of the process fluid is

a. 1 kg/s

b. 1.5 kg/s

c. 1.78 kg/s

d. 0.8 kg/s

Solution   The amount of heat to be delivered to the process fluid per unit mass is

$$\Delta i_c = c_c(t - t_1) = 2,500 \times (70 - 20) = 125 \text{ kJ/kg}$$

Thus, to process 1 kg/s, the heat to be delivered is

$$Q = 1 \times 125 = 125 \text{ kW}$$

The steam flow that must be fed to the jacket is

$$W_h = \frac{Q}{\lambda}$$

And taking $\lambda = 2.115 \times 10^6$ J/kg (which shall be considered constant in the entire problem), the result will be

$$W_h = \frac{125 \text{ kJ/s}}{2,115 \text{ kJ/kg}} = 0.059 \text{ kg/s}$$

The temperature difference required to transfer this heat is

$$\Delta T = \frac{Q}{UA} = \frac{125 x 10^3}{350 \times 10} = 35.7 °C$$

Then the temperature inside the jacket must be

$$T = 35.7 + 70 = 105.7$$

The steam pressure corresponding to this temperature is 123 kPa. Then, since the pressure in the distribution header is 3 bar (300 kPa), to heat up 1 kg/s of process liquid, the steam flow must be throttled in the valve so as to have a pressure drop of $300 - 123 = 177$ kPa. The following table shows the calculation for the other process conditions requested in the example.

Case	b	c	d
W_c	1.5	1.78	0.8
$Q(W) = 125,000 \times W_c$	187,500	222,500	100, 000
W_h (kg/s) $= Q/2.115 \times 10^6$	0.088	0.105	0.047
$\Delta T = Q/(350 \times 10)$	53.5	63.5	29
$T = 70 + \Delta T$	123.5	133.5	99
P_{steam}	220	300	98

We shall analyze the results of this table. We see that in order to move from condition (a) to condition (b) with a higher process liquid flow rate, it was necessary to open the steam valve to increase the jacket pressure. If we want to move to condition (c) with a higher process flow, it can be seen that a jacket pressure of 3 bar (300 kPa) is necessary. But this pressure equals the header pressure. This means that the valve pressure drop must be zero. This is a limit condition that ideally happens when the valve is completely open. It is obviously the maximum pressure that can be established into the jacket, so process condition (c) determines the maximum capacity of the system.

If it is desired to increase the operating capacity further, it will be necessary to increase the heat transfer area of the vessel. According to Eq. (10-6-7), this allows us to increase the heat flow with a lower ΔT.

If a steam-heated vessel has its admission valve completely open and cannot reach the desired process condition, the steam pressure downstream of the control valve must be measured. If this pressure does not differ from the header pressure, there is some problem in vessel design (the heat transfer area is too small, the heat transfer coefficients are different from expected, or fouling is too high).

On the other hand, if the valve downstream pressure is lower than the header pressure, it means that the valve is too small because even at its maximum opening its pressure drop for the required steam flow is too high.

Let's now see what happens if we attempt to reduce the process flow to 0.8 kg/s (condition d). We see that to reduce the capacity so much, it would be necessary to have inside the jacket a very low temperature fluid that corresponds to a subatmospheric pressure. In this condition, the condensate cannot discharge from the trap. Thus it builds up into the jacket, reducing the heat transfer surface exposed to the steam. This reduction in A allows us to decrease the heat transfer rate without making a subatmospheric pressure into the jacket necessary. But this type of operation can lead to fluctuating process conditions.

Possible solutions to this problem may be

1. To install a condensate receiver below the vessel level with a barometric leg that provides the necessary height to overcome the negative Δp.

2. To install an inverted check valve to allow admission of air into the jacket when pressure falls below atmospheric. In this way, the steam partial pressure is reduced, and it is possible to have a lower condensation temperature with a total pressure equal to atmospheric.

GLOSSARY

A = area (m^2)

a_s = shell-side flow area (m^2)

B = condensate layer width (m)

B = baffle spacing (m)

c = specific heat [J/(kg $\cdot$ K)]

c = clearance between tubes (m)

D = diameter (m)

D_{eq} = equivalent diameter (m)

D_s = shell diameter (m)

f = friction factor

g = gravity acceleration (m/s^2)

G = mass flow density [kg/(s $\cdot$ m^2)]

G' = flow per unit length at the bottom of a vertical condensate layer [kg/(m $\cdot$ s)]

G'' = flow per unit length on a horizontal condensate layer [kg/(m $\cdot$ s)]

H = parameter of Eq. (10-3-72)

h = height (m)

h = heat transfer coefficient (generic) [W/(m^2 $\cdot$ K)]

h_f = heat transfer coefficient for the condensation of a vapor on a film without considering the effect of vapor velocity [W/(m^2 $\cdot$ K)]

h'_f = heat transfer coefficient for the condensation of a vapor on a film corrected for the effect of vapor velocity

h_G = single-phase heat transfer coefficient [W/(m^2 · K)]

i = specific enthalpy (J/kg)

J = mass flow density [kg/(s · m^2)]

k = thermal conductivity [W/(m · K)]

L = tube length (m)

M = mole weight

N = number of tubes

n = number of passes

p = pressure (N/m^2)

P_t = separation between tube axes in a row (m)

Pr = Prandtl number

Q = heat flow (J/s)

q = heat-flow density [J/(s · m^2)]

R = parameter in Eq. (10-3-71)

Re = Reynolds number

R_f = fouling resistance [(m^2 · K)/W]

T = hot-fluid temperature (°C)

t = cold-fluid temperature (°C)

U = overall heat transfer coefficient [W/(m^2 · K)]

U' = combined heat transfer resistance including the condensate layer, fouling, and refrigerant film coefficient [W/(m^2 · K)]

v = velocity (m/s)

W = mass flow (kg/s)

W_s = shell-side mass flow (kg/s)

W' = mass flow toward the film surface (kg/s)

W_f = mass flow at the bottom of the condensate layer (kg/s)

x = longitudinal coordinate (m)

y = longitudinal coordinate (m)

z = longitudinal coordinate (m)

λ = heat of condensation (J/kg)

δ = film thickness (m)

δ_f = film thickness at the film bottom (m)

μ = viscosity [kg/(m · s)]

ρ = density (kg/m^3)

τ = shear stress (N/m^2)

Subscripts

1 = inlet

2 = outlet

c = cold

h = hot

i = internal

K = Kern's method

L = liquid

m = mean

o = external

R = refrigerant

T = total (liquid + vapor)

V = vapor

w = wall

x = position

REFERENCES

1. Nusselt W.: "Surface Condensation of Water Vapor," *Zeitschr Ver Deutsch Ing* 60(27):541–546, 1916.
2. Kern D.: *Process Heat Exchange.* New York: McGraw-Hill, 1950.
3. Dukler A. E.: "Dynamics of vertical falling film systems,": *Chem. Eng. Progr.* 55(10):62–67, 1959.
4. Collier J.: *Convective Boiling and Condensation.* London: McGraw-Hill, 1967.
5. Boyko I. D., Krushilin G. N.: "Heat Transfer and Hydraulic Resistance during Condensation of Steam in a Horizontal Tube and a Bundle of Tubes," *Int. J. Heat Mass Transfer* 10:361–373, 1967.
6. Carpenter E. F., Colburn A. P.: "The Effect of Vapor Velocity on Condensation Inside tubes," *Inst. Mech. Engrs Proc. General Discussion on Heat Transfer.* New York: ASME, 1951, pp. 20–26.
7. Mueller A. C.: in *Heat Exchanger Design Handbook.* New York: Hemisphere Publishing, 1983.
8. Wallis G. B.: "Flooding Velocities for Air and water in Vertical Tubes," U.K.A.E.A Rep. AEEW-R-123, 1961.
9. Othmer D. F.: *Ind Eng Chem* 21:576, 1929.

CHAPTER 11
BOILING

11-1 MECHANISMS OF HEAT TRANSFER TO BOILING LIQUIDS

The change of phase from liquid to vapor can take place following two different mechanisms: evaporation and boiling. Let's assume that we have a subcooled liquid (at a temperature below the equilibrium temperature at the existing pressure). If this liquid comes in contact with a heating surface at a higher temperature, heat will flow from the heating surface to the liquid. The heat flux can be expressed as

$$q = h(T_W - t_F)$$

where t_F is the liquid temperature (away from the surface), T_W is the temperature of the surface, and h is the convection heat transfer coefficient corresponding to the hydrodynamic conditions of the system, which, in natural convection, is a function of the temperature difference $(T_W - t_F)$.

For low heat transfer rates (or low wall temperatures), the only effect of the convection heat flux is to progressively increase the liquid temperature. When heat flux exists, there is a temperature profile, as shown in Fig. 11-1.

As long as the liquid is being heated and the temperature at the free surface of the liquid increases, the vapor pressure of the liquid at the liquid-vapor interface also increases, and liquid molecules abandon the liquid phase. This phenomenon is called *evaporation,* and its characteristic is that the transition from liquid to vapor state occurs only at the free surface of the liquid. Evaporation takes place at any liquid temperature, and the driving force for this mechanism is the partial pressure difference of the volatile component between the liquid surface and the vapor phase.

Let's assume that now it has been decided to increase the heat transfer rate in the system of Fig 11-1. To achieve this, it will be necessary to increase the temperature of the heating surface. It is then possible that the solid-surface temperature T_W will be increased above the liquid-saturation temperature T_{SAT}. Since the liquid temperature varies continuously from T_W to t_F, part of the liquid also will be above the saturation temperature T_{SAT}.

In this case it is possible that vapor bubbles begin to appear near the wall. This process is called *boiling.* While the bulk temperature of the liquid t_F (away from the solid wall) is still below the saturation temperature, once the bubbles formed at the heating surface depart from the surface and contact the colder liquid, collapse, and yield their latent heat of condensation to the liquid. This phenomenon is called *subcooled boiling* and may be described as formation of vapor bubbles at the heating surface with further condensation of them when they depart from the surface.

If the heat transfer to the liquid continues, its temperature will increase, and finally, all the liquid will reach the saturation temperature. In this case, the bubbles do not collapse any more and can reach the liquid surface. This is called *saturated boiling.* The three mechanisms are shown in Fig. 11-2.

11-1-1 Formation of Vapor Bubbles in Boiling Liquids: Equilibrium Condition of a Bubble

For a better understanding of the different boiling heat transfer mechanisms, it is necessary to review some topics dealing with the formation of bubbles over a solid-liquid interface. We shall first consider the equilibrium condition of a vapor bubble submerged into a liquid and, for the moment, far from any solid surface.

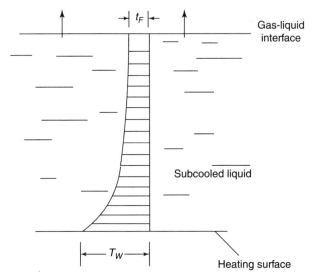

FIGURE 11-1   Temperature profile in a heated liquid.

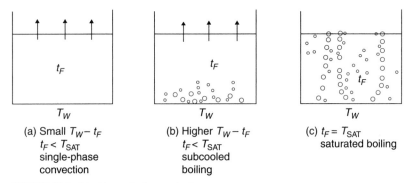

FIGURE 11-2   Liquid vaporization mechanisms.

The study of the equilibrium condition of bubbles is a classic in physics books, and it can be easily demonstrated that the necessary condition for the existence of a bubble is that the pressure inside the bubble be higher than the exterior pressure by an amount given by $2\sigma/r$, where σ is the surface tension and r is the bubble radius. We shall apply this concept to the case of a vapor bubble in thermodynamic equilibrium with the surrounding liquid. Let's assume that we have a system at a constant temperature T_G consisting of a single-component liquid pool and a spherical vapor nucleus with radius r.

Since we assume that the system is in thermodynamic equilibrium, the pressure above the liquid-vapor interface, which is also the pressure inside the bubble, shall be the equilibrium vapor pressure at temperature T_G. We shall call this pressure p_G.

Note: In fact, the vapor pressure over a curve interface is slightly different from the value obtained from the saturation curve. This is a simplification of the problem, and for a more rigorous treatment, ref. 1 may be consulted.

As stated previously, there must be a pressure difference between the interior and the exterior of the bubble, so if p_F is the exterior pressure, then

$$p_G - p_F = \frac{2\sigma}{r} \qquad (11\text{-}1\text{-}1)$$

Let T_{SAT} be the saturation temperature corresponding to p_F. This means that the pairs (p_G, T_G) and (p_F, T_{SAT}) determine two points belonging to the vapor-pressure curve of the liquid. Since $p_G > p_F$, it is also true that $T_G > T_{SAT}$. This means that the equilibrium condition of the bubble requires that the system temperature T_G be higher than the saturation temperature corresponding to the imposed liquid pressure. In other words, the liquid must be superheated, where $T_G - T_{SAT}$ is the value of superheat required by the bubble to exist in equilibrium.

If the temperature of the system is increased above T_G, the bubble grows, whereas if the temperature decreases below T_G, the bubble collapses. The slope of the saturation curve of a pure liquid can be expressed by the Clausius Clapeyron equation:

$$\frac{dp}{dT} = \frac{\lambda}{T(v_V - v_L)} \tag{11-1-2}$$

where λ is the heat of vaporization and v_V and v_L are the specific volumes of vapor and liquid, respectively. T is the absolute temperature (K).

Since $v_V \gg v_L$, and assuming that the ideal gas equation of state can be applied to the vapor phase,

$$Mpv_V = RT \tag{11-1-3}$$

Equation (11-1-2) then can be rewritten as

$$\frac{dp}{p} = \frac{\lambda M}{RT^2} dT \tag{11-1-4}$$

Remembering that the points (p_F, T_{SAT}) and (p_G, T_G) belong to the saturation curve, Eq. (11-1-4) can be integrated between these points, resulting in

$$\ln \frac{p_G}{p_F} = \frac{\lambda M}{R} \left(\frac{1}{T_{SAT}} - \frac{1}{T_G} \right) \tag{11-1-5}$$

It also must be considered that the pressure difference between the interior and exterior of the bubble is small in comparison with the absolute value of the pressure p_F, and since $\ln(1 + x) \cong x$ for $x \ll 1$, we get

$$\ln \frac{p_G}{p_F} = \ln \left(1 + \frac{p_G - p_F}{p_F} \right) = \frac{p_G - p_F}{p_F} = \frac{2\sigma}{rp_F} \tag{11-1-6}$$

Combining Eqs. (11-1-5) and (11-1-6) and rearranging, we get

$$(T_G - T_{SAT}) = \frac{RT_{SAT}T_G}{\lambda M} \frac{2\sigma}{rp_F} \tag{11-1-7}$$

Since the product $T_{SAT} \times T_G$ approximately equals T_{SAT}^2, we have

$$T_G - T_{SAT} = \Delta T_{SAT} = \frac{2\sigma R T_{SAT}^2}{\lambda M p_F r} \tag{11-1-8}$$

This equation expresses the superheat necessary for the existence of a vapor bubble with radius r in equilibrium with the liquid.

11-1-2   Nucleation into Boiling Liquids

We shall analyze what the necessary conditions are for the production of vapor bubbles into a heated liquid. This phenomenon is called *nucleation*. There are two different nucleation mechanisms: homogeneous and heterogeneous.

The first takes place into the bulk of a liquid that is superheated above its saturation temperature. The second, which is much more common, is the one that takes place over a solid surface.

Homogeneous Nucleation.   It can be seen in Eq. (11-1-8) that the smaller the bubble radius, the higher is the superheat required for the bubble to grow into a homogeneous mass of liquid. If the value of the superheat is high, there is a small but finite probability of a cluster of molecules with energy similar to that of the vapor phase coming together at a certain point to form a vapor embryo of the size of the equilibrium nucleus over which the bubble can grow.

The number of nuclei that may form per unit time depends on the liquid temperature T_G and on the nature of the liquid, and some theoretical models to predict the nucleation rate were developed. In the case of water, it could be estimated that for a significant nucleation to take place, degrees of superheating as high as 220°C would be necessary. However, in practice, any strange body or the walls of the recipient normally provide a great number of active sites that provoke the start of boiling with much lower superheat.

It is possible, deliberately working with very pure fluids and highly polished and clean surfaces, to minimize the effects of the walls and delay the start of nucleation. However, it was never possible to reach values of superheat as great as those mentioned. For this reason, it is of much higher interest to study the second mechanism, namely, heterogeneous nucleation.

Heterogeneous Nucleation, or Nucleation Over Solid Surfaces.   It can be demonstrated that when a bubble forms over a solid surface, the necessary superheat for the existence of this bubble in equilibrium with the liquid is different from that calculated with Eq. (11-1-8). In this case, the necessary superheat can be obtained by multiplying the value from Eq. (11-1-8) by a factor Φ that depends on the contact angle θ between the solid surface and the liquid-vapor interface, always measured through the liquid[30] (Fig. 11-3).

The shape of the function Φ is shown in Fig. 11-4. When a liquid completely wets the surface, the value of the contact angle θ is 0, and the situation is that of Fig. 11-5. In this case, the superheat required for the existence of the bubble with radius r in equilibrium is the same as predicted with Eq. (11-1-8) because the correction factor Φ equals 1. It is as if the solid were not present.

However, for any other substance that forms an angle $\theta > 0$, the necessary value of $T_G - T_{SAT}$ for the existence of a nucleus in equilibrium is smaller than that calculated with Eq. (11-1-8) because the correction factor is smaller than 1.

When a liquid does not wet the surface at all, then $\theta = 180$ degrees. In this case, Φ equals 0, which means that no superheat is required. The equilibrium condition is that the liquid temperature coincides with the saturation temperature. This is the situation on a planar interface.

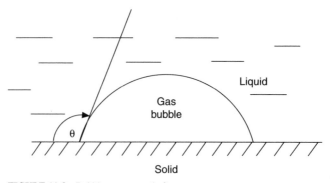

FIGURE 11-3   Bubble contact angle θ.

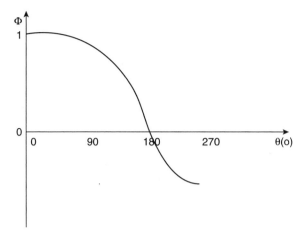

FIGURE 11-4   Superheat correction factor.

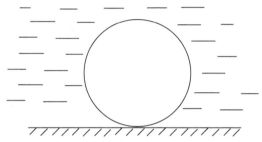

FIGURE 11-5   Case of $\theta = 0$.

One may think that it is not possible to have contact angles over an interface that are higher than 180 degrees. However, normally, all surfaces have pits or cavities, and a vapor nucleus can exist in them, as shown in Fig. 11-6.

In a cavity of this type, with an irregular shape, the curvature direction of the interface can be inverted. In this case, the contact angle can be higher than 180 degrees, so the correction factor Φ is negative. This means that the superheat is also negative, and a vapor nucleus can exist even at temperatures lower than the saturation temperature. This explains the reason why the pores that normally are present on the solid surface can act as nucleation centers because even at low temperatures it is possible for a vapor embryo to be present in them and this vapor embryo can give origin to a bubble when the temperature is increased.

The way in which a bubble grows is shown in Fig. 11-7. Let's imagine a vapor nucleus in the interior of a pit, represented by the situation indicated as 1 in the figure.

We have already explained that this nucleus can exist even when the temperature of the system is lower than the saturation temperature. If now the surface temperature is raised above the saturation temperature, the bubble will grow and will move to position 2 with a positive value of Φ because now $\theta < 180$ degrees.

If the bubble continues growing and moves to position 3, the curvature radius of the bubble decreases with respect to that of situation 2. Since the superheat required for the equilibrium condition of the bubble is inversely proportional to the radius, the result is that situation 3 requires a higher superheat than situation 2. This means that to achieve the bubble growth from situation 2 to situation 3, it is necessary to increase the temperature.

The same can be said to pass from situation 3 to situation 4. However, once the bubble radius equals that of the cavity mouth (situation 4), to attain further growth and to pass into situation 5, the curvature radius of the bubble must start increasing, and the necessary overheating for the equilibrium of the system decreases.

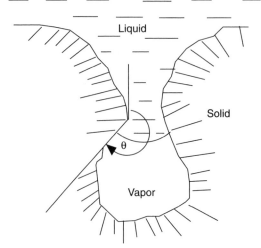

FIGURE 11-6 Vapor embryo into a cavity.

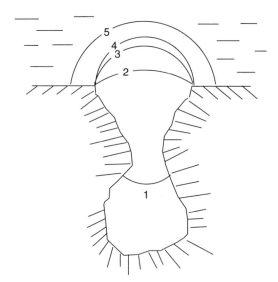

FIGURE 11-7 Bubble growth.

Since the degree of superheat that the system has reached up to that moment is higher than the bubble equilibrium temperature, the bubble starts growing spontaneously and finally departs from the surface.

This means that in the process of bubble growth, there is a critical instant, represented by situation 4 in Fig. 11-7, after which bubble growth is spontaneous and continues up to bubble departure. If the temperature of the system reaches or surpasses the superheat required by this critical instant, the bubble will be released. This critical instant is reached when the bubble has a radius equal to that of the cavity mouth.

It must be noted that when moving from situation 3 to situation 4, the contact angle θ also decreases. Since the reduction in θ produces an opposite effect to a reduction in r, the combination of both effects causes the critical bubble radius to be somewhat different than the mouth-cavity radius. However, this difference is not taken into account in the majority of the models.

Nucleation in a Temperature Gradient. The preceding discussion of vapor nucleation considered only uniform temperature in all the system. However, this situation is not real when the solid surface over which the bubbles are formed acts as a heating surface.

In this case, if the surface temperature is T_{W1}, for heat transfer to exist, it is required that this temperature be higher than the liquid temperature. This means that a temperature gradient exists in the liquid phase. Figure 11-8 represents a heating surface with an active nucleation site.

Well away from the wall, the liquid temperature is t_F. In this case, t_F has been represented below T_{SAT}. This corresponds to a situation where the bulk of the fluid is subcooled, but the conclusions are also valid for saturated boiling with $t_F = T_{SAT}$.

The change in the liquid temperature from T_{W1} to t_F takes place in a zone near the solid, which is called the *boundary layer.* A usual simplification is to assume a linear temperature profile through the boundary layer. Let's assume that this profile is represented by line I in Fig. 11-8.

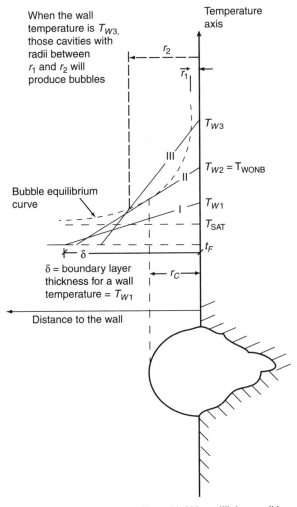

FIGURE 11-8 Temperature profiles and bubble equilibrium condition.

If h is the heat transfer coefficient, the heat-flux density at the interface is

$$k \frac{T_{W1} - t_F}{\delta} = q \tag{11-1-9}$$

where δ is the distance in which the temperature varies from T_{W1} to t_P

Since

$$q = h(T_{W1} - t_F) \tag{11-1-10}$$

it follows that

$$\delta = \frac{k}{h} \tag{11-1-11}$$

Typically, for the case of stagnant water, the boundary-layer thickness is about 0.1 mm.

The dashed line in Fig. 11-8 also represents Eq. (11-1-8):

$$T_G = T_{SAT} + \frac{RT_{SAT}^2 \, 2\sigma}{\lambda M p_F r} \tag{11-1-12}$$

This equation represents the equilibrium temperature for a bubble of radius r. The plotted curve is the function $T_G = f(r)$ given by Eq. (11-1-12).

It was explained previously that during bubble growth, there is a critical instant when the size of the bubble reaches that of the cavity mouth. The following theory was postulated by Hsu[2] and establishes that the bubble will grow and release if the temperature of the liquid surrounding the top of the bubble, when this critical instant is reached, exceeds the temperature required for the nucleus to remain in equilibrium given by Eq. (11-1-12).

Let's consider the behavior of the cavity represented in Fig. 11-8. Figure 11-9 represents the situation in more detail.

The temperature at the top of the bubble can be obtained from the line representing the temperature profile at a distance $r = r_1$ from the wall (r_1 = cavity radius). This temperature is T_b. According to Hsu[2], the temperature required for bubble growth is given by the equilibrium curve. This value is T_a.

In the situation represented in Fig. 11-9, since T_a is higher than T_b, the bubble cannot progress, and this situation is the same for any cavity radius because for any value of r, the temperature profile is below the equilibrium temperature. Then none of the cavities, whatever their radii, have reached the necessary temperature to produce bubbles.

Coming back to Fig. 11-8, let's assume that the surface temperature is increased up to T_{W2}. In this case, the temperature profile changes, and it becomes tangent to the curve that represents the equilibrium condition at a point corresponding to a radius r_c. In this situation, the cavities whose radii are equal to r_c would have reached the condition required for nucleation, and they will start producing bubbles. T_{W2} is the wall temperature at which the production of bubbles starts. It is called the *onset of nucleation boiling temperature* (T_{WONB})

Let's assume that the wall temperature is increased further up to a value T_{W3} so that the temperature profile in the system corresponds to line III in Fig 11-8 (see also Fig. 11-10). It can be seen that now there is a range of r values for which the system temperature is higher than that required for bubble growth. This range corresponds to r values between r_1 and r_2. It means that when the wall temperature is T_{W3}, all the cavities whose sizes are between r_1 and r_2 will be active, producing bubbles. Thus, as the wall temperature is being raised, the number of active sites increases and the boiling rate also increases.

This model, then, allows us to calculate the wall temperature at which nucleate boiling starts (T_{WONB}), which corresponds to T_{W2} in Fig. 11-8. This temperature can be obtained by finding the T_W value that makes the temperature profile tangent to the bubble equilibrium curve. It can be demonstrated geometrically that the abscissa of the tangency point is half the abscissa at which the tangent intersects the asymptote of the equilibrium curve given by $T = T_{SAT}$.

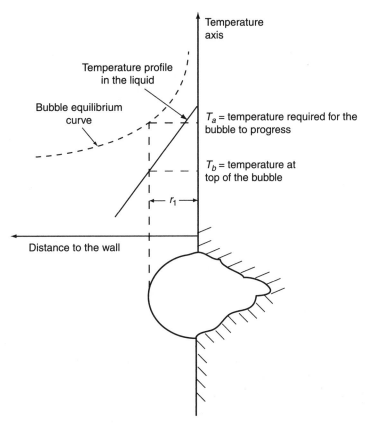

FIGURE 11-9 Wall superheat required for bubble growth.

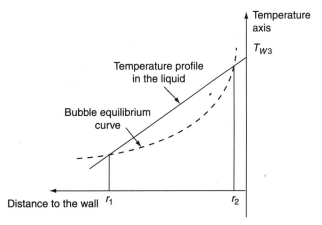

FIGURE 11-10 Range of active cavity sizes at a certain wall superheat.

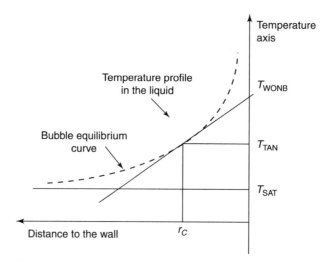

FIGURE 11-11   In saturated boiling $T_{\text{WONB}} - T_{\text{TAN}} = 1/2\,(T_{\text{TAN}} - T_{\text{SAT}})$.

With relation to Fig. 11-11, if the temperature corresponding to the tangency point is T_{TAN}, the result will be

$$T_{\text{WONB}} - T_{\text{TAN}} = T_{\text{TAN}} - T_{\text{SAT}} = \frac{1}{2}(T_{\text{WONB}} - T_{\text{SAT}}) \tag{11-1-13}$$

According to Eq. (11-1-8), at the tangency point,

$$T_{\text{TAN}} - T_{\text{SAT}} = \frac{2\sigma R T_{\text{SAT}}^{2}}{\lambda M p_{F} r_{C}} \tag{11-1-14}$$

While the equation of the line representing the temperature profile can be written

$$T_{\text{WONB}} - T_{\text{TAN}} = \frac{q r_{C}}{k} \tag{11-1-15}$$

Multiplying Eqs. (11-1-14) and (11-1-15) together and considering Eq. (11-1-13), we get

$$\frac{1}{4}(T_{\text{WONB}} - T_{\text{SAT}})^{2} = \frac{2\sigma R T_{\text{SAT}}^{2} q}{\lambda M k p_{F}} \tag{11-1-16}$$

Since p_{F} and T_{SAT} define a point on the saturation curve, the specific volume corresponding to this state will be called v_{VSAT}. And applying the ideal gas state equation, we get

$$v_{\text{VSAT}} = \frac{R T_{\text{SAT}}}{M p_{F}} \tag{11-1-17}$$

And Eq. (11-1-16) can be rewritten as

$$q = \frac{k}{4B}(T_{\text{WONB}} - T_{\text{SAT}})^{2} \tag{11-1-18}$$

where

$$B = \frac{2\sigma T_{\text{SAT}} v_{\text{VSAT}}}{\lambda}$$ (11-1-19)

A more rigorous deduction gives the following expression[4] for B:

$$B = \frac{2\sigma T_{\text{SAT}} (v_V - v_L)_{\text{SAT}}}{\lambda}$$ (11-1-20)

Equation (11-1-18) gives a relationship between the heat-flux density and the wall temperature at the onset of nucleate boiling.

Since up to that moment the heat transfer regime was single-phase convection, defined by

$$q = h(T_W - t_F)$$ (11-1-21)

if a correlation for h is available (the usual correlations for convection in liquids can be used), Eqs. (11-1-18) and (11-1-21) can be solved, and it is possible to calculate the values of q and T_{WONB} for the onset of nucleate boiling.

Some authors consider that the presence of a bubble produces a distortion of the temperature profile so that the shape of the isotherm is as shown in Fig. 11-12. In this case, if r_c is the radius of the first bubble reaching the equilibrium condition, the isotherm that passes near its top is, away from the bubble, at a distance nr_c from the solid surface, where n is a factor greater than 1.

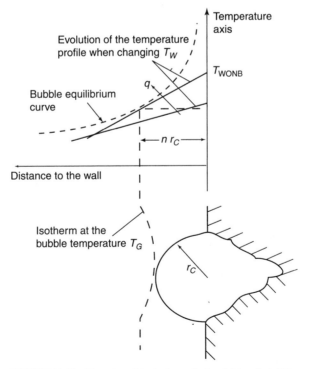

FIGURE 11-12 Distortion of the isotherms in the vicinity of a bubble.

Some authors suggest a value between 1 and 2 for n. Apparently, the value $n = 1$ reproduces with good agreement experimental results for the boiling of water. Han and Griffith[3] suggested to assume $n = 1.5$ based on theoretical considerations. With this model, Eq. (11-1-18) can be modified as

$$q = \frac{k}{4B}\left(\frac{T_{\text{WONB}} - T_{\text{SAT}}}{n}\right)^2 \tag{11-1-22}$$

One deficiency of the previously exposed theory is that it assumes a continuous distribution of cavity sizes from $r = 0$ to $r = \infty$. However, for a cavity to act as a nucleation site, it is necessary that its size is not too big. In such cases, the liquid completely fills the cavity, and if a vapor embryo is not present, the cavity cannot generate a bubble. This will be better understood with the following example.[4]

Example 11-1   Calculate the critical radius and wall temperature required for the onset of nucleation in the case of water at atmospheric pressure and at its boiling temperature (saturated boiling).

Solution   1. *Obtaining the temperature profile in the liquid.* Up to the start of boiling, the natural convection film coefficient for a planar surface in contact with a liquid is given by

$$h = 0.14\frac{k_L}{D}\left(\frac{\beta_L g \Delta T \rho_L^2 D^3}{\mu_L^2}\frac{c_L \mu_L}{k_L}\right)^{1/3}$$

For water at 100°C,

$\beta_L = 7.2 \times 10^{-4}\ K^{-1}$ $\mu_L = 2.97 \times 10^{-4}\ kg/(m \cdot s]$
$\rho_L = 960\ kg/m^3$ $k_L = 0.68\ W/(m \cdot K)$
$Pr_L = 1.9$ $g = 9.8\ m/s^2$

$$h = 0.14 \times 0.68\left[\frac{7.2 \times 10^{-4} \times 9.8 \times 960^2 \times 1.9}{(2.97 \times 10^{-4})^2}\right]^{0.33} \Delta T^{0.33} = 453\Delta T^{0.33} = 453(T_W - T_{\text{SAT}})^{0.33}$$

Since in this case $t_F = T_{\text{SAT}}$. The boundary-layer thickness is

$$\delta = \frac{k}{h} = \frac{0.68}{453(T_W - T_{\text{SAT}})^{0.33}} = 1.5 \times 10^{-3}(T_W - T_{\text{SAT}})^{-0.33}$$

This means that the boundary-layer thickness is a function of T_W and for each T_W, the temperature profile is given by a line with a slope $(T_W - T_{\text{SAT}})/\delta$. Figure 11-13 shows the temperature profile for three different values of the wall temperature.

2. *Bubble equilibrium curve.* The bubble equilibrium curve is given by

$$T_G - T_{\text{SAT}} = \frac{RT_{\text{SAT}}^2}{\lambda M p_F r}2\sigma = \frac{B}{r}$$

For water at 100°C,

$\sigma = 0.058\ N/m$
$M = 18$
$T_{\text{SAT}} = 373\ K$
$\lambda = 2.257 \times 10^6\ J/kg$
$p_F = 1.01 \times 10^5 N/m^2$

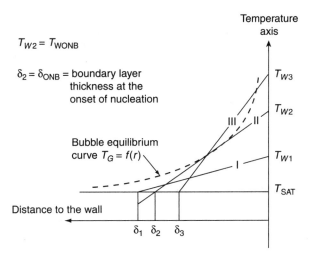

FIGURE 11-13 Boundary-layer thickness at different wall superheats.

The ideal gas constant is $R = 8{,}307$ J/(kg · mol · K). With these values, we get $B = 3.26 \times 10^{-5}$. The bubble equilibrium curve then can be written as

$$T_G - T_{SAT} = \frac{3.26 \times 10^{-5}}{r}$$

This is the equation of a hyperbole with asymptotes $T = T_{SAT}$ and $r = 0$.

3. *Determination of the wall temperature for the start of boiling.* We must calculate the wall temperature at which the temperature profile is tangent to the bubble equilibrium curve. It has been demonstrated that in that condition, Eqs. (11-1-18) and (11-1-21) must be satisfied simultaneously. Then

$$q = \frac{k}{4B}(T_{WONB} - T_{SAT})^2 = \frac{0.68}{4 \times 3.26 \times 10^{-5}}(T_{WONB} - 373)^2$$

and

$$q = h(T_{WONB} - T_{SAT}) = 453(T_{WONB} - T_{SAT})^{4/3}$$

This is a two-equations system with two unknowns. Solving, we get $T_{WONB} = 373.02$ K and $q = 2.2$ W/m^2. The value of r_c then can be found as

$$r_C = \frac{\delta_{ONB}}{2} = \frac{1.5 \times 10^{-3}}{2}(T_{WONB} - T_{SAT})^{-0.33}$$
$$= 0.75 \times 10^{-3}(373.02 - 373)^{-0.33} = 2.76 \times 10^{-3}\,\text{m} = 2.76\,\text{mm}$$

This means that according to the theoretical prediction, with a wall temperature only 0.02 K above the saturation temperature, boiling already starts. However, it can be found experimentally that a few degrees of superheat are always necessary for the start of boiling. The discrepancy lies in the fact that not all the cavities or pores of the heating surface can act as nucleation sites. If the size of a cavity is large, once the first vapor bubble detaches, the liquid completely fills the cavity, and it cannot continue acting as a nucleation site.

Note that according to these calculations, the critical radius is 2.76 mm, which hardly can be accepted as a pore. It is normally accepted that only cavities whose radii are 5 μm or less (5×10^{-6} m) can act as nucleation sites. Some authors reduce this figure to 1 μm or even less. If we accept 5 μm as the minimum pore radius capable of generating bubbles, this means that the wall temperature has to be raised up to the value that makes the liquid temperature at 5 μm from the wall reach the required super-heat predicted by Eq. (11-1-8) for a cavity $r = 5$ μm. This is illustrated in Fig. 11-14. Since in this case $T_G \cong T_W$, then

$$T_G \cong T_W = T_{SAT} + \frac{3.26 \times 10^{-5}}{r} = 373 + \frac{3.26 \times 10^{-5}}{5 \times 10^{-6}} = 378.9 \text{ K}$$

If we accept that the minimum active-site radius is 1 μm, then the required wall temperature for the start of boiling is 405.6 K.

11-1-3   Bubble Diameter at Detachment

As explained previously, once a bubble has reached its critical radius, it continues growing spontaneously. For a certain amount of time, owing to surface-tension forces, the bubble remains adhered to the solid surface, but as it continues growing, the buoyancy forces, which depend on the third power of the radius, overcome the surface-tension forces, and the bubble detaches from the solid.

Several models were proposed to predict the growth velocity of bubbles and the detachment diameter. Detachment diameter is obtained by a balance between surface-tension and buoyancy forces. The following expression is obtained[31]

$$\text{Detachment diameter} = D_d = 0.0208\theta \left[\frac{\sigma}{g(\rho_L - \rho_V)} \right]^{1/2} \tag{11-1-23}$$

where θ is the contact angle of the bubble at detachment (in degrees).

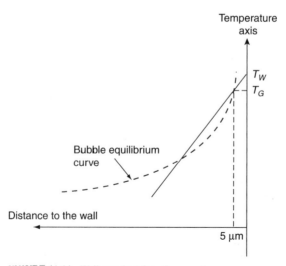

FIGURE 11-14   Wall superheat for a 5-μm cavity.

11-2 POOL BOILING

To study the boiling mechanism, it is necessary to distinguish between the boiling of stagnant liquids (*pool boiling*) and that of moving liquids, which is called *forced convective boiling*. We shall first study the case of pool boiling.

11-2-1 Pool-Boiling Curve

We shall consider the case of a liquid at its saturation temperature T_{SAT} contained in a vessel. This liquid is being heated by means of a heating surface at a temperature T_W, so $T_W > T_{SAT}$. The temperatures difference $T_W - T_{SAT}$ is called ΔT_{SAT}. The results of investigations about heat transfer rates in pool boiling are usually plotted on a graph of surface heat flux against heater-wall surface temperature or, more frequently, against the wall superheat ΔT_{SAT} rather than the wall temperature itself. The resulting curve—the *boiling curve*—is like that shown in Fig. 11-15. (When the relationship between q and ΔT is not linear, the definition of a heat transfer coefficient h is not particularly useful, and this is why it is preferred to present the results directly as a plot of q against ΔT.) It can be observed on the curve in the figure the existence of several regions.

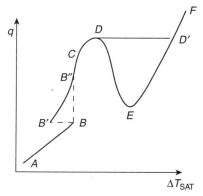

FIGURE 11-15 Typical pool-boiling curve.

In the region corresponding to low wall temperatures (curve AB), heat transfer takes place without bubble formation. Heat is removed by natural convection because the temperature required for the onset of nucleate boiling, T_{WONB}, has not been reached. In this region, all the correlations for natural convection are valid.

When point B is reached, the wall superheat becomes sufficient to cause vapor bubble formation at the heating surface. This was analyzed in detail in the preceding section, and we also explained how to predict the value of this temperature.

When bubble production starts, the heat transfer coefficient suddenly increases as a consequence of the liquid agitation produced by the bubbles. This explains the discontinuity presented by the curve at this point.

The boiling curve in Fig. 11-15 can be constructed in two ways:

1. Increasing the wall temperature slowly and continuously and measuring the resulting heat flux
2. Increasing the heat flux slowly and continuously and measuring the resulting wall temperature (This is the procedure when heat is supplied by an electric resistance. The electric power delivered to the resistance is increased progressively, and this provokes an increase in the wall temperature.)

If the boiling curve is constructed by increasing the wall temperature, it is observed that when point B is reached, a slight increase in T_W provokes a jump in heat flux to point B''. If, on the other hand, the curve is constructed by increasing the heat flux, when point B is reached, a slight increase in q causes a decrease in the wall temperature to point B' because, owing to the increase in the heat transfer coefficient, a smaller ΔT is required to maintain the heat flux.

Region $B'CD$ corresponds to the nucleate boiling regime; this means with bubble production. It can be observed now that the slope of the boiling curve is higher than that of region AB, indicating an increase in the film boiling coefficient. The agitation effect produced by the detachment of the bubbles explains the high values of the heat transfer coefficients in this region.

As long as the difference $T_W - T_{SAT}$ is being increased, the heat flux also increases, but a moment is reached when a change begins in the hydrodynamic regime of the system. The bubble production is so high that they begin to agglomerate before detaching, and a vapor blanket forms over the surface. This vapor blanket prevents the liquid from flowing onto the surface.

As long as ΔT is increased further, more and more heating surface is being covered by this blanket, and the heat transfer coefficient falls. This corresponds to region DE of the boiling curve. In this region, a quite unique phenomenon in heat transfer takes place, consisting of a reduction in heat flux with an increase in the driving force.

Point D is called the *burnout point*. This is so because when the experimental construction of the boiling curve is performed, increasing the heat flux as the independent variable and measuring the resulting wall temperature, once point D is reached, a further increase in the heat flux moves the system to situation D', with a sudden increase in the temperature of the heating element (normally an electric resistance). In this case, the resistance may burn out, and that is the reason for the expression.

In region EF, the heating surface is covered by a stable vapor film, and the heat has to flow to the liquid through it. In this region, the wall temperature is high enough for the radiation effects to become significant, and heat flux begins again to increase with the increase in ΔT. This mechanism is known as *film boiling*. This effect sometimes can be observed if a drop of water falls onto a very hot iron plate. In this case, sometimes it can be seen that the drop takes a certain amount of time to evaporate and remains, for a while, moving rapidly over the surface up to its final consumption. This is due to the formation of a vapor film that insulates the drop from the surface, and the heat transfer is through this vapor film.

11-2-2 Boiling-Curve Correlations

Several correlations have been proposed for each region of the boiling curve. They represent the experimental results with more or less success. It was not possible to find general correlations that adjust the results of different authors. The explanation may be in the fact that boiling is a very complex phenomenon and depends on a large number of variables. In the following sections, we shall present the best-known correlations.

Single-Phase Convection (Zone AB).   In this zone, the usual single-phase natural convection correlations may be used. For example, for natural convection in a turbulent regime from a horizontal plane, a well-known correlation is

$$\frac{hD}{k_L} = 0.14 \left(\frac{\beta g \Delta T D^3 \rho_L^2}{\mu_L^2} \frac{c_L \mu_L}{k_L} \right)^{1/3}$$ (11-2-1)

where D is the diameter of the surface.

Onset of Nucleation Boiling (Point B).   The model based on Hsu's theory to predict the onset of nucleate boiling was already explained (see Sec. 11-1).

Nucleate Boiling Zone (Zone B'D).   This region of the boiling curve is the most important for design because most heat transfer equipment normally operates in it. Many researchers tried to correlate the heat transfer coefficients, and many expressions have been proposed. Most of these correlations look completely different and are based on different variables. This is confirmation that a good understanding of this phenomenon is still lacking, and a large amount of research is still to be done. In what follows we shall present the best-known correlations.

Even the usual form of correlation is to express the heat transfer coefficient as $h = a\Delta T^b$. Substituting h with $q/\Delta T$ or ΔT with q/h, it is possible to present the correlations as $h = f(q)$ or $q = f(\Delta T)$. We shall use all these forms.

(a) *Rohsenow Correlation.*   Rohsenow[7] proposed a correlation of the type

$$\mathrm{Nu} = a\,\mathrm{Re}^b\,\mathrm{Pr}_L^c$$ (11-2-2)

For definitions of Nusselt and Reynolds numbers, the bubble detachment diameter predicted by Eq. (11-1-23) is used.

The velocity in Reynolds number is taken as the superficial velocity of liquid toward the heating surface. If q is the heat-flux density and λ is the latent heat of vaporization, the vaporization rate on the surface (kg/s) is given by q/λ, and the velocity of the liquid toward the surface to replace the vaporized mass will be $q/\lambda\rho_L$. Hence the Reynolds number will be

$$\mathrm{Re} = \frac{q}{\lambda\rho_L}\left[\frac{\sigma}{g(\rho_L - \rho_V)}\right]^{1/2}\frac{\rho_L}{\mu_L} \tag{11-2-3}$$

It must be noted that the contact angle θ of the Eq. (11-1-23) has not been included. This is due to the fact that this variable is very difficult to measure, and it was decided not to include it in the correlation,
The Nusselt number then will be

$$\mathrm{Nu} = \frac{hD_d}{k_L} = \frac{h}{k_L}\left[\frac{\sigma}{g(\rho_L - \rho_V)}\right]^{1/2} \tag{11-2-4}$$

The correlation proposed by Rohsenow is

$$\mathrm{Nu} = \frac{1}{C_{SF}}\mathrm{Re}^{(1-n)}\mathrm{Pr}^{-m} \tag{11-2-5}$$

which can be arranged to give

$$\frac{c_L\Delta T_{\mathrm{SAT}}}{\lambda} = C_{SF}\left\{\frac{q}{\mu_L\lambda}\left[\frac{\sigma}{g(\rho_L - \rho_V)}\right]^{1/2}\right\}^{n}\left(\frac{c_L\mu_L}{k_L}\right)^{m+1} \tag{11-2-6}$$

The original equation had $n = 0.33$ and $m = 0.7$. Afterwards, Rohsenow recommended that m be changed to 0 for the case of water. Values of C_{SF} for several solid-liquid combinations are included in Table 11-1. A more detailed study of C_{SF} values was performed by Vachon and colleagues.[6]
The author's recommendation, in case a specific value for the particular fluid surface combination is not available, is to assume as a first approximation $C_{SF} = 0.013$. It is not indicated how to later improve this first approximation.

(b) McNelly Equation.   McNelly[13] proposed the following equation:

$$h = 0.225\left(\frac{qc_L}{\lambda}\right)^{0.69}\left(\frac{pk_L}{\sigma}\right)^{0.31}\left(\frac{\rho_L}{\rho_V} - 1\right)^{0.33} \tag{11-2-7}$$

where p is the system pressure.

(c) Forster and Zuber Equation.   These authors[8] proposed a correlation of the type

$$\mathrm{Nu} = \frac{hr_b}{k_L} = 0.0015\mathrm{Re}_b^{0.62}\mathrm{Pr}_L^{0.33} \tag{11-2-8}$$

TABLE 11-1   Values of the C_{SF} Constant in Rhosenhow's Equation

Fluid-Surface Combination	C_{SF}	Fluid-Surface Combination	C_{SF}
Water-nikel	0.006	n-Butylic alcohol–copper	0.003
Water–stainless steel	0.015	CO_3K_2 50%–copper	0.0027
Water-copper	0.013	CO_3K_2 35%–copper	0.0054
Carbon tetrachloride–copper	0.013	Isopropylic alcohol–copper	0.0022

where

$$r_b = \frac{\Delta T}{\lambda \rho_V} \left(\frac{2\pi k_L \rho_L c_L \sigma}{\Delta p_{SAT}} \right) \left(\frac{\rho_L}{\Delta p_{SAT}} \right)^{0.25} \qquad (11\text{-}2\text{-}9)$$

$$\mathrm{Re}_b = \frac{\pi k_L c_L}{\mu_L} \left(\frac{\rho_L \Delta T_{SAT}^2}{\rho_V \lambda} \right)^2 \qquad (11\text{-}2\text{-}10)$$

Substituting, we get

$$q = 0.00122 \left(\frac{k_L^{0.79} c_L^{0.45} \rho_L^{0.49}}{\sigma^{0.5} \mu_L^{0.29} \lambda^{0.24} \rho_V^{0.24}} \right) \Delta T_{SAT}^{1.24} \Delta p_{SAT}^{0.75} \qquad (11\text{-}2\text{-}11)$$

where Δp_{SAT} is the difference between the vapor pressures at the heating surface and liquid temperatures.

(d) Mostinsky Equation. Mostinsky[12] proposed the following dimensional correlation:

$$h = 0.00417 p_c^{0.69} q^{0.7} \left[1.8 \left(\frac{p}{p_c} \right)^{0.17} + 4 \left(\frac{p}{p_c} \right)^{1.2} + 10 \left(\frac{p}{p_c} \right)^{10} \right] \qquad (11\text{-}2\text{-}12)$$

where p_c (critical pressure) must be expressed in kPa and q in W/m^2 to get h in W/(m$^2 \cdot$ K). If p_c is expressed in N/m^2, the value of the constant is 3.7×10^{-5}.

(e) Stephan and Abdelsalam Correlations. These researchers[14] made a regression of experimental results obtained from different authors. Using dimensional analysis techniques, they obtained the following dimensionless groups to express the solution of the problem:

$$X1 = \left(\frac{q D_d}{k_L T_{SAT}} \right) \quad X2 = \left(\frac{\alpha^2 \rho_L}{\sigma D_d} \right) \quad X3 = \left(\frac{c_L T_{SAT} D_d^2}{\alpha^2} \right) \quad X4 = \left(\frac{\lambda D_d^2}{\alpha^2} \right)$$

$$X5 = \left(\frac{\rho_V}{\rho_L} \right) \quad X6 = \left(\frac{c_L \mu_L}{k_L} \right) \quad X7 = \left(\frac{\rho_w c_w k_w}{\rho_L c_L k_L} \right) \quad X8 = \left(\frac{\rho_L - \rho_V}{\rho_L} \right) \qquad (11\text{-}2\text{-}13)$$

It is interesting to note that the dimensionless group $X7$ is a ratio of physical properties of the heating-surface material and liquid properties. In these expressions, α is the liquid thermal diffusivity ($k/\rho c$), and D_d is the bubble detachment diameter given by Eq. (11-1-23). T_{SAT} is the absolute saturation temperature.

The authors correlated the Nusselt number ($h D_d/k_L$) as a function of these dimensionless groups. It was not possible to find a single correlation to adjust the experimental data for all types of compounds. The authors, then, divided the data in four groups corresponding to the following four types of substances:

1. Water
2. Hydrocarbons
3. Cryogenic fluids
4. Refrigerants

Since in the experimental information analyzed by these authors there were no measurements of the contact angle θ included in the definition of D_d, they correlated the data using typical values of θ for each of the four groups of substances. The values of θ, as well as the correlations they obtained, are shown in Table 11-2.

Boiling of Mixtures. All the previous correlations are recommended by their authors for boiling of single-component fluids. In the case of mixtures, the heat transfer rate is considerably reduced by the existence of diffusive resistances. In these cases, the film coefficients must be corrected as indicated in Sec. 11-4-2.

TABLE 11-2   Stephan and Abdelsalam Correlations

Group	Correlation	Range (p/p_c)	θ, degrees	Equation
Water	$Nu = 0.24 \times 10^7 X1^{0.67} X4^{-1.58} X3^{1.26} X8^{5.22}$	$10^{-4} < p/p_c < 0.88$	45	(11-2-14)
Hydrocarbons	$Nu = 0.054 X5^{0.34} X1^{0.67} X8^{-4.33} X4^{0.248}$	$5 \times 10^{-3} < p/p_c < 0.9$	35	(11-2-15)
Cryogenics	$Nu = 4.82 X1^{0.624} X7^{0.117} X3^{0.374} X4^{-0.329} X5^{0.257}$	$4 \times 10^{-3} < p/p_c < 0.97$	1	(11-2-16)
Refrigerants	$Nu = 207 X1^{0.745} X5^{0.581} X6^{0.533}$	$3 \times 10^{-3} < p/p_c < 0.78$	35	(11-2-17)

Maximum Heat Flux.   For the maximum heat-flux density (point D of the boiling curve), the Zuber[18,19] equation is normally accepted:

$$q_{max} = 0.18 \rho_V \lambda \left[\frac{(\rho_L - \rho_V)\sigma g}{\rho_V^2} \right]^{1/4} \qquad (11\text{-}2\text{-}14)$$

This equation is valid for planar surfaces facing upward. It can be extended to single tubes immersed in a big mass of liquid. However, it cannot be used for tube bundles without correction. This will be explained later.

Another correlation owing to Mostinsky, also valid for a single tube immersed in a boiling liquid, is

$$q_{max} = 367 p_c \left(\frac{p}{p_c} \right)^{0.35} \left(1 - \frac{p}{p_c} \right)^{0.9} \qquad (11\text{-}2\text{-}15)$$

where pressures are expressed in k_{Pa} and q_{max} in W/m^2.

Transition Flux Zone (Zone DE).   This is a very unstable zone, and no correlations for it are available at present.

Point of Minimum Heat Flux (Point E).   The theoretical correlation developed by Zuber[19] can be applied:

$$q_{min} = 0.125 \rho_V \lambda \left[\frac{(\rho_L - \rho_V)\sigma g}{\rho_L^2} \right]^{1/4} \qquad (11\text{-}2\text{-}16)$$

Afterwards, Kutateladze[20] proposed to change the coefficient to 0.09.

Stable Film Boiling Zone (Zone EF).   The Bromley correlation[9] can be used:

$$h = 0.62 \left[\frac{k_V^3 (\rho_L - \rho_V)\rho_V \lambda g}{\mu_V D_o \Delta T} \right]^{1/4} \qquad (11\text{-}2\text{-}17)$$

where D_o is the tube diameter.

11-2-3   Comparison among the Several Nucleate Boiling Correlations

The nucleate boiling region is that of highest interest because most process equipment is designed to operate in this zone. We can see that different physical variables are used for different authors to correlate the same phenomena. It is quite obvious, then, that different results will be obtained depending on the correlation employed. The differences can be important, as will be seen in the following examples.

Example 11-2   Calculate the values of the nucleate boiling heat transfer coefficients for water, ammonia, and benzene at atmospheric pressure corresponding to the following values of ΔT_{SAT} ($=T_W - T_{SAT}$): 4.3°C, 9.1°C, and 13°C using Rohsenow, Forster and Zuber, Mostinsky, McNelly, and Stephan correlations. (We have chosen these values of ΔT because we had at hand experimental data for water.) The following table includes the physical properties necessary for the calculations:

	Water	NH_3	Benzene
T_{SAT}, K	373	244.2	353
c_L, J/(kg · K)	4,180	4,472	1,985
λ, J/kg	2.25×10^6	1.37×10^6	3.97×10^5
ρ_L, kg/m^3	958	689	820
ρ_V, kg/m^3	0.597	0.843	2.82
μ_L, kg/(m · s)	0.275×10^{-3}	0.24×10^{-3}	0.3×10^{-3}
k_L, W/(m · K)	0.688	0.502	0.129
σ, N/m	0.0588	0.0325	0.021
p_c, N/m^2	221×10^5	112×10^5	48.9×10^5
θ, °	45	35	35
$\Delta p_{SAT}/\Delta T_{SAT}$	3,906	5,479	3,306

The last row of the table is the slope of the vapor pressure curve, necessary for Forster and Zuber correlations.

Substituting the physical properties into the correlations, it is possible to obtain expressions of the type

$$q = a\Delta T_{SAT}{}^b$$

And we come to

Correlation	Water	Ammonia	Benzene
Rohsenow	$154.2\Delta T_{SAT}{}^3$	$50.59\Delta T_{SAT}{}^3$	$1.6\Delta T_{SAT}{}^3$
Mostinsky	$46.45\Delta T_{SAT}{}^{3.33}$	$14.12\Delta T_{SAT}{}^{3.33}$	$3.4\Delta T_{SAT}{}^{3.33}$
McNelly	$21\Delta T_{SAT}{}^{3.22}$	$46.2\Delta T_{SAT}{}^{3.22}$	$16.3\Delta T_{SAT}{}^{3.22}$
Stephan	$64.2\Delta T_{SAT}{}^{3.05}$	$0.044\Delta T_{SAT}{}^{3.92}$	$8\Delta T_{SAT}{}^3$
Forster and Zuber	$1.68\Delta T_{SAT}{}^{1.24}\Delta p_{SAT}{}^{0.75}$ $= 830\Delta T_{SAT}{}^{1.99}$	$1.66\Delta T_{SAT}{}^{1.24}\Delta p_{SAT}{}^{0.75}$ $= 1057\Delta T_{SAT}{}^{1.99}$	$0.508\Delta T_{SAT}{}^{1.24}\Delta p_{SAT}{}^{0.75}$ $= 221\Delta T_{SAT}{}^{1.99}$

And calculating the values of q for the selected temperature differences, we get

1. Water at atmospheric pressure (values of q in W/m^2):

ΔT_{SAT}	Rohsenow	Mostinsky	McNelly	Stephan	Forster & Zuber	Experimental
4.3	11,200	5,900	2,300	5,490	15,100	55,000
9.1	1.16×10^5	7.25×10^4	2.57×10^4	5.4×10^4	6.72×10^4	2.36×10^5
13	3.38×10^5	2.37×10^5	8.11×10^5	1.60×10^5	1.36×10^5	5.17×10^5

2. Ammonia at atmospheric pressure (values of q in W/m^2):

ΔT_{SAT}	Rohsenow	Mostinsky	McNelly	Stephan	Forster & Zuber
4.3	4,022	1,816	5,060	13.3	19,200
9.1	3.8×10^4	2.2×10^4	5.66×10^4	252	8.56×10^4
13	1.11×10^5	7.23×10^4	1.78×10^5	1,023	1.74×10^5

3. Benzene at atmospheric pressure (values of q in W/m²):

ΔT_{SAT}	Rohsenow	Mostinsky	McNelly	Stephan	Forster & Zuber
4.3	130	440	1,780	640	4,030
9.1	1,200	5,300	19,900	6,000	17,900
13	3,500	17,400	62,900	17,500	36,400

Note the important dispersion in the calculated results.

Expression for h as a Function of the Heat-Flux Density. Nucleate boiling correlations are of the type

$$q = a\Delta T^b \tag{11-2-18}$$

where $\Delta T = T_W - t$ (we use the small t for cold-fluid temperature). Combining this expresion with

$$q = h_c\Delta T \tag{11-2-19}$$

(we added the subscript c to identify the cold fluid, which is the boiling fluid), we get

$$h_c = cq^d \tag{11-2-20}$$

where $d = (b-1)/b$ and $c = a^{1/b}$. As we shall see, when the correlations are expressed this way, the deviations in the calculated h at same q are much smaller than those compared at the same ΔT. This will be shown in the following example.

Example 11-3 For the same correlations and fluids of Example 11-2, calculate the boiling film coefficient for similar q values.

Solution We shall first write all the correlations using the transformation equations explained above

Correlation	Water	Ammonia	Benzene
Rohsenow	$h = 5.35q^{0.66}$	$h = 3.69q^{0.66}$	$h = 1.16q^{0.66}$
Mostinsky	$h = 3.16q^{0.7}$	$h = 2.21q^{0.7}$	$h = 1.44q^{0.7}$
McNelly	$h = 2.56q^{0.69}$	$h = 3.27q^{0.69}$	$h = 2.37q^{0.69}$
Stephan	$h = 3.05q^{0.66}$	$h = 0.45q^{0.75}$	$h = 1.99q^{0.66}$
Forster and Zuber	$h = 29.3q^{0.5}$	$h = 33q^{0.5}$	$h = 15q^{0.5}$

For each fluid of Example 11-2, we shall calculate h_c for different q values:

	Water			Ammonia			Benzene		
q (W/m²)	1,000	10,000	50,000	1,000	10,000	50,000	1,000	10,000	50,000
Rohsenow	532	2,460	7,200	350	1,610	4,660	110	510	1,460
Mostinsky	397	2,000	6,150	279	1,400	4,300	181	910	2,800
McNelly	300	1,470	4,470	384	1,880	5,700	280	1,360	4,140
Stephan	389	1,800	4,930	80	450	1,500	190	500	2,500
Forster and Zuber	926	2,930	6,550	1,040	3,300	7,400	470	1,500	3,350

It can be seen that except for some particular cases (e.g., the big dispersion showed by the Stephan correlation for the case of ammonia), when the results are presented this way, the differences among the correlations look smaller.

Effect of the Controlling Resistances.   The heat-flux density q depends not only on the boiling side coefficient but also on the fouling resistance and heating medium coefficient. This is

$$q = U(T - t) \tag{11-2-21}$$

where

$$U = (1/h_h + 1/h_c + R_f)^{-1} \tag{11-2-22}$$

The boiling heat transfer coefficients are usually very high, so frequently the controlling resistance for heat transfer lies in the combined effect of fouling and the hot-fluid coefficient. This means that independent of the differences that the boiling correlations may present, their influence in the calculation of q is normally dampened.

Thus, when calculation of the boiling h is performed with expressions of the type $h = cq^d$, the deviations are flattened, and the results obtained with different correlations are not very dissimilar. This is not an indication of the quality of the correlations but rather of the low influence that the selection of a good correlation has in the final equipment design. The concept is illustrated in the following example.

Example 11-4   Calculate, using the previously studied correlations, the area of a coil to evaporate 500 kg/h of benzene (0.138 kg/s) at atmospheric pressure (boiling temperature = 80°C) using 140°C steam as the heating medium. Assume that the combined resistance $(1/h_h + R_f)$ corresponding to the heating steam and fouling is 0.0005 $(m^2 \cdot K)$/W. We shall assume that the benzene is fed at its boiling temperature, so only the latent heat of vaporization has to be delivered.
The following procedure is proposed:

1. Assume a value for q as a first trial.
2. With the boiling correlations ($h_c = cq^d$), calculate h_c.
3. Calculate $1/U = 1/h_c + (1/h_h + R_f)$.
4. Calculate $q = U(T - t) = U(140 - 80)$, and compare with the assumed q. With the calculated q, go back to step 1 up to convergence.
5. The required area then will be $A = W\lambda/q = (0.138 \times 3.97 \times 10^5)/q$.

The final calculations are shown in the following table:

Correlation	Q (W/m^2)	h_c	U	$U(140–80)$	A (m^2)
Rohsenow	5×10^4	1,470	847	5.08×10^4	1.1
Mostinsky	8×10^4	3,895	1,321	7.92×10^4	0.69
McNelly	9×10^4	6,211	1,512	9.07×10^4	0.61
Stephan	7.5×10^4	3,280	1,242	7.45×10^4	0.73
Forster and Zuber	8.1×10^4	4,270	1,360	8.17×10^4	0.67

The damping effect of the other resistances can be observed. The results obtained look more presentable.

11-3   FLOW BOILING IN TUBES

A case of great interest in process engineering is the flow boiling of liquids in a tube. To study this case, it is necessary to analyze some aspects of fluid mechanics dealing with two-phase flow.

11-3-1   Two-Phase Flow Patterns

When a liquid-vapor mixture flows in a tube, the flow pattern can assume different configurations depending on the relative amount of each phase, their physical properties, and the flow rates. The flow patterns that can exist are different for vertical and horizontal flow. We shall analyze the case of two-phase vertical upward flow because it is the case that most interests us. A description of the different two-phase flow patterns in horizontal tubes can be found in ref. 4.

The different flow patterns that will be described next follow an order of increasing vapor quality x ($x =$ vapor mass flow/total mass flow).

Bubbly Flow. This pattern appears when the vapor-phase mass flow is small. The vapor phase is distributed as discrete bubbles in a continuous liquid phase (Fig. 11-16a).

Slug Flow. In slug flow, the vapor bubbles are approximately the diameter of the pipe. The length of these bubbles can vary considerably, and they are separated from the pipe wall by a slowly moving liquid film. The liquid flow is contained in liquid slugs that separate successive gas bubbles (Fig. 11-16b).

Churn or Semiannular Flow. Churn flow is formed by the breakdown of large vapor bubbles in slug flow. The vapor flows in a somewhat chaotic manner through the liquid, which is mainly displaced to the tube wall. The flow has an oscillatory or time-varying character, hence the descriptive name *churn flow.* This region is also called *semiannular flow* (Fig. 11-16c).

Wispy Annular Flow. The flow takes the form of a relatively thick liquid film on the walls of the pipe carrying small vapor bubbles, together with a considerable amount of liquid entrained in a central vapor core. This entrained liquid phase appears as large droplets that have agglomerated into long, irregular filaments or wisps (Fig. 11-16d).

Annular Flow. A liquid film forms at the pipe wall with a continuous central gas or vapor core. Large-amplitude waves usually are present on the surface of the film, and the continued breakup of these waves forms a source for droplet entrainment that occurs in the central core. As distinct from the wispy annular pattern, the droplets are separated rather than agglomerated (Fig. 11-16e).

Mist Flow. The liquid is only present in the form of small droplets entrained in the vapor phase that occupy the entire pipe (Fig. 11-16f).

Frequently, the distinction among the different flow patterns is not so clear. Some authors describe other intermediate patterns, and no uniform criterion exists in the definition of one pattern or the other.

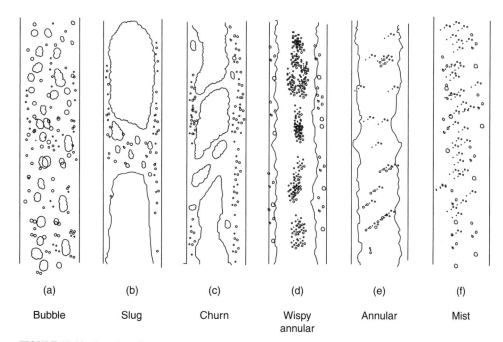

|(a)|(b)|(c)|(d)|(e)|(f)|
|Bubble|Slug|Churn|Wispy annular|Annular|Mist|

FIGURE 11-16 Two-phase flow patterns.

When a fluid flows in a heated tube, as long as the fluid receives heat, it experiences a boiling process, and the vapor quality x increases along its path. This means that the two-phase flow pattern changes from one type to the following as in the system of Fig. 11-18.

11-3-2 Flow-Pattern Characterization

The type of flow pattern existing in a two-phase system depends on the fluid mass flow and the relative amounts of vapor and liquid present. The vapor quality of the circulating fluid x, can be defined as

$$x = \frac{W_V}{W_V + W_L} \tag{11-3-1}$$

where W_V = vapor mass flow rate (kg/s)
 W_L = liquid mass flow rate (kg/s)
$W_V + W_L = W$ = total mass flow rate (kg/s)

The mass velocity G is the mass flow rate divided by the flow area:

$$G = \frac{W}{a_t} \tag{11-3-2}$$

For characterization of the different flow patterns that may be present in a section of the tube, some graphs known as *flow-pattern maps* have been developed. The Hewit and Roberts[32] flow-pattern map for vertical flow is shown in Fig. 11-17a. This map is developed as a function of two variables, which are the vapor and liquid mass velocities, defined as

$$G_L = G(1 - x) \qquad \text{and} \qquad G_V = Gx$$

Another flow-pattern map for vertical upflow was presented by Fair[16] and is shown in Fig. 11-17b. The author restricts the validity of this map to the sizing of vertical thermosiphon reboilers.

Liquid and Vapor Volume Fractions.   Other parameters used frequently in the analysis of two-phase systems are the liquid or vapor volume fractions (R_L or R_V). These are the fraction of volume or cross section of tube occupied by the liquid or vapor, respectively. These parameters cannot be calculated directly from the values of x and the liquid and vapor densities due to the fact that both phases normally flow with different velocities.

The first authors to study two-phase flow systems and correlate the different flow parameters were Lockhardt and Martinelli.[15] Different authors later proposed modifications to their original correlations. Fair[16] presents a correlation to calculate R_L as a function of a parameter X_{tt} defined as

$$X_{tt} = \left(\frac{1-x}{x}\right)^{0.9} \left(\frac{\rho_V}{\rho_L}\right)^{0.5} \left(\frac{\mu_L}{\mu_V}\right)^{0.1} \tag{11-3-3}$$

This parameter is known as the *Lockhardt-Martinelli parameter* for turbulent flow in both phases. The correlation proposed by Fair is

$$R_L = \left(\frac{1}{1 + 21X' + X'^2}\right)^{0.5} \tag{11-3-4}$$

where $X' = 1/X_{tt}$.

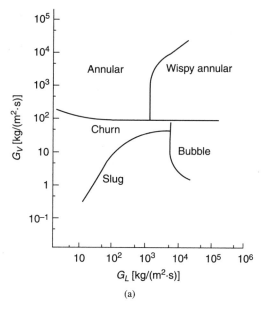

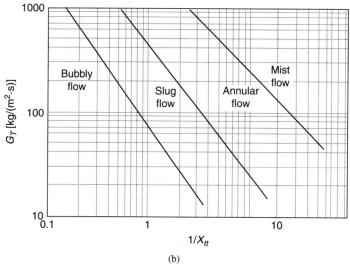

FIGURE 11-17   (*a*) Hewit and Roberts map for two-phase vertical upflow. (*b*) Fair map
for two-phase vertical upflow.

11-3-3 Frictional Pressure Drop

The frictional pressure drop for a two-phase system circulating in a pipe can be calculated as

$$\Delta p_{TP} = \phi^2 \Delta p_L \tag{11-3-5}$$

where Δp_{TP} = frictional pressure drop of the two-phase system
Δp_L = frictional pressure drop if only the liquid phase circulates in the pipe

ϕ^2 is a correction factor (the fact that ϕ^2 rather than ϕ is used has no special significance and is only due to historic reasons).

ϕ^2 was originally correlated by Lockhardt and Martinelli. The original correlation was later modified by other authors. Fair[16] proposed the following:

$$\phi^2 = 1 + 21X' + X'^2 \tag{11-3-6}$$

11-3-4   Heat Transfer Mechanisms for Boiling Inside Vertical Tubes

Let's consider a vertical tube fed at its base with subcooled liquid with enthalpy i_1 and temperature t_1. Let W be the mass flow rate. The tube is externally heated, so there is a heat flux through the wall to the circulating fluid.

Depending on the boundary condition for heat transfer, we can have different situations. For example, it is usual in laboratory conditions to work with electrically heated tubes. In this case, if a constant electrical power per unit length is imposed, the boundary condition is constant heat flux for the whole tube surface. Then the tube wall temperature will vary from point to point to deliver this heat flux according to

$$q = \text{constant} = h_i(T_{W(z)} - t_{F(z)}) \tag{11-3-7}$$

where h_i is the heat transfer coefficient and $T_{W(z)}$ and $t_{F(z)}$ are the local temperatures of the wall and circulating fluid corresponding to a tube section at coordinate z. The heat transfer coefficient h_i, as well as $t_{F(z)}$ and $T_{W(z)}$ are a function of z.

A more usual situation in industrial processes is to have the tube heated by an external fluid. For example, let's assume that the heating medium at the external side of the tube is condensing steam at temperature T. Let h_o be its condensation heat transfer coefficient. The wall temperature of the tube at every section must be such that

$$h_o(T - T_{W(z)}) = h_i(T_{W(z)} - t_{F(z)}) \tag{11-3-8}$$

or

$$T_{W(z)} = \frac{h_o T + h_i t_{F(z)}}{h_i + h_o} \tag{11-3-9}$$

Figure 11-18 shows the variation of fluid temperature $t_{F(z)}$ and wall temperature $T_{W(z)}$ for a heated tube along the z axis.

The origin of z coordinate is at the tube entrance. During the first stage of its evolution, the liquid temperature increases owing to the sensible heat transfer received (zone A in Fig. 11-18). As the fluid temperature increases, the wall temperature also increases according to Eq. (11-3-9).

Somewhere a condition is reached in which the wall and fluid temperatures are such that vapor bubbles begin to form at the tube wall. This point corresponds to the onset of nucleate boiling. Note that even though the wall temperature at this location must be higher than the saturation temperature, the bulk of the fluid still has not reached this saturation temperature. This means that the liquid is still subcooled. When the vapor bubbles produced at the tube wall depart and come in contact with the colder fluid, they recondense and deliver their heat of condensation to the liquid. From the point where nucleate boiling begins, the flow pattern starts being two-phase flow.

To the right of the tube in Fig. 11-18, the different flow patterns are indicated, and farther to the right are the different heat transfer regions. Region B corresponds to subcooled boiling. At the point where nucleate boiling starts, the microconvective effect created by bubble detachment produces an increase in the heat transfer coefficient h, the consequence of which is a sudden decrease in the wall temperature according to Eq. (11-3-9).

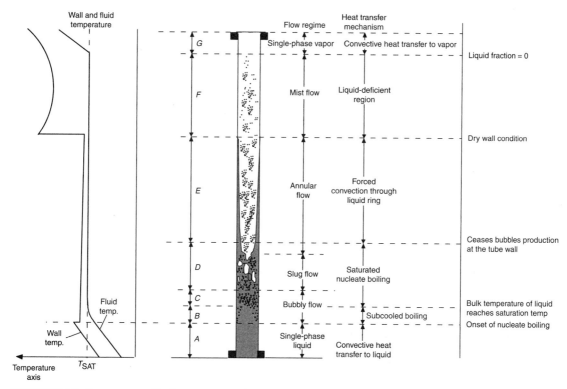

FIGURE 11-18 Temperature profile, flow patterns, and heat transfer regions in a heated tube.

As the temperature of the liquid continues increasing, at a certain location it reaches the saturation temperature. This happens at a distance from the inlet section z_{SAT} given by

$$W(i_{SAT} - i_1) = \int_0^{z_{SAT}} q\,dA \tag{11-3-10}$$

where i_{SAT} is the saturation enthalpy of the liquid stream per unit mass. Note that i_{SAT} is the average enthalpy of the stream at section z_{SAT}. However, since there is a radial temperature profile, and since in the region near the tube wall the fluid temperature is higher than the saturation temperature, there must exist at the center of the tube a region with fluid temperatures below t_{SAT}. Nevertheless, it is usual to accept Eq. (11-3-10) as a valid criterion to define the start of saturated boiling (start of region C in Fig. 11-18).

After the point where saturated boiling starts, the most characteristic variable to represent the condition of flow is the vapor quality or vapor mass fraction, defined as

$$x = \frac{\int_{z_{SAT}}^{z} q\,dA}{W\lambda} = \frac{i_z - i_{z_{SAT}}}{\lambda} \tag{11-3-11}$$

This vapor quality or vapor mass fraction represents the ratio between the vapor mass flow and the total mass flow as expressed by Eq. (11-3-1).

As long as x increases, the flow pattern experiences further transformations and may turn into slug flow or annular flow, as shown in Fig. 11-18. If the circulating fluid is a pure single component, its temperature remains almost constant during the entire saturated boiling region (except for the change in boiling temperature as the pressure changes along the tube). The heat transfer coefficient h_i increases as long as the vaporization proceeds owing to the increase in velocity caused by the reduction in the fluid density. Hence the wall temperature decreases slowly along the z axis (even though this is difficult to appreciate in Fig. 11-18).

After the annular flow pattern starts, there is a change in the heat transfer mechanism consisting of suppression of the nucleate boiling mechanism. This means that bubble production at the tube wall stops, and heat must flow by forced convection through the liquid film to the liquid-vapor interface, where evaporation takes place. In this region we cannot speak about boiling any more. It is called the *two-phase force convection region* (region E in Fig. 11-18).

If heat transfer continues, the liquid film may disappear completely, and the flow pattern changes from annular to mist flow. This transition is known as *tube-wall dryout*. In this region the heat transfer is directly from the tube wall to the vapor. The effect is a sudden reduction in the heat transfer coefficient and, consequently, an increase in the wall temperature, as shown in Fig. 11-18.

In this region, indicated as F in the figure, the liquid droplets evaporate, and finally, a single vapor phase will be present. If heat transfer continues further, there is an increase in the fluid temperature, and according to Eq. (11-3-9), the wall temperature also increases (zone G).

11-3-5   Correlations for Boiling Heat Transfer Inside Tubes

We shall consider the correlations that can be used to calculate the heat transfer coefficient in each region.

Single-Phase Liquid Convection Region.   In this region the convective heat transfer coefficient can be calculated with the usual single-phase convection correlations. For example, the Dittus-Boelter correlation can be used for turbulent flow:

$$h_i = 0.023 \, \mathrm{Re}^{0.8} \, \mathrm{Pr}^{0.33} \left(\frac{k}{D_i} \right) \left(\frac{\mu}{\mu_W} \right)^{0.14} \tag{11-3-12}$$

Onset of Subcooled Nucleate Boiling.   This point can be predicted by Hsu's theory, which was already explained for pool boiling. Figure 11-19 represents the radial temperature profiles in the liquid $t = f(r)$ corresponding to different tube sections and superimposed on the same diagram the bubble equilibrium curve given by Eq. (11-1-8) is represented.

At a certain tube section, before the start of boiling, the temperature profile is represented by line AB. As long as the fluid and wall temperatures increase in upper sections, the temperature profile moves upward until it becomes tangent to the bubble equilibrium curve.

This situation corresponds to the tube section where nucleation starts. In this section, Eq. (11-1-18) or Eq. (11-1-22) applies, and it is possible to calculate the wall temperature combining Eqs. (11-1-21), (11-1-22), (11-3-9), and (11-3-12). There must be taken into consideration the restrictions explained in Example 11-1 regarding the maximum size of cavities that can act as active nucleation sites.

Subcooled Nucleate Boiling Region.   It is generally accepted that in boiling heat transfer, the heat flux can be calculated as the addition of two mechanisms: a macroscopic two-phase forced convection heat transfer and a microconvective effect caused by the agitation produced by bubble detachment at the solid wall. This approach has been proposed by several authors:

$$q = q_c + q_n \tag{11-3-13}$$

The microconvective effect can be explained by a relationship of the type

$$q_n = \varphi(T_W - T_{\mathrm{SAT}})^n \tag{11-3-14}$$

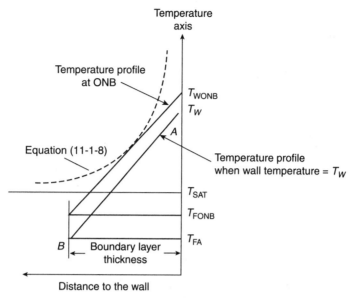

FIGURE 11-19   Start of boiling in a heated tube.

This is a correlation of the same type as those studied for saturated nucleate boiling. But it must be noted that in the expression of ΔT, the saturation temperature, rather than the actual fluid temperature, is used, even though the liquid is subcooled. The exponent n ranges from 2 to 4.

The second mechanism, forced-convection heat transfer, is calculated the same way as single-phase liquid heat transfer:

$$q_c = h_L(T_W - t_{F(z)}) \qquad (11\text{-}3\text{-}15)$$

where h_L is calculated with Eq. (11-3-12). A conservative simplification normally assumed in reboiler design is to ignore the microconvective effect and consider this region and the single-phase liquid region as a single region.

Start of Saturated Boiling.   The transition from subcooled boiling to saturated boiling is defined from a thermodynamic viewpoint. It is the point at which the liquid reaches the saturation temperature found on the basis of simple heat-balance calculations. In this region the bubbles formed at the tube wall, once they detach from the surface, remain in the fluid without recondensation.

The wall temperature can be calculated using Eq. (11-3-9) with $t_{F(z)} = T_{SAT}$. If, as a simplification, it is assumed that the pressure may be considered constant along the tube, the saturation temperature also will be constant. If, on the other hand, the pressure drop due to friction and hydrostatic head effects cannot be neglected, the saturation temperature will change along tube length z.

For example, consider the tube in Fig. 11-20. At the inlet section, the pressure is p_1. If a pressure gradient dp/dz exists in the system, we shall have a curve like curve I, representing the pressure along the tube length. Curve II represents the saturation temperature at the pressure corresponding to each z value. This means that points as A and A' indicate conditions p and t corresponding to the saturation curve.

Curve III represents the evolution of liquid temperature along the tube length. The inlet temperature is t_1. From t_1, the temperature increases as long as the liquid is heated up (subcooled region). The evolution is represented by segment BC of curve III.

At point C, the fluid temperature reaches the saturation temperature, and saturated boiling starts. From then on, the evolution of the fluid temperature follows curve CD corresponding to saturation. Please note

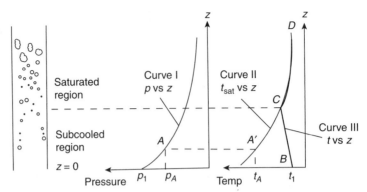

FIGURE 11-20   Pressure and temperature variation along the tube.

that in Fig. 11-18, on the contrary, it has been supposed that the pressure drop is negligible, so the saturated boiling region is represented by an essentially constant fluid temperature.

Heat Transfer in the Saturated Boiling Region.   From the section where saturated boiling starts, the vapor bubbles formed at the tube wall remain in the system without recondensing, and the vapor mass fraction begins to increase along tube length z. Since the total mass flow is constant, the average fluid velocity increases when the fluid density decreases as a consequence of this vaporization. This also produces an increase in the heat transfer coefficient with the progress of vaporization.

As in the case of the subcooled region, the heat transfer mechanism can be considered as the superimposition of two effects: the macroconvective effect and a second effect originated by bubble detachment from the heat transfer surface. The difference is that in the subcooled case, the ΔT values acting as driving forces are different for both mechanisms. They are $T_W - T_{SAT}$ for the nucleation mechanism and $T_W - t_{F(z)}$ for the macroconvection. In the case of saturated boiling, since the saturation temperature and fluid temperature coincide along the whole region, both mechanisms occur with the same driving force. Thus

$$q = q_C + q_N = (h_N + h_{TP})(T_W - t_F) \qquad (11\text{-}3\text{-}16)$$

where h_N is the coefficient corresponding to the microconvective effect of bubble detachment and h_{TP} is the two-phase macroconvective coefficient.

As a difference with the subcooled boiling region, where the macroconvective coefficient was calculated with Eq. (11-3-12) assuming only liquid flow, in the saturated boiling region, where the vapor generated remains in the system, the velocity in the Reynolds number will be a function of the mass vapor fraction.

It is usually accepted that h_{TP} is related to the single-phase convection coefficient by

$$h_{TP} = f_{(X_{tt})} h_L \qquad (11\text{-}3\text{-}17)$$

where h_L is the coefficient for the total flow assumed to be liquid and X_{tt} is the Lockhardt-Martinelli parameter. Many correlations of the type in Eq. (11-3-17) were proposed. One of the best known is due to Dengler and Addoms:[11]

$$h_{TP} = 3.5(X')^{0.5} h_L \qquad (11\text{-}3\text{-}18)$$

where X' is the reciprocal of X_{tt} ($X' = 1/X_{tt}$).

Calculation of h_N to complete Eq. (11-3-16) can be performed with the usual expressions for nucleate boiling presented in Sec. 11-3. Chen[17] has developed a correlation that has been recommended for the saturated nucleate boiling region, as well as for the two-phase force convection region. This will be studied later .

Suppression of Saturated Nucleate Boiling.   As was explained earlier, when the heat transfer coefficient h_{TP} increases because of the increase in the vapor mass fraction, a condition called *suppression of saturated nucleate boiling* can be reached. In this condition, the mechanism of bubble production disappears. This transition normally takes place in the annular flow region, but it can occur in any of the flow patterns described in Sec. 11-3.

The suppression of nucleate boiling can be explained using the same model employed to predict the onset of nucleate boiling. In the entire saturated boiling region, if no pressure gradient exists, the fluid temperature $t_{F(z)}$ remains constant. Hence the heat-flux density increases with the vapor mass fraction as a consequence of the increase in h_{TP}. Figure 11-21 represents the evolution of the temperature profile in the liquid phase while the heat flux increases. It can be seen that with the increase in q, the boundary layer thickness decreases because the coefficient h_{TP} is given by

$$h_{TP} = \frac{k_L}{\delta}$$
(11-3-19)

The curve representing the bubble equilibrium condition given by Eq. (11-1-8) is also shown in the figure.

Curve A corresponds to a condition where nucleate boiling exists because there are cavities with superheat higher than necessary to act as nucleation sites. However, when the temperature profile moves from B to C, the nucleate boiling disappears. This is so because at a very short distance from the wall, less than a bubble radius, the temperature has decreased to a value that is lower than that required for the bubble to grow and detach. Then curve B corresponds to the limit of nucleate boiling.

Accepting that coefficient h_{TP} is given by Eq. (11-3-18), it can be demonstrated that the heat-flux density at the point of nucleate boiling suppression is

$$q_S = \frac{49 B h_L^2}{k_L X_{tt}}$$
(11-3-20)

where B is defined by Eq. (11-1-20).

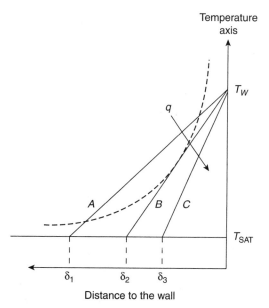

FIGURE 11-21   Suppression of nucleate boiling.

Two-Phase Force-Convection Region.   This region can be associated to any of the two-phase flow patterns, but if fluid velocities are not extremely high, the usual situation is that this region is associated with the annular flow. Heat is transferred by conduction and convection through the liquid film, and vapor is being generated continuously at the interface. It is normally considered that the vapor is at its saturation temperature corresponding to the local pressure. However, the existence of a heat flux from the interface requires the existence of, at least, a small temperature drop. This is normally neglected, but it can be important in certain cases, particularly at low reduced pressures ($p_r < 0.001$).

It is then assumed that the total temperature drop ($T_W - T_{SAT}$) takes place in the liquid film, and the smaller the liquid film thickness and the higher its turbulence, the smaller this temperature drop will be. The heat transfer coefficient can reach extremely high values in this region. Values for water as high as 200,000 W/(m^2 · K) have been reported. Following Martinelli recommendations, several authors have correlated their two-phase convection heat transfer data in the form of Eq.(11-3-17). These correlations can be found in the ref. 21.

In what follows, we shall consider in more detail one of the most frequently used correlations, which can be used not only for the two-phase convective region but also can be extended for the nucleate boiling region, either saturated or subcooled.

Chen Correlation.[17]   Chen performed research work comparing about 600 experimental points with the results predicted by several authors' correlations. None of these correlations could be considered satisfactory in the whole range. Chen therefore proposed a new correlation, and it proved very successful in correlating all the available data.

The proposed correlation covers both the saturated nucleate boiling region and the two-phase forced-convection region for pure liquids, either water or organics. Some authors suggest extending it to the subcooled nucleate boiling region following a procedure that will be explained later.

As was mentioned earlier, the basic assumption is that the saturated convective boiling coefficient results from the superposition of two additive mechanisms: the two-phase convection plus a microconvective effect associated with nucleation and the detachment of bubbles. Thus

$$h = h_N + h_{TP} \tag{11-3-21}$$

However, there is an interaction between both mechanisms that causes the individual effects to be modified. The equations describing these mechanisms thus also must be modified as follows.

The convective component can be represented by a Dittus-Boelter type of equation:

$$h_{TP} = 0.023 \, Re_{TP}^{0.8} \, Pr_{TP}^{0.4} \frac{k_{TP}}{D} \tag{11-3-22}$$

where the thermal conductivity and the Reynolds and Prandtl numbers are effective values associated with the two-phase fluid. However, since heat is transferred to a liquid film in annular and dispersed flow, Chen argued that it is reasonable to use the liquid thermal conductivity in Eq. (11-3-22). Similarly, the values of the Prandtl modulus for liquid and vapor are normally of the same magnitude, and it may be expected that the two-phase value will be close to this value.

Regarding Re, a parameter F is defined such that

$$F = \left(\frac{Re_{TP}}{Re_L} \right)^{0.8} = \left[\frac{Re_{TP}}{G(1-x)D/\mu_L} \right]^{0.8} \tag{11-3-23}$$

Equation (11-3-22) now may be rewritten as

$$h_{TP} = 0.023 \, Re_L^{0.8} \, Pr_L^{0.4} \left(\frac{k_L}{D} \right) F \tag{11-3-24}$$

As can be appreciated in Eq. (11-3-23), Re_L is the Reynolds number that would exist if only the liquid phase circulates in the system.

Regarding the calculation of F, since this ratio is a flow parameter only, it may be expected that it can be expressed as a function of the Lockhardt-Martinelli factor X_{tt}. This was found to be the case. Figure 11-22 is a representation of F versus X_{tt}.

The analysis of Forster and Zuber, developed for pool boiling, was taken as a basis for evaluation of the nucleate boiling component, and this can be expressed by Eqs. (11-2-8) through (11-2-11). These equations are based on the premise that the Reynolds number for the nucleate boiling mechanism is governed by the bubble growth rate. The value of ΔT_{SAT} included in the expression of the Reynolds number [Eq. (11-2-10)] is the difference $T_W - T_{SAT}$. This difference appears in the Forster and Zuber model as the bubble superheat because it is assumed that the bubble forms in a region where the temperatures is approximately equal to T_W (close to the tube wall).

However, in some cases, the actual superheat of a bubble can be quite different from ΔT_{SAT}. Figure 11-23 represents the actual superheat of a bubble ΔT_e, which is the difference between the temperature in the region where the bubble grows and the saturation temperature. This difference depends on the shape of the temperature profile existing in the boundary layer.

As can be appreciated in the figure, the actual superheat of the fluid in which the bubble grows is lower than the wall superheat ΔT_{SAT} ($\Delta T_{SAT} = T_W - T_{SAT}$). The difference between this lower actual superheat ΔT_e and the wall superheat is small for the case of pool boiling and was neglected by Forster and Zuber, but it is significant in the forced-convection case because the boundary layer is much thinner and the temperature gradients much steeper, so the bubble temperature may be quite lower than the wall temperature.

Equation (11-2-11) then must be written in terms of effective ΔT and Δp:

$$h_N = 0.00122 \left(\frac{k_L^{0.79} c_L^{0.45} \rho_L^{0.49}}{\sigma^{0.5} \mu_L^{0.29} \rho_V^{0.24} \lambda^{0.24}} \right) \Delta T_e^{0.24} \Delta p_e^{0.75} \qquad (11\text{-}3\text{-}25)$$

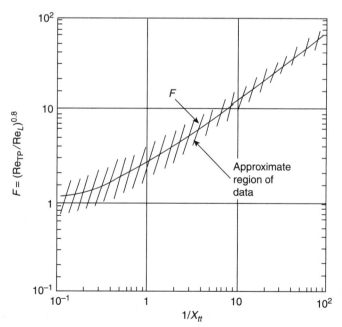

FIGURE 11-22 *F factor versus* $1/X_{tt}$ *(Reprinted with permission from Chen JC: "Correlation for Boiling Heat Transfer to Saturated Fluids in Convective Flow," I&EC Process Design and Developments 5(3):326, 1966. Copyright by the American Chemical Society.)*

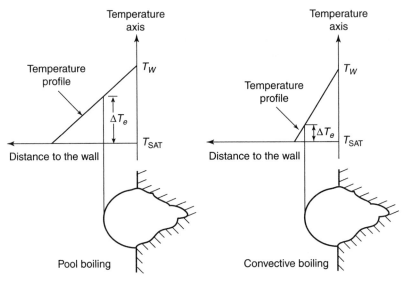

FIGURE 11-23   Temperature profiles in pool and convective boiling.

Chen defines a *suppression factor S* as

$$S = \left(\frac{\Delta T_e}{\Delta T_{SAT}} \right)^{0.99}$$ (11-3-26)

The exponent 0.99 allows S to appear to the first power in the final equation. Using the Clausius-Clapeyron equation for the vapor pressure, the result is

$$S = \left(\frac{\Delta T_e}{\Delta T_{SAT}} \right)^{0.24} \left(\frac{\Delta p_e}{\Delta p_{SAT}} \right)^{0.75}$$ (11-3-27)

with which we finally get

$$h_N = 0.00122 \left(\frac{k_L^{0.79} c_L^{0.45} \rho_L^{0.49}}{\sigma^{0.5} \mu_L^{0.29} \lambda^{0.24} \rho_V^{0.24}} \right) \Delta T_{SAT}^{0.24} \Delta p_{SAT}^{0.75} S$$ (11-3-28)

It might be expected that S would approach unity at low flows and zero at high flows. Chen suggested that S can be represented as a function of the local two-phase Reynolds number Re_{TP}, which was confirmed experimentally.

The function S versus Re_{TP} is shown in Fig. 11-24. An analysis of Figs. 11-22 and 11-24 may contribute to a better comprehension of this phenomenon. F increases with increasing values of $1/X_{tt}$, thus indicating that the turbulence associated with the two-phase flow increases with higher vapor mass fractions.

At low mass flows and low vapor mass fractions, S approaches unity, indicating that the microconvective mechanism plays a more important role, whereas at high mass velocities and vapor mass fractions, the macroconvective mechanism becomes preponderant. Thus it is logical that S approaches 0.

Chen used the Forster and Zuber equation as the basis for his analysis. However, it is possible to use any other of the previously mentioned correlations to calculate the nucleate boiling contribution. At present,

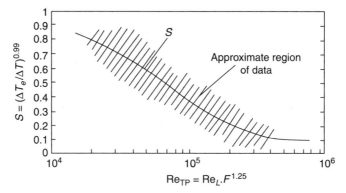

FIGURE 11-24 *S* factor versus Re_{TP}. *(Reprinted with permission from Chen JC: "Correlation for Boiling Heat Transfer to Saturated Fluids in Convective Flow," I&EC Process Design and Developments 5(3):326, 1966. Copyright by the American Chemical Society.)*

some authors prefer the Mostinsky equation [Eq. (11-2-12)] owing to its simplicity. The expression for h_N then could be written as

$$h_N = 0.00417 p_c^{0.69} q^{0.7} \left[1.8 \left(\frac{p}{p_c} \right)^{0.17} + 4 \left(\frac{p}{p_c} \right)^{1.2} \right] S \qquad (11\text{-}3\text{-}29)$$

with pressures in kPa and q in W/m² to get h in W/(m² · K). (The third term of the polynomial has been neglected.)

Equations (11-3-28) and (11-3-29) can only be applied to the boiling of single-component fluids or fluids with a small boiling range. In the case of mixtures, diffusion resistances appear at the vapor-liquid interface, and h_N has to be corrected with a factor F_c (see Eq. 11-4-17).

This effect is similar to what we explained in condensation of vapor mixtures. In the case of boiling, since the lighter components preferably vaporize, the composition of the liquid near the interface is richer in heavy components, and the lighter components in the liquid mixture must diffuse through this film to reach the interface.

It also was found that in the case of mixtures, h_{TP} also must be corrected for diffusion effects, but the correction factors are difficult to predict. We shall not include them, but in high-boiling-range mixtures this may have an effect on the calculated results.

The Chen correlation can be extended for use in the subcooled boiling region by applying the principle of mechanisms superposition:

$$q = h_N (T_W - T_{SAT}) + h_{TP} (T_W - T_{F(z)}) \qquad (11\text{-}3\text{-}30)$$

In this case, h_{TP} is obtained from Eq. (11-3-24) with $F = 1$ (this means that it coincides with h_L), and h_N is evaluated from Eq. (11-3-29), with the value of S obtained from the single-phase-liquid Reynolds number. This extension of the Chen correlation has been tested against experimental information with satisfactory results.

Limits to the Convective Boiling Correlations.

Mist Flow Region. As was explained earlier, in cases where the vapor mass fraction at the outlet region of the tube is very high, it is possible to experience a transition from annular to mist flow. In heat transfer equipment design, this is an undesirable circumstance because there is a substantial deterioration in the performance of the unit. When in a thermosiphon reboiler mist flow exists near the tube outlet, the local wall temperature approaches the heating medium temperature, indicating a very low value of the heat transfer coefficient.

This condition may be predicted as a function of the total mass velocity and the Lockhardt-Martinelli parameter calculated for the outlet conditions. Fair proposed to calculate a critical mass velocity for mist flow as

$$G \text{ (mist flow)} = 2441 X_{tt} \qquad (11\text{-}3\text{-}31)$$

where G is in kg/(m^2 · s). If the mass velocity at the tube outlet is higher than that calculated with Eq. (11-3-31), it is possible to have mist flow at the tube outlet, and the Chen correlation is not valid.

Upper Limit to the Heat-Flux Density.   The curve representing the heat-flux density versus temperature difference for flow boiling is similar to that for pool boiling and also presents a maximum. Several correlations have been proposed to predict the maximum heat-flux density. A correlation[21] that has proven to be successful for a wide variety of fluids, including water, alcohols, and hydrocarbons, is

$$q_{max} = 23,600 \left(\frac{D_i^2}{L} \right)^{0.35} p_c^{0.61} \left(\frac{p}{p_c} \right)^{0.25} \left(1 - \frac{p}{p_c} \right) \qquad (11\text{-}3\text{-}32)$$

where q_{max} = maximum heat-flux density (W/m^2)
   p_c = critical pressure (kPa)
   p = absolute pressure (kPa)
   D_i = internal tube diameter (m)
   L = tube length (m)

And the previously studied correlations (such as that of Chen) cannot be used when they predict heat-flux densities higher than that calculated with Eq. (11-3-32).

In-Tube Film Boiling.   For in-tube boiling, it is also possible, if temperature differences are high, to have a stable film-boiling regime in which heat transfer to the liquid takes place through a vapor film. The ΔT value must be high enough to move away from the transition region (where q decreases with an increase in ΔT). An approximate value for the minimum ΔT required for film boiling can be calculated[28] as

$$\Delta T_{fb} = 0.555 \left[52 \left(1 - \frac{p}{p_c} \right) + 0.04 \left(\frac{p}{p_c} \right)^{-2} \right] \qquad (11\text{-}3\text{-}33)$$

where ΔT_{fb} is expressed in °C.
   The film boiling heat transfer coefficient can be estimated[21,29] by means of the following equation:

$$h_{fb} = 105 p_c^{0.5} \Delta T^{-0.33} \left(\frac{p}{p_c} \right)^{0.38} \left(1 - \frac{p}{p_c} \right)^{0.22} \qquad (11\text{-}3\text{-}34)$$

with p_c in kPa and h_{fb} in W/(m^2 · K). This equation must be considered only as a conservative estimation because the available information is scarce.

The Critical Pressure.   Many of the correlations presented in this chapter are based on the use of critical pressure. The *critical pressure* of a single-component fluid is the maximum pressure at which two phases in equilibrium can exist. At this pressure, liquid and vapor densities became equal. Any fluid at a pressure higher than this presents a unique phase, called *supercritical,* whose density decreases continuously with an increase in temperature without presenting a two-phase state. For pure substances, the critical pressures of fluids are tabulated[27] and can be found easily.
   In the case of mixtures, the situation is more complex. The critical pressure of a mixture can be obtained by calculating with a process simulator the bubble and dew-point temperatures at different pressures. The

pressure at which both temperatures coincide corresponds to the critical state. Some process simulators include the calculation of critical properties of all the streams routinely.

For a multicomponent mixture with known composition, an approximation consists of calculating a pseudo-critical pressure, defined as

$$p_{sc} = \sum p_{ci} x_i \tag{11-3-35}$$

where x_i is the mole fraction of component i and p_{ci} is its critical pressure.

For petroleum fractions, the method proposed by the American Petroleum Institute (API)[25] and reproduced in Gas Processors Suppliers Association (GPSA) manuals[26] allows the calculation of pseudocritical properties as a function of the average boiling temperature and specific gravity.

11-4   REBOILERS

11-4-1   Different Types

The purpose of a reboiler is to vaporize the bottoms of a distillation column, thus producing the necessary amount of vapor required for the operation of the column, which is determined by the reflux ratio adopted by the designer. For a brief description of the distillation process, App. A can be consulted.

A reboiler consists of a heat exchanger, heated by a hot fluid, which can be steam, hot oil, or any available process stream. The hot stream transfers heat to the column bottoms liquid, which vaporizes into the reboiler. There are several types of reboilers, and they will be described below.

Forced-Circulation Reboiler.   The reboiler shown in Fig. 11-25 is called a *forced-circulation reboiler* because the liquid is fed into it by means of a pump. The liquid coming from the lowest tray gets into the bottom end of the column, which acts as an accumulator. From there, it is taken up by the pump and is forced through the reboiler (stream W), whereas a fraction B is withdrawn as bottom distillation product.

The device indicated as LC is a level controller that opens or closes the product outlet valve, thus preventing the column bottom from getting flooded or dried out. The liquid flow rate fed into the reboiler must be higher than the required vapor production, so the reboiler outlet stream is a liquid-vapor mixture. The vapor V passes to the lowest tray, and the liquid L comes back to the column bottoms from where it is recirculated.

For stable system operation, without pressure fluctuations or pulsating flows, it is recommended to keep a circulation ratio $W{:}V$ of at least 3:1. The higher this ratio, the more stable will be the hydraulic behavior of the system.

Another reason to work with high circulation ratios is that in this way the fouling problems on the tube surfaces are reduced. The bottom part of a distillation column is where high-boiling-point impurities accumulate. If the reboiler feed were completely vaporized, nonvolatile compounds (salts or dissolved substances) potentially could precipitate or deposit on the tube walls.

This reboiler type has the particularly attractive feature of allowing any desired circulation rate because the rate is defined by the capacity of the pump and is practically independent of reboiler design. It is a very convenient alternative in cases where the column bottom product is very viscous.

The principal disadvantage is that it requires installation of a pumping system in a location that is problematic for the following reasons:

1. The column bottom liquid is at its boiling temperature. The pump is then prone to cavitation, and it is necessary to raise the column bottom above the pump foundation level by a value dictated by the required net positive suction head (NPSH). This usually means 3 or 4 m.

2. Temperatures at column bottoms are high, and the pump and mechanical seals are normally subject to severe service conditions that may result in frequent maintenance.

For these reasons, natural-circulation units, as described below, are normally preferred.

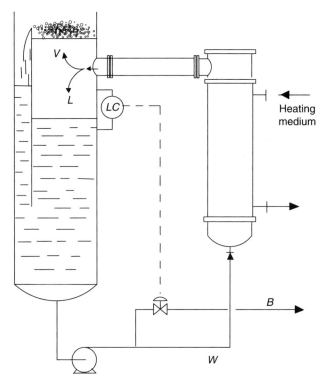

FIGURE 11-25 Forced-circulation reboiler.

Vertical Thermosiphon Reboiler. The thermosiphon reboiler is the most common type in the process industry. The basic idea consists of eliminating the circulation pump and using the difference between liquid and vapor densities as the driving force for fluid circulation. A schematic diagram of a vertical thermosiphon reboiler can be seen in Fig. 11-26.

The unit is a vertical heat exchanger in which the liquid coming from the column bottoms flows into and through the tubes, where it is vaporized. In the vaporization process, a vapor-liquid mixture is generated into the tubes. The average density of this mixture is lower than that of the liquid.

With reference to Fig. 11-26, between points A and B, the hydrostatic pressure difference is given by $\rho_L g z_1$. Between points C and B, the hydrostatic pressure difference will be the product of an average density corresponding to the vapor-liquid mixture and $g z_2$. Since this density is lower than that of the liquid, it results that the hydrostatic pressure difference is higher for segment AB than for segment BC. Then a fluid circulation is established between the column and the reboiler.

As the liquid in the reboiler partially vaporizes, a vapor-liquid mixtures exits the unit from the top outlet nozzle and flows back into the column, where both phases separate as in a forced-circulation reboiler. The circulation rate that results in the system will be such that the frictional pressure drop balances the hydrostatic pressure difference. Since normally $z_1 \cong z_2 = z$, this hydrostatic pressure difference will be given by

$$gz(\rho_L - \rho_{\text{average liquid-vapor}})$$

This means that the higher z (= reboiler length) is, the higher will be the circulation rate. This has an effect on the heat transfer coefficient that improves with the increase in fluid velocity.

In well-designed thermosiphon reboilers, it is possible to achieve very high circulation rates and hence good values of the heat transfer coefficient. The piping configuration also is very simple, which simplifies plant layout. Cleaning of this type of reboiler is also a simple task because the process fluid is inside

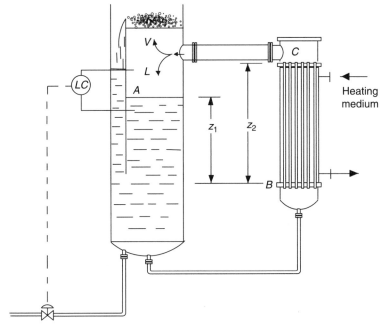

FIGURE 11-26   Vertical thermosiphon reboiler.

the tubes. The shell-side heating fluid is normally a clean fluid, so low-cost fixed tubesheet units can be employed.

The main disadvantage is that the column has to be raised to achieve the necessary height for fluid circulation. The distillation column is itself a tall piece of equipment, and an additional height increase may have a significant impact on the mechanical design and cost.

An important aspect that must be considered in design is the temperature increase required by the fluid to achieve the desired vaporization fraction. For example, with reference to Fig. 11-27, let T_A be the temperature of the bottom tray liquid. This liquid falls down through the downcomer to the column bottom, where it mixes with the liquid fraction coming back from the reboiler, resulting in a mixture temperature T_C.

If the liquid consists of a mixture of several components with different boiling points, because the lighter components vaporize preferably, the reboiler outlet temperature T_B will be higher than the inlet temperature T_C. If the circulation ratio (total inlet flow/vaporized flow) is high, the vaporization process does not produce a significant change in the liquid-phase composition; thus the outlet bubble-point temperature T_B will not be very different from the inlet bubble-point temperature T_C. If, on the other hand, the circulation ratio is low, this temperature change can rise up by several degrees and has to be considered for the following reasons:

1. For temperature-sensitive fluids, undesirable transformations may occur.

2. The temperature increase in the boiling fluid reduces the available ΔT between the hot and cold fluids. In certain cases, when the temperature of the heating fluid is very close to that of the column bottom product, this may have a significant impact on the required heat transfer area. In other cases, this can be beneficial because it moves the operating condition away from the critical heat flow.

Another aspect that may be considered a disadvantage of the two reboiler types considered up to now (Figs. 11-25 and 11-27) is related to their efficiency from the mass-transfer point of view. The objective of a distillation is always to obtain the bottoms product as heavy as possible (this means rich in high-boiling-point components).

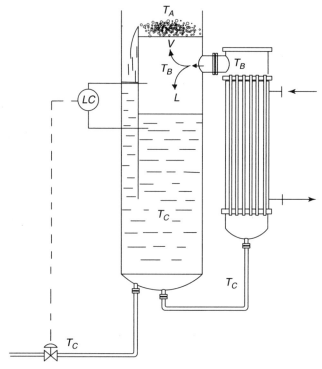

FIGURE 11-27   Temperatures in a reboiler system.

If we consider the unit shown in Fig. 11-27, we can see that the stream richest in heavy components corresponds to the liquid fraction exiting the reboiler. But this fraction is mixed in the column bottom with the liquid coming down from the lowest tray, which is lighter. Hence the distillation bottom product, which is a mixture of both streams, does not correspond to the heaviest composition. The quality of the product has been degraded with this mixture.

Sometimes the configuration shown in Fig. 11-28 is used. In this case, the liquid from the bottom tray is extracted completely and send to the reboiler. The outlet vapor-liquid mixture is separated into the column, so the liquid fraction is withdrawn as distillation bottoms without being recirculated.

This is called *once-through circulation*. It can be observed that the vapor generated in the reboiler is in thermodynamic equilibrium with the bottom product of the distillation. This means that the reboiler behaves as an additional column tray, and from the mass-transfer point of view, this is more efficient than the recirculation configuration.

This configuration is also particularly attractive in services subject to polymerization or product decomposition caused by high temperatures because it minimizes the residence time in the reboiler and avoids repeated contact of the fluid with the heating surface as it is the case in recirculation units.

If the distillation column is a packed-type column instead of a tray-type, this configuration becomes more complex because it is necessary to add a liquid drawoff tray. The other disadvantage of this design is that both the liquid bottom product and the reboiler vapor product are always external data for the reboiler designers, so they have no chance to handle the recirculation rate as a design variable to improve the heat transfer coefficients

Additionally, it was explained earlier that it is always convenient to work with a vaporization ratio lower than 30 percent to avoid fouling problems. This type of reboiler, then, should not be selected when the vapor generation is higher than 3/7 of the bottom-product flow rate.

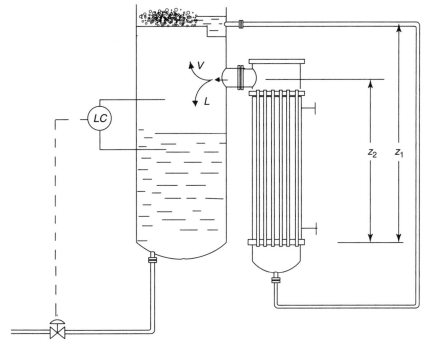

FIGURE 11-28 Once through the vertical reboiler.

The maximum temperature in this system is the reboiler outlet temperature, which coincides with the bottom-product temperature. This means that with this type of unit it is not necessary to reach temperatures higher than the bubble point of the bottom product, as was the situation with recirculation reboilers.

Figure 11-29 shows a configuration pretending to combine the advantages of the recirculation and at the same time reduce the efficiency loss originating in the mixture of the liquid fraction coming back from the reboiler with the bottom-tray liquid. In this case, the liquid fed to the reboiler has approximately the composition of the bottom tray, and the bottom product of the distillation has a similar composition to that of the reboiler outlet liquid fraction.

Horizontal Thermosiphon Reboiler. In this type of reboiler, the vaporization takes place in the shell side while the heating fluid circulates through the tubes. Since normally the process fluid presents more fouling issues than the heating medium, in certain cases this design may not be very attractive because of the difficult cleaning.

This type of reboiler requires more plot area than a vertical unit, and recirculation ratios are not so high. To reduce the frictional pressure drop, the double-split configuration (TEMA H shell) shown in Fig. 11-30 is frequently adopted. As in the case of a vertical thermosiphon, it is possible to connect this unit in a once-through circulation diagram.

Kettle Reboilers. The kettle reboiler is shown in Fig. 11-31, and the way it is connected to the distillation column is shown in Fig. 11-32. It is a TEMA BKT or BKU heat exchanger. The liquid feed gets into the unit through a nozzle (3) and fills the shell side up to a level determined by the height of an overflow weir (4). The liquid pool contained into the shell receives heat from the tubes and vaporizes. The amount of vapor produced obviously depends on the heat transfer area and the temperature of the heating medium. The fraction of the feed that is not vaporized flows over the weir and exits through a nozzle (5) as distillation bottom product.

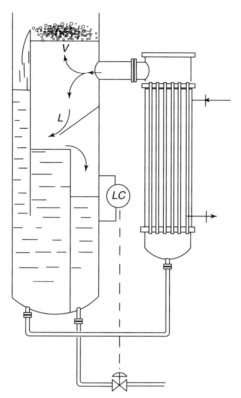

FIGURE 11-29 An improved column bottom arrangement.

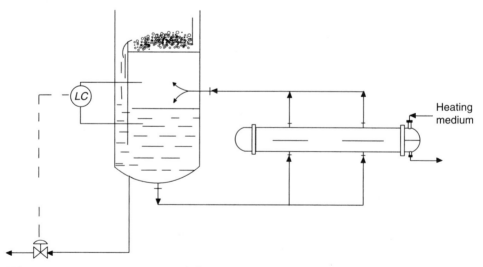

FIGURE 11-30 Horizontal thermosiphon reboiler.

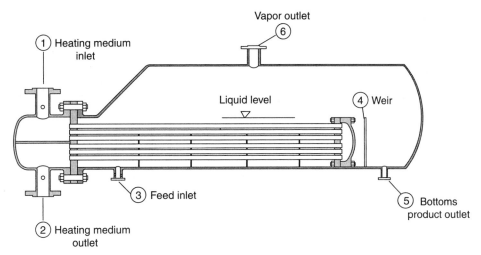

FIGURE 11-31   Kettle reboiler.

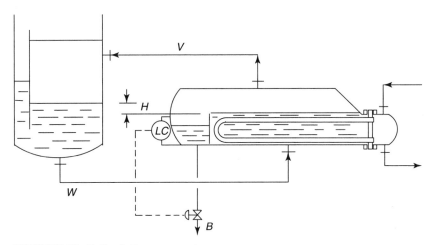

FIGURE 11-32   Kettle reboiler arrangement.

The purpose of the vapor space above the liquid level is to allow separation of the vapor and entrained liquid droplets so that only vapor exits through the outlet nozzle (6). The tube bundle is a simple floating-head type or U tubes, and the heating fluid circulating in the tubes may be steam or any other fluid.

The difference between this unit and the thermosiphon or forced-circulation reboiler is that there is no recirculation or liquid coming back from the reboiler to the column. It can be observed that the vapors generated into the unit are in thermodynamic equilibrium with the distillation bottom product extracted after the weir through the nozzle (5). This means that in this configuration the reboiler behaves as an additional equilibrium stage from the mass-transfer point of view.

The fluid velocities in this type of reboiler are very low, and frictional pressure drops are negligible. The only pressure drops in the system are due to friction in the inlet and outlet piping. These must be overcome by the difference in liquid levels between the column bottom and the reboiler (indicated as H in Fig. 11-32). This difference is very small, and the reboiler and the column bottom behave as communicating vessels. It is necessary neither to raise the column to allow liquid circulation, as is the case of thermosiphon units, nor to install a circulation pump.

As a counterpart, the reboiler is more expensive, especially in cases where the operating pressure is high because the big shell diameters make it necessary to use greater plate thickness in its construction. These units also require more plot area than vertical types. However, they have the advantage, in comparison with thermosiphon units, that the thermal design is not conditioned by the hydraulics of the system. This gives the designer more freedom to modify the relative location between the column and the reboiler. This is particularly important in the case of big units.

When the amount of distillation bottom product is small in comparison with the amount of vapor, this type of reboiler has a tendency to act as an impurities concentrator. Since these units do not have the high circulation rates of thermosiphons, it is convenient to limit the vaporization to not more than 80 percent of the feed rate. This means that the 20 percent constituting the bottoms product acts as a purge, thus eliminating the concentrated impurities.

Internal Reboiler. In certain less frequent cases, when the amount of heat to be transferred in the reboiler is small, the construction shown in Fig. 11-33 may be used. In this case, a tube bundle is installed directly in the bottom of the column. The cost of the shell is avoided, and there is no connecting piping. However, the column must be provided with the internal supports for the bundle and a big flange. The tube bundle is comparatively expensive because it is necessary to employ more tubes with smaller lengths.

Other Applications of Boiling Heat Transfer Equipment. Among all the process equipment in which boiling heat transfer takes place, the most common is the distillation-column reboiler. This is why in the following topics we shall only make reference to this application. But there are other process equipment apart from reboilers in which fluids are vaporized.

However, the calculation methods that can be employed in their design are the same as we shall study for reboilers. This means that even though we shall always make reference to the distillation process, the same treatment can be extended to other applications as well.

For example, let us consider the case of Fig. 11-34. It represents a waste-heat-recovery boiler of a chemical plant. In this plant, an exothermic vapor-phase reaction takes place, and the high-temperature reaction products must be cooled down before passing to the fractionation section.

The waste heat of the gases is recovered in the equipment shown in the figure to generate steam that is later used in other sectors of the plant. The gases circulate through the shell side of the exchanger while water is vaporized within the tubes.

A steam-water mixture exits the outlet nozzle into the separator *S*. There, both phases are separated, and the steam is withdrawn from the upper nozzle while the liquid is mixed with the fresh feed (which replaces the vaporized mass) and recirculates through the tubes. This recirculation is caused by the difference in

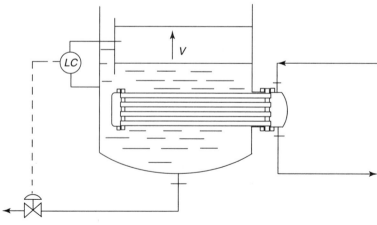

FIGURE 11-33 Internal reboiler

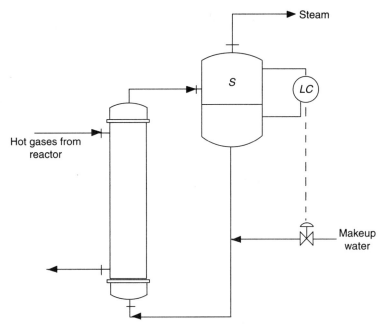

FIGURE 11-34   Thermosiphon steam generator.

densities of the water-steam mixture in the tubes and that of the liquid in the return leg. It is evident that this unit can be assimilated to a thermosiphon reboiler and can be designed with the same techniques that will be studied for this case.

Let us analyze the case of the chiller represented in Fig. 10-24. In this case, a process stream is cooled down using as cooling medium a refrigerant fluid such as ammonia, propane, or a halogenated hydrocarbon and follows the refrigerating cycle shown in the figure. In the chiller, the refrigerant receives heat from the process stream and vaporizes. This type of unit operates as a kettle reboiler and must be designed as such.

11-4-2 General Aspects of the Thermal Design of Reboilers

It has already been mentioned that the prediction of boiling heat transfer coefficients presents many difficulties, and research work has led to the development of a variety of correlations that, when compared, reveal considerable differences. It is unavoidable, then, that these uncertainties extend to the design of boiling heat transfer equipment, especially to the case of reboilers. Several authors have published design methods that proved to be useful for a particular application or fluid but had little success when extended to other situations.

In recent years, with the appearance of commercial software, it became more difficult to find systematic design methods for this type of equipment in the open literature. This section is based on the information quoted most frequently in the references, but it is recommended that one verifies manual calculations with commercial programs and vice versa to be sure that the results are interpreted adequately.

The following general considerations must be taken into account:

1. There are critical phenomena that impose limits to the heat transfer rate (e.g., the transition from nucleate to film boiling or the mist flow in convective boiling) that must be checked. High temperature differences between hot and cold fluids can lead to these effects and generally are not convenient.

2. Except for these critical effects, the heat transfer coefficients for boiling liquids are usually very high, and frequently, the controlling resistance is fouling. However, when the unit is clean, the wall temperature in contact with the boiling liquid is higher than that assumed for design, and a critical condition may be reached. This condition should be checked if it is not possible to adjust the temperature of the heating medium.

11-4-3 Kettle Reboiler Design

Heat Balance of a Kettle Reboiler.   With reference to Fig. 11-32, the liquid from the column bottom is at the bubble-point temperature corresponding to the column operating pressure. Let this temperature be t_F. When this liquid goes into the reboiler, it receives a heat flow Q, increasing its enthalpy and suffering a partial vaporization. Two phases are then formed, V and B, that are, respectively, the vapor going back to the column and the distillation bottom product. Since it is considered that the reboiler is an ideal stage, these two streams will be at thermodynamic equilibrium at temperature t, which is the temperature inside the reboiler.

If the liquid coming down from the bottom plate is a pure component (which roughly may be the situation in binary distillations), since the column and reboiler pressures are practically the same, the temperature t coincides with the temperature of stream W. This means that the operation of the unit is practically isothermal. Only latent heat is transferred, and the enthalpy balance is reduced to

$$Q = V\lambda \tag{11-4-1}$$

If, instead, the column bottom product is a multicomponent mixture, the vaporization process will generate two phases with different composition as well as different from the composition of W. Since B and W are two liquid streams at their respective bubble points but with different compositions, their temperatures also will be different. This means that the liquid W receives sensible and latent heat in the reboiler and it increases its temperature and is partially vaporized.

It is generally accepted that when the liquid feed enters the reboiler, its temperature is instantaneously increased as the result of the mixture with a big mass of fluid at temperature t, so the boiling fluid temperature in all the reboiler interiors may be considered constant and equal to the outlet temperature. This means that the heating medium must deliver all the heat (sensible and latent) to a fluid at temperature t. If the heating medium is, in turn, isothermal (e.g., in the case of condensing steam) at temperature T, the temperature difference for heat transfer will be

$$\Delta T = T - t \tag{11-4-2}$$

The value of the heat flux Q required for the distillation normally is obtained as a result of the process simulation when the heat balance of the column is solved and the enthalpies of streams W, V, and B are calculated. This balance can be expressed as

$$Q = Vi_V + Bi_B - Wi_W \tag{11-4-3}$$

Even though the calculation of enthalpies is performed directly by thermodynamic models employed by the simulator, some reboiler design methods require separation of the heat flux Q into a fraction of *sensible heat* and a fraction of *latent heat*. By making use of a mean liquid specific heat c_L, we can define

$$Q_S = Wc_L(t - t_F) \qquad \text{(sensible heat)} \tag{11-4-4}$$

$$Q_L = Q - Q_S \qquad \text{(latent heat)} \tag{11-4-5}$$

Calculation of the Reboiler Area.   The design of this type of reboiler follows a method originally published by Palen and Small.[22] The method assumes that the boiling heat transfer coefficient can be calculated by the addition of two contributions. One of them is the macroscopic convection as consequence of the two-phase fluid circulation through the tube bundle. This is a convective mechanism in which the fluid moving across the bundle is liquid and vapor. The convection heat transfer coefficient that allows the calculation of this contribution will be called h_{TP}.

Superimposed to this mechanism, there exists a microagitation effect caused by bubble nucleation and detachment at the heat transfer surface. The coefficient associated with this mechanism will be called h_{NB}. Then

$$h_o = h_{NB} + h_{TP} \qquad (11\text{-}4\text{-}6)$$

In a kettle reboiler, the liquid velocities are small, and the effect of h_{NB} is much more important than that of h_{TP}. High precision in the calculation of h_{TP} thus is not required, and normally, it is assumed to have a value in the range of 250–500 W/(m^2 · K) without further verification. h_{NB}, in turn, is calculated with the methods and correlations mentioned in Sec. 11-2. However, it must be considered that those correlations are valid only for simple geometries such as, for example, a single tube submerged into a boiling liquid. In the case of tube bundles, some corrections are necessary because the vapor bubbles produced on the surface of a tube affect the tubes above it.

The general expression is

$$h_{NB} = h_{NB1} F_b F_c \qquad (11\text{-}4\text{-}7)$$

where h_{NB1} = nucleate boiling coefficient for a single tube submerged into a fluid with similar properties to that in the reboiler and submitted to the same heat-flux density
F_c = boiling-range correction factor
F_b = bundle-geometry correction factor

In what follows, we shall analyze how to calculate these terms.

Calculation of h_{NB1}.   In Sec. 11-2 we studied some expressions for calculation of the boiling heat transfer coefficient for a single tube submerged into a fluid. Some of these equations were expressed in the form

$$q = a\Delta T^b \qquad (11\text{-}4\text{-}8)$$

If we consider that

$$q = h_{NB1}\Delta T \qquad (11\text{-}4\text{-}9)$$

then it can be written

$$h_{NB1} = a\Delta T^{b-1} \qquad (11\text{-}4\text{-}10)$$

or

$$h_{NB1} = cq^d \qquad (11\text{-}4\text{-}11)$$

where

$$c = a^{1/b} \qquad \text{and} \qquad d = (b-1)/b$$

When the only mechanism present is the nucleate boiling, all these expressions are equivalent and any one of these forms is valid. However, if the heat-flux density results from the addition of two effects, Eq. (11-4-9) is no longer valid, and the transformation from one type of equation into another cannot be done.

This subject is not very clear in the references, and some confusion may arise. Thus it is very important to remark that h_{NB1} must be evaluated at the same heat-flux density as that of the reboiler. This means that an equation like Eq. (11-4-11) must be used, where q has to be taken as the total heat-flux density considering all the mechanisms present.

It was explained in Sec. 11-2 that important deviations may be found among the results of the different nucleate boiling correlations available. Among all the possible correlations, most authors prefer the Mostinsky correlation [Eq. (11-2-12)], very probably because of its simplicity. This is expressed as

$$h_{NB1} = 0.00417 p_c^{0.69} q^{0.7} F_p \qquad (11\text{-}4\text{-}12)$$

where F_p is a pressure correction factor, expressed as a function of p/p_c (pressures in kPa), that is,

$$F_p = \left[1.8\left(\frac{p}{p_c}\right)^{0.17} + 4\left(\frac{p}{p_c}\right)^{1.2} + 10\left(\frac{p}{p_c}\right)^{10} \right] \qquad (11\text{-}4\text{-}13)$$

For reboilers design, Palen[21] suggested considering only the first term of the polynomial.

Calculation of F_b, the Correction Factor for Tube Bundle Geometry. The correlations studied for a single tube cannot be employed for a tube bundle because the bubbles produced on the surface of a tube affect the neighboring tubes. In one of his first papers, Palen[22] considered that the boiling heat transfer coefficient of a tube bundle is smaller than that of a single tube (about 30 percent). Some years later, in another paper, the same author explained that the bundle heat transfer coefficient is about two or three times higher than the coefficient of a single tube. It is remarkable that between the first and second values there is 10 times difference that remained hidden for years. This gives one an idea on how difficult is boiling heat transfer cofficients experimental measurement.

Afterwards, in successive papers,[21] Palen proposed different correlations for the calculation of F_b. One of them is

$$F_b = 1 + 0.1\left[(1-\psi)/\psi \right]^{0.75} \qquad (11\text{-}4\text{-}14)$$

where

$$\psi = (\pi D_b L/A) \qquad (11\text{-}4\text{-}15)$$

D_b is the tube bundle diameter (diameter of the minimum circle completely containing all the tubes). To evaluate it, it is necessary to perform the tubes distribution.

Another correlation proposed by Taborek[24] is

$$F_b = 1 + 0.1\left[\frac{0.785 D_b}{C_1 (Pt/D_o)^2 D_o} - 1 \right]^{0.75} \qquad (11\text{-}4\text{-}16)$$

where C_1 is unity for 90- and 45-degree tube arrays and 0.866 for triangle arrays. All these equations require one to know the geometry of the tube bundle. In early design stages, when this information is not available, it is suggested to adopt $F_b = 2$ for kettle reboilers. For the case of horizontal thermosiphons (whose calculation method is similar), it is suggested to adopt $F_b = 2.3$ for preliminary calculations.

Calculation of F_c, Boiling-Range Correction Factor. This factor is important when the liquid presents a boiling range. This means that the reboiler feed temperature is different from the fluid temperature into the reboiler. All the correlations studied previously for nucleate boiling are valid for single-component fluids. In the case of mixtures, the heat transfer coefficients can be considerably smaller owing to diffusion effects that create additional resistances to heat transfer, as was the case in multicomponent condensers.

Several correlations were proposed to correct the heat transfer coefficient to include diffusion effects. The following was proposed by Palen and Young,[21] and we adopt it because of its simplicity:

$$F_c = 1/(1 + 0.023 q^{0.15} BR^{0.75}) \qquad (11\text{-}4\text{-}17)$$

where BR is the boiling range of the mixture in °C (difference between dew point and bubble point of the reboiler feed at the operating pressure). Please note that since the vaporization in the reboiler is not total, BR normally is higher than the temperature change that the fluid actually experiences. In Eq. (11-4-17), q is in W/m².

Maximum Heat-Flux Density and Film Boiling.   The previously studied correlations are valid only for low ΔT between the tube wall and boiling fluid. Otherwise, it is possible to reach the transition to film boiling regime with a consequent reduction of the heat transfer coefficient. It is then necessary to have a method to establish the validity limit of the correlations. Normally, this is handled through calculation of q_{max} (maximum heat-flux density for the tube bundle).

Mostinsky[12] proposed a correlation to calculate the maximum heat-flux density for a single tube:

$$q_{1,\,max} = 367 p_c \left(\frac{p}{p_c} \right)^{0.35} \left(1 - \frac{p}{p_c} \right)^{0.9} \tag{11-4-18}$$

p_c is expressed in kPa and $q_{1,\,max}$ in W/m².

In a tube bundle, the maximum heat-flux density is lower than for a single tube because the vapor generated in the lower tubes increases the amount of vapor in the upper tubes, thus making access of liquid to the tube wall more difficult. Palen and Small[22] suggested correcting the maximum heat-flux density for a single tube by means of

$$q_{b,\,max} = q_{1,\,max} \phi_b \tag{11-4-19}$$

where ϕ_b is a correction factor related to the geometry of the tube bundle. Several expressions were proposed for the calculation of ϕ_b. The simplest[21] is

$$\phi_b = 3.1\psi \tag{11-4-20}$$

with the restriction

$$\phi_b \leq 1 \tag{11-4-21}$$

where ψ is the same parameter defined in Eq. (11-4-15). It must be noted that for a single tube, $\psi = 1$.

The equation tells us that essentially it must be considered that $\phi_b = 1$ for all cases with ψ higher than 1/3. In these cases, which correspond to small tube bundles, the behavior is similar to that of a single tube. The bigger the tube bundle and the more compact its design, the smaller are the values of ψ and ϕ_b. This means that blockage of the heat transfer surface by the vapor occurs at lower heat-flux densities.

A certain safety margin with respect to these values must be adopted. It can be considered that when the heat-flux density at which the reboiler operates is higher than $0.5q_{b,\,max}$, there is a potential risk of vapor interference. In these cases it is recommended to modify the tube bundle geometry, increasing ψ (e.g., increasing the separation between the tubes). A heat-flux density higher than $0.7q_{b,\,max}$ never should be used for design.

Kern's Recommendations for Maximum Heat Flux.   In the 1950s, Donald Kern suggested upper limits to the maximum heat-flux density that should be adopted for reboiler design. These values were 94,600 W/m² [30,000 Btu/(h · ft²)] for aqueous substances and 37,800 W/m² [12,000 Btu/(h · ft²) for organics. These recommendations were not supported by any theory but were obtained by industrial experience. The engineering practice around the world followed Kern's recommendations for decades. Since these heat-flux densities are rather low, they normally were the limiting factor in design, and usually, the heat transfer area simply was adopted as the quotient between Q and this heat-flux density.

At present, the theories that we have presented predict maximum heat-flux densities generally higher than Kern's values. However, owing to the weakness of the theoretical models and the complexity of the problem, even today, Kern's limits are used widely, and some commercial design programs still maintain them as a user option.

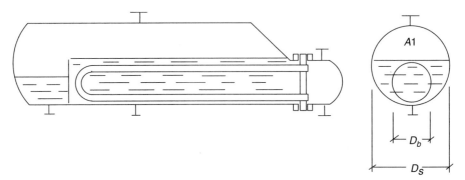

FIGURE 11-35 Kettle reboiler section.

Shell Sizing. The shell diameter must be larger than the tube bundle diameter D_b to allow vapor-liquid separation. Usually the shell diameter is approximately twice the tube bundle diameter (Fig. 11-35). It is recommended to calculate the shell cross-sectional area above the level of the weir (A_1 in the figure) using a critical vapor velocity defined as

$$v_c = k_1 \sqrt{\frac{\rho_L - \rho_V}{\rho_V}} \qquad (11\text{-}4\text{-}22)$$

with $k_1 = 0.03$ to 0.09 m/s.
If $V_V = W_V / \rho_V = $ m³/s of vapor at the outlet, then it must be

$$A_1 \geq \frac{V_V}{v_c N_N} \qquad (11\text{-}4\text{-}23)$$

where N_N is the number of vapor outlet nozzles (if the tube length is higher than 5 m, it is usual to have two vapor outlet nozzles).

Example 11-5   Reboiler with boiling range.   A kettle reboiler for a distillation column operating at 400 kPa pressure must be designed. The column bottom product entering the reboiler has a flow rate of 10,000 kg/h and a temperature of 104.5°C (essentially equal to its bubble-point temperature). Fifty percent of the feed is vaporized into the reboiler, and the other 50 percent is extracted as liquid product. The outlet temperature (temperature inside the reboiler) is 113.3°C. The dew point of the feed at 400 kPa is 119°C. The required heat flux obtained from the process simulation is 470 kW. The heating media is steam. The available steam is at 717 kPa. To perform this service, it is proposed to use a reboiler with the following characteristics:
TEMA type AKU

$N = $ number of tubes: 44
$n = $ number of tube passes: 2

Tube diameters: $D_o = 0.019$ mm; $D_i = 0.0144$ mm

$P_t = $ tube pitch: 0.0238-mm triangle

Tube length: 3.3 m

Fouling resistances of 0.0001 for the tube side and 0.0003 for the shell side must be considered. Verify the suitability of this unit.
Physical properties: Cold fluid (properties at the outlet temperature 113.3°C):

Vapor

$c_V = 2,100$ J/(kg · K)

$\mu_V = 0.0083$ cP

$\rho_V = 11.7$ kg/m^3

$k_V = 0.0213$ W/(m · K)

$p_c = 3214$ kPa (critical pressure)

Liquid

$c_L = 2,700$ J/(kg · K)

$\mu_L = 0.142$ cP

$\rho_L = 564$ kg/m^3

$k_L = 0.087$ W/(m · K)

Solution We shall use the correlations for the prediction of q_{max}, even though we know that this may not be the most conservative approach. The available steam pressure is 717 kPa. If a pressure drop for the steam control valve of 100 kPa is assumed, and considering a 50-kPa pressure drop in the reboiler, the inlet and outlet steam pressures will be 617 and 567 kPa, respectively. At these pressures, the corresponding condensation temperatures are 160 and 155.5°C. Thus we shall assume a mean temperature for the condensing steam of 157.7°C.

The temperature difference for heat transfer thus will be $\Delta T = 157.7 - 113.3 = 44.4$°C (remember that for the boiling fluid, the outlet temperature must be used). For the steam side, we shall assume $h_{io} = 8500$ W/(m^2 · K). The heat transfer area is

$A = N\pi\, D_o L = 44 \times \pi \times 0.019 \times 3.3 = 8.66$ m^2

The heat-flux density thus will be

$q = Q/A = 470,000/8.66 = 54,272$ W/m^2

We shall verify this value using Mostinsky correlations [Eqs. (11-4-12) to (11-4-17)]:

$h_{NB1} = 0.00417(3214)^{0.69} \times 54,272^{\,0.7}F_p = 2260F_p$

$F_p = 1.8(p/p_c)^{0.17} + 4(p/p_c)^{1.2} + 10(p/p_c)^{10}$

$\qquad = 1.8(400/3214)^{0.17} + 4(400/3214)^{1.2} + 10(400/3214)^{10} = 1.59$

$h_{NB1} = 2260 \times 1.59 = 3593$ W/(m^2 · K)

Correction factor for tube bundle geometry: For calculation of this parameter, it is necessary to analyze the tube-sheet layout. A TEMA AKU shell with two tube passes is considered. 44 tubes, 19 mm OD, with 23.8-mm pitch in a 30-degree triangular distribution corresponds to a tube bundle diameter $D_b = 240$ mm.

With $D_b = 0.240$, $P_t = 0.0238$, $D_o = 0.019$, $C_1 = 0.866$, from Eq. (11-4-16), we obtain $F_b = 1.4$.
Correction factor for boiling-range F_c: The boiling range of the mixture is BR $= 119 - 104.5 = 14.5$°C. Then

$F_c = 1/(1 + 0.023q^{0.15}\mathrm{BR}^{0.75}) = 1/(1 + 0.023 \times 54,272^{0.15} \times 14.5^{0.75}) = 0.532$
$h_{NB} = h_{NB1}F_bF_c = 3,593 \times 1.4 \times 0.532 = 2,670$ W/(m^2 · K)

The shell-side coefficient then will be

$h_o = h_{NB} + h_{TP} = 2,670 + 500 = 3,170$ W/(m^2 · K)

The total heat transfer coefficient will be

$U = (1/h_o + 1/h_{io} + R_f)^{-1} = (1/3,170 + 1/8,500 + 0.0004)^{-1} = 1,200$ W/(m^2 · K)

Verification

$q = U\Delta T = 1,200 \times 44.4 = 53,300$ W/(m^2 · K)

This value is in good agreement with the 54,272 W/m^2 adopted for calculation of the boiling h. Otherwise, h_o should be recalculated. The area required by the unit will be

$A = Q/q = 470,000/53,300 = 8.8$ m^2

The unit is slightly undersized. Since the difference is very small, a new calculation is not considered necessary.

Verification of the maximum heat-flux density: The preceding calculations are only valid if the maximum nucleate boiling heat-flux density is not exceeded. Calculating $q_{1,\text{max}}$ with Eq. (11-4-18) with $p = 400$ kPa and $p_c = 3214$ kPa, we get

$$q_{1,\text{max}} = 504{,}000 \text{ W/m}^2$$

The geometric correction factor is $\phi_b = 3.1\psi$. Thus

$$\psi = \pi D_b \, L/A = \pi \times 0.24 \times 3.3/8.66 = 0.29$$

$$\phi_b = 0.89$$

Then

$$q_{b,\text{max}} = 0.89 \times 504{,}000 = 448{,}500 \text{ W/m}^2$$

Since this value is much higher than the design heat-flux density, the geometry of the unit is satisfactory.

Example 11-6 Design a reboiler for the same process conditions as Example 11-5 but with a vapor production of 85,100 kg/h. Analyze the convenience of using a heating steam of higher temperature.

Solution The heat duty will be proportional to the preceding case, that is,

$$Q = (85{,}100/5{,}000) \times 470 = 8{,}000 \text{ kW}$$

In the preceding example, we noticed that the maximum heat-flux density was much higher than that used in the design. One could think that it is possible to work with a higher-temperature steam, which would allow a higher heat-flux density. However, since it is a bigger unit, the geometric factor ϕ_b and hence the maximum heat-flux density will be smaller. Let's maintain the design with $3/4$-in (19-mm) tubes with a 15/16-in (23.8-mm) triangular pitch and a tube length of 3.3 m. To study the impact of the scale factor in the design, the following procedure will be used:

1. A value for q is assumed.
2. Calculate $A = Q/q$.
3. Calculate the number of tubes $N = A/\pi D_o L$.
4. Draw the tube distribution, and estimate the tube bundle diameter D_b.
5. Calculate ψ and ϕ_b.
6. Calculate $q_{b,\text{max}} = q_{1,\text{max}}\phi_b$.
7. Assuming a safety factor of 0.7, if $q > 0.7q_{b,\text{max}}$, a smaller value for the heat-flux density must be assumed. Otherwise, the assumed heat-flux density is acceptable.
8. Calculate the film coefficients and the total U. Then, with the heat-flux density resulting from the preceding steps, calculate the required ΔT as $\Delta T = q/U$.
9. The required steam temperature will be $T_h = T_c + \Delta T = 113 + \Delta T$.

The results of the final iterations corresponding to steps 1 to 7 are shown in the following table:

q	A	N	D_b, mm	ψ	ϕ_b	$q_{b,\text{max}}$	$0.7q_{b,\text{max}}$
100,000	80	405	598	0.0774	0.24	121,100	84,700
65,000	123	624	685	0.057	0.176	89,056	62,300 OK

This means that the design must be based on a maximum heat-flux density of about 62,300 W/m², and a 624-tube reboiler is necessary.

Calculation of the heat transfer coefficients: Recalculating h_{NB1} and F_c with $q = 62,300$ W/m^2, the result is $h_{NB1} = 3,957$ W/(m$^2 \cdot$ K), and $F_c = 0.527$. The geometric factor F_b is calculated with $D_b = 0.685$ using Eq. (11-4-16), and the result is $F_b = 1.94$.

$$h_{NB} = h_{NB1}F_bF_c = 3,957 \times 1.94 \times 0.527 = 4,045 \text{ W/(m}^2 \cdot \text{K)}$$

$$h_o = h_{NB} + h_{TP} = 4,045 + 500 = 4,545 \text{ W/(m}^2 \cdot \text{K)}$$

$$U = (1/h_{io} + 1/h_o + R_f)^{-1} = 1,355 \text{ W/(m}^2 \cdot \text{K)}$$

Then

$$\Delta T = q/U = 62,300/1,355 = 47°C$$

This means that the maximum steam temperature that can be used is $113.3 + 47 = 160.3°C$.

11-4-4   Horizontal Thermosiphon Reboiler

The heat transfer mechanism in a horizontal thermosiphon reboiler is essentially the same as that in a kettle reboiler, and Eq. (11-4-6) is applied. However, in this case it may be expected that the convective contribution (h_{TP}) is incremented as a consequence of the higher liquid circulation velocity. Several models with different degrees of complexity were proposed to take into account the convective effects. One of the best known is that of Fair and Klip.[23] These models require a significant number of calculations and the available information to support their use is scarce, so a conservative recommendation is to design them as kettle reboilers.

11-4-5   Axial-Flow Reboilers

This type includes forced-circulation and vertical thermosiphon reboilers. In these units, boiling takes place into the tubes, and both the boiling heat transfer coefficient and the frictional pressure drops depend on the two-phase Reynolds number $Re_{TP} = GF^{1.25}(1 - x)D_i/\mu_L$ (see Eq. 11-3-23).

In a forced-circulation reboiler, the flow rate is imposed by the circulation pump, but in a thermosiphon unit, as explained in Sec. 11-4, the flow rate results from the hydraulic balance in the circuit, where the circulation driving force is the density difference between the reboiler liquid feed and the vapor liquid mixture in the tubes. But the density of the vapor-liquid mixture depends, in turn, on the heat transfer capacity of the reboiler. This means that the thermal and hydraulic calculations are coupled and cannot be solved independently.

The usual approach is to assume a total flow rate and calculate the required heat transfer area to reach the desired vaporization for this flow rate. The hydraulic balance then is verified, and based on the results, the assumed flow rate is modified, repeating the calculations up to final agreement.

It must be noted that the vapor mass fraction x included in the Reynolds number Re_{TP} changes as the vaporization progresses. Hence both Re_{TP} and the frictional pressure drop per unit length, as well as the density of the vapor-liquid mixture, change along the tube length.

The calculation method requires that the reboiler be divided into intervals small enough to assume that the fluid conditions are uniform, and for each interval, the heat transfer surface and frictional pressure drops are calculated. The total value results from summation of the individual values for all the intervals. The detailed procedure is explained in the following section.

Vertical Thermosiphon Reboiler Design.   Assume that it is necessary to design a thermosiphon reboiler to vaporize W_V (kg/h) of a liquid mixture at a distillation column bottom. The column bottom conditions correspond to a pressure p_A. At this pressure, the liquid is at its bubble-point temperature T_A.

The liquid-feed mass flow to the reboiler W will be higher than the vaporization rate W_v owing to the recirculated fraction. It is usual to design with circulation ratios W/W_v of at least 4:1. We assume a mass flow rate W. As already explained, it must be verified that the reboiler design allows this circulation rate.

The vapor mass fraction at the reboiler outlet thus will be

$$x_E = \frac{W_v}{W} \tag{11-4-24}$$

If the liquid is a multicomponent mixture, the boiling temperature increases as long as vaporization progresses. It is then necessary to have a vaporization curve showing the amount of heat to be delivered to the circulating fluid for different vapor mass fractions between 0 and x_E to achieve that vapor fraction, as well as the temperatures corresponding to each point. This curve can be calculated with a process simulator and can be presented in a table such as Table 11-3.

TABLE 11-3   The Boiling-Curve Data

x	T	ΔQ
0	t_A	0
.	.	.
.	.	.
.	.	.
x_E	t_E	
		$Q = \Sigma \Delta Q$

We must remark that the pressure also changes along the tube length. However, for the calculation of this table, it is usual to neglect the effects of pressure changes, and the table is calculated at a constant pressure equal to the column bottom pressure

In certain cases, if the circulation ratio is high or the boiling range of the product is small, the table may not be necessary, and it can be assumed that the fluid vaporizes at constant temperature. In this case, the boiling curve will be a linear function of x and can be expressed as

$$\Delta Q = \lambda W \Delta x \tag{11-4-25}$$

where λ is a mean heat of vaporization.

As in other design cases, it is necessary to assume a geometric configuration and then verify that it allows the required heat flux. We shall then assume that the following variables were adopted: tube diameter, number of tubes, shell diameter, heating-fluid flow rate, and temperatures. With this information, the shell-side heat transfer coefficient can be calculated. To simplify the explanation, it will be assumed that the heating medium is isothermal condensing steam at temperature T.

Liquid Single-Phase Region.   We refer to Fig. 11-36. The liquid at the distillation column bottom is at p_A and T_A, where T_A is the bubble-point temperature corresponding to p_A. The liquid pressure at the lower tube sheet level will be p_B, and neglecting the frictional pressure drop in the reboiler inlet piping, this is

$$p_B = p_A + \rho_L g z_1 \tag{11-4-26}$$

Since p_B is higher than p_A, the bubble-point temperature at p_B is also higher than T_A, meaning that the reboiler liquid feed is subcooled. The first reboiler region thus will be liquid single-phase heat transfer.

As was explained in Sec. 11-3 (see Fig. 11-18), at a certain reboiler section, the bubble production at the tube wall starts, and the heat transfer mechanism turns into a subcooled nucleate boiling regime. From then on, the liquid continues increasing its temperature, whereas the pressure decreases, up to the section in

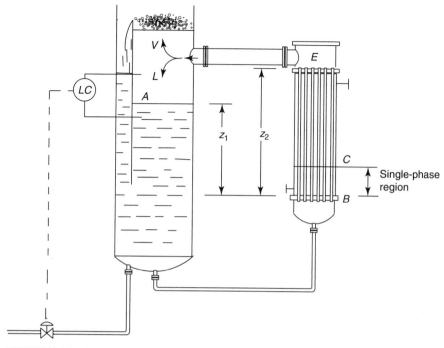

FIGURE 11-36 Thermosiphon reboiler with single-phase region.

which the mean liquid temperature reaches the bubble point at the prevailing pressure. There the saturated boiling regime starts.

Coming back to Fig. 11-36, let's assume that section C corresponds to the start of saturated boiling. In this section, pressure p_c and temperature T_c correspond to a point on the vapor-pressure curve. This was already explained in Sec. 11-3 (see Fig. 11-20). The pressure-temperature conditions corresponding to point C can be obtained as the intersection between the saturation curve of the fluid (vapor pressure versus temperature) and the curve representing the pressure-temperature evolution of the fluid as it flows up through the tube.

If $t_{(z)}$ and $p_{(z)}$ represent the temperature and pressure condition of the fluid corresponding to a coordinate z, this evolution is represented by the equation

$$t_{(z)} - t_B = \frac{\dfrac{dt_{(z)}}{dz}}{\dfrac{dp_{(z)}}{dz}} (p_{(z)} - p_B) \qquad (11\text{-}4\text{-}27)$$

And the saturation curve is

$$t_{\text{SAT}(z)} - t_A = \left(\frac{dt}{dp}\right)_S (p - p_A) \qquad (11\text{-}4\text{-}28)$$

where the derivative is the slope of the saturation curve (bubble-point temperature versus pressure). This also can be calculated with the help of a process simulator. In the equation, p is the saturation pressure at

$t_{\text{SAT}(z)}$. At point C, the last two equations must be satisfied, and considering that t_A equals t_B because there is no heating between A and B, combining both expressions, we have

$$\frac{p_B - p_C}{p_B - p_A} = \frac{\left(\dfrac{dt}{dp}\right)_S}{-\dfrac{dt/dz}{dp/dz} + \left(\dfrac{dt}{dp}\right)_S} \tag{11-4-29}$$

The derivatives of temperature and pressure with respect to z can be calculated easily because it is a liquid single-phase region. (Even though part of this region corresponds to subcooled boiling, it is conservative to calculate it as a liquid single-phase region.) Then, with the total mass flow rate W, it is possible to calculate the total heat transfer coefficient U_{SP} for this single-phase region, and it is

$$Wc_L dt = (\pi D_i N dz) U_{\text{SP}} (T - t_A) \tag{11-4-30}$$

$$\therefore \frac{dt}{dz} = \frac{\pi D_i N U_{\text{SP}} (T - t_A)}{Wc_L} \tag{11-4-31}$$

and the pressure drop is

$$dp = -\rho_L g dz - dF - v dv \tag{11-4-32}$$

The last two terms represent the frictional and acceleration pressure drops. For this single-phase region, these terms can be neglected, resulting in

$$\frac{dp}{dz} = -\rho_L g \tag{11-4-33}$$

Then, with the preceding equations, it is possible to calculate the pressure and temperature corresponding to point C where saturated boiling starts.

The heat flux corresponding to the single-phase region is

$$Q_{\text{SP}} = Wc_L (T_C - T_A) \tag{11-4-34}$$

And the heat transfer area corresponding to this region may be calculated as

$$A_{\text{SP}} = \frac{Q_{\text{SP}}}{U_{\text{SP}} (T - t_A)} \tag{11-4-35}$$

The length of this single-phase region will be

$$\Delta z_{\text{SP}} = A_{\text{SP}} / (\pi D_o N) \tag{11-4-36}$$

We must remark that this length is usually not significant with respect to the total area of the reboiler. Thus it is often neglected, and calculations can be started directly with the saturated boiling region assuming that $T_C = T_A$ and $p_C = p_B$.

Saturated Boiling Region. As was explained in Sec. 11-3, the boiling heat transfer coefficient in this region is calculated with the Chen correlation. For this purpose, we shall divide the unit into intervals, as indicated in Table 11-3, and for each interval with the corresponding value of x, we shall calculate Re_{TP}, and with it we shall calculate X_{tt} and S, which are the necessary parameters for the calculation of h_N and h_{TP}. Then, for each one of the intervals, we can calculate

$$h_i = h_N + h_{TP} \tag{11-4-37}$$

$$h_{io} = h_i \frac{D_i}{D_o} \tag{11-4-38}$$

$$U = \left(\frac{1}{h_{io}} + \frac{1}{h_o} + R_f \right)^{-1} \tag{11-4-39}$$

It is then possible to calculate the area and length of each interval as

$$\Delta A = \frac{\Delta Q}{U(T - t_{(z)})} \tag{11-4-40}$$

$$\Delta z = \frac{\Delta A}{\pi D_o N} \tag{11-4-41}$$

Knowing ΔZ, it is possible to calculate the frictional pressure drop corresponding to the interval. It can be obtained with Eqs. (11-3-5) and (11-3-6).

For each interval, the Reynolds number for liquid flowing alone $[Re_L = W(1 - x)D_i/a_t\mu_L]$ is calculated, and with it, the Fanning friction factor f_L can be obtained. Then the frictional pressure drop for the interval can be calculated as

$$\Delta p_f = \phi^2 \Delta p_L = 4\phi^2 f_L \frac{\Delta z}{D_i} \frac{1}{2\rho_L} \left[\frac{W(1-x)}{a_t} \right]^2 \tag{11-4-42}$$

The two-phase mixture density in each interval can be calculated as

$$\rho_{TP} = \rho_L R_L + \rho_V (1 - R_L) \tag{11-4-43}$$

where R_L is obtained with Eqs. (11-3-3) and (11-3-4).

Verification of the Recirculation Rate. With reference to Fig. 11-36, the pressure p_B is given by Eq. (11-4-26) (neglecting the frictional pressure drop in the inlet piping). The mass flow W circulating through the circuit must be such that p_B is the sum of

1. Frictional pressure drop in the boiler
2. Hydrostatic pressure owing to the liquid-vapor column inside the tubes
3. Acceleration pressure drop
4. Outlet pressure

We shall see how to calculate these terms.

Frictional Pressure Drop.   The frictional pressure drops correspond to the sum of

1. Pressure drop in the single-phase region *BC*. This can be calculated as

$$4f \frac{\Delta z_{SP}}{D_i} \frac{W^2}{2\rho_L a_t^2} \tag{11-4-44}$$

where f is the friction factor for liquid flow in the single-phase region. This pressure drop is not significant and often is neglected.

2. Frictional pressure drop in the double-phase region. This is calculated as the summation $\Sigma\Delta pf$ corresponding to all the intervals.

3. Outlet frictional pressure drop. This can be calculated as

$$\frac{W^2}{a_E^2\rho_L}(1-x_E)^2\phi_E^2 \tag{11-4-45}$$

where a_E is the area of the outlet nozzle and ϕ_E^2 is calculated with Eq. (11-3-6) for the outlet condition.

Hydrostatic Pressure Corresponding to the Liquid-Vapor Column. This will be the sum of that corresponding to the liquid single-phase region and all the double-phase intervals. This is $\rho_L g\Delta z_{SP} + \Sigma\rho_{TP}g\Delta z$.

Acceleration Pressure Drop. Since the velocity of the vapor-liquid mixture exiting the reboiler is much higher than the liquid velocity at the reboiler inlet, there is an acceleration term that must be considered (this is analogous to the $\Delta\rho V^2$ term in the Bernouilli equation). This term can be expressed as

$$\frac{W^2}{a_t^2}\left[\frac{x_E^2}{\rho_V R_{VE}}+\frac{(1-x_E)^2}{\rho_L R_{LE}}-\frac{1}{\rho_L}\right] \tag{11-4-46}$$

where subscript E refers to the exit condition.

Outlet Pressure. This pressure is p_A.

Then the hydraulic verification requires the following equation to be satisfied:

$$\rho_L g z_1 = 4f\frac{\Delta z_{SP}}{D_i}\frac{W^2}{2\rho_L a_t^2}+\Sigma\rho_{TP}g\Delta z+\Sigma\Delta p_f +$$
$$+\frac{W^2}{a_t^2}\left[\frac{x_E^2}{\rho_V R_{VE}}+\frac{(1-x_E)^2}{\rho_L R_{LE}}-\frac{1}{\rho_L}\right]+\frac{W^2}{a_E^2\rho_L}(1-x_E)^2\phi_E^2 \tag{11-4-47}$$

Limitations to the Heat-Flux Density. As was pointed out in Sec. 11-3, in flow boiling it is also necessary to establish limits to the validity of the correlations. These limits correspond to the beginning of the transition boiling regime. A simple correlation to predict the maximum heat-flux density is expressed by Eq. (11-3-32). Thus, if the maximum heat-flux density resulting from the calculation is higher than that obtained with Eq. (11-3-32), it will be necessary to decrease the temperature of the heating medium.

Again, in this case the values suggested by Kern (94,600 W/m² for aqueous fluids and 37,800 W/m² for organics) can be adopted as the upper limit of the heat-flux density. However, at present, these values are considered excessively conservative and may lead to oversized designs.

When working at low pressures, if ΔT is increased above a certain value, another phenomenon may appear, consisting of a hydraulic instability that provokes a pulsating flow in the reboiler circuit. This effect was studied by Blumenkrantz,[24] and the conclusion is that this limitation can be stabilized by increasing the frictional resistance in the reboiler inlet piping either by the addition of a valve or by using a smaller diameter. However, at moderate pressures, the critical temperature corresponding to the start of transition boiling is reached before the occurrence of this hydraulic instability, so the limit to ΔT is given by Eq. (11-3-32).

If either the heat-flux density or the mass vapor fraction at the reboiler outlet is too low, it is also possible to have an unsatisfactory operation. This is so because in these cases the thermosiphon effect can be too small to have a predictable circulation. Thus it is not recommended to design with heat-flux densities lower than 5,000 W/m². If this condition cannot be achieved because of a very low available ΔT, it is preferable to use a forced-circulation or kettle reboiler.

The other limit that must be verified is that the outlet vapor fraction is not so large as to have mist flow. A correlation that was presented to predict the mist-flow condition is given by Eq. (11-3-31). The recommendation, then, is to calculate the mass flow density, and if it is higher than the value predicted by

Eq. (1-3-31) for the outlet value of X_{tt}, the design must be adjusted so as to work with higher recirculation rates. This is another reason why it is not recommended to design with circulation rates lower than 3:1 for hydrocarbons or 10:1 for water.

Example 11-7   It is desired to design a vertical thermosiphon reboiler to vaporize 4,138 kg/h of the bottom liquids of a distillation column. The column bottom temperature is 142.1°C, and the pressure is 1,261 kPa. The following data corresponding to the boiling curve was obtained with a process simulator at 1,261 kPa, assuming a liquid inlet flow to the reboiler of 12,143 kg/h (which roughly corresponds to a recirculation ratio about 3:1).

Q (kW)	Temperature (°C)	Vapor Mass Fraction x (kg/kg)
0	142	0
82	146.2	0.0682
164	148.1	0.1364
246	151.28	0.2046
328	153.2	0.2728
410	156.4	0.34

The following physical properties of the vapor and liquid may be assumed constant in the whole operating range:

	Vapor	Líquid
Specific heat, J/(kg · K)	2,400	3,010
Viscosity, cP	0.01	0.1
Thermal conductivity, W/(m · K)	0.027	0.055
Density, kg/m³		510
Critical pressure, kPa		3,167
Molecular weight	70	70
Latent heat, J/kg	234	

The slope of the vapor-pressure curve of the column bottom liquids (bubble-point curve) is 0.039°C/kPa

BR = boiling range of the mixture (difference between dew point and bubble point) = 18°C

With the molecular weight, the vapor density may be calculated:

$$\rho_V = Mp/zRT = 25.635 \text{ kg/m}^3$$

The heating media will be steam. A condensing temperature of 187°C will be assumed. We shall use 19-mm OD and 14.4-mm ID tubes. A fouling factor of 0.00027 (m² · K)/W will be adopted. Design the unit and calculate the necessary liquid height to achieve the assumed recirculation rate.

Note: A downloadable spreadsheet with the calculations of this example is available at http://www.mhprofessional.com/product.php?isbn=0071624082

Solution   We shall adopt 42 tubes, and we shall calculate their necessary length. The mass flow density will be

$$G = W/(N\pi D_i^2/4) = 12,143/[3,600 \times 42 \times 0.0144^2 \times (3.14/4)] = 493 \text{ kg/(s} \cdot \text{m}^2)$$

As a first approach, we can neglect the liquid single-phase region. This implies assuming that the liquid enters into the reboiler at its bubble-point temperature, and from the beginning, a saturated boiling regime exists. We shall adjust the results later.

To calculate the heat transfer coefficient h_N using Eq. (11-3-29), it would be necessary to know in advance the value of q. Since this value is not known, an iterative procedure is necessary. We shall first assume a value for q, and with it, we shall calculate h_N. At the end of the calculation, this can be verified and the calculations repeated if necessary. We shall start with $q = 45,000 \text{ W/m}^2$. Also with

$q = 45,000$ W/m^2, the correction factor for boiling range $F_c = 0.499$ [(Eq. (11-4-17)] is calculated. The following table contains the calculation results for all the intervals:

Interval	1	2	3	4	5
ΔQ, kW	82	82	82	82	82
t	144.1	147.14	149.7	152.2	154.8
$\Delta T = (T - t = 187 - t)$	42.8	39.85	37.5	34.75	32.2
x	0.0341	0.1023	0.1705	0.2387	0.3064
X_{tt} [Eq. (11-3-3)]	5.722	1.993	1.722	0.8016	0.588
F (Fig. 11-22)	1.170	1.835	2.463	3.105	3.875
$Re_L = [G(1 - x)D_i/\mu_L]$	68,623	63,778	58,932	54,087	49,278
$Re_{TP} = (Re_L F^{1.25})$	83,516	136,237	181,867	222,981	260,207
h_{TP}, W/(m$^2 \cdot$ K) [Eq. (11-3-24)]	1,348	1,995	2,514	2,959	3,348
S (Fig. 11-24)	0.408	0.28	0.217	0.179	0.154
h_N, W/(m$^2 \cdot$ K) [Eq. (11-3-29) with $q = 45,000$ W/m^2]	1,236	848	658	543	467
$h_N F_c$	617	423	328	271	233
h_i, W/(m$^2 \cdot$ K) [Eq. (11-4-37)]	1,966	2,419	2,842	3,231	3,582
h_{io}, W/(m$^2 \cdot$ K) [Eq. (11-4-38)]	1,490	1,833	2,154	2,448	2,715
U [Eq. (11-4-39)]	945	1,072	1,174	1,256	1,323
ΔA, m^2 [Eq. (11-4-40)]	2.02	1.92	1.87	1.88	1.93

The total area will be $A = \Sigma \Delta A = 9.62$ m^2. The tube length should be

$$L = A/(\pi N D_o) = 9.62/(3.14 \times 42 \times 0.019) = 3.84 \text{ m}$$

The average heat-flux density is $410/9.62 = 42,618$ W/m^2. It is noteworthy that the importance of h_N in the calculation of h_i is not significant, so it is not necessary to feed back the calculated value in place of the assumed 45,000 W/m^2.

Verification of the maximum heat-flux density: The maximum heat-flux density will be verified with Eq. (11-3-32):

$$q_{max} = 23,600(0.0144^2/3.84)^{0.35} \times 3,167^{0.61}(1,261/3,167)^{0.25}(1 - 1,261/3,167)$$
$$= 49,123 \text{W/m}^2 > 42,618 \qquad \text{OK}$$

Verification of mist flow at reboiler outlet: We shall verify that mist flow does not exist at the unit outlet using Eq. (11-3-31):

$$G = 2,441 X_{ttE} = 2,441 \times 0.512 = 1,251 > 493 \qquad \text{OK}$$

Calculation of the height of the liquid leg required to achieve the desired circulation rate: We shall first calculate the frictional pressure drop and the hydrostatic pressure of the liquid vapor column:

Interval	1	2	3	4	5
Δz [Eq. (11-4-41)]	0.808	0.766	0.747	0.750	0.768
X_{tt} [Eq. (11-3-3)]	5.722	1.993	1.722	0.8016	0.588
$X' = 1/X_{tt}$	0.174	0.501	0.580	1.247	1.700
R_L [Eq. (11-3-4)]	0.461	0.291	0.225	0.186	0.159
ρ_{TP} [Eq. (11-4-43)]	249	166	135	116	103
$\rho_{TP}g\Delta z$	1,971	1,252	988	852	773
ϕ^2 [Eq. (11-3-6)]	4.7	11.78	19.64	28.73	39.55
f [Eq. (7-4-9)]	0.0059	0.0060	0.0061	0.0063	0.0064
Δp_f [Eq. (11-4-42)]	1,055	2,206	3,117	3,934	4,705

The total frictional pressure drop will be $\Sigma \Delta p_{fi} = 15,016$ Pa. The hydrostatic pressure will be $\Sigma \rho_{TP} g \Delta z =$

5,836 Pa. The friction loss at the outlet nozzle can be calculated with Eq. (11-4-45) where a_E is the area of the reboiler outlet nozzle. Assuming a 0.2-m (8-in) diameter, $a_E = 0.0314$ m². With Eq. (11-4-45), we get 451 Pa. The acceleration pressure drop is calculated with Eq. (11-4-46), where subscript E refers to the outlet conditions (corresponding to a mass vapor fraction = 0.34). The calculated value is 2,217 Pa. Then

Hydrostatic pressure of the liquid vapor column	5,836 Pa
Friction loss at the outlet	451 Pa
Acceleration loss	2,217 Pa
Friction loss in tubes	15,016 Pa
Total	23,521 Pa

Then, according to Eq. (11-4-47), the required height is $z_1 = 23,521/g\rho_L = 23,521/9.8 \times 510) = 4.71$ m.

If the reboiler were built with a 3.84-m tube length as was calculated, to achieve the desired circulation rate, it would be necessary to operate with a liquid level into the column 0.9 m above the upper tubesheet level. In this situation, the liquid would be obstructing the reboiler outlet nozzle. Thus it is necessary to modify the design.

With the help of a spreadsheet, it is easy to modify the design by changing the number of tubes to 50. The following table contains the calculation results:

Interval	1	2	3	4	5
F (Fig. 11-22)	1.170	1.835	2.463	3.105	3.875
$Re_L = [G(1 - x)D/\mu_L]$	57,643	53,573	49,503	45,433	41,393
$Re_{TP} = (Re_L F^{1.25})$	70,153	114,439	152,768	187,304	218,573
h_{TP}, W/(m² · K) [Eq. (11-3-24)]	1,173	1,735	2,186	2,574	2,912
S (Fig. 11-24)	0.458	0.322	0.253	0.211	0.182
$h_N F_c$, W/(m² · K) [Eq. (11-3-29) with $q = 45,000$ W/m²]	644	454	357	297	257
h_i, W/(m² · K) [Eq. (11-4-37)]	1,817	2,189	2,543	2,871	3,169
h_{io}, W/(m² · K) [Eq. (11-4-38)]	1,377	1,659	1,928	2,176	2,402
U [Eq. (11-4-39)]	898	1,010	1,103	1,180	1,244
ΔA, m² [Eq. (11-4-40)]	2.13	2.04	1.99	2.00	2.05

The required area is then $A = \Sigma\Delta A = 10.12$ m². The tube length necessary for heat transfer is $10.12/(50 \times \pi \times 0.019) = 3.42$ m. The hydraulic calculation results are

Hydrostatic pressure liquid + vapor	5,194
Friction loss at outlet	451
Acceleration loss	1,564
Friction loss in tubes	9,854
Total	17,063

Required liquid height for circulation is $17,063/(9.8 \times 512) = 3.41$ m, which is acceptable.

Example 11-8 Calculate the length of the liquid single-phase region for Example 11-7 (with the 50 tubes).

Solution Using Eq. (11-4-29) and the nomenclature of Fig. 11-36, we get

$$\frac{p_B - p_C}{p_B - p_A} = \frac{\Delta z_{SP}}{z_1}$$

$$\Delta z_{SP} = 3.14m \frac{(dt/dp)s}{-\dfrac{dt/dz}{dp/dz} + (dt/dp)s} = \frac{0.039}{-\dfrac{dt/dz}{dp/dz} + 0.039}$$

In the single-phase region, h_i is calculated with Eq. (11-3-24) with $F = 1$. We get $h_i = 1{,}002$ W/(m$^2 \cdot$ K).

$h_{io} = 759$ W/(m$^2 \cdot$ K)

$U = (1/759 + 1/8{,}500 + 0.00027)^{-1} = 586$ W/(m$^2 \cdot$ K)

$dt/dz = \pi D_o NU(T - t)/Wc_L = \pi \times 0.019 \times 50 \times 586 \times (187 - 142)/[(12{,}143/3{,}600) \times 3{,}010] = 7.74°$C/m

$dp/dz = -\rho_L g = -510 \times 9.8 = -4{,}998$ Pa/m $= -4.99$ kPa/m

$\Delta z_{SP} = 3.14(0.039)/[(7.74/4.99) + 0.039] = 0.077$ m

The temperature at which boiling starts will be $t_C = t_B + dt/dz \cdot \Delta z_{SP} = 142.6°$C.
We see that the length of the single-phase region is not significant with respect to the total tube length, which confirms our previous assumption. Otherwise, it would be necessary to add this length to the value calculated for the convective boiling region.

In a rigorous way, it would be necessary to recalculate the convective boiling region later because the temperature at the section where boiling starts is different from the reboiler inlet temperature. However, the differences are very small, and the effort is not necessary. In units working at very low pressures (vacuum service), the single-phase region may be important and must be considered.

GLOSSARY

a_t = flow area of a tube (m)

B = constant defined in Eq. (11-1-19) (K $\cdot$ m)

C_{SF} = coefficient for Eq. (11-2-6) (dimensionless)

D = diameter (m)

D_d = bubble detachment diameter (m)

F = correction factor in Chen's Eq. (11-3-23)

f = friction factor

G = mass flow density or mass velocity (kg/s)

g = gravity acceleration (m/s^2)

i = enthalpy per unit mass (J/kg)

h = film coefficient [W/(m$^2 \cdot$ K)]

h_L = single-phase film coefficient calculated assuming all the fluid at liquid state [W/(m$^2 \cdot$ K)]

$h_N = h_{NB}$ = nucleation contribution to the film coefficient [W/(m$^2 \cdot$ K)]

h_{NB1} = nucleation coefficient for a single tube

h_{TP} = two-phase convection contribution to the film coefficient [W/(m$^2 \cdot$ K)]

k = thermal conductivity [W/(m $\cdot$ K)]

M = molecular weight (kg/kmol)

m = exponent in Eq. (11-2-6) (dimensionless)

n = exponent in Eq. (11-2-6) (dimensionless)

Nu = Nusselt number (dimensionless)

p = pressure (N/m^2)

p_c = critical pressure (N/m^2) (in Mostinsky dimensional equations, kPa)

p_G = equilibrium saturation pressure at temperature T_G (N/m^2)

p_F = system pressure described in Sec. 11-1 (N/m^2)

Pr = Prandtl number (dimensionless)

q = heat-flux density at the interface (W/m^2)

q_N = contribution to q owing to nucleation effect (W/m²)

q_C = contribution to q owing to convection effect (W/m²)

r = bubble radius (m)

r_c = critical radius of the cavity (m)

r_b = characteristic radius defined by Eq (11-3-8) (m)

R = Ideal gases constant [(N · m)/(K · mol)]

Re = Reynolds number (dimensionless)

Re_L = Reynolds number for the liquid flowing alone (dimensionless)

Re_{TP} = Reynolds number for the two-phase flow (dimensionless)

R_L = volumetric fraction of the tube occupied by the liquid (dimensionless)

R_V = volumetric fraction of the tube occupied by the vapor (dimensionless)

S = Chen's supression factor (dimensionless)

T = temperature (generic or of the hot fluid medium) (K)

T_G = system temperature defined in Sec. 11-1 (K)

t_F = temperature in the bulk of the fluid (K)

T_{SAT} = saturation temperature corresponding to the system pressure (K)

T_{WONB} = wall temperature at which boiling starts (K)

v_V = vapor specific volume (m³/kg)

v_L = liquid specific volume (m³/kg)

v_{VSAT} = saturated vapor specific volume (m³/kg)

x = vapor quality (vapor mass flow/total mass flow)

X_{tt} = Lokhardt-Martinelli parameter [Eq. (11-3-3)] (dimensionless)

X' = reciprocal of X_{tt} (dimensionless)

ΔT_{SAT} = difference between wall temperature and saturation temperature (K)

Δp_{SAT} = difference between the vapor pressure at wall temperature and the vapor pressure at saturation temperature (K)

σ = surface tension (N/m)

λ = heat of vaporization (J/kg)

δ = boundary layer thickness (m)

θ = bubble contact angle (degrees or radians)

ϕ^2 = correction factor defined by Eq. (11-3-5)

Φ = correction factor for bubble contact angle

ρ = density (kg/m³)

μ = viscosity (kg/ms)

β = expansion coefficient (1/K)

Subscripts

L = liquid

V = vapor

TP = two phase

i = internal

o = external

REFERENCES

1. Thomson W.: *Phil Mag* 42(4):448, 1871.

2. Hsu Y. Y.: "On the size of range of active nucleation cavities on a heating surface," *Trans ASME J Heat Transfer* 84:207, 1962.

3. Han C. Y., Griffith P.: "The mechanism of heat transfer in nucleate pool boiling. Pt 1: Bubble initiation growth an departure," *Int J Heat Mass Transfer* 8(6):887–904, 1965.

4. Collier J.: *Convective Boiling and Condensation.* London: McGraw-Hill, 1972.

5. Fritz W.: "Berechnung des Maximal Volume von Dampfblasen," *Phys Z* 36:379, 1935.

6. Vachon R.: Paper presented at U.S. National Heat Transfer Conference, Seattle, WA, paper 67-HT-34, 1967.

7. Rohsenow W.: "A method of correlating heat transfer data for surface boiling of liquids," *Trans ASME* 74:969, 1952.

8. Forster H., Zuber N.: "Bubble dynamics and boiling heat transfer," *AIChE J* 1:532, 1955.

9. Bromley L. A.: "Heat transfer in stable film boiling," *Chem Eng Progr* 46:221–227, 1950.

10. Alves G.: *Chem Process Eng* 50(9):449–456, 1964.

11. Dengler C. E., Addoms J. N.: "Heat transfer mechanism for vaporization of water in a vertical tube," *Chem Eng Progr Symp Series* 57(18):95–103, 1956.

12. Mostinsky I. L.:" Application of the rule of corresponding states for the calculation of heat transfer and critical heat flux," *Teploenergetika* 4:66, 1963 *English Abstr. Br Chem Eng* **8**(8):586, 1963.

13. McNelly M. J.: "A correlation of the rates of heat transfer to nucleate boiling liquids," *J. Imp Coll. Chem Eng Soc* 7:18, 1953.

14. Stephan K., Abdelsalam M.: "Heat transfer correlations for natural convection boiling," *Int J Heat Mass Transfer* 25:73–85, 1979.

15. Lokhardt R. W., Martinelli R.: "Proposed correlation of data for isothermal two phase two components flow in pipes," *Chem Eng Progr* 45(1):39–48, 1949.

16. Fair J. R.: "What you need to design thermosiphon reboilers," *Petroleum Refiner* 39(2):105–123, 1960.

17. Chen J.: "Correlation for boiling heat transfer to saturated fluids in convective flow," *I&E.C Process Design and Development* 5(3):322–329, July 1966.

18. Zuber N.: "On the stability of boiling heat transfer," *Trans ASME* 80:711, 1958.

19. Zuber N., Tribus M., Westwater J. W.: "The hydrodinamic crisis in pool boiling of saturated and subcooled liquids," *Int Dev Heat Transfer* 2:230, 1961.

20. Kutateladze S. S.: "A hydrodinamic theory of changes in the boiling process under free convection," *Izv Akad Nauk SSSR Otd Tekh Nauk* 4:529, 1951.

21. Palen J. W., in *Heat Exchangers Design Handbook.* New York: Hemisphere Publishing, 1983.

22. Palen J. W., Small W. M.: "A new way to design kettle and internal reboilers," *Hydrocarbon Proc* 43(11):199, 1964.

23. Fair J. R., Klip A.: "Thermal design of horizontal reboilers," *Chem Eng Progr* March 1983: 83–96.

24. Blumenkrantz A., Taborek J.: "Application of stability analysis for design of natural circulation boiling systems and comparison with experimental data," *AIChE Symp Series* 68(118) 1971.

25. Procedures 4D4.1 and 4D3.1. *Technical Data Book: Petroleum Refining.* Washington: American Petroleum Institute, 1980.

26. *Engineering Data Book,* 11th ed: Gas Processors Suppliers Association, Tulsa, Oklahoma, 1998.

27. Perry: *Chemical Engineers Handbook.* New York: McGraw-Hill.

28. Cichelli M. T., Bonilla C. F.: "Heat Transfer to Boiling Liquids under Pressure," *Trans AIChe* 41:755, 1945.

29. Frederking, Clark: "Natural Convection Film Boiling on a Sphere." *Adv Cryogen Eng* 8:501, 1963.

30 Bankoff S. G.: "Ebullition from solid surfaces in the absence of a pre existing gaseous phase," *Trans ASME* 79:735, 1957.

31 Fritz W.: "Berechnung des Maximal Volume von Dampfblasen," *Phys. Z* 36:379, 1935.

32 Hewitt G. F., Roberts D. N.: "Studies of two phase flow patterns by simultaneous X rays and flash photography," A.E.R.E.-M 2159 HMSO, 1969.

CHAPTER 12
THERMAL RADIATION

12-1 HEAT TRANSFER BY RADIATION THROUGH TRANSPARENT MEDIA

12-1-1 Characteristics of the Radiant Energy

Radiation is a heat transfer mechanism that, unlike conduction and convection, does not require a physical medium to propagate. Even more, if the propagation medium contains certain gases as water vapor or CO_2, part of the energy will be absorbed by these gases, and a smaller amount of energy will reach the receptor. Radiant energy propagates better in a vacuum than through any other medium, either gas, liquid, or solid.

Radiant energy is sometimes envisioned as transported by electromagnetic waves and at other times as transported by photons, whose energy depends on the associated wavelength. It is also known that radiation travels with the speed of light.

Radiant energy can be emitted by any body, solid, liquid, or gas. For a body to emit radiant energy, its atoms or molecules need to be previously excited by the absorption of any form of energy. Then, when the body emits radiation, the energy of its atoms or molecules decreases to the previous level. These changes in the energy level may consist of a modification of the vibration or rotation modes of the molecules, electrons jumping from one orbital to another with a lower energy level, or a decrease in the energy of the atomic nucleus.

Owing to this variety of mechanisms, radiation may consist of photons having very different energy; thus radiant energy encompasses a wide range of wavelengths, and the body can emit a continuous spectrum. The electromagnetic spectrum does not extend uniformly from $\lambda = 0$ to $\lambda =$ infinite because the radiant energy is concentrated in a certain region.

The wavelengths of the complete electromagnetic spectrum extend from $\lambda = 10^{-13}$ m (cosmic rays) to values of λ as high as 1,000 m (long-wavelength radio waves). There is a relatively narrow zone of this spectrum which is associated with the thermal energy of the emitting body. (This is the internal energy related to the macroscopic observable parameter that is temperature.) This wavelength interval ranges from 0.1–100 μm. All bodies, as a consequence of their temperature, emit radiation within this wavelength range. The higher the temperature of the body, the higher is the amount of energy emitted within this wavelength range. (Also, the distribution of energy within the wavelength spectrum changes with temperature.) The portion of the radiant energy emitted by a body as a consequence of its thermal level is called *thermal radiation*.

Within this wavelength interval, there is a very narrow band (from 0.4 to 0.7 μm) corresponding to radiation that can affect the human optic nerve. This is the *visible radiation*. When a certain body, as a consequence of its temperature, emits an important amount of radiation within the visible region, it appears luminous (e.g., the filament of an incandescent light bulb, a flame, a very hot iron, etc.).

Cold bodies, on the other hand, emit little energy within the visible range, and they lack luminosity. They can, however, reflect the light emitted by other hot sources. When we observe an object illuminated by a light source (e.g., solar light), the radiation we see is not that emitted by the object, but the radiation from the source reflected by the object.

The inverse process to emission takes place when a radiant energy beam is incident on the surface of a body. In this case, it may result in an increase in the energy level of atoms or molecules of the body by absorption of part of the incident radiation. In the case of solid surfaces, this absorption takes place within a very thin layer under the surface. In the case of liquids, the radiant energy can travel deeper into the medium, and this distance can be even much higher with gases.

Not all gases absorb thermal radiation. Elemental gases (N_2, H_2, and O_2) do not absorb thermal radiation to a significant degree. Some other gases, such as water vapor and CO_2, are capable of absorbing in very definite bands of the spectrum. Solids and liquids, on the other hand, can absorb energy of all wavelengths.

In the same way that the effect of radiant energy emission is to decrease the temperature of an emitting body (unless it is simultaneously receiving energy from another source), the effect of radiant energy absorption is to increase the temperature of the absorber. If a body is at a certain temperature in a steady state, this is so because it has reached a thermal level such that the energy it is emitting equals the energy it is absorbing from other sources.

If two bodies are facing each other, there will be a net heat transmission from the warmer body to the colder one. If the two bodies reach the thermal equilibrium, the energy emitted by each equals the energy each absorbs from the other. There is no net heat transmission between them. This condition is reached when the temperatures of both bodies are equal. Thus it is evident that bodies capable of emitting a higher amount of energy at a certain temperature also must be capable of absorbing energy in a higher proportion because this is the only way in which they can reach thermal equilibrium with other bodies that are not as good emitters.

A body can reach a steady state within a system when it absorbs the same amount of energy than it emits. For example, a body exposed to solar radiation interchanges energy with the sun, which acts as a source, and also with the rest of the universe that surrounds it. It is possible that the body reaches a temperature that is higher than the surroundings (e.g., a piece of iron remaining under the sun in summer). In this system, the sun acts as the energy source (it emits more energy than it absorbs), the body reaches a steady state (emitting the same amount that it absorbs), and the rest of the world acts as a final receptor of energy.

There are some other wavelengts of the electromagnetic spectrum capable of producing a temperature increase in an absorbing body. For example, microwaves used for cooking correspond to wavelengths of about 10^{-1} m. However, this type of radiation is not produced by a hot source but rather by electronic devices. The waves do not belong to the field of thermal radiation, respond to different mechanisms, and will not be studied here. This chapter will deal only with radiation resulting directly from thermal excitation, and we shall refer to it hereafter merely as *radiation*.

12-1-2 Absorption, Reflection, and Transmission Coefficients

When radiation is incident on a surface, a portion of the total energy is absorbed in the material, a portion is reflected from the surface, and the remainder is transmitted through the body. The fraction absorbed is called the *absorptivity* or *absorption coefficient* α, the fraction reflected is the *reflectivity* or *reflection coefficient* ρ, and the fraction transmitted is the transmissivity or *transmission coefficient* τ. These magnitudes are related by

$$\alpha + \rho + \tau = 1 \qquad (12\text{-}1\text{-}1)$$

The relative values of these coefficients depend on the material of the body and the state of its surface. In an opaque body, an incident beam can penetrate only a very short distance into the body. This distance is on the order of a micron for metals. The absorbed energy increases the temperature of the zone immediately adjacent to the surface, and then heat penetrates into the body by conduction. We may think, then, that all the process takes place at the body surface, and then, for an opaque solid, it can be considered that

$$\alpha + \rho = 1 \qquad (12\text{-}1\text{-}2)$$

A body with $\alpha = \rho = 0$ is called a *white body* (then $\tau = 1$). A material that behaves roughly as a white body is glass. A body with $\rho = 1$ is a mirror. Glass is not a perfect white body. The fact that it heats up when exposed to solar radiation demonstrates that its absorption coefficient is not nil.

A black body is one that absorbs all the incident radiation. Then, for a black body, $\alpha = 1$. For all real bodies, $\alpha < 1$. This indicates that the body reflects part of the incident radiation.

The absorptivity or absorption coefficient depends on the nature of the body, its temperature, and the nature of the incident radiation. This means that any body will present different values of the absorption coefficient when changes the source from which it is absorbing radiant energy.

12-1-3 Kirchoff's Law

Let's consider a big cavity into which there is a small body a whose external area is A_a (Fig. 12-1). This body is in thermal equilibrium with the cavity walls, which are assumed to be insulated, and it is also assumed that there is nothing else in the cavity other than body a. Let t be the temperature of the system in equilibrium. Both body a and the internal walls of the cavity emit radiant energy. The energy emitted by the body per unit time and per unit area is what we shall call W_a (emissive power of body a).

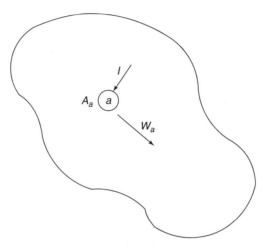

FIGURE 12-1 Body in an infinite cavity.

Let I be the irradiation over the surface of the body (which is the incident energy per unit area and per unit time). Then the surface of the body emits an amount of energy $A_a W_a$ and absorbs an amount of energy $A_a \alpha_a I$, where α_a is the absorption coefficient of the body for the radiation coming from the cavity. Since we have assumed that the system is in thermal equilibrium,

$$A_a W_a = A_a \alpha_a I \tag{12-1-3}$$

This means that

$$I = \frac{W_a}{\alpha_a} \tag{12-1-4}$$

If now we replace body a with another body c, whose area is A_c, at the same temperature as the cavity (we reserve the letter b for the blackbody), we have

$$I = \frac{W_c}{\alpha_c} \tag{12-1-5}$$

Since we have assumed that the dimensions of the cavity are very large in comparison with the bodies, we can assume that the presence of these bodies does not affect the behavior of the cavity. Then irradiation I remains constant when we change the bodies. Then

$$\frac{W_a}{\alpha_a} = \frac{W_c}{\alpha_c} = \frac{W_i}{\alpha_i} = I \tag{12-1-6}$$

This relationship is valid as long as the bodies a, c, and i are at the same temperature as the cavity. In other words, the bodies and the cavity are in *thermal equilibrium*. This expression is known as *Kirchhoff's law*, and it says that the quotient between the emissive power of a body and the coefficient of absorption exhibited by that body for the radiation coming from a big cavity at the same temperature is a constant. This means that a body that is capable of absorbing a higher amount of radiation also must be the best emitter because it must get rid of all the absorbed energy to reach equilibrium with other bodies. In particular, if we place into the cavity a body capable of absorbing all the incident energy (blackbody), we can write

$$I = \frac{W_a}{\alpha_a} = \frac{W_c}{\alpha_c} = \frac{W_b}{1} \qquad (12\text{-}1\text{-}7)$$

Since for any body other than a blackbody α is always lower than 1, Eq. (12-1-7) shows that the blackbody is the maximum emitter at the considered temperature.

The emissivity ε of a body is the quotient between its emissive power and the emissive power of the blackbody at the same temperature. For a generic body a, this is

$$\varepsilon_a = \frac{W_a}{W_b} \qquad (12\text{-}1\text{-}8)$$

By comparing Eqs. (12-1-7) and (12-1-8), we get

$$\varepsilon_a = \alpha_a \qquad (12\text{-}1\text{-}9)$$

This is another expression of Kirchhoff's law, and it establishes that the emissivity of a body is equal to the absorption coefficient of that body for the radiation coming from a big cavity in thermal equilibrium with it.

The emissivity of a body is a function of its temperature, but the absorption coefficient depends not only on its temperature but also on the characteristics of the incident radiation. If the temperature or the nature of the source with which the body is exchanging energy is modified, the value of the absorption coefficient will change. Equation (12-1-9) is only valid for the case of a body in thermal equilibrium with a big cavity, and it cannot, in principle, be generalized for other situations. Equation (12-1-7) indicates that the irradiation I in a cavity of large dimensions is equal to the emissive power of the blackbody at the same temperature.

12-1-4 The Cavity as Blackbody

The blackbody is a theoretical concept. No real body absorbs all the energy it receives. For experimental purposes, though, a blackbody can be simulated by a cavity such as a hollow sphere maintained at a uniform temperature with a small hole in its wall (Fig. 12-2). Any radiation entering through this hole has little probability of reaching the orifice again and escaping because it repeatedly strikes and reflects on the interior surface, and part of the energy is absorbed in each reflection. When the original radiation beam finally reaches the hole again and escapes, it has been so weakened by repeated reflection that the energy leaving the cavity is negligible. Thus, for any external observer, the orifice behaves as a blackbody because it absorbs all the radiation that is incident on it.

We have explained that the irradiation within the cavity equals the emissive power of the blackbody. A small hole in the cavity wall will not modify the situation, so the energy coming out of the hole equals the emissive power of the blackbody at the temperature of the cavity. Thus a small hole placed in the wall of a big enclosure behaves as a blackbody from the point of view of either absorption or emission of radiant energy, and it is a way to approximate the characteristics of an element of an area belonging to a blackbody.

A big cavity also acts as a blackbody surface for any body placed into it because all the energy emitted by the body is being absorbed in successive reflections at the cavity walls with little probability to reach back the surface of the emitter.

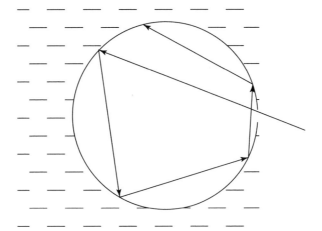

FIGURE 12-2 Radiation coming into a big cavity is absorbed completely.

12-1-5 Monochromatic Emissive Power

Any body emits energy within a range of wavelengths. The *monochromatic emissive power* at a certain wavelength is the amount of energy emitted per unit time, per unit area, and per unit wavelength interval. For example, if W_{50} is the energy emitted in the interval from 0 to 50 μm and W_{51} is the energy emitted in the interval from 0 to 51 μm, always per unit time and per unit area, then the monochromatic emissive power at the considered wavelength (in this case 50 μm) is

$$W_\lambda = \frac{W_{51} - W_{50}}{1 \text{ μm}} \tag{12-1-10}$$

Any body shows an emission spectrum that is characteristic of the nature of the body and its temperature. The emission spectrum is the distribution $W_\lambda = f(\lambda)$ (Fig. 12-3). The total emissive power of the body is then

$$W = \int_0^\infty W_\lambda \, d\lambda \tag{12-1-11}$$

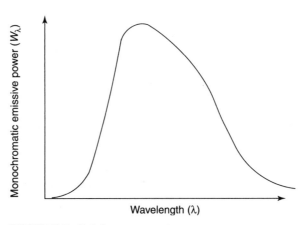

FIGURE 12-3 Emission spectrum.

The unit for W (total emissive power) is W/m², whereas the unit for W_λ (monochromatic emissive power) is W/(m²·μm).

12-1-6   Spectral Radiant Energy Distribution for a Blackbody

A relationship showing how the emissive power of a blackbody is distributed among the different wavelengths was derived by Max Plank[3] in 1,900 through his quantum theory. According to this theory,

$$W_{b\lambda} = \frac{2\pi C_1 \lambda^{-5}}{e^{C_2/\lambda T} - 1} \tag{12-1-12}$$

where

$$C_1 = hc^2$$
$$C_2 = hc/k$$

where h and k are the Planck and Boltzman constants, respectively, and c is the speed of light in a vacuum = 2.9979×10^8 m/s.

Figure 12-4 is a plot of Eq. (12-1-12) showing $W_{b\lambda}$ as a function of λ at given parameters of the body temperature. From the graph it is possible to extract some conclusions:

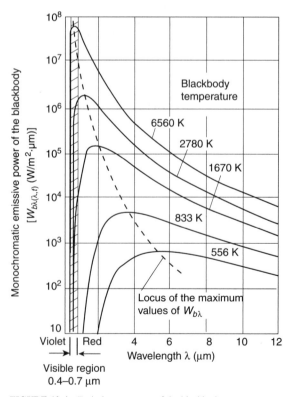

FIGURE 12-4   Emission spectrum of the blackbody.

1. The total energy emitted per unit time and per unit area over the entire wavelength spectrum, which is the emissive power of the blackbody, increases with the temperature of the body. This value can be obtained as the total area under a curve corresponding to any temperature. It can be demonstrated that

$$W_b = \int_0^\infty \frac{2\pi C_1 \lambda^{-5} \, d\lambda}{e^{C_2/\lambda T} - 1} = \sigma T^4 \qquad (12\text{-}1\text{-}13)$$

where σ is known as the *Stephan Boltzman constant,* and its value can be obtained by calculating the integral in Eq. (12-1-13), resulting in

$$\sigma = 5.672 \times 10^{-8} \; W/(m^2 \cdot K^4)$$

2. When the temperature increases, the wavelength corresponding to the peak of emission moves toward the smaller wavelength's zone. The relationship between the wavelength λ_{max} at which $W_{b\lambda}$ is a maximum and the absolute temperature is called *Wien's displacement law.* Thus

$$\lambda_{max} T = 2897.6 \; \mu m \cdot K \qquad (12\text{-}1\text{-}14)$$

where λ_{max} is the wavelength for which the monochromatic emissive power is a maximum at temperature T. The dashed line in Fig. 12-4 is the locus of all points represented by Eq. (12-1-14).

The shaded portion of the graph shows the region of the spectrum corresponding to visible light. It can be noticed that at temperatures about 1,000 K, a certain part of the emitted energy falls within this range, and the human eye begins to detect radiation. At this temperature, the object glows with a dull-red color (within the visible spectrum, the higher wavelengths correspond to red). As the temperature is increased, the peak of emission moves toward smaller wavelengths, and the color changes becoming nearly white at about 1,500 K.

Equation (12-1-14) made it possible to calculate the temperature of the sun. Solar radiation was resolved, and it was observed that the maximum emissive power corresponds to a wavelength of 0.25 μm. Then, by application of Eq. (12-1-14), it can be concluded that the temperature of the sun is 6,366 K.[1]

12-1-7 Emission from Real Surfaces

Any real surface (not black) emits energy with a spectral distribution different from that of the blackbody. With reference to Fig. 12-5, if at a certain temperature curve b represents the emissive power of the blackbody and curve a represents that of any other body at the same temperature, we can define the monochromatic emissivity of that body at the wavelength and temperature considered as

$$\varepsilon_\lambda = \frac{W_\lambda}{W_{b\lambda}} = f_{(\lambda, T)} \qquad (12\text{-}1\text{-}15)$$

The last part of the equation merely means that the monochromatic emissivity is a function of the wavelength and temperature.

While the total emissivity of the body at the considered temperature is

$$\varepsilon = \frac{W}{W_b} = \frac{\int_0^\infty W_\lambda \, d\lambda}{\int_0^\infty W_{b\lambda} \, d\lambda} = \frac{\int_0^\infty W_{b\lambda} \varepsilon_\lambda \, d\lambda}{W_b} = f_{(T)} \qquad (12\text{-}1\text{-}16)$$

since usually ε_λ is not a constant for all wavelengths, when dealing with real bodies, not only is the amount of energy emitted at a certain temperature smaller to that of the blackbody, but also the spectral distribution is different from that of the blackbody.

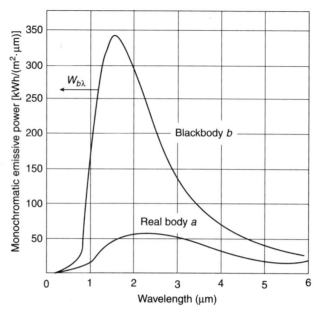

FIGURE 12-5 Emission spectrum of real surfaces.

12-1-8 Gray Bodies

Gray bodies are bodies whose surfaces present a constant emissivity for any wavelength and any temperature. In this case, the curve representing the emissive power is similar to that of the blackbody, with the difference being that all the ordinates are reduced by a constant factor, which is the emissivity of the gray surface (Fig. 12-6).

For a gray surface, the total emissivity becomes

$$\varepsilon = \frac{\int_0^\infty W_{b\lambda}\varepsilon_\lambda\, d\lambda}{\int_0^\infty W_{b\lambda}\, d\lambda} = \varepsilon_\lambda = \text{constant}$$

Real surfaces do not behave as gray surfaces, but the concept of gray body greatly simplifies the problems of radiant heat transfer and is an approximation used frequently in technical calculations.

12-1-9 Monochromatic Absorption Coefficient

When a surface absorbs radiant energy, it preferably exhibits a higher capacity to absorb radiation from certain wavelengths. We shall define the monochromatic absorption coefficient for a specific wavelength as the fraction of the total incident monochromatic energy that is absorbed by the surface. This coefficient is denoted as α_λ. Then, if the monochromatic energy of a certain wavelength that is incident on a surface per unit area and per unit time is I_λ, we get

$$\alpha_\lambda = \frac{\text{energy absorbed}}{I_\lambda} = f(\lambda, T)$$

This monochromatic absorption coefficient is a function of the temperature of the surface and the wavelength under consideration.

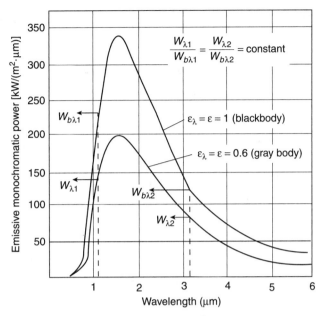

FIGURE 12-6   Emission spectrum of gray surfaces.

The total absorption coefficient defined in Sec. 12-1-2 is the fraction of the energy that is absorbed with respect to the total incident energy of all wavelengths. Then

$$\alpha = \frac{\int_0^\infty \alpha_\lambda I_\lambda \, d\lambda}{\int_0^\infty I_\lambda \, d\lambda} \tag{12-1-17}$$

This coefficient is a function of the temperature of the body but also depends on the distribution of the incident radiation $I_\lambda = f(\lambda)$, as can be seen in Eq. (12-1-17).

According to Kirchhoff's law, when a body absorbs energy from a cavity in equilibrium with it, $\varepsilon = \alpha$. Since a big cavity behaves as a blackbody thermally, we can say that Kirchhoff's law expresses the equality between the emissivity at a certain temperature and the absorption coefficient for the radiation emitted by a blackbody at the same temperature as that of the considered surface. It then follows that the equality between emissivity and absorption coefficient requires that the incident radiation have the same spectral distribution as that of the blackbody. In these conditions, the absorption coefficient is

$$\alpha = \frac{\int_0^\infty \alpha_\lambda I_{b\lambda} \, d\lambda}{\int_0^\infty I_{b\lambda} \, d\lambda} \tag{12-1-18}$$

where the energy distribution $I_b\lambda$ corresponds to the black body.

If the source with which the body exchanges energy is a gray body, the absorption coefficient wil be the same, since

$$\alpha = \frac{\int_0^\infty \alpha_\lambda (\varepsilon I_{b\lambda}) \, d\lambda}{\int_0^\infty (\varepsilon I_{b\lambda}) \, d\lambda} \tag{12-1-19}$$

where ε (= constant) is the emissivity of the source. Then the absorption coefficient of a real surface is the same, whether the surface absorbs energy from a black source or from a gray source at equal temperatures.

12-1-10 Generalization of Kirchhoff's Law for Gray Surfaces

We shall demonstrate that the absorption coefficient of a gray body is constant and independent of the nature and temperature of the source from which it is absorbing energy. Let's consider a gray body absorbing energy coming from a black source in equilibrium with it. Under these conditions, $\alpha = \varepsilon$. This is true whatever the temperature is.

However, since ε remains constant when the temperature changes (definition of gray surface), it follows that α also must be constant. Thus a gray body absorbing energy from a black source exhibits the same absorption coefficient for any temperature of the source. However, since changing the temperature of the source also changes the spectral distribution of the incident energy, if the gray body remains insensible to this change, we can conclude that the gray surface has a constant monochromatic absorption coefficient.

If α_λ is constant, the total absorption coefficient of a gray body for the radiation coming from any source, defined by Eq. (12-1-17), also will be constant and independent of the spectral distribution of the incident energy or, which is the same, independent of the temperature and nature of the source.

We can see that the gray-body concept results in a simplification much more powerful than that suggested by the initial definition. A gray body thus is a body whose emissivity is constant and equal to its absorption coefficient whatever the temperatures of the body and the source are and whether or not thermal equilibrium exists between them.

12-1-11 Intensity of Radiation

The intensity of radiation or intensity of emission is a magnitude used to indicate the total amount of radiation leaving a surface that propagates in a certain direction. Figure 12-7 represents an element of area dA belonging to an energy-emitting body surrounded by a hemisphere with radius r constructed around the radiating surface. All the energy emitted by dA will reach the hemisphere, whose area is $2\pi r^2$. Each

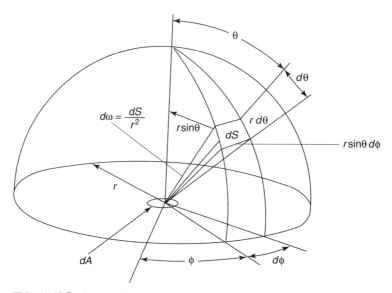

FIGURE 12-7 Geometry for the definition of the intensity of radiation.

element of area belonging to the hemisphere subtends a solid angle $d\omega$ with center at dA. The unit of the solid angle is the steradian. In the same way as the ratio between the length of an elemental arc belonging to a circumference and the radius of the circumference is a measure of the subtended angle in radians; the measure of a solid angle in steradians is the ratio between the differential area belonging to the hemisphere that subtends the angle and the square of the radius. In the hemisphere surrounding dA, there are therefore 2π steradians.

We can see in the figure that if θ is the zenith angle and ϕ is the azimuthal angle, the area of the surface element dS is

$$dS = (r\sin\theta\, d\phi)(r d\theta) \tag{12-1-20}$$

and

$$d\omega = \sin\theta\, d\phi\, d\theta \tag{12-1-21}$$

The intensity of radiation emitted by the element of area dA in a certain direction is the energy emitted in that direction per unit time, per unit solid angle, and per unit area of emitting surface projected onto a plane normal to the direction of emission. Thus

$$i = \frac{dQ}{d\omega(dA\cos\theta)} \tag{12-1-22}$$

where dQ = energy emitted per unit time within the solid angle $d\omega$
    $d\omega$ = element of solid angle
    dA = element of emitting surface
    θ = angle formed by the direction of emission and the normal to dA (Fig. 12-8)

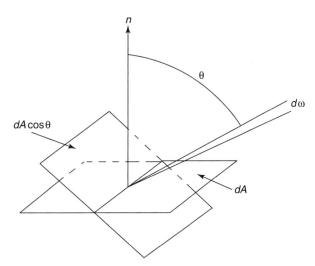

FIGURE 12-8   Nomenclature for Eq. (12-1-22).

The reason to include $\cos\theta$ is because it results in simpler final expressions. This is due to the fact that for the case of a blackbody, the intensity of radiation thus defined is independent of the direction of emission, as will be explained below. For the case of real surfaces, the intensity of radiation will be a function of θ.

The total energy emitted by the surface per unit time and per unit area is the emissive power, which, according to Eq. (12-1-22), is

$$W = \frac{1}{dA} \int dQ = \int i \cos\theta \, d\omega = \int_{\phi=0}^{\phi=2\pi} \int_{\theta=0}^{\theta=\pi/2} i \cos\theta \, \sin\theta \, d\phi \, d\theta \qquad (12\text{-}1\text{-}23)$$

12-1-12   Monochromatic Intensity of Radiation

It is also possible to define the monochromatic intensity of radiation as the energy emitted per unit time, per unit projected area, per unit solid angle, and per unit wavelength interval. The relationship between the total intensity of radiation defined in the preceding section and the monochromatic intensity of radiation is the same as that which exists between the total emissive power and the monochromatic emissive power. This means that it can be interpreted that the total intensity of radiation corresponds to the integration of the monochromatic intensity of radiation over the entire range of wavelengths.

The monochromatic intensity of radiation depends on the characteristics of the emitting surface, on its temperature, on the direction of emission, and on the wavelength. This is

$$i_\lambda = f_{(\lambda,\theta,\phi,T)} \qquad (12\text{-}1\text{-}24)$$

whereas the total intensity of radiation is

$$i = \int_0^\infty i_\lambda \, d\lambda = f_{(T,\theta,\phi)} \qquad (12\text{-}1\text{-}25)$$

We shall explain in what follows that for the case of a blackbody, the intensity of radiation depends only on the temperature and not on the direction of emission.

12-1-13   Spacial Distribution of a Blackbody Emission

Let's assume that we have two elements of area dA_1 and dA_2, both being black surfaces and in thermal equilibrium. Let dA_2 belong to a spherical surface whose center is occupied by dA_1 (Fig. 12-9). If both

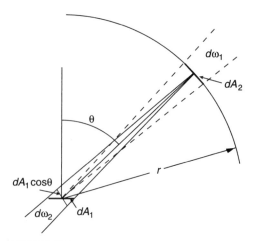

FIGURE 12-9   Heat exchange between two elements of area belonging to blackbodies. dA_1 is located at the center of a sphere to which dA_2 belongs.

elements are in thermal equilibrium, the amount of energy that each element absorbs from the other must be the same. However, if both surfaces are black, the energy that each element absorbs is the total that is incident on it. The energy emitted by dA_1 that strikes dA_2 and is absorbed; thus can be expressed as

$$dQ_{1\to2} = i_\theta dA_1 \cos\theta\, d\omega_1 \qquad (12\text{-}1\text{-}26)$$

where i_θ is the intensity of radiation emitted by dA_1 in the direction θ.

The solid angle is

$$d\omega_1 = \frac{dA_2}{r^2} \qquad (12\text{-}1\text{-}27)$$

Then

$$dQ_{1\to2} = \frac{i_\theta dA_1 dA_2 \cos\theta}{r^2} \qquad (12\text{-}1\text{-}28)$$

In turn, the energy emitted by dA_2 and reaching dA_1 is

$$dQ_{2\to1} = i_n dA_2 d\omega_2 \qquad (12\text{-}1\text{-}29)$$

where i_n is the intensity of radiation emitted by dA_2 in the direction connecting both elements of area, which is the direction normal to dA_2 [for the same reason, the cosine does not appear in Eq. (12-1-29)]. But $d\omega_2$ can be obtained as

$$d\omega_2 = \frac{dA_1 \cos\theta}{r^2} \qquad (12\text{-}1\text{-}30)$$

(It must be remembered that to obtain the value of the solid angle, we must imagine a hemispherical surface with its center at dA_2, intersect it with the solid angle, and divide the intersection area by the square of the radius.)

Then

$$dQ_{2\to1} = \frac{i_n dA_1 dA_2 \cos\theta}{r^2} \qquad (12\text{-}1\text{-}31)$$

Since thermal equilibrium exists, $dQ_{1\to2} = dQ_{2\to1}$, and it can be deduced from Eqs. (12-1-28) and (12-1-31) that

$$i_\theta = i_n \qquad (12\text{-}1\text{-}32)$$

This means that the intensity of radiation emitted by a blackbody in the normal direction n is the same as that emitted in an arbitrary direction θ. In other words, the intensity of radiation from a blackbody depends only on its temperature and not on the direction of emission.

12-1-14 Directional Emissive Power: Cosine Law—Real Surface Emission Distribution

It is possible to define a directional emissive power as the amount of energy emitted per unit time, per unit solid angle, and per unit area of emitting surface (not projected). The directional emissive power, denoted as E_θ, is related to the intensity of radiation by

$$E_\theta = \frac{dQ}{dA\,d\omega} = i_\theta \cos\theta \qquad (12\text{-}1\text{-}33)$$

For a blackbody, since i_θ is independent of θ, we get

$$E_{b\theta} = i_b \cos\theta \qquad (12\text{-}1\text{-}34)$$

where i_b depends only on the temperature.

This expression is known as *Lambert's law* or the *cosine law*, and the surfaces that follow it are called lambertians. Usually, real surfaces do not follow this law, and the intensity of emitted radiation varies with the direction. Figure 12-10 is a plot in polar coordinates showing the characteristics of directional emission for several surfaces.

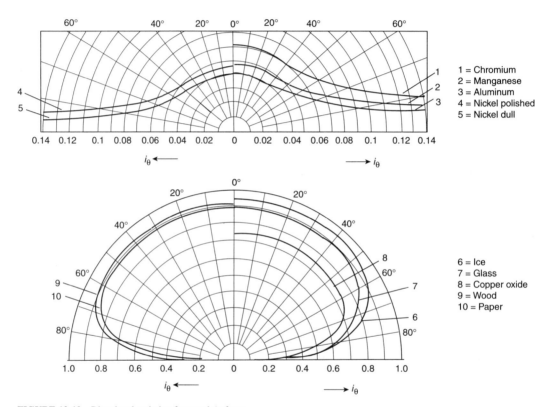

1 = Chromium
2 = Manganese
3 = Aluminum
4 = Nickel polished
5 = Nickel dull

6 = Ice
7 = Glass
8 = Copper oxide
9 = Wood
10 = Paper

FIGURE 12-10   Directional emission from real surfaces.

Usually, for electric conductors, the intensity of emission is relatively constant for small values of θ (angle between the normal to the surface and the direction of the radiant beam) and then increase when viewed from wider angles.[7] For electric nonconductors, the intensity shows a maximum in the direction normal to the surface, and this value is practically constant up to $\theta = \pi/4$. For higher values, the intensity decreases, and it is nil for $\theta = \pi/2$.

12-1-15   Intensity of Radiation and Emissive Power of a Blackbody

Since in the case of a blackbody the intensity of emission does not depend on θ, Eq. (12-1-23) can be easily integrated to find the total emissive power:

$$W_b = \int_{\phi=0}^{\phi=2\pi} \int_{\theta=0}^{\theta=\pi/2} i_b \cos\theta \sin\theta \, d\theta \, d\phi = 2\pi i_b \int_0^{\pi/2} \sin\theta\cos\theta \, d\theta$$

$$= 2\pi i_b \int_0^1 1/2 \, d(\sin^2\theta) \tag{12-1-35}$$

$$\therefore W_b = \pi i_b$$

This expression relates the emissive power to the intensity of emission for a blackbody.

12-1-16 Radiant Heat Transfer between Blackbodies: View Factor

Let's consider two black surfaces such that part of the radiation emitted by each one can reach the other (Fig. 12-11). Let these surfaces be A_1 and A_2. On each surface we can consider an element of area. Let them be dA_1 and dA_2, respectively. The radiant beams that leave each element of area and are intercepted by the other form angles θ_1 and θ_2 with the respective normals.

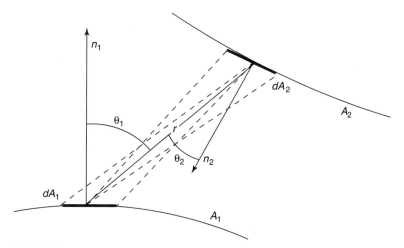

FIGURE 12-11   Heat exchange between two area elements of blackbodies.

The heat flow leaving dA_1 and reaching dA_2 is

$$dQ_{1\to2} = i_{b_1} dA_1 \cos\theta_1 \frac{dA_2 \cos\theta_2}{r^2} \tag{12-1-36}$$

We wrote directly i_{b1} without any other subscript because in the case of blackbodies it is not a function of the angle. Subscript 1 denotes the temperature T_1.

By integrating this expression over both surfaces, we obtain the total heat flow leaving A_1 and striking A_2. Since A_2 is a blackbody, this energy is absorbed completely. Then

$$Q_{1\to2} = \int_{A_1} \int_{A_2} \frac{W_{b_1} \cos\theta_1 \cos\theta_2}{\pi r^2} dA_1 dA_2 \tag{12-1-37}$$

The total energy that leaves A_1 and expands through all the half-space is

$$Q_{1\to\infty} = W_{b_1} A_1 \tag{12-1-38}$$

It is possible to define a magnitude F_{12} as

$$Q_{1\to2} = W_{b_1} A_1 F_{12} \qquad (12\text{-}1\text{-}39)$$

where F_{12} is called the *view factor* between A_1 and A_2 evaluated on the basis of area A_1. It represents the fraction of the energy leaving A_1 and striking A_2 with respect to the total energy leaving A_1. (Some authors prefer the expression *shape factor* rather than *view factor*.) In this particular case, since both bodies are black, the energy that reaches A_2 is absorbed completely.

Now we can perform the same analysis for the heat flow from A_2 to A_1:

$$dQ_{2\to1} = i_{b_2} dA_2 \cos\theta_2 \frac{dA_1 \cos\theta_1}{r^2} \qquad (12\text{-}1\text{-}40)$$

Integrating and defining a view factor for the heat flow from A_2 to A_1, we get

$$Q_{2\to1} = W_{b_2} \int_{A_1} \int_{A_2} \frac{\cos\theta_1 \cos\theta_2}{\pi r^2} dA_1 dA_2 = W_{b_2} A_2 F_{21} \qquad (12\text{-}1\text{-}41)$$

The total heat flux leaving A_2 is

$$Q_{2\to\infty} = W_{b_2} A_2 \qquad (12\text{-}1\text{-}42)$$

Equation (12-1-41) defines the view factor F_{21} as the fraction of the energy emitted by A_2 that reaches A_1. It is called the *view factor* between A_1 and A_2 evaluated on the basis of area A_2. Comparing Eqs. (12-1-37), (12-1-39), and (12-1-41), we see that

$$A_1 F_{12} = A_2 F_{21} = \int_{A_1} \int_{A_2} \frac{dA_1 dA_2 \cos\theta_1 \cos\theta_2}{\pi r^2} \qquad (12\text{-}1\text{-}43)$$

Then the net rate of radiant heat transfer between A_1 and A_2 is

$$Q_{1\rightleftarrows2} = W_{b_1} A_1 F_{12} - W_{b_2} A_2 F_{21} \qquad (12\text{-}1\text{-}44)$$

or

$$Q_{1\rightleftarrows2} = A_1 F_{12}(W_{b_1} - W_{b_2}) = A_1 F_{12}\sigma(T_1^4 - T_2^4) \qquad (12\text{-}1\text{-}45)$$

It can be noted that if both surfaces are at the same temperature, the net rate of heat transfer is obviously nil.

The integrals defining the view factors, of the type of Eq. (12-1-43), have been calculated for the most common geometries, and they can be found in the references in the form of graphs or tables.[6] Figures 12-12, 12-13, and 12-14 are graphs of the view factors between parallel planes, between a differential element of area and a parallel plane, and between perpendicular planes. Figure 12-15 is a graph of the view factor between a plane and a bank of tubes parallel to it. Since the view factor F_{12} represents the fraction of the energy emitted by surface A_1 that strikes surface A_2, it is evident that the sum of the vision factors of a body with respect to all other bodies in the universe is unity.

Example 12-1   Calculate the net rate of radiant heat transfer between two black surfaces consisting of two parallel 1×2 m rectangles separated by 1 m and at 800 and 300 K, respectively.

Solution

$$Q_{1\rightleftarrows2} = A_1 F_{12}(W_{b_1} - W_{b_2}) = A_1 F_{12}\sigma\left(T_1^4 - T_2^4\right)$$

F_{12} is obtained from the graph in Fig. 12-12. In curve 3, we get $F_{12} = 0.3$.

$$Q_{1\rightleftarrows2} = 2 \times 0.3 \times 5.672 \times 10^{-8} \times (800^4 - 300^4) = 13{,}663 \text{ W}$$

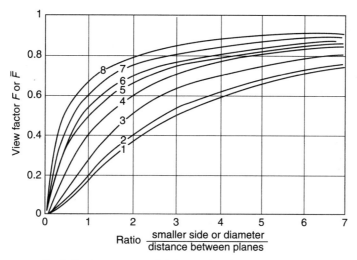

FIGURE 12-12 View factors for parallel squares, rectangles, and disks. (*From Hottel HC: in McAdams WH (ed): Heat Transmission. New York: McGraw-Hill, 1954. Reproduced with permission of the editors.*)

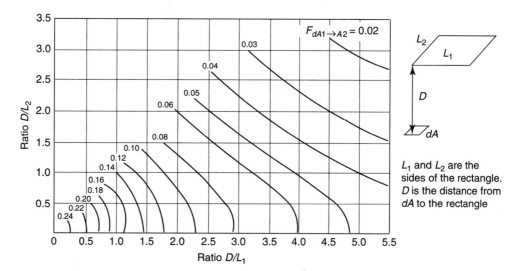

FIGURE 12-13 View factor for a differential area located below one corner of a finite rectangular plane $F_{dA1 \to A2}$. (*From Hottel HC: in McAdams WH (ed): Heat Transmission. New York: McGraw-Hill, 1954. Reproduced with permission of the editors.*)

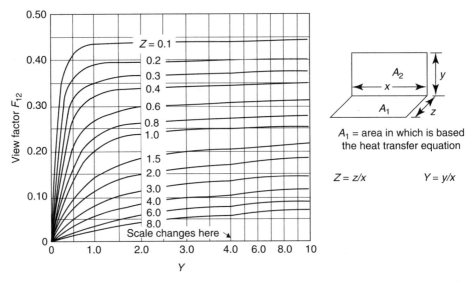

FIGURE 12-14 View factor for perpendicular planes with a common edge. (*From Hottel HC: in McAdams WH (ed): Heat Transmission. New York: McGraw-Hill, 1954. Reproduced with permission of the editors.*)

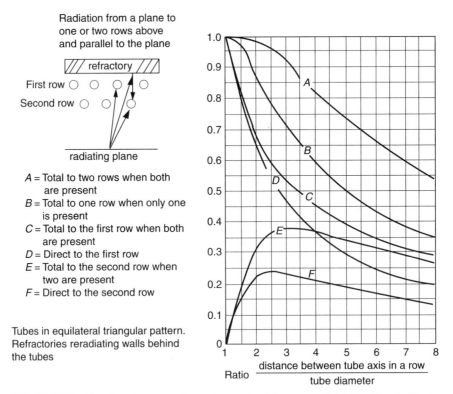

FIGURE 12-15 View factor between a plane and one or two tube rows parallel to it, with and without a refractory wall. (*From Hottel HC: in McAdams WH (ed): Heat Transmission. New York: McGraw-Hill, 1954. Reproduced with permission of the editors.*)

12-1-17 View-Factor Algebra

It is sometimes possible to calculate the view factor for a particular geometric arrangement by substracting or adding other known view factors corresponding to simpler arrangements. This procedure is known as *view-factor algebra*. It is illustrated in the following examples.

Example 12-2   A 0.01-m² radiant heater at 1,000 K, oriented as shown in Fig. 12-16, is used to heat a floor 6 m wide and 8.5 m long. Assuming that the surface of the floor behaves as a blackbody, calculate the heat flux that it receives from the source.

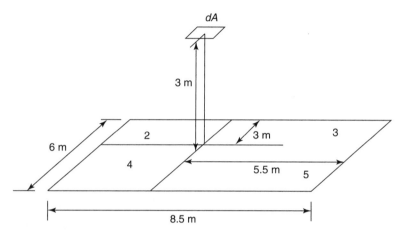

FIGURE 12-16   Figure for Example 12-2.

Solution   Let 1 designate the heater. Owing to its small size, it can be considered as a differential surface element when compared with the size of the floor. Let the floor be divided into surfaces 2, 3, 4, and 5, as shown in the figure. Then

$$Q_{1\rightleftarrows(2+3+4+5)} = Q_{1\rightleftarrows2} + Q_{1\rightleftarrows3} + Q_{1\rightleftarrows4} + Q_{1\rightleftarrows5}$$

$$Q_{1\rightleftarrows2} = A_1 F_{12} W_{b_1}$$

$$Q_{1\rightleftarrows3} = A_1 F_{13} W_{b_1}$$

$$Q_{1\rightleftarrows4} = A_1 F_{14} W_{b_1}$$

$$Q_{1\rightleftarrows5} = A_1 F_{15} W_{b_1}$$

$$Q_{1\rightleftarrows\text{floor}} = A_1 W_{b_1} (F_{12} + F_{13} + F_{14} + F_{15})$$

Observe that

$$F_{1\rightleftarrows\text{floor}} = \sum F_{1\rightleftarrows i}$$

From Fig. 12-13, we get $F_{12} = F_{14} = 0.14$ and $F_{13} = F_{15} = 0.164$. Thus

$$Q_{1 \rightleftarrows \text{floor}} = 0.01 \times 5.672 \times 10^{-8} \times 1,000^4 \times (0.14 + 0.164 + 0.14 + 0.164) = 345 \text{ W}$$

It was assumed that the heating element is also a blackbody.

Example 12-3   Calculate the view factors between surfaces A_1 and A_3 in Fig. 12-17.

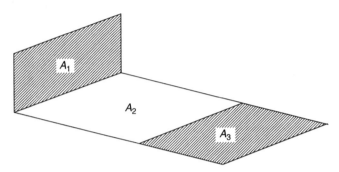

FIGURE 12-17   Figure for Example 12-3.

Solution   We can write

$$F_{1(2,3)} = F_{12} + F_{13}$$

Factors $F_{1(2,3)}$ and F_{12} can be obtained from Fig. 12-14, and then F_{13} can be easily calculated by difference.

12-1-18   Electric Analogy

Equation (12-1-45) can be interpreted as the equation describing an electric circuit where there are two opposite electromotive forces (W_{b1} and W_{b2}) and a resistance $1/A_1 F_{12}$. The electric current in the circuit represents the heat flux $Q_{1 \leftrightarrow 2}$ (Fig. 12-18). In this simple case, this electric analogy does not look particularly useful, but its advantage will be appreciated when we deal with more complex systems.

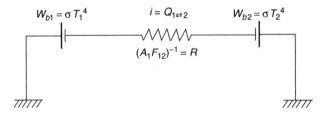

FIGURE 12-18   Electric analogy for the heat exchange between two blackbodies.

12-1-19   Radiant Heat Exchange among Three Blackbodies

Let us consider a system consisting of three blackbodies exchanging heat among themselves (Fig. 12-19). The energy exchanged between 1 and 2 can be expressed as

$$Q_{1 \rightleftarrows 2} = A_1 F_{12}(W_{b_1} - W_{b_2}) \tag{12-1-46}$$

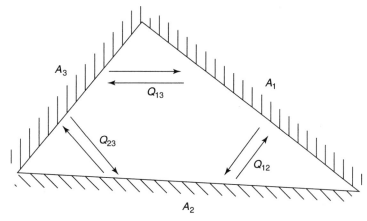

FIGURE 12-19   Heat exchange among three blackbodies.

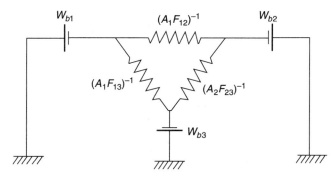

FIGURE 12-20   Electric analogy for the heat exchange among three black-bodies.

The energy exchanged between 2 and 3 is

$$Q_{2\rightleftarrows3} = A_2 F_{23}(W_{b_2} - W_{b_3})$$

(12-1-47)

And the energy exchanged between 1 and 3 is

$$Q_{1\rightleftarrows3} = A_1 F_{13}(W_{b_1} - W_{b_3})$$

(12-1-48)

Each one of these equations represents a loop of the electric network of Fig. 12-20. The current through each resistor represents one of the binary heat fluxes of Eqs. (12-1-46) through (12-1-48).

The amount of energy that body 1 is exchanging with the rest of the system is the algebraic sum of $Q_{1\leftrightarrow2}$ and $Q_{1\leftrightarrow3}$. This value is represented in the circuit by the current circulating through the electromotive force W_{b1}. Depending on the direction of this current, body 1 would be delivering or absorbing energy to or from the system. The same can be said about surfaces A_2 and A_3.

12-1-20   Insulating Walls or Refractories

Let us consider the system represented in Fig. 12-21, where surface A_1 is a heat source, A_2 is a heat sink, and the A_3's are walls insulated from the exterior. For example, the figure can be a representation of an

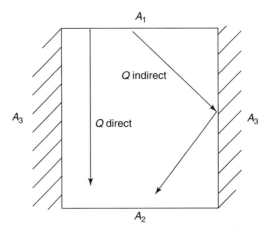

FIGURE 12-21 Heat exchange between two bodies in the presence of refractory reradiating walls.

electric oven where A_1 is the surface of the heating electrical resistor, A_2 is the object that is being heated, and the A_3's are the oven walls. Since the walls A_3 are insulated, it is not possible for energy to be delivered to or withdrawn from the system through them. This is normally the case with refractory walls of ovens and furnaces, so we shall refer to this type of surface either as *insulating walls* or *refractories.*

When the system reaches its steady state, the energy emitted by A_1 must be totally absorbed by A_2. (Note that for this system to reach a steady state it is necessary that energy be withdrawn continuously through A_2. For example, A_2 can be the surface of a coil internally refrigerated by a circulating fluid.)

In a system such as this, with a hot source and a cold energy receptor, the insulating walls necessarily must reach an intermediate temperature between the source and the receptor. Thus there must be an energy exchange between the walls and surfaces A_1 and A_2, and owing to the adiabatic character of the walls, the energy absorbed by them must be equal to the energy they emit.

From the total energy that is incident on A_3, a portion is reflected and a portion is absorbed. (A_3 is not necessarily a black surface.) But all the energy absorbed is emitted again toward the rest of the system. This behavior is similar to that of a blackbody at a temperature such that its emissive power is equal to the energy that is incident on it.

We then can represent this system by an electric network such as that in Fig. 12-20, to which we shall impose the additional condition that $Q_{1\leftrightarrow3} = Q_{3\leftrightarrow2}$. In this case, there is no current circulation through the branch containing W_{b3}, so it can be eliminated from the network without modification of the current distribution in the rest of the circuit. The resulting network is shown in Fig. 12-22. The interpretation of the circuit is as follows: From the total energy leaving A_1, a portion strikes A_2 directly. This fraction is the current circulating through the upper branch and is given by

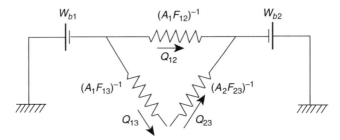

FIGURE 12-22 Electric circuit representing the heat exchange between two blackbodies in the presence of refractory reradiating walls.

$$Q_{1 \rightleftarrows 2} = A_1 F_{12}(W_{b_1} - W_{b_2})$$ (12-1-49)

where F_{12} is the view factor representing the direct vision between the two surfaces.

Another fraction of the energy leaving A_1 reaches the refractory wall, where it is absorbed and reemitted again to finally reach A_2. This energy is represented by the current in the lower branch. We see that the refractory offers an alternative path for the heat transmission between A_1 and A_2.

The triangle of resistances in Fig. 12-22 can be solved easily to find the total resistance (first adding the two resistances in series and then solving the parallel with the upper branch). The total resistance will be called $\left(A_1 \overline{F_{12}}\right)^{-1}$.

The equivalent circuit is that of Fig. 12-23, where the expression for the total resistance is

$$A_1 \overline{F_{12}} = A_1 F_{12} + \frac{A_1 F_{13} A_2 F_{23}}{A_1 F_{13} + A_2 F_{23}}$$ (12-1-50)

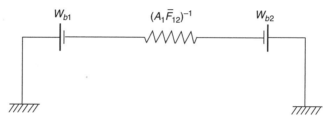

FIGURE 12-23   Equivalent resistance of the circuit in Fig. 12-22.

This means that the effective view factors $\overline{F}$ can be calculated from the binary view factors between the several surfaces of the system.

The graph in Fig. 12-15 shows curves representing the view factors $\overline{F_{12}}$ for the case of radiant heat transfer from an emitting planar surface to a bank of tubes, including, in some of the curves, the presence of refractory walls behind the tubes. Also, curves 5, 6, 7, and 8 in Fig. 12-12 represent the view factors $\overline{F_{12}}$ for parallel planes connected by reradiant refractory walls.

The main difference that exists between the circuits in Figs. 12-20 and 12-22 is that in the latter case, the refractory wall has an emissive power (electric potential) that "floats" between the emissive powers W_{b1} and W_{b2} at a level that depends on the magnitudes of W_{b1} and W_{b2} and on the thermal resistances participating in the circuit.

Usually a refractory wall is a real surface (not black). However, we explained that with the purpose of calculating the heat flux between A_1 and A_2, it is possible to replace the refractory wall with a blackbody having a fictitious temperature such that its emissive power is equal to the incident irradiation. In the particular case where the refractory is a blackbody, this is exactly the situation, and the temperature, determined by

$$T_3 = \sqrt[4]{\frac{W_{b_3}}{\sigma}}$$ (12-1-51)

coincides with the real temperature of the refractory wall. However, in a more general situation, the temperature determined by Eq. (12-1-51) will be the ficitious temperature we mentioned, and it will not coincide with the real temperature of the refractories.

12-1-21   Case of Several Insulating Walls

For Eq. (12-1-50) to be valid, it is necessary that all the insulating walls can be represented by a single node in the circuit and have the same electrical potential. This requires the hypothesis that all the refractory walls are at the same temperature. If the geometry of the systems is such that we presume there will be areas at different temperatures, they must be treated as different surfaces.

For example, let's consider the case of heat transfer between two surfaces A_1 and A_2 connected by refractory walls A_S and A_R, as shown in Fig. 12-24. In this case, the equivalent electric circuit is that shown in the right side of the figure which also can be solved to find the effective view factor $\overline{F}_{12}$.

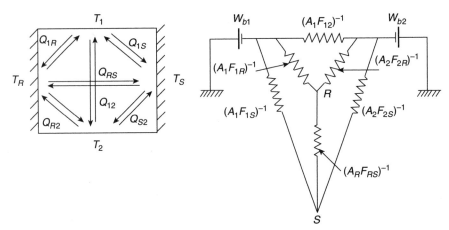

FIGURE 12-24 Heat transfer between two blackbodies with two reradiating refractory walls.

As a general case, where a higher number of refractory surfaces are involved, the system can be solved as a quotient of determinants of the following form (see ref. 2):

$$A_1\overline{F}_{12} = \frac{\begin{vmatrix} A_1F_{12} & A_2F_{2R} & A_2F_{2S} & \cdots \\ A_1F_{1R} & A_RF_{RR}-A_R & A_RF_{RS} & \cdots \\ A_1F_{1S} & A_RF_{RS} & A_SF_{SS}-A_S & \cdots \\ \cdots & \cdots & \cdots & \cdots \end{vmatrix}}{\begin{vmatrix} A_RF_{RR}-AR & A_RF_{RS} & \cdots \\ A_RF_{RS} & A_SF_{SS}-A_S & \cdots \\ \cdots & \cdots & \cdots \end{vmatrix}} \tag{12-1-52}$$

F_{RR} is the view factor of the wall R with respect to itself (concave surfaces can intercept part of the radiation emitted by themselves). All the view factors of a body are related by

$$F_{11}+F_{12}+F_{1i}=1 \tag{12-1-53}$$

If we apply the general solution to the more simple case of a single refractory wall, we get

$$A_1\overline{F}_{12} = \frac{\begin{pmatrix} A_1F_{12} & A_2F_{2R} \\ A_1F_{1R} & A_RF_{RR}-A_R \end{pmatrix}}{A_RF_{RR}-A_R} = A_1F_{12}-\frac{A_1A_2F_{1R}F_{2R}}{A_RF_{RR}-A_R} \tag{12-1-54}$$

Additionally, the summation of F_{Ri} values equals 1; thus

$$(F_{R1}+F_{R2}+F_{RR})A_R=A_R \tag{12-1-55}$$

Thus

$$A_R F_{RR} - A_R = -(A_R F_{R1} + A_R F_{R2})$$ (12-1-56)

Substituting in the solution of the determinant

$$A_1 \overline{F}_{12} = A_1 F_{12} + \frac{A_1 F_{1R} A_2 F_{2R}}{A_1 F_{1R} + A_2 F_{2R}}$$ (12-1-57)

which is the same expression we get with the equivalent electric circuit.

12-1-22   Enclosures with Gray Surfaces

In all the cases considered up to now, all the energy striking a surface was absorbed completely because we considered only black surfaces. In the case of real surfaces, not all the incident energy is absorbed because it is partly reflected. The reflected energy can come back to the emitter or to other bodies in the universe according to the view factors of the reflecting surface. This energy, in turn, is partly reflected by those other bodies. Thus the steady state is reached through a process of successive reflections and absorptions in which the energy is being transmitted among the different bodies participating in the system.

The solution of this type of system is extremely complex, but it is possible to simplify the mathematical treatment by assuming that the bodies behave as gray surfaces. To solve the problem, we must perform the following analysis: Let's consider a gray body that is exchanging energy with the ambient. Since it is not a blackbody, it will reflect part of the incident energy. Then the radiation leaving its surface per unit time and per unit area, which is called the *radiosity J,* is the sum of the radiation emitted and reflected (Fig. 12-25).

If I is the irradiation to which this surface is submitted (density of incident energy), the absorbed fraction of it is α. The reflected energy is

$$I(1-\alpha) = I\rho$$

Then

$$J = W + I\rho$$ (12-1-58)

where W is the emissive power, given by

$$W = W_b \varepsilon$$ (12-1-59)

From Eq. (12-1-58), we get

$$I = \frac{J - W}{\rho}$$ (12-1-60)

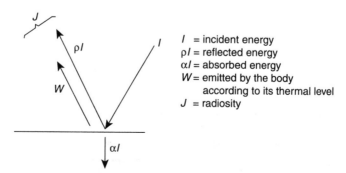

I = incident energy
ρI = reflected energy
αI = absorbed energy
W = emitted by the body
    according to its thermal level
J = radiosity

FIGURE 12-25   Behavior of a gray surface.

The net energy flux at the body surface (heat exchanged with the exterior) is

$$Q_{1\rightleftarrows\text{ext}} = A_1(J - I) \tag{12-1-61}$$

Substituting I, the result is

$$Q_{1\rightleftarrows\text{ext}} = A_1\left(J - \frac{J - W}{\rho}\right) \tag{12-1-62}$$

Considering Eq. (12-1-59), and since $(1 - \rho) = \varepsilon$ because the body is gray and opaque, the result is

$$Q_{1\rightleftarrows\text{ext}} = \frac{A_1\varepsilon}{1-\varepsilon}(W_b - J) \tag{12-1-63}$$

If the surface were a blackbody, the energy flux emitted per unit area would be W_b.
The energy flux per unit area coming from the gray surface is J, which is related to W_b by

$$W_b - J = Q_{1\rightleftarrows\text{ext}}\frac{1-\varepsilon}{A_1\varepsilon} \tag{12-1-64}$$

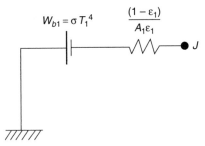

FIGURE 12-26   Representation of a gray surface.

We see that the higher is the energy exchange between the surface and the rest of the system, the higher is the difference between W_b and J. If we analyze this from the point of view of the electric analogy, we could think that the gray body acts as an electromotive force W_b, but with an internal resistance that provokes a voltage drop, so the electric potential for an external observer is J (see Fig. 12-26). We shall call this resistance *gray-to-black transformation resistance*.

Let's consider the case of two gray surfaces with areas A_1 and A_2 at temperatures T_1 and T_2 and with emissivities ε_1 and ε_2. Let's transform these bodies from gray to black. The blackbody emissive powers at these temperatures are

$$W_{b_1} = \sigma T_1^4 \quad \text{and} \quad W_{b_2} = \sigma T_2^4 \tag{12-1-65}$$

We then add the gray-to-black transformation resistances.
From the side of body A_1 we have

$$\frac{1-\varepsilon_1}{A_1\varepsilon_1} \tag{12-1-66}$$

And from the side of body A_2 we have

$$\frac{1-\varepsilon_2}{A_2\varepsilon_2} \tag{12-1-67}$$

With the addition of these resistances, the blackbody emissive powers are transformed into the radiosities J_1 and J_2 (Fig. 12-27). From here on, we have the same circuits that we analyzed earlier for the case of black surfaces. We shall consider some examples:

1. Heat exchange between gray bodies in a vacuum or surrounded by nonabsorbing gases (Fig. 12-28), where F_{12} is the view factor, as before

2. Heat exchange between gray bodies surrounded by refractory walls at a uniform temperature (Fig. 12-29)

3. Heat exchange between two gray bodies in the presence of refractory walls at different temperatures (in this case, two walls) (Fig. 12-30)

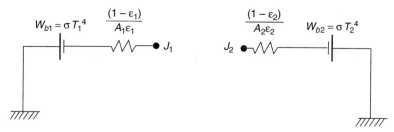

FIGURE 12-27 Radiosities of two gray surfaces.

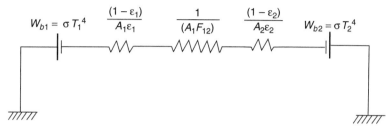

FIGURE 12-28 Electric circuit representing heat exchange between two gray bodies.

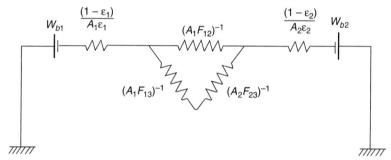

FIGURE 12-29 Electric circuit representing heat exchange between two gray bodies in the presence of reradiating refractory walls at uniform temperature.

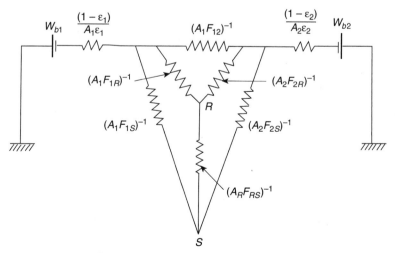

FIGURE 12-30 Electric circuit representing heat exchange between two gray bodies in the presence of reradiating refractory walls at two different temperatures.

In all these cases, the network is solved by means of a determinant, and finally, it reduces to the case shown in Fig. 12-31, where $\bar{F}_{12}$ designes the view factor including the effect of the refractory walls. The total resistance of the circuit shown in Fig. 12-31 will be called $1/A_1\mathcal{F}_{12}$, and the equivalent circuit is that shown in Fig. 12-32. Here, we made reference to A_1, but A_2 also can be used, and the resistance also may be called $1/A_2\mathcal{F}_{21}$.

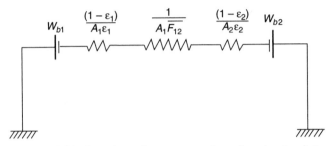

FIGURE 12-31 General case for two gray surfaces where the view factor includes the presence of refractory walls.

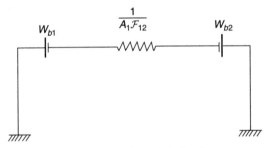

FIGURE 12-32 Total view factor including the presence of refractory walls and the effect of surface emissivities.

12-1-23 Heat Transfer between a Gray Body and a Body That Completely Surrounds It

Let's consider the case of a gray body at temperature T_1, completely surrounded by a second body at temperature T_2. The situation is that shown in Fig. 12-33. If the view factor of body 1 with respect to itself is nil, then $F_{12} = 1$, and with reference to Fig. 12-28, the total resistance of the circuit is

$$\frac{1}{A_1\mathcal{F}_{12}} = \frac{1-\varepsilon_1}{A_1\varepsilon_1} + \frac{1}{A_1} + \frac{1-\varepsilon_2}{A_2\varepsilon_2} \tag{12-1-68}$$

This means that

$$\frac{1}{\mathcal{F}_{12}} = \frac{1-\varepsilon_1}{\varepsilon_1} + 1 + \frac{1-\varepsilon_2}{\varepsilon_2}\frac{A_1}{A_2} \tag{12-1-69}$$

And if the quotient A_1/A_2 is small (e.g., in the case of a small body in an infinite ambient), Eq. (12-1-69) reduces to

$$\mathcal{F}_{12} = \varepsilon_1 \tag{12-1-70}$$

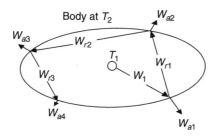

W_1: Energy emitted by A_1.
This energy strikes on A_2 and is partially absorbed (W_{a1}). The nonabsorbed fraction is reflected (W_{r1}). If area A_2 is much larger than A_1, the reflected fraction strikes again on other region of A_2, where another fraction of the energy is absorbed. Since there is a very low probability that this energy comes back to A_1, it is finaly completely absorbed in a multiple reflections process.
A_2 behaves as a blackbody since it absorbs all the energy it receives

FIGURE 12-33   Heat transfer between a gray body and another body that completely surrounds it.

Then the heat lost by body A_1 can be calculated as

$$Q = A_1 \varepsilon_1 (W_{b_1} - W_{b_2}) = A_1 \varepsilon_1 \sigma \left(T_1^4 - T_2^4 \right) \tag{12-1-71}$$

This expression was already used in previous chapters to calculate the radiant heat loss from a body to an infinite ambient. We were in debt for its justification. It is interesting to note that the emissivity ε_2 does not appear in Eq. (12-1-71). This is so because since we assume that A_2 has infinite dimensions, it behaves like a blackbody because all the energy received from A_1 is being absorbed in successive reflections until it is completely exhausted, with very little probability that part of this energy comes back to body A_1 (see Fig. 12-33).

Example 12-4   Two parallel plates 1 m wide and 1 m in length are separated by a distance of 0.5 m between them. One plate is at $T_1 = 1,000$ K, and the other is at $T_2 = 500$ K. Both plates are immersed in an ambient at 27°C (300 K) that can be considered infinite with respect to the dimensions of the plates (Fig. 12-34). Calculate the heat exchanged by each plate and by the ambient. Consider that both surfaces are gray, with emissivities $\varepsilon_1 = 0.5$ and $\varepsilon_2 = 0.3$.

Solution   The ambient can be treated as a third body, and the equivalent electric network is represented in the Fig. 12-34. The view factor F_{12} is obtained from Fig. 12-12 (curve 2). We get $F_{12} = F_{21} = 0.4$. Since each plate only has a view of the other plate and the ambient, the result is

$$F_{13} = 1 - F_{12} = 0.6$$
$$F_{23} = 1 - F_{21} = 0.6$$

And since A_1 and A_2 are unity, the resistances are

$$(A_1 F_{12})^{-1} = 1/0.4 = 2.5 \qquad (A_1 F_{13})^{-1} = (A_2 F_{23})^{-1} = 1/0.6 = 1.66$$

$$\frac{1-\varepsilon_1}{\varepsilon_1 A_1} = \frac{1-0.5}{0.5} = 1 \qquad \frac{1-\varepsilon_2}{\varepsilon_2 A_2} = \frac{1-0.3}{0.3} = 2.33$$

$$W_{b_1} = 5.672 \times 10^{-8} \times 1,000^4 = 56,720 \text{ W/m}^2$$

$$W_{b_2} = 5.672 \times 10^{-8} \times 500^4 = 3,570 \text{ W/m}^2$$

$$W_{b_3} = 5.672 \times 10^{-8} \times 300^4 = 459 \text{ W/m}^2$$

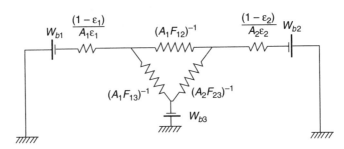

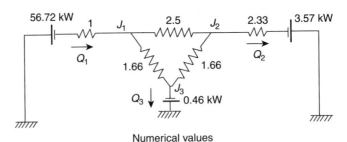

FIGURE 12-34   Figure for Example 12-4.

The electric network can be solved by any method used for electric circuit resolution. A simple way to solve it is to consider that the sum of all the currents (heat fluxes) coming to a node must be zero. Then

$$\frac{56.72 - J_1}{1} = \frac{J_1 - J_2}{2.5} + \frac{J_1 - 0.46}{1.66}$$

$$\frac{J_2 - 3.57}{2.33} = \frac{J_1 - J_2}{2.5} + \frac{0.46 - J_2}{1.66}$$

We have two equations with two unknowns. Solving, we get

$$J_1 = 30.45 \qquad \text{and} \qquad J_2 = 9.796$$

By knowing the electric potentials of these points, we can find the currents in the network. We get

$$Q_1 = 26.27 \text{ kW} \quad Q_2 = 2.67 \text{ kW} \quad Q_3 = 23.6 \text{ kW} \quad Q_{1-2} = 8.264 \text{ kW} \quad Q_{1-3} = 18.0 \text{ kW} \quad Q_{2-3} = 5.6 \text{ kW}$$

This means that most of the radiant energy emitted by both plates is transferred to the ambient. Only 2.67 kW from a total of 26.26 kW emmitted by plate 1 is absorbed by plate 2.

12-1-24   Radiant Shields

One possible way to reduce the radiant heat transfer between two surfaces is to interpose a screen or shield between them. We know from our personal experience that if we do not want to feel the effects of solar radiation, we must stay under a shadow or that if we interpose a screen between a quartz infrared heater and ourselves, we do not feel its radiation.

Let's consider the case of two infinite parallel planes at temperatures T_1 and T_2, one acting as a source and the other acting as a sink. Here again, the view factor $F_{12} = 1$, and then Eq. (12-1-68) is also valid. Let's consider now that we interpose between both planes a screen A_3. Since this screen can exchange energy only with the planes A_1 and A_2, it will finally reach steady state when the energy received from A_1 equals that transmitted to A_2.

The energy exchange between plane A_1 and plane A_3 can be represented by the left loop in Fig. 12-35, whereas the energy exchange between plane A_3 and plane A_2 corresponds to the right loop. Since $Q_{(1\leftrightarrow2)} = Q_{(2\leftrightarrow3)}$ (plane 3 does not supply or withdraw energy to or from the system), both branches can be joined and we get the equivalent circuit of Fig. 12-36.

FIGURE 12-35   Heat exchange between two parallel planes with a radiant screen separating them.

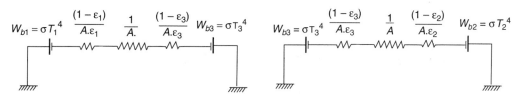

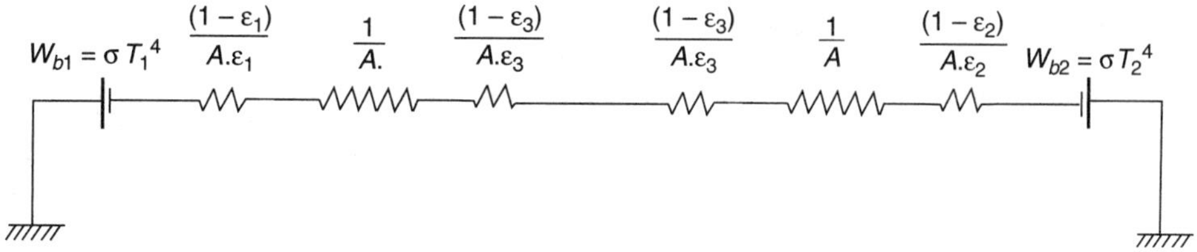

FIGURE 12-36   Simplification of the circuit in Fig. 12-35.

Example 12-5   Calculate the radiant heat-flux density exchanged between two black infinite planes at 900 and 400 K, respectively, and in what proportion the heat flux is reduced if a radiant screen with emissivity 0.4 is interposed between them.

Solution   The view factors for infinite parallel planes are unity. In the case of the two planes facing each other without a screen, the heat exchanged is

$$\frac{Q}{A} = \sigma\left(T_1^4 - T_2^4\right) = 5.672 \times 10^{-8}(900^4 - 400^4) = 35,760 \text{ W/m}^2$$

When the screen is added, the system can be represented by Fig. 12-36. The total resistance of the circuit is

$$\frac{1}{A} + \frac{1-\varepsilon_3}{\varepsilon_3 A} + \frac{1-\varepsilon_3}{\varepsilon_3 A} + \frac{1}{A} = \frac{2}{A}\left(1 + \frac{1-0.4}{0.4}\right) = \frac{5}{A}$$

Then

$$\frac{Q}{A} = \frac{\sigma\left(T_1^4 - T_2^4\right)}{5} = \frac{5.672 \times 10^{-8}(900^4 - 400^4)}{5} = 7,152 \text{ W/m}^2$$

We see that the net energy exchanged is reduced to one-fifth by the simple addition of the screen. This is greatly due to its relatively low emissivity, which makes the screen reflect most of its incident energy.

We shall see what would be the situation if we consider that the screen is a blackbody. In this case, the total resistance of the network is $2/A$; then

$$\frac{Q}{A} = \frac{\sigma\left(T_1^4 - T_2^4\right)}{2} = \frac{5.672 \times 10^{-8}\left(900^4 - 400^4\right)}{2} = 17,880 \ \text{W/m}^2$$

This means that in this case the heat exchanged is reduced to 50 percent from the original.

It can be demonstrated easily that if additional screens are added in the radiation path, the net heat exchanged is

$$\frac{Q}{A} = \frac{\sigma\left(T_1^4 - T_2^4\right)}{2(n+1)} \tag{12-1-72}$$

where n is the number of screens.

12-2   RADIANT HEAT TRANSMISSION IN ABSORBING MEDIA

12-2-1   Radiation Absorbing Gases and Inert Gases

Up to now, we considered the transmission of radiant energy, assuming that this energy propagates in a vacuum or in the presence of a gas completely transparent to radiation. Certain gases, however, are capable of absorbing and emitting energy.

Up to a temperature about 2,800 K, the energy emitted by a gas is almost exclusively due to changes in the vibration or rotation of molecules. Those gases whose molecules are formed by atoms of the same chemical element are not capable of absorbing or emitting radiant thermal energy. These gases (N_2, O_2, and H_2) are transparent to thermal energy. On the other hand, gases with asymmetric molecules, such as CO_2, CO, H_2O, etc., can absorb and emit energy in the range of wavelengths corresponding to thermal radiation.

12-2-2   Description of the Gas Energy-Absorption Mechanism: Beer's Law

Let dA be an element of a surface belonging to a blackbody surrounded by a gas capable of absorbing energy. Let $i_{b\lambda}^o$ be the monochromatic intensity of radiation of a certain wavelength emitted by the surface in the direction θ (Fig. 12-37).

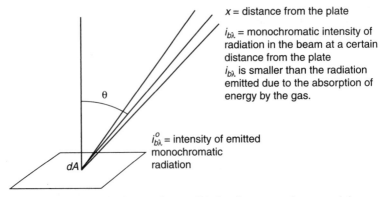

x = distance from the plate

$i_{b\lambda}$ = monochromatic intensity of radiation in the beam at a certain distance from the plate
$i_{b\lambda}$ is smaller than the radiation emitted due to the absorption of energy by the gas.

$i_{b\lambda}^o$ = intensity of emitted monochromatic radiation

θ

dA

FIGURE 12-37   Radiation from an element of black surface to a gas that surrounds it.

The absorption of energy by the gas is a volume-dependent process. In a length dx, the intensity of radiation decreases in a value that, according to the Beer's law, is given by

$$-di_{b\lambda} = k_\lambda i_{b\lambda} dx \qquad (12\text{-}2\text{-}1)$$

where k_λ is a function of the concentration of absorbent molecules of the gas (in other words, a function of the partial pressure of the components of the gas capable of absorbing radiant energy).

The boundary conditions for the integration of Eq. (12-2-1) are

$$\text{For } x = 0, \ ib_\lambda = ib_\lambda^o$$

$$\text{For } x = x, \ ib_\lambda = ib_{\lambda(x)}$$

Then

$$\int_{i_{b\lambda}^o}^{i_{b\lambda(x)}} \frac{di_{b\lambda}}{i_{b\lambda}} = \int_0^x -k\, dx \qquad (12\text{-}2\text{-}2)$$

As a result of the integration, we obtain the following expression for the intensity of radiation as a function of the distance to the surface element:

$$i_{b\lambda(x)} = i_{b\lambda}^o e^{-k_\lambda x} \qquad (12\text{-}2\text{-}3)$$

This relationship indicates that the intensity of radiation reduces exponentially with the distance through the gas.

Let's assume that x tends to 0; then

$$i_{b\lambda(x)} = i_{b\lambda}^o \qquad (12\text{-}2\text{-}4)$$

This means that if the width is small, the gas behaves as transparent. If dQ_λ is the monochromatic radiation emitted by the element of area per unit time within the solid angle $d\omega$, we have

$$dQ_{\lambda,\,\text{emitted}} = i_{b\lambda}^o dA \cos\theta\, d\omega \qquad (12\text{-}2\text{-}5)$$

At a certain distance x from the surface, the intensity of the beam has been weakened according to Eq. (12-2-3), and then the amount of energy remaining in the beam is

$$i_{b\lambda}^o dA \cos\theta\, d\omega e^{-k_\lambda x} \qquad (12\text{-}2\text{-}6)$$

and the absorbed energy is given by

$$dQ_{\lambda,\,\text{absorbed}} = i_{b\lambda}^o dA \cos\theta\, d\omega (1 - e^{-k_\lambda x}) \qquad (12\text{-}2\text{-}7)$$

We can define the monochromatic absorption factor of the gas up to a certain distance from the emitting surface as the fraction of the energy that is absorbed by the gas mass contained in the solid angle up to the considered distance. This fraction is taken with respect to the total energy emitted by the surface in that direction:

$$\alpha_\lambda = \frac{i_{b\lambda}^o dA \cos\theta\, d\omega (1 - e^{-k_\lambda x})}{i_{b\lambda}^o dA \cos\theta\, d\omega} = (1 - e^{-k_\lambda x}) \qquad (12\text{-}2\text{-}8)$$

By analyzing Eq. (12-2-8), we see that

1. The monochromatic absorption factor is independent from the direction because k_λ is not a function of θ.
2. The monochromatic absorption factor is a function of the distance to the emitting surface. This means that it depends on the gas mass crossed by the beam (it is not exclusively a function of the thermodynamic

parameters of the gas). If the distance is very high, α tends to 1, meaning that at a big distance to the surface the gas has absorbed all the monochromatic energy emitted by the surface.

3. The monochromatic absorption factor depends on the wavelength. This dependence is very strong. The emission and absorption spectra of gases are in general discontinuous. This means that there are wavelength bands in which the gas absorbs and some others in which it is completely transparent.

12-2-3   Hemispheric Monochromatic Absorption Factor

Let's assume again that we have an element of black surface surrounded by a hemisphere of absorbing gas with radius L (Fig. 12-38).

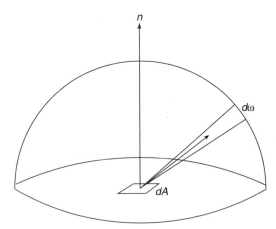

FIGURE 12-38   Geometry for the definition of gas radiation parameters.

The monochromatic energy emitted by the elemental surface within the differential solid angle $d\omega$ is given by Eq. (12-2-5). To calculate the total monochromatic energy emitted by dA, we must integrate Eq. (12-2-5) in all the half-space. We get

$$\int i_{b\lambda}^o \cos\theta \, dA d\omega = \pi i_{b\lambda}^o \, dA \tag{12-2-9}$$

(For details about this integration, see Sec. 12-1-14.) But all the energy absorbed within the solid angle and up to a radius L is, according to Eq. (12-2-7),

$$dQ_{\lambda,\text{absorbed}} = i_{b\lambda}^o dA \cos\theta \, d\omega (1 - e^{-k_\lambda L}) \tag{12-2-10}$$

The term within parentheses is the monochromatic absorption factor corresponding to the radius L. Since this is independent of θ, the integration of Eq. (12-2-10) is performed in the same way as that of Eq. (12-2-9). It then results that the total amount of energy absorbed in the entire hemisphere of radius L is

$$dQ_{\lambda,1\rightarrow G} = \int i_{b\lambda}^o \cos\theta \, dA d\omega \alpha_\lambda = \int_{\phi=0}^{\phi=2\pi} \int_{\theta=0}^{\theta=\pi/2} i_{b\lambda}^o \alpha_\lambda \cos\theta \sin\theta \, dA d\varphi d\theta \tag{12-2-11}$$

$$= \pi i_{b\lambda}^o \alpha_\lambda dA$$

Since $\pi i_{b\lambda}^o = W_{b\lambda 1}$ (monochromatic emissive power of the blackbody at the surface temperature T_1) :

$$dQ_{\lambda,1 \to G} = \alpha_\lambda W_{b\lambda 1} dA \qquad (12\text{-}2\text{-}12)$$

The fraction of the absorbed energy with respect to the energy emitted by the surface can be calculated by dividing Eqs. (12-2-11) and (12-2-9), giving

$$\frac{\text{Energy absorbed}}{\text{Energy emitted}} = \frac{\pi i_{b\lambda}^o \alpha_\lambda dA}{\pi i_{b\lambda}^o dA} = \alpha_\lambda \qquad (12\text{-}2\text{-}13)$$

For this reason, the absorption factor defined by Eq. (12-2-8) is also called the *hemispheric monochromatic absorption factor* because its definition can be expressed by Eq. (12-2-13). The hemispheric monochromatic absorption factor represents the fraction of the monochromatic energy of the considered wavelength that is absorbed by the gas contained in a hemisphere of radius L with respect to the total monochromatic energy emitted by a surface element located at its center.

Since $\alpha_\lambda = 1 - e^{-k_\lambda L}$, it is clear that this factor depends on L. This means that it depends on the volume of gas and also, through k_λ, on the nature of the gas, its pressure, temperature, and wavelength considered.

12-2-4 Monochromatic Hemispheric Emissivity

In the same way that gases absorb energy, they are capable of emitting it. As with absorption, the emission of energy by a gas is a volumetric phenomenon taking place in the entire volume of the gas. To study the radiant energy emission by a gas, the same geometry used for definition of the absorption coefficient will be used: a hemispherical gas volume with a blackbody element of area located at its center.

As a consequence of its temperature, the gas emits radiation. Part of the emitted radiation reaches the element of area. If the monochromatic energy reaching the element of area per unit time is called $dQ_{\lambda G \to 1}$, it is possible to define monochromatic hemispheric emissivity as

$$\varepsilon_\lambda = \frac{dQ_{\lambda G \to 1}}{dA\, W_{b\lambda G}} \qquad (12\text{-}2\text{-}14)$$

where $W_{b\lambda G}$ is the monochromatic emissive power of the blackbody at the gas temperature.

It is important to note that the gas emissivity is defined with reference to the receptor area and not the emitter area (because no area can be associated with the gas). Since dA is a black surface, all the incident energy is absorbed, and then the net rate of energy exchange between the gas and the element of area is

$$dQ_{\lambda,1 \rightleftarrows G} = dQ_{\lambda,1 \to G} - dQ_{\lambda,G \to 1} \qquad (12\text{-}2\text{-}15)$$

and combining Eqs. (12-2-12) and (12-2-14), we get

$$dQ_{\lambda,1 \rightleftarrows G} = dA(\varepsilon_\lambda W b_{\lambda G} - \alpha_\lambda W_{b\lambda 1}) \qquad (12\text{-}2\text{-}16)$$

It's important to realize that both ε_λ and α_λ are a function of the gas volume. The higher this volume, the higher will be ε_λ and α_λ, and the higher will be the heat flux in both directions.

12-2-5 Total Emissivity and Total Absorption Factor

We have explained that the monochromatic absorption factor and the monochromatic gas emissivity are almost noncontinuous functions of the wavelength. Unlike the case of solid bodies, the gas absorption and emission mechanisms exist only in narrow bands of the electromagnetic spectrum. In each wavelength interval where the gas is absorbent, both the absorption factor and the emissivity are functions like

$$\alpha_\lambda = 1 - e^{-k_\lambda L} \qquad (12\text{-}2\text{-}17)$$

where k_λ depends on the wavelength, as well as on system pressure and temperature.

However, in technical calculations, we need to have total absorption factors and emissivities for the entire range of thermal radiation. These total coefficients and their dependence on the gas parameters were defined and correlated by Hottel[4,5] for the cases of CO_2 and water vapor. The definitions of the total absorption factor and total emissivity are based on the same geometry used for the hemispheric monochromatic factors. We must consider an element of area located at the center of a gas hemisphere of radius L containing the gas.

If $dQ_{G \to 1}$ is the amount of energy (corresponding to all the wavelengths) reaching the area element per unit time, the total emissivity of the gas is defined as

$$dQ_{G \to 1} = \varepsilon W_{bG} dA \tag{12-2-18}$$

where W_{bG} is the emissive power of the blackbody at the gas temperature. This means that

$$W_{bG} = \sigma T_G^4 \tag{12-2-19}$$

In turn, if $dQ_1 \to G$ is the energy emitted by dA and absorbed by the gas, the total absorption factor of the gas can be defined as

$$dQ_{1 \to G} = \alpha W_{b_1} dA \tag{12-2-20}$$

where W_{b1} is the emissive power of the blackbody at the temperature of the surface. This is

$$W_{b_1} = \sigma T_1^4 \tag{12-2-21}$$

The net heat exchange between the gas and the surface element then can be calculated as

$$dQ_{1 \rightleftarrows G} = dA\, \sigma \left(\varepsilon T_G^4 - \alpha T_1^4 \right) \tag{12-2-22}$$

12-2-6   Graphs for Gas Emissivity and Absorption Factor

Total gas emissivities and absorption factors were correlated for the cases of CO_2 and water vapor, which are the gases present in the products of all combustion. The dependence of these total coefficients on the parameters of the system differs completely from Eq. (12-2-17), which is only valid for the monochromatic radiation.

Calculation of the emissivity and absorption factor of a gas at temperature T_G contained in a hemisphere of radius L exchanging energy with an element of black surface located at its center can be calculated with graphs, as explained in what follows. For CO_2, if p_c is the partial pressure of CO_2 in the gas mixture, the emissivity can be obtained from Fig. 12-39 as a function of the gas temperature at given parameters of the product $p_c L$.

This graph is only valid for a total pressure of 1 atm (101.3 kPa). For other conditions, a correction factor C_c must be obtained from Fig. 12-40 as a function of the total pressure and the product $p_c L$. The emissivity then is calculated as

$$\varepsilon_{CO_2} = eC_c \tag{12-2-23}$$

where e is the value obtained from Fig. 12-39.

For water vapor, the procedure is quite similar. If p_w is the partial pressure of water vapor in the gas mixture, it is possible to obtain from Fig. 12-41 an emissivity e as a function of T_G and $p_w L$. This graph is valid for a total pressure of 101.3 kPa and for p_w tending to 0. For other conditions, a correction factor obtained from Fig. 12-42 must be used. It must be observed that the abscissa is in this case $(p + p_w)^2$. The emissivity of water vapor is then the product.

$$\varepsilon_{H_2O} = eC_w \tag{12-2-24}$$

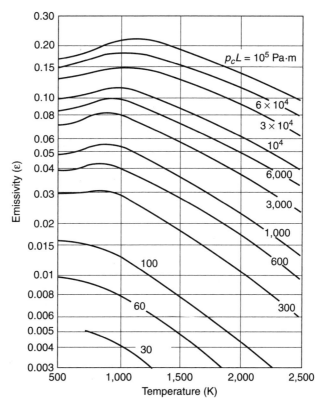

FIGURE 12-39 Emissivity of CO_2 at 1 atm (abs) total pressure. (*From H.C Hottel: in W.H McAdams: Heat Transmission. New York: McGraw-Hill, 1954 Reproduced with permission of the editors.*)

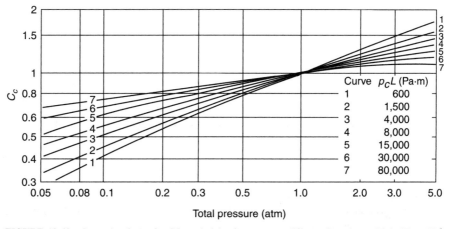

FIGURE 12-40 Correction factor for CO_2 emissivity for pressures different from 1 atm (abs). (*From H.C Hottel: in W.H McAdams: Heat Transmission. New York: McGraw-Hill, 1954 Reproduced with permission of the editors.*)

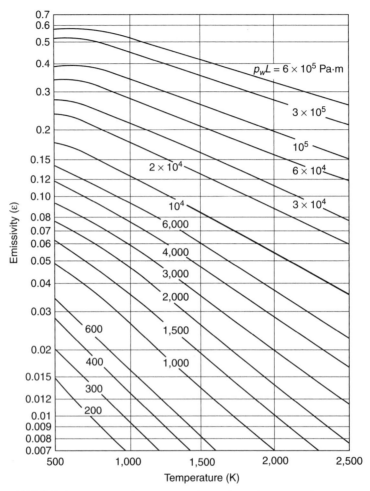

FIGURE 12-41 Emissivity of water vapor at 1 atm (abs) total pressure and water vapor partial pressure tending to 0. (*From H.C Hottel: in W.H McAdams:* Heat Transmission. *New York: McGraw-Hill 1954 Reproduced with permission of the editors.*)

For a mixture containing CO_2 and water vapor together, an additional correction is necessary. Since the emission bands of water vapor and CO_2 present a certain overlap, when both gases are present, part of the energy emitted by one of them is absorbed by the other. Then the mixture emissivity, which in case of no overlap would be the sum of the emissivities of both components, is reduced. The mixture emissivity then can be calculated as

$$\varepsilon = \varepsilon_{CO_2} + \varepsilon_{H_2O} - \Delta\varepsilon \qquad (12\text{-}2\text{-}25)$$

where $\Delta\varepsilon$ is obtained from Fig. 12-43.

The absorption factor can be obtained in a similar way. The absorption factor of a mixture containing CO_2 and water vapor is

$$\alpha = \alpha_{CO_2} + \alpha_{H_2O} - \Delta\alpha \qquad (12\text{-}2\text{-}26)$$

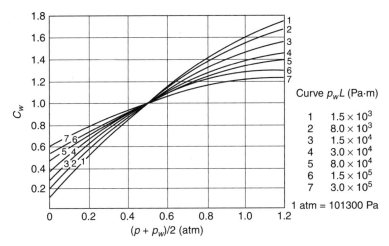

FIGURE 12-42   Correction factor for water vapor emissivity. (*From H.C Hottel: in W.H McAdams : Heat Transmission. New York: McGraw-Hill, 1954 Reproduced with permission of the editors.*)

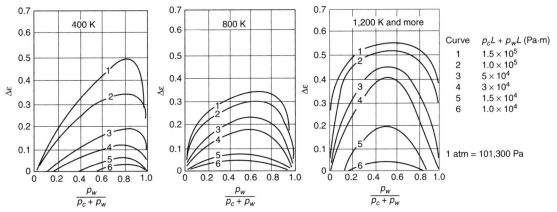

FIGURE 12-43   Emissivity correction for mixtures of CO_2 and water vapor. (*From H.C Hottel: in W.H McAdams: Heat Transmission. New York: McGraw-Hill, 1954 Reproduced with permission of the editors.*)

where

$$\alpha_{CO_2} = e'C_c'\left(\frac{T_G}{T_1}\right)^{0.65} \tag{12-2-27}$$

Where e' is obtained from Fig. 12-39 but entering the graph with the surface temperature T_1 as abscissa instead of gas temperature and calculating the parameter as $p_cL(T_1/T_G)$ instead of p_cL.

For water vapor,

$$\alpha_{H_2O} = e'C_w'\left(\frac{T_G}{T_1}\right)^{0.45} \tag{12-2-28}$$

where e' and C_w are obtained from Figs. 12-41 and 12-42. To find e', we enter the graph in Fig. 12-41 with T_1 as abscissa and with $p_w L(T_1/T_G)$ as the parameter. The correction coefficients are obtained in the same way as for emissivity calculation. A detailed calculation is explained in Example 12-6.

12-2-7 Radiant Heat Exchange between a Gas and a Finite Surface

Up to now, we only considered the case of heat exchange between a gas hemisphere and an element of area located at its center. For other geometries, the problem is solved by defining an equivalent mean hemispherical beam length. The equivalent mean hemispherical beam length is a fictitious length characteristic of the considered geometry that allows one to obtain, using the graphs in Figs. 12-39 through 12-43, emissivities and absorption factors for the gas such that the net heat exchanged between the gas and the surface can be expressed as

$$Q_{G \rightleftarrows 1} = A(\varepsilon W_{bG} - \alpha W_{b1}) \tag{12-2-29}$$

or

$$Q_{G \rightleftarrows 1} = A\sigma \left(\varepsilon T_G^4 - \alpha T_1^4 \right) \tag{12-2-30}$$

The more usual case is that of a gas exchanging energy with the walls of the enclosure in which it is contained. In these cases, Table 12-1 allows one to obtain the equivalent mean hemispherical beam length for simple geometries.

TABLE 12-1 Equivalent Mean Hemispherical Beam Length[6]

Shape	L
Infinite cylinder	Diameter
Circular cylinder with height = diameter	$2/3 \times$ diameter
Clearance between infinite parallel planes	$2 \times$ distance between planes
Cube	$2/3 \times$ edge

For geometries not included in Table 12-1, an approximate value can be calculated as

$$L = 3.4 \frac{\text{volume}}{\text{surface area}}$$

12-2-8 Equivalent Gray Gas Emissivity

Making an extension of the gray-body concept, we can define a gray gas as that whose emissivity equals the absorption factor. In this case, Eq. (12-2-29) reduces to

$$Q_{G \rightleftarrows 1} = A\varepsilon(W_{bG} - W_{b1}) \tag{12-2-31}$$

As we have seen in the preceding section, gases such as CO_2 and water vapor are very far from being gray. However, we can keep the form of Eq. (12-2-31) if we define an equivalent gray gas emissivity as

$$\varepsilon_g = \frac{\varepsilon W_{bG} - \alpha W_{b1}}{W_{bG} - W_{b1}} \tag{12-2-32}$$

In this way, the net heat exchange between the gas and the surface can be expressed as

$$Q_{G \rightleftarrows 1} = A\varepsilon_g (W_{bG} - W_{b1}) \tag{12-2-33}$$

And the equation becomes simpler because instead of ε and α, it requires a single parameter ε_g. Of course, in order to use this expression, both ε and α have to be calculated anyway. So this approach does not look to offer any particular advantage. However, by using a single parameter, the equation can be represented by the equivalent electric network of Fig. 12-44a, and this will be useful when dealing with more complex cases, as we shall later see. If the solid surface is supposed to have an emissivity ε_1, the gray-to-black transformation resistance must be included, as shown in Fig. 12-44b.

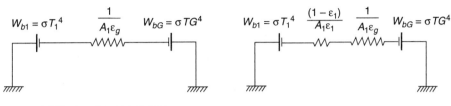

(a) Heat exchange between
a gas and a blackbody

(b) Heat exchange between a gas
and a gray body

FIGURE 12-44   Heat exchange between a gas and a solid surface.

Example 12-6   The exhaust gases of an internal combustion motor are discharged into the atmosphere through an exhaust pipe with 0.1-m internal diameter. The pressure within the pipe is 125 kPa. The components mole fractions are

H_2O: 0.096
CO_2: 0.085
N_2: 0.747
O_2: 0.072

(*Note:* This composition corresponds to the combustion gases when a fuel with a molecular composition $CH_{1.8}$ is burned with 20 percent excess air. In Chap. 13, the stoichiometry of combustion will be explained in detail, and this will allow you to calculate the composition of the products of any combustion.)

If the wall temperature is assumed to be at 700 K, and the emissivity of the walls is 0.8, calculate the net radiant heat exchanged between the gas and the wall.

Solution   The partial pressures of the absorbing gases (CO_2 and H_2O) are

For CO_2: $p_c = 0.085 \times 125 = 10.62$ kPa
For H_2O: $p_w = 0.096 \times 125 = 12$ kPa

The equivalent mean hemispherical beam length, assuming that the pipe is an infinite cylinder, is equal to the diameter according to Table 12-1. This is 0.1 m. Then

$$p_c \times L = 10.62 \times 0.1 = 1.062 \text{ kPa} \cdot \text{m}$$
$$p_w \times L = 12 \times 0.1 = 1.2 \text{ kPa} \cdot \text{m}$$

CO_2 *emissivity:* From Fig. 12-39, with $T_G = 1,500$ K and $p_c L = 1,062$ Pa·m (0.0105 atm·m), we get $e = 0.037$.
The correction factor from Fig. 12-40 with $p = 125$ kPa (1.23 atm) and $p_c L = 1,062$ Pa·m is $C_c = 1.05$. Thus

$$\varepsilon_{CO_2} = 0.037 \times 1.05 = 0.0388$$

Water vapor emissivity: With $p_w L = 1{,}200$ Pa·m (0.0118 atm·m) and $T_G = 1{,}500$ K, from Fig. 12-41, we get $e = 0.015$. The correction factor from Fig. 12-42 is obtained by entering with an abscissa of $\frac{1}{2}(1.23 + 0.118) = 0.673$ atm and 1,200 Pa·m as the parameter. We obtain $C_w = 1.27$. Thus

$$\varepsilon_{H_2O} = 0.015 \times 1.27 = 0.019$$

Mixture emissivity: The correction factor from Fig. 12-43 is negligible. Thus the emissivity of the mixture will be the sum

$$\varepsilon = \varepsilon_{CO_2} + \varepsilon_{H_2O} = 0.0388 + 0.019 = 0.0578$$

CO_2 absorption factor: Entering Fig. 12-39 with abscissa $T_1 = 700$ K, and calculating the value of the parameter as

$$P_c L(T_1/T_G) = 1{,}062(700/1{,}500) = 495 \text{ Pa·m } (0.00488 \text{ atm·m})$$

we get $e = 0.035$. The correction factor $C'_c = C_c = 1.05$. Then

$$\alpha_{CO_2} = 0.035 \times 1.05 \times (1{,}500/700)^{0.65} = 0.060$$

Water vapor absorption factor: Entering Fig. 12-41 with an abscissa of 700 K and a parameter value of

$$p_w L(T_1/T_G) = 1{,}200 \ (700/1{,}500) = 560 \text{ Pa·m } (0.0052 \text{ atm·m})$$

we get $e = 0.024$. The correction factor is $C'_w = C_w = 1.27$. Then

$$\alpha_{H_2O} = 0.024 \times 1.27 \times (1{,}500/700)^{0.45} = 0.043$$

Mixture absorption factor: Again, neglecting the mixture correction, the absorption factor will be the sum of

$$\alpha_{CO_2} + \alpha_{H_2O} = 0.060 + 0.043 = 0.102$$

The equivalent gray gas emissivity thus will be

$$\varepsilon_g = \frac{\varepsilon W_{bG} - \alpha W_{b1}}{W_{bG} - W_{b1}} = \frac{0.0578 \times 1{,}500^4 - 0.102 \times 700^4}{1{,}500^4 - 700^4} = 0.055$$

The area of the pipe per meter length is $A_1 = \pi \times D = 0.314$ m²/m. Thus the total resistance of the electric circuit represented in Fig. 12-44b is

$$\frac{1}{A_1 \varepsilon g} + \frac{1 - \varepsilon_1}{A_1 \varepsilon_1} = \frac{1}{0.314 \times 0.055} + \frac{1 - 0.8}{0.314 \times 0.8} = 58.7$$

and the net heat exchanged is

$$Q_{G_1} = \frac{W_{bG} - W_{b1}}{58.7} = \frac{\sigma\left(T_G^4 - T_1^4\right)}{58.7} = \frac{5.678 \times 10^{-8}(1{,}500^4 - 700^4)}{58.7} = 4{,}664 \text{ W}$$

12-2-9 Radiant Heat Transfer between Surfaces in the Presence of Absorbing Gases

Let's first consider the case of two planar parallel plates at temperatures T_1 and T_2 exchanging energy when a vacuum or a transparent gas exists between them. The heat exchange rate can be expressed as

$$Q_{1 \rightleftarrows 2} = A_1 F_{12}(W_{b1} - W_{b2}) \tag{12-2-34}$$

This expression has its electric analogy in a circuit where the resistance is $(A_1 F_{12})^{-1}$. If the surfaces are gray, instead of F_{12}, we use $\mathcal{F}_{12}$. If now we introduce a nontransparent gas between the plates, the heat exchanged will not be the same. The gas will absorb energy according to its coefficient α, and it will only allow the passage of a fraction $(1 - \alpha) = \tau_{G12}$.

We could think that the resistance to the heat transmission between plates 1 and 2 has increased in a factor $1/\tau_{G12}$ owing to the presence of the gas. The gas, in turn, can exchange energy with both surfaces. The radiant energy exchanged between the gas and plate 1 is given by $A_1 \varepsilon_{g1}(W_{bG} - W_{b1})$, and the energy exchanged between the gas and plate 2 is $A_2 \varepsilon_{g2}(W_{bG} - W_{b2})$. The electric circuit in Fig. 12-45 represents this system.

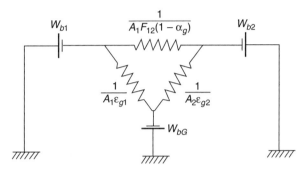

FIGURE 12-45   System with two solid surfaces and a gas.

If the gas is enclosed in the space between the plates without delivering or withdrawing energy to or from the system, it will reach a steady-state temperature in between those of plates 1 and 2. This steady state is reached when the gas temperature is such that all the energy received from the hot source is yielded to the cold one. Then we can eliminate in the electric circuit the branch containing W_{bG} without modifying the current distribution in the rest of the network. The network represented in Fig. 12-46 is obtained. We see that in this case the final result is a net heat transmission from plate 1 to plate 2.

The energy flow from plate 1 to plate 2 is the sum of two fractions. The first one comes directly to plate 2 through the resistance of the upper branch. This is the energy flow allowed by the gas according to its transparency coefficient.

The rest of the energy emitted by plate 1 is absorbed by the gas. But the gas reemits it, and it finally reaches A_2. This is the current circulating through the lower resistances. The total energy exchanged is the sum of both currents. If temperatures T_1 and T_2 are known, the network can be solved and the current distribution found. It is then possible to calculate the potential of point G (W_{bG}), and the gas temperature can be obtained from

$$T_G = \sqrt[4]{\frac{W_{bG}}{\sigma}} \qquad (12\text{-}2\text{-}35)$$

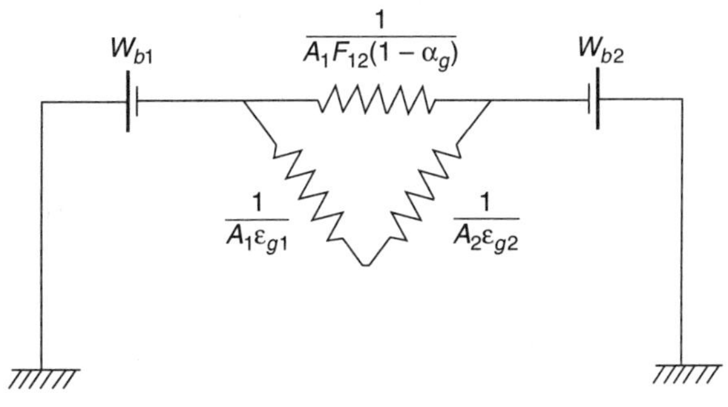

FIGURE 12-46   Gas enclosed between plates.

GLOSSARY

A = area (m^2)

F = direct view factor

$\overline{F}$ = view factor including refractories

$\mathcal{F}$ = view factor between gray surfaces

I = intensity of radiation (W/m^2)

Q = heat flux (W)

r = radial coordinate (m)

T = temperature (K)

W = emissive power (W/m^2)

W_λ = monochromatic emissive power [W/(m$^2 \cdot$ m)]

α = absorption coefficient

ε = emissivity

λ = wavelength (m)

ω = solid angle

ρ = reflection coefficient

σ = Stephan Boltzman constant

τ = transmissivity

θ, ϕ = angular coordinates

Subscripts

b = blackbody

n = normal direction

θ = direction θ

λ = wavelength λ

REFERENCES

1. Kreith F.: *Principles of Heat Transfer.* New York: Internacional Textbook Company, 1968.
2. Hottel H. C.: in McAdams W.H : *Heat Transmission.* New York: McGraw-Hill, 1954.
3. Planck M.: *The theory of Heat Radiation.* New York: Dover, 1959.
4. Hottel H. C.: "Radiant Heat Transmission," *Mech Eng* 52:699–704, 1930.
5. Hottel H. C., Egbert: "Radiant Heat Transmission from Water Vapor," *AIChE Trans* 38:531–565, 1942.
6. Rohsenow M., Hartnett J. P.: *Handbook of Heat Transfer.* New York: McGraw-Hill, 1973.
7. Schmidt E., Eckert E.: "Über die Richtunsverteilun der Warmestrahlung von Oberflachen," *Forsch Geb Ingenieurwes* 6:175–183, 1935.

CHAPTER 13
PROCESS FIRED HEATERS

13-1 SIMPLIFIED SCHEMATIC

A process furnace or fired heater consists of a combustion chamber or firebox with one or more burners and containing a tubular coil where a process fluid circulates. The chamber walls normally are metallic, interiorly covered by a refractory lining. A gaseous or liquid fuel is injected in the burners with the necessary combustion air. The combustion reaction that takes place in the burners generates hot flue gases that are the product of the combustion. Part of the energy liberated in the combustion process is used to heat up the product circulating into the coil. Another part is lost through the stack (because the flue gas is still hot when it leaves the system), and there also is a conduction heat loss through the walls of the chamber. A schematic is shown on Fig. 13-1.

Before proceeding with a description of the several elements constituting the furnace, we shall review some aspects related to the process of combustion.

13-2 COMBUSTION

13-2-1 Stoichiometry

Combustion is an exothermic chemical reaction between oxygen and a fuel. Usually the fuel is a hydrocarbon and the oxygen is that contained in air. If the amount of oxygen is enough to convert all the fuel to CO_2 and H_2O (the principal products of combustion), it is said that the combustion is *complete*. If the amount of oxygen in the burner is not enough for complete burning of the fuel, the combustion gases will contain unburned hydrocarbon and CO.

The combustion reactions of the most important hydrocarbons are

Methane: $CH_4 + 2O_2 = CO_2 + 2H_2O$
Ethane: $C_2H_6 + \frac{7}{2}O_2 = 2CO_2 + 3H_2O$
Propane: $C_3H_8 + 5O_2 = 3CO_2 + 4H_2O$
Ethylene: $C_2H_4 + 3O_2 = 2CO_2 + 2H_2O$

Any sulfur present in the fuel will react according to

$$S + O_2 = SO_2$$

One mole of oxygen is required for each carbon atom, 0.25 mole of oxygen for every hydrogen atom, and 1 mole for every sulfur atom. Thus, for a fuel containing n carbon atoms, m hydrogen atoms, and s sulfur atoms, the reaction can be written

$$C_nH_mS_s + (n + 0.25m + s)O_2 = nCO_2 + 0.5mH_2O + sSO_2$$

In industrial burners, the amount of oxygen employed is not stoichiometric because an oxygen excess is always necessary to achieve complete burning of the fuel.

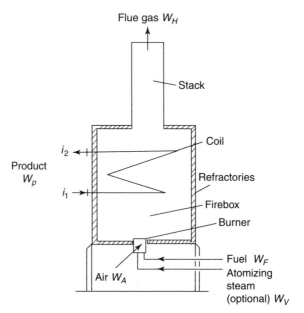

FIGURE 13-1 Schematic of a process fired heater.

When gaseous fuels are employed (natural gas or refinery fuel gas), 10–15 percent excess air normally is used. Liquid-fuel burners normally require a higher air excess because it is more difficult to obtain a good mixture between the air and the fuel, especially if high-viscosity fuel oil is used. In these cases, the excess air can be 25 percent or even more.

Most liquid-fuel burners require the addition of steam, which is injected into the burner and mixed with the fuel oil to atomize it at the burner outlet nozzle. The fuel exits the nozzle at high velocity, divided into small droplets that can burn easily.

If e is the oxygen excess (e = excess air/stoichiometric air), and considering that the atmospheric nitrogen:oxygen ratio is 3.762:1, the combustion reaction for a molecule with a chemical formula $C_nH_mS_s$ can be written

$$C_nH_mS_s + (1 + e)(n + 0.25m + s)O_2 + 3.762(1 + e)(n + 0.25m + s)N_2 = nCO_2 + 0.5mH_2O + sSO_2$$
$$+ e(n + 0.25m + s)O_2 + 3.762(1 + e)(n + 0.25m + s)N_2$$

To this equation we should add the atomizing steam, if used. Even though the atomizing steam does not take part in the combustion reaction, it is important to consider it among the combustion products because it increases the partial pressure of the water vapor, and this is important for the calculation of gas emissivities and the enthalpy balance.

It also should be considered that atmospheric air always contains some water vapor. However, we shall neglect this contribution in the calculations. Fuel gas also can contain nitrogen and CO_2. These components will incorporate directly into the combustion gases. If, additionally, some CO is present in the fuel gas, these molecules will react according to

$$CO + \tfrac{1}{2}(1 + e)O_2 + 1.88(1 + e)N_2 = CO_2 + \tfrac{1}{2}eO_2 + 1.88(1 + e)N_2$$

Example 13-1 Calculate the amount of air and the composition of the combustion gases when a fuel with the following molar composition (in mole fractions)is burnt with 15 percent excess air:

$$CH_4: 0.85; \ C_2H_6: 0.08; \ C_2H_4: 0.03; \ C_3H_8: 0.02; \ C_3H_6: 0.02$$

Solution For every 100 moles of fuel gas, we shall have

85 moles of CH_4	$n = 85$	$m = 340$
8 moles of C_2H_6	$n = 16$	$m = 48$
3 moles of C_2H_4	$n = 6$	$m = 12$
2 moles of C_3H_6	$n = 6$	$m = 12$
2 moles of C_3H_8	$n = 6$	$m = 18$
	$\Sigma n = 119$	$\Sigma m = 430$

We shall need $(1 + e)(n + 0.25m) = 1.15 \times (119 + 107.5) = 260.48$ moles of oxygen. The associated nitrogen is $3.762 \times 260.48 = 979.9$ moles. The combustion will produce 119 moles of CO_2 and $430/2 = 215$ moles of H_2O.

The combustion gases will contain the excess oxygen, which is $0.15 \times (119 + 107.5) = 33.98$ moles. The total moles thus will be $119 + 215 + 979.9 + 33.98 = 1,347.88$. The mole fractions will be

$N_2 = 979.9/1,347.88 = 0.727$

$O_2 = 33.98/1,347.88 = 0.0252$

$H_2O = 115/1,347.88 = 0.1595$

$CO_2 = 119/1,347.88 = 0.0953$

13-2-2 Heating Value

The *heating value* of a fuel is the amount of heat liberated during complete combustion with the stoichiometric air. For example, in the case of methane, it will be the heat of the reaction

$$CH_4 + 2O_2 = CO_2 + 2H_2O$$

The heating value can be expressed in kcal/mol of fuel or in kcal/kg or kcal/(Sm³). The heats of reaction normally are defined at a constant temperature, usually 15°C. This means that the heat of reaction is the amount of heat liberated during the combustion of 1 mol of CH_4 initially at 15°C, with air at the same temperature, once the combustion products are cooled again to 15°C.

However, we can see that one of the reaction products is water. To complete the definition, it is necessary to clarify if in the reaction products water is considered to be in the vapor or liquid state. If the liquid state is assumed, the water produced in the reaction must be condensed and then will deliver its heat of condensation. This amount of heat thus will be part of the heat of reaction. The heating value thus defined is called the *gross heating value*. If, on the other hand, the water produced in the combustion remains in the vapor state, the heat of condensation will not be part of the heat of reaction, which is then called the *net heating value*.

For example, in the case of methane, the gross heating value is 13,252 kcal/kg, and the net heating value is 11,932 kcal/kg. Normally, in combustion equipment (e.g., a process heater), the water produced in the reaction exits through the stack as vapor, so the net heating value must be used in calculations.

Domestic gas distribution companies normally use the gross heating value in their invoices. This implies the assumption that the users can install a heat-recovery system in their stove burners to make use of the enthalpy of the water vapor contained in the combustion gas. Table 13-1 lists the heating values of some common gases.

Heats of Combustion of Fuel Mixtures.

Gases. Normally, fuel gases consist of a mixture of hydrocarbons. Usually a fuel-gas analysis is available. If the fuel composition is known, it is possible to calculate the heating value of the mixture as a molar average of the individual heating values.

Liquid Fuels. Unlike gaseous fuels, the heating value of a liquid fuel is not usually calculated on a chemical analysis basis. Instead, the heating value of a liquid fuel can be correlated as a function of the specific gravity of the oil with sufficient accuracy for most engineering calculations. Sometimes the American

TABLE 13-1 Heating Values of Some Common Gases

	Gross Heating Value (kcal/kg)	Gross Heating Value (kcal/kmol)	Net Heating Value (kcal/kg)	Net Heating Value (kcal/kmol)
Carbon monoxide (CO)	2,413	67,564	2,413	67,564
Hydrogen (H$_2$)	33,912	67,824	28,654	57,308
Methane (CH$_4$)	13,252	212,032	11,932	190,912
Ethane (C$_2$H$_6$)	12,391	371,730	11,333	339,990
Propane (C$_3$H$_8$)	12,028	529,232	11,066	486,904
n-Butane (C$_4$H$_{10}$)	11,835	686,430	10,923	633,534
n-Pentane (C$_5$H$_{12}$)	11,710	843,120	10,828	779,616
n-Hexane (C$_6$H$_{14}$)	11,638	1,000,868	10,777	926,822
Ethylene (C$_2$H$_4$)	12,009	336,252	11,254	315,112
Propylene (C$_3$H$_6$)	11,683	490,686	10,928	458,976
Butylene (C$_4$H$_8$)	11,575	648,200	10,820	605,920
Benzene (C$_6$H$_6$)	10,093	787,254	9,686	755,508
Toluene (C$_7$H$_8$)	10,269	944,748	9,809	902,428
p-Xylene (C$_8$H$_{10}$)	10,343	1,096,358	9,844	1,043,464
Acetylene (C$_2$H$_2$)	11,935	310,310	11,528	299,728
Naftalene (C$_{10}$H$_8$)	9,604	1,229,312	9,274	118,7072
Ammonia (NH$_3$)	5,366	91,222	4,433	75,361
Hydrogen sulfide (SH$_2$)	3,939	133,926	3,628	123,352

Petroleum Institute (API) gravity is used. The API gravity of a hydrocarbon is a parameter related to its density by

$$°API = \frac{141.4}{specific\ gravity} - 131.5$$

(Specific gravity is at 15°C.)

The following equations allow the prediction of liquid hydrocarbon heating values as a function of API gravity:

Gross heating value (kcal/kg) $= 9,886 + 34.477(°API) + 0.009989(°API)^2 - 0.004275(°API)^3$

Net heating value (kcal/kg) $= 9,263 + 45.61(°API) - 0.6561(°API)^2 + 0.003878(°API)^3$

13-2-3 Enthalpy Balance

We refer again to Fig. 13-1. A fuel mass flow W_F and an air mass flow W_A enter the combustion chamber. Within the heater, the combustion reaction takes place, and a mass of hot flue gases is produced. Part of the heat liberated in the combustion is absorbed by the fluid circulating in the coil. The heat transfer between the hot gases and the surface of the coil is mainly by radiation. If Q is the heat flux (J/s or kcal/h) withdrawn by the process fluid, then

$$Q = W_P(i_2 - i_1) \tag{13-2-1}$$

We can write a heat balance for the unit as

$$W_F i_F + W_A i_A + W_V i_V - W_H i_H - Q - Q_P = 0 \tag{13-2-2}$$

where i_F, i_A, i_H, and i_V are the enthalpies of the fuel, air, flue gas, and atomizing steam, respectively, and Q_p represents the energy lost by conduction through the heater walls.

To make an enthalpy balance in a system where a chemical reaction as the combustion takes place, not only must the definition of the reference state for enthalpies calculation include the reference temperature, but it also must specify which are the chemical species to which the zero enthalpy is assigned at the reference temperature. For example, if the reference state is defined as the combustion products at 15°C, then we must include the heat of reaction in the fuel enthalpy.

Then we could write the enthalpy balance as

$$W_F\left[c_F(T_F - T_0) + \Delta i_{\text{comb}}\right] + W_A c_A(T_A - T_0) + W_V c_V(T_v - T_0) = W_H c_H(T_H - T_0) + Q + Q_P \quad (13\text{-}2\text{-}3)$$

Since we choose as the reference state for water the vapor state, Δi_{comb} must be the net heating value (NHV). The preceding equation can be combined with Eq. (13-2-1) to give

$$Q = Wp(i_2 - i_1) = W_A c_A(T_A - T_0) + W_F c_F(T_F - T_0) + W_F(\text{NHV}) + W_V c_V(T_V - T_0)$$
$$-W_H c_H(T_H - T_0) - Q_p \quad (13\text{-}2\text{-}4)$$

The heat lost through the heater walls depends on the quality of the thermal insulation. Its evaluation can be done by calculating the combined radiation and convection coefficients for both the interior and exterior sides of the walls and knowing the thermal conductivities and thickness of insulation and refractory materials. However, since this magnitude is not very significant in the heat balance, it is enough for our present purposes to estimate it as a percentage of the heat liberated in the combustion chamber, whose order is between 2 and 5 percent. We shall call this fraction γ.

It is useful to write Eq. (13-2-3) per unit mass of fuel, and it results

$$\frac{Q}{W_F} = \frac{W_A}{W_F} c_A(T_A - T_0) + \frac{W_V}{W_F} c_V(T_V - T_0) + c_F(T_F - T_0) + \text{NHV}\left(1 - \frac{\gamma}{100}\right) - \frac{W_H}{W_F} c_H(T_H - T_0) \quad (13\text{-}2\text{-}5)$$

The left member of the equation represents the heat transferred to the process per kilogram of burned fuel. This is an important parameter related to the thermal efficiency of the process.

The quotients included in the right side member depend on the stoichiometry of the combustion and can be calculated as explained previously. The ratio W_V/W_F is information supplied by the burner vendor (for burners that use atomizing steam), and its value is about 0.3 kg steam/kg fuel.

Among the terms on the right side of the equation, it is very important that corresponding to the flue gas because most of the heat not delivered to the process is lost as enthalpy in the flue gases leaving the stack. Note that in order to reduce this heat loss, it is necessary to achieve a low value of T_H. This means that the outlet temperature of the flue gas should be as low as possible.

We shall explain later that the flue gas outlet temperature is roughly equal to the mean gas temperature into the combustion chamber. Since the heat-flux density on the coil surface is a function of the fourth power of the gas temperature, a low gas temperature means a low heat flux, thus requiring more heat transfer surface—in other words, a bigger and more expensive heater.

This means that the outlet temperature of the flue gas is an important variable whose value defines the thermal efficiency of the heater. This value should be adopted in the design stage looking for a proper balance between fuel consumption versus heater cost.

The ratio W_H/W_F, included in the last term of the right side of the equation, also must be analyzed. The flue-gas stream contains all the combustion products and also the excess air. If a high amount of excess air is used, the flue-gas mass flow will be high, and this means a higher amount of hot gases leaving the stack. Then, according to Eq. (13-2-5), a higher amount of fuel will be necessary.

If the chemical composition of the fuel is known, the ratio W_H/W_F can be calculated, as we did in the Example 13-1. In the case of liquid fuels, where a chemical analysis is not normally available, it is not possible to calculate the mass of flue gases using stoichiometry. Charts such as that in Fig. 13-2 then are

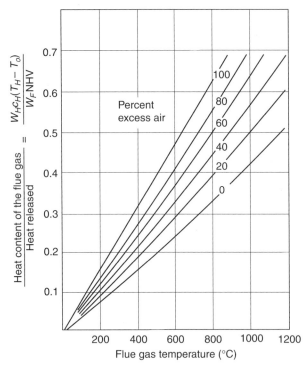

FIGURE 13-2 Fraction of energy lost with the combustion products. (*From Wimpress N: "Generalized Method Predicts Fired Heaters Performance,"* Chem Eng *May 22:pp 95–102, 1978. Reproduced with permission.*)

useful for this purpose. The graph in the figure represents the ratio between the enthalpy of the combustion products and the total heat delivered to the heater, which is

$$\frac{W_H c_H (T_H - T_0)}{W_F (\text{NHV})}$$

as a function of the flue gas temperature and the excess air for typical refinery fuels. The graph can be used for either gaseous or liquid fuels. It cannot be used for highly aromatic fuels or other unusual fuels.

Fuel Efficiency

Fuel efficiency is the ratio between the heat delivered to the process and the total heat liberated in the combustion. It is obtained dividing Eq. (13-2-5) by NHV:

$$\eta = \frac{Q}{W_F(\text{NHV})} = \frac{W_A c_A (T_A - T_0)}{W_F(\text{NHV})} + \frac{W_V c_V (T_V - T_0)}{W_F(\text{NHV})} + \frac{c_F (T_F - T_0)}{\text{NHV}} + \left(1 - \frac{\gamma}{100}\right) - \frac{W_H c_H (T_H - T_0)}{W_F(\text{NHV})} \quad (13\text{-}2\text{-}6)$$

Example 13-2 If the combustion gases generated when burning the fuel in the condition of Example 13-1 leave the stack at 700°C after delivering heat in the process heater, what was the fuel efficiency of the unit?

Note: A downloadable spreadsheet with the calculations of this example is available at http://www.mhprofessional.com/product.php?isbn=0071624082

Solution We shall assume that the fuel gas and the air enter the heater at ambient temperature (approximately equal to the reference temperature). We then can neglect their sensible enthalpy. If we assume that the heat loss through the heater walls is about 2 percent of the heat liberated, Eq. (13-2-6) reduces to

$$\eta = 0.98 - \frac{W_H c_H (T_H - T_0)}{W_F (\text{NHV})}$$

The combustion was performed with 15 percent excess air. For 700°C and 15 percent excess air, from Fig. 13-2, we get an ordinate of 0.33. Then $\eta = 0.98 - 0.33 = 0.65$.

We shall perform the same calculation analytically. We can calculate the heating value of the fuel with the chemical composition and using Table 13-1. On the basis of 100 kmol of fuel, we have

Component	N_i = No. of Moles	NHV, kcal/kmol	$N_i \times$ NHV, Kcal
CH_4	85	190,912	16,227,500
C_2H_6	8	339,990	2,719,900
C_2H_4	3	315,112	945,300
C_3H_6	2	458,976	918,000
C_3H_8	2	486,904	973,800
		Σ	21,784,500

The enthalpy of the combustion gases can be calculated with the help of Table I-6 of the App. I. The molar heat capacity is expressed as a polynomial of the type

$$c = a + bT + cT^2 + dT^3$$

where T is in K. Then, to calculate the enthalpy at 700°C (973 K), we must calculate the integral

$$\int_{288}^{973} (a + bT + cT^2 + dT^3)\,dT = a(973 - 288) + \frac{b}{2}(973^2 - 288^2) + \frac{c}{3}(973^3 - 288^3) + \frac{d}{4}(973^4 - 288^4)$$

$$= 685a + 431,892b + 2.99 \times 10^8 c + 2.223 \times 10^{11} d$$

The combustion products are 979.9 kmol nitrogen, 49.5 kmol oxygen, 115 kmol water, and 119 kmol CO_2 (see Example 13-1). The calculation is shown in the following table:

	N = Number of Moles	a	b	c	d	I	$N \times I$, kcal
N_2	979.9	7.44	-3.24×10^{-3}	6.400×10^{-6}	-2.79×10^{-9}	4,990	4,889,701
H_2O	215	7.701	4.59×10^{-4}	2.53×10^{-6}	-8.59×10^{-10}	6,038	1,298,170
O_2	33.98	6.713	-8.79×10^{-7}	4.17×10^{-6}	-2.544×10^{-9}	5,279	179,380
CO_2	119	4.728	1.754×10^{-2}	-1.338×10^{-5}	4.097×10^{-9}	7,723	919,037

The enthalpy of the combustion gases thus will be 4,889,701 + 1,298,170 + 179,380 + 919,037 = 7,286,288 kcal. The fuel efficiency will be

$$\eta = 0.98 - \frac{W_H c_H (T_H - T_0)}{W_F (\text{NHV})} = 0.98 - \frac{7,286,288}{21,784,500} = 0.64$$

which reasonably agrees with the previous result.

Radiant Convective Arrangement.   As we have seen, the fired heater shown in Fig. 13-1 has a very low thermal efficiency owing to the energy lost with the hot flue gases, which leave the stack at a very high temperature. Nowadays, most fired heaters recover heat from the flue gases with the installation of a tube bank in the flue-gas path before the stack entrance. Since the combustion gases already have cooled down, the predominant mechanism in this zone is convection, and it is called the *convection section.*

For a better use of the temperature differences, the process fluid enters first to the convection section and then passes to the radiant section, thus circulating countercurrently to the flue gas, as shown in Fig. 13-3. This arrangement gives higher furnace efficiency (lower stack-gas temperature) than would be the case with co-current flow. The coil sections corresponding to the transition from the convection to the radiant sections are called *crossovers.*

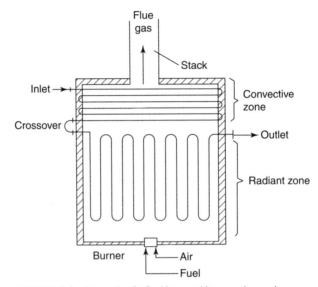

FIGURE 13-3   Schematic of a fired heater with convection section.

An illustration of a radiant convective heater showing the distribution of the tubes in radiant and convection sections can be seen in Fig. 13-4. In this type of fired heater, efficiencies are usually 85–90 percent.

13-3   APPLICATIONS OF FIRED HEATERS

The principal applications of fired heaters in process industry are as follows:

Fractionating column-feed preheaters. A typical example of this service is the feed heater for the atmospheric distillation column in a petroleum refinery. Here, crude oil entering the fired heater as a 220°C liquid, might exit near 360–370°C with about 60 percent of the charge being vaporized. The outlet mixture then is fed to a fractionating column. Other refinery applications can be found in vacuum distillation units or the fractionation section of cracking units.

Reactor-feed preheaters. Fired heaters in this application heat up the charge stock to the necessary temperature for controlling a chemical reaction taking place in the reactor vessel. An example of this is a catalytic reformer unit, where a single-phase mixture of hydrogen and vaporized hydrocarbon react to produce aromatics or high-octane gasoline. In this case, the temperature rise in the heater is about 100°C, and operating pressures are in the range of 20–40 bar. Another example is the heating of mixtures of liquid hydrocarbons and recycled hydrogen gas for reaction in a refinery hydrocracker.

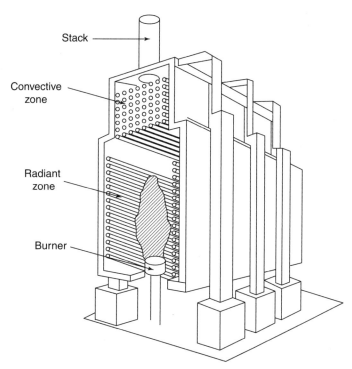

FIGURE 13-4   Cross section of a radiant convective heater.

Fluid temperatures typically run from 320°C at the inlet to 450°C at the outlet. Operating pressures can reach 200 bar.

Column reboilers. In this application, the charge stock coming from a distillation column is a recirculating liquid that is partially vaporized in the fired heater. The mixed vapor-liquid stream reenters the column. The heater acts as a reboiler, delivering the vaporization heat, with low temperature differences between inlet and outlet. Typically, 50 percent or more of the charge is vaporized.

Heat supplied to a heat transfer medium. Many plants furnish heat to individual users by means of an intermediate heat transfer medium, which is normally mineral or synthetic oil. A fired heater is used to elevate the temperature of this heat transfer medium. The fluid normally remains in the liquid state from inlet to outlet.

Heat supplied to viscous fluids. Heaters are used to increase the temperature of heavy oils that must be pumped in order to reduce viscosity and facilitate pumping or reduce pumping power. These types of heaters can be found in the pumping stations of pipelines

Fired reactors. This category includes heaters in which a chemical reaction occurs within the tube coil. These heaters normally are constructed with the fired-heater industry's most sophisticated technology. They normally operate at very high temperatures and pressures and require high-quality materials and accurate control of the temperature in every zone of the heater. An example is a steam-hydrocarbon reformer heater, in which the tubes of the combustion chamber function individually as vertical reaction vessels filled with nickel catalyst. The outlet temperatures are about 800–900°C. Another example is a pyrolysis heaters, used to produce olefins from gaseous feedstocks such as ethane and propane and from liquid feedstocks such as naphtha and gas oil. The tubes and burners are arranged so as to ensure pinpoint firing control. Outlet temperatures are also in the range of 800–900°C.

13-4 HEATER TYPES

Process furnaces generally are important and large pieces of equipment. To achieve a high heat-flux density, a high gas emissivity is required. As we have studied, gas emissivity is a function of volume, so a large firebox is necessary. Typical refinery heaters have the size of a regular room.

The operation of a fired heater must be monitored carefully owing to the large amount of energy normally handled and the potential risks that direct-fired units present. For this reason, these units are built under strict codes and safety standards.

Standard API 560/ISO 13705 ("Fired Heaters for General Refinery Services")[6] is the best known standard, and usually, fired-heater specifications require adherence to it. This standard defines criteria for thermal and mechanical design of the heater and construction details of the several heater components, such as tubes, burners, structure, thermal insulation, stack, etc.

There are several types of process heaters, but normally, all of them are variations on the three basic types shown in Fig. 13-5. In what follows, we shall describe the most common geometries.

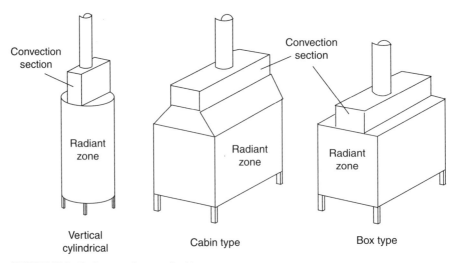

FIGURE 13-5 Basic types of process fired heaters.

Vertical Cylindrical Fired Heaters.   These are probably the most common fired heaters in use today and are used for heat duties up to about 40 million kcal/h. The radiant section is a cylindrical chamber, and the coil tubes are standing or hanging vertically in a circle around the floor-mounted burners. Thus the flame extends parallel to the tubes. The furnace floor must be elevated about 2 m with respect to the ground to allow maintenance and access to the burners (Fig. 13-6).

One of the most important parameters in heater design is the heat-flux density received by the surface of the tubes in the radiant chamber. Although heater size can be reduced by increasing the radiant heat-flux density, this also will increase the operating temperature of the tube material and reduce its service life. There are therefore upper-limit values for the average heat-flux density in a radiant chamber (see Table 13-2).

However, not only must the average heat flux on the tube surface be limited, but it is also important that the heat-flux density be as uniform as possible over the whole tube length because otherwise some sectors of the tubes may suffer localized damage. This is why API Standard 560 defines maximum values for the ratio between tube length and other dimensions of the radiant chamber because for very long tubes it would be difficult to maintain a uniform heat density, and the tube zone closer to the burners would suffer more deterioration.

In the case of cylindrical vertical heaters, the maximum L/D ratio allowed by the standard is 2.75 (L is tube length and D is diameter of the tube circle).

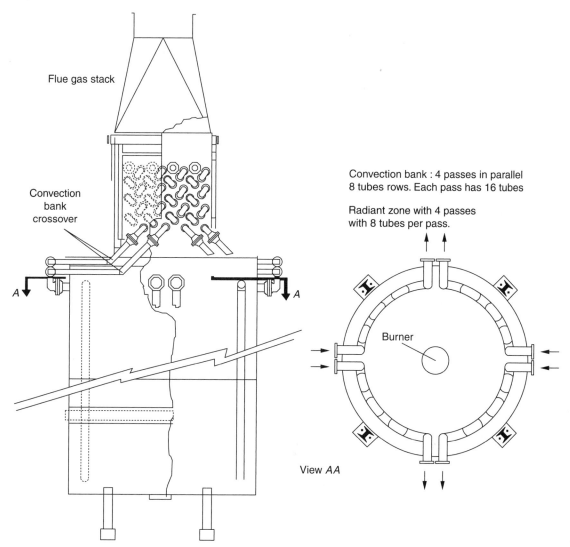

Flue gas stack

Convection bank crossover

A

A

Convection bank : 4 passes in parallel 8 tubes rows. Each pass has 16 tubes

Radiant zone with 4 passes with 8 tubes per pass.

Burner

View *AA*

FIGURE 13-6 Vertical cylindrical fired heater.

TABLE 13-2 Typical Values of the Average Heat-Flux Density in the Radiant Zone

Type of Heater	Average Heat-Flux Density (W/m^2)
Crude oil heaters in atmospheric distillation	31,400–44,000
Crude oil heaters in vacuum distillation	25,000–31,400
Reboilers	31,400–37,600
Hot oil heaters	25,000–34,500
Catalytic reforming furnaces	23,500–37,600
Delayed coker heaters	31,400–34,599
Propane deasphalting units	25,000–28,000
Charge heaters in hydrotreating units	31,400
Charge heaters in catalytic cracking units	31,400–34,500
Steam superheaters	28,200–40,000
Gasoline plant heaters	31,400–37,600

Cylindrical vertical heaters have the convection section in the upper part. The convection section consists of a bank of tubes whose length is approximately equal to the diameter of the radiant section. The two lowest tube rows can "see" the radiant chamber, so they receive radiant heat from it. These tubes are normally considered part of the radiant section and are called *shield tubes* because they protect the upper tube rows from radiation.

The fluid circulating into the tubes suffers a pressure drop that must be below the allowable value specified by process considerations. To avoid an excessive pressure drop, it is sometimes necessary to divide the process stream in two or more parallel paths that receive the name of *passes.* (It is important to note that the concept of pass is different from that used in heat exchanger terminology. In fired heaters, passes are parallel branches, whereas in heat exchangers, a pass is each one of the fluid passages through the unit.) For example, Fig. 13-6 represents a heater with a four-pass arrangement.

It is important to have a symmetric pass arrangement so that each pass can receive the same flow rate and the same heat flux. This is particularly important in vaporizing heaters because an uneven heat distribution may lead to a very different hydraulic behavior in different passes. This is why it is always recommended to design heaters to work in a two-phase regime with as low a number of passes as possible.

Figure 13-7 shows some variations of the vertical cylindrical heater. That in Fig. 13-7a is an all-radiant furnace (without a convection section). It is inexpensive, but since the temperature of the gases leaving the furnace is high (800–1,000°C), it has a very low efficiency. It is used in low-duty applications where efficiency is not important. The heater in Fig. 13-7b is also used for low capacities. It cannot be employed when more than one parallel pass is necessary. The construction in Fig. 13-7c is the most common and is the same type represented in Fig. 13-6.

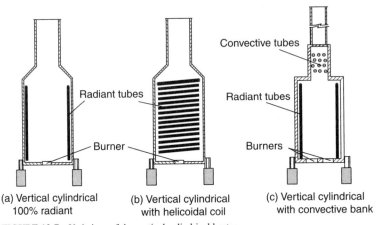

FIGURE 13-7 Variations of the vertical cylindrical heater.

Constant-Cross-Section Heaters (Cabin and Box Types). Cabin- and box-type heaters have a constant cross section, and the tubes are most times installed in horizontal position. The radiant section of a cabin-type heater includes tubes on the sidewalls and on the sloping roof of the cabin, which is called the *arch.* In box-type heaters, the arch is horizontal. The convection section extends the entire length of the heater, so it is possible to use longer tubes than in vertical cylindrical heaters.

In small sizes (up to about 40 million kcal/h), cylindrical vertical heaters are more economical than cabin- or box-type heaters; however, these types have some advantages as:

• Tubes can be completely drained.

• The problems originating in two-phase flow are less severe (vertical tube heaters are prone to slug flow owing to the accumulation of liquid in the ascending flow tubes).

• Even though the usual design is with floor-mounted burners, it is also possible to install the burners in the lateral walls so that the heater floor can be at ground level.

For horizontal-tube heaters with floor-mounted burners, API Standard 560 also defines upper and lower limits to the height:width ratio in order to achieve a uniform heat-flux density on the tube surface. These ratios depend on the heater duty and are indicated in the following table:

Heater Capacity (MW)	Maximum H/W	Minimum H/W
Up to 3	2	1.5
3–6	2.5	1.5
Above 6	2.75	1.5

H is the vertical distance from the floor to the arch, and W is the distance between refractory walls. For horizontal-tube heaters with wall-mounted burners (such as that in Fig. 13-9a), the maximum tube length allowed by the standard is 12 m.

Figures 13-8 through 13-10 show some possible variations on these heaters. The heater designs in Fig. 13-8b and c allow use of the same furnace for two different services because it is possible to regulate the fire in each radiant chamber. However, since both chambers influence each other, when one of them is

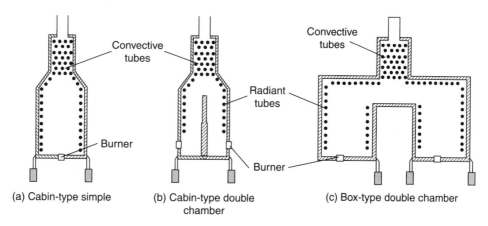

(a) Cabin-type simple   (b) Cabin-type double chamber   (c) Box-type double chamber

FIGURE 13-8   Constant-cross-section heaters.

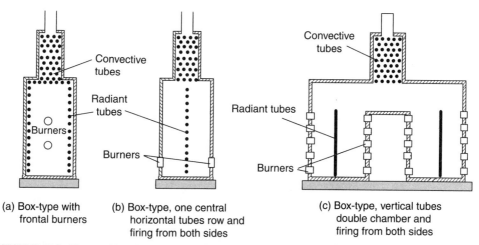

(a) Box-type with frontal burners   (b) Box-type, one central horizontal tubes row and firing from both sides   (c) Box-type, vertical tubes double chamber and firing from both sides

FIGURE 13-9   Heaters with wall-mounted burners.

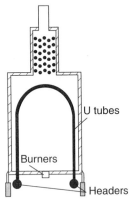

FIGURE 13-10 Arbor-type heater.

working at maximum capacity, it is not possible to reduce the heat input to the other below 50–75 percent (depending on whether both services share the convection section). Nor is it possible to interrupt the flow circulation in one chamber while the other is operating because this would damage the tubes.

The heaters in Fig. 13-9 have wall-mounted burners. They do not require maintenance access below the heater floor, and the foundation is simpler. In reaction furnaces, it is very important to have an even heat distribution around the tube circumference. In this case, it is usual to locate the tubes in the central part of the heater and install lateral burners so that the tubes are heated from both sides. This can be appreciated in Fig. 13-9*b* and *c*. The heater design in Fig. 13-9*b* has horizontal tubes, whereas that in Fig. 13-9*c* has vertical tubes.

Figure 13-10 represents a special design for high process flow rates when a low pressure drop is required. Each arc is a parallel branch, which allows low mass velocities owing to the high number of tubes in parallel. It is also possible to install interior walls to divide the heater in sections, thus allowing several services simultaneously, but as mentioned earlier, it is necessary to carefully study the effect that variations on heat load in one section must have in the others. Other geometries are possible, and some designs may be patent-protected.

13-5 HEATER COMPONENTS

13-5-1 Tubes

The diameters of the radiant-section tubes can range from 2 to 10 in, The more usual diameters are 4 and 6 in. API Standard RP 530[9] defines the tube materials that can be used, the procedures to perform the mechanical calculations, and the corrosion allowances according to the material selected. The most common tube material is carbon steel in all services where oxidation and corrosion are not very severe. Alloy steels, used for high-temperature applications, contain chromium, molybdenum, or silicon. Molybdenum is added to improve the mechanical resistance; chromium is added to reduce graphitization and improve resistance to oxidation, the same as silicium. Austenitic steels (chrome-molybdenum stainless steels) are used for severe corrosive or oxidative conditions and resist very high temperatures.

The material selection must be based in the metal temperature, which is higher than the process-fluid temperature. Additionally, it must be considered that this temperature difference may increase with time owing to the formation of coke deposits on the interior surface of the tubes. The following table lists the maximum design temperatures for various tube materials:

Material	Type or Grade	Maximum Temperature
Carbon steel	B	538°C
C-$\frac{1}{2}$Mo	T1 or P1	593°C
1$\frac{1}{4}$Cr-$\frac{1}{2}$Mo	T11 or P11	593°C
2$\frac{1}{4}$Cr-1Mo	T22 or P22	649°C
5Cr-$\frac{1}{2}$Mo	T5 or P5	649°C
7Cr-$\frac{1}{2}$Mo	T7 or P7	704°C
9Cr-1Mo	T9 or P9	704°C
18Cr-8Ni	304 or 304H	815°C
16Cr-12Ni-2Mo	316 or 316H	815°C
18Cr-10Ni-Ti	321 or 321H	815°C
18Cr-10Ni-Cb	347 or 347H	815°C
Ni-Fe-Cr	Alloy 800H	982°C
25Cr-20Ni	HK-40	1,010°C

Plain tube type are used in the radiation section, but in the convection section, extended-surface tubes normally are used to compensate for the low flue-gas heat transfer coefficient. These may consist of studs welded to the tubes or transverse fins (Fig. 13-11). The studs must have a minimum diameter of $1/2$ in and a maximum height of 1 in.

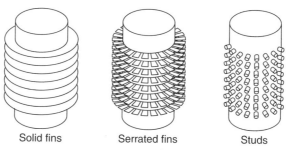

Solid fins   Serrated fins   Studs

FIGURE 13-11   Types of fins.

The most important consideration when selecting the type of extended surface is the ease of cleaning. Gaseous fuels (natural gas or refinery fuel gas) are considered clean fuels, and they do not produce appreciable fouling on the exterior surface of the tubes. On the other hand, in the case of fuel oil, soot deposition is quite normal. To clean the tubes, soot blowers sometimes are installed, but in this case, the fins height must be smaller and the fin spacing higher to facilitate the operation. The maximum fin height allowed by the API standard is 1 in for gaseous fuels and $3/4$ in for fuel oil. The maximum fin density is 197 fins/m for gaseous fuels and 118 fins/m for liquid fuels. Studs are cleaned more easily but are more expensive, resulting in about a 5 percent increase in the total cost of the heater.

The fins used in fired heaters are thicker than those used for air coolers and can be solid or serrated. Serrated fins are manufactured by effecting transverse cuts on a metallic ribbon, leaving an uncut portion. When the ribbon is wrapped around the tube, the individual segments separate at the outer edge, forming individual rectangular fins. Since the fins are neither stretched nor compressed while the ribbon is being wrapped around the tube, the granular structure, physical properties, corrosion resistance, and fin thickness remain equal to that of the original ribbon. After being wrapped, the ribbon is welded by means of a high-frequency process.

In a solid fin, these transverse cuts are not made, and during the wrapping process, the outer metal fibers are stretched, and the inner fibers are compressed. This results in a slight slimming of the fins at the external edge and fattening or waving at the base.

Connections from one tube to the other are made with U bends. In the radiant section, the bends are inside the radiant chamber. In the convection section, tubesheets such as that shown in Fig. 13-12 are used.

API Standard RP 560 specifies how the tubes must be supported in both the radiant and the convection sections and gives all the necessary criteria for the mechanical design of these supports. Sometimes, instead of using U bends, special pieces called *plug headers* are used. These are provided with removable plugs that allow interior cleaning of the tubes, but this also increases the heater cost.

Nowadays, when in-tube coke deposition is anticipated, decoking systems are installed. The traditional steam-air decoking systems are based on alternate circulation of air and steam into the tubes for coke deposit burning and scale removal. Most modern systems employ the circulation of cleaning balls (normally called *pigs*) in the tubes with the help of an auxiliary fluid.

13-5-2   Burners

The burners can fire gas or liquid fuels or both. The burners must be selected so that their maximum heat release is a certain percentage higher than the normal heat duty of the heater. API Standard RP 560 defines this excess as a function of the number of burners of the heater. These values range from 15 to 25 percent.

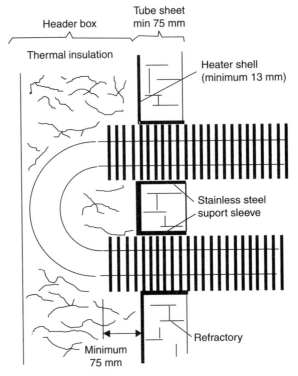

FIGURE 13-12 Tubesheets.

Another API publication (API RP 535, "Burners for Fired Heaters") deals with all aspects of the design and operation of heater burners.

Usually the burners have pilots, which are small auxiliary burners that provide ignition to the main burners or reignition in case of flame-off. Gas burners are broadly categorized into two types: natural-draft or forced-draft. In the first category, atmospheric air is induced through the burner either by the negative pressure inside the firebox or by fuel-gas pressure that suctions the air through a venturi. These burners are provided with adjustable slots that allow manual regulation of the fuel-air ratio.

Forced-draft burners are supplied with combustion air at a positive pressure by a combustion air fan. In this case, an air distribution box is installed in the burner's zone. This box is called a *plenum.*

Sometimes it is specified that forced-draft systems be designed to be capable of operating in natural draft at reduced capacity. Natural draft may be required when the air fan fails. Air doors at the plenum should open automatically to provide a source of ambient air on fan failure.

In the burner's flame, there is a zone where the temperature reaches a maximum or peak value. The location of this zone and the maximum temperature depend on how the mixing of fuel gas and air is done.[8] In some type of burners, called *premix burners,* air and fuel gas are mixed prior to entering the combustion zone. A premix burner is represented in Fig. 13-13.

A raw-gas burner is a gas burner in which both components enter the combustion zone separated. Combustion starts only when the mixture reaches the flammability range. This type of burner presents a longer flame, and peak temperatures are lower. Lower temperatures reduce the NO_x production.

One of the most important subjects to be considered in a burner's specification is the production of contaminants, especially nitrogen oxides. These are compounds that form in the zones of the flame where temperatures are higher by reaction of the fuel-bound nitrogen with oxygen. The main component of the NO_x is NO. This is a gas that rapidly reacts with atmospheric oxygen to form NO_2.

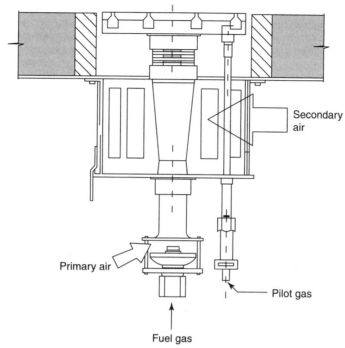

FIGURE 13-13 Premix burner.

NO$_2$ is a very reactive gas, and when it comes in contact with water, it produce highly corrosive nitrous and nitric acid. Thus, when NO$_2$ comes in contact with rain, it produces the *acid rain* that is highly harmful not only for vegetation but also for buildings, structures, etc.

To control the NO$_x$ production in a burner, it is important to avoid high-temperature zones in the flame. For this purpose, stage-air burners are usually employed. These burners complete the combustion in two separate combustion zones. All fuel is injected into the primary combustion zone with only a portion of the total air. Much of the fuel does not ignite because insufficient air is available. This incomplete combustion results in a lower flame temperature. The flame loses heat as it radiates to the surroundings. NO$_x$ production is limited by the lower flame temperature and the low oxygen content, and the nitrogen atoms present in the fuel form molecular N$_2$ rather than NO$_x$. Combustion is completed in the secondary combustion zone, where the remaining air is injected. Flame temperatures will not approach those in a standard burner because heat already has been lost to the surroundings during the initial combustion stage. A stage-air burner is shown in Fig. 13-14.

The most important aspect of burner selection is to achieve an even heat distribution within the firebox, thus avoiding localized overheating of the tubes and direct impingement of the flame on the metal surface. The burner vendor normally indicates the size of the flame for the maximum heat release, and the heater configuration must be such that the flame does not touch the tube surface. API Standard RP 560 defines minimum distances between burner and tubes as a function of the burner heat release. The art and expertise in fired heater design are mainly related to proper selection of the number and type of burners and their distribution in the heater to satisfy all these objectives.

13-5-3 Stack and Draft

The pressure within the firebox is always below atmospheric. This depression is called *draft* and is created by the buoyant effect of the hot flue gases in the stack that draws air into the combustion zone. The draft

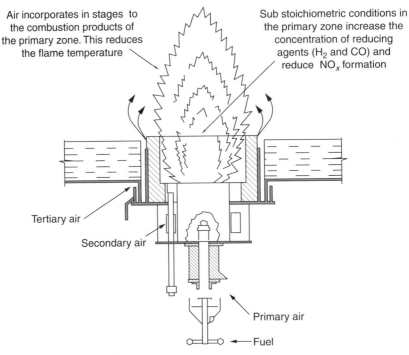

Air incorporates in stages to the combustion products of the primary zone. This reduces the flame temperature

Sub stoichiometric conditions in the primary zone increase the concentration of reducing agents (H_2 and CO) and reduce NO_x formation

Tertiary air

Secondary air

Primary air

Fuel

FIGURE 13-14 Stage-air burner.

produced by a hot-gas column depends on the density difference between the atmospheric air and the hot gas and can be calculated as

$$\Delta p = Lg(\rho_{air} - \rho_{flue\ gas}) \qquad (13\text{-}5\text{-}1)$$

approximately[3]

$$\Delta p\ (\text{Pa}) = 3,465L\left(\frac{1}{T_{air}} - \frac{1}{T_{flue\ gas}}\right) \qquad (13\text{-}5\text{-}2)$$

where temperatures are in K and L is the stack height in meters. Owing to heat losses, the temperature of the flue gas at the top of the stack is quite a bit lower than at the stack base. This difference can be estimated conservatively to be 40°C.

To estimate the stack diameter, a reasonable criterion is to define it so as to have a mass velocity of the flue gases of about 3.5–5 kg/(m²·s). It is recommended that the stack design create a negative pressure of 1.3 mm H_2O at the entrance to the convection section. Additionally, the draft must compensate the frictional pressure drop through the convection section and the stack itself. The frictional pressure drop in the stack can be calculated[4] as

$$\Delta p_{stack}\ (\text{Pa}) = 2.76 \times 10^{-5}\left(\frac{G_H^2 T}{D}\right)L \qquad (13\text{-}5\text{-}3)$$

where G_{II} is the mass velocity of the flue gas in kg/(s·m²), T is the absolute temperature in kelvins, D is the stack diameter in meters, and L is the height.

The other frictional pressure drops can be expressed in terms of velocity heads ($G_H^2/2\rho$) at the section under consideration. For plain tubes in the convection section, a rough estimation is 0.2 velocity heads per tube row. For finned tubes, one velocity head per tube row can be considered, or the correlations in Sec. 13-6-5 can be used for a better prediction. For the stack entrance, 0.5 velocity heads must be considered. One velocity head is the loss at the stack outlet, and 1.5 velocity heads is the loss at the draft regulating damper.[10]

13-5-4   Heat Recovery

Sometimes, to increase the thermal efficiency of the process, part of the residual heat of the flue gas is recovered to preheat the combustion air. This requires rather complex and expensive installations that are cost-effective only in high-capacity fired heaters. An example is shown in Fig. 13-15.

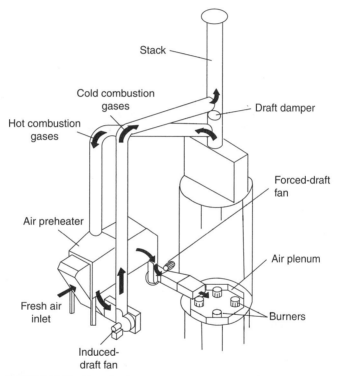

FIGURE 13-15   Air preheating system.

The combustion gases at the outlet of the convection zone are taken by an induced-draft fan that causes them to circulate through a heat exchanger, where they preheat the fresh air going into the heater. This fresh air, in turn, is circulated through the preheater and the burners by a forced-draft fan. After the heat exchange, the cold flue gases enter the stack to be evacuated.

13-5-5   Control Systems

The amount of fuel fed to the heater is normally regulated as a function of the outlet temperature of the process stream. In heaters with natural-draft burners, the amount of air induced by the burners cannot be measured, and the air must be manually adjusted by moving the air registers of the burners. For good regulation, it is necessary to have an analysis of the flue gas, which can be performed manually with an Orsat apparatus or similar device or with an inline gas analyzer. The color of the flame is another parameter that allows one to know if the air:fuel ratio is deficient, in which case the flame is yellow (normal color for a good air-fuel mixture tends to blue). The other element that must be monitored is the fuel efficiency, measured by fuel consumption. A low fuel efficiency can indicate a high excess of air. In forced-draft heaters, it is possible to install an air flowmeter between the fan and the burners, and an automatic control valve also can be installed to keep the air:fuel ratio within design values even with a variable heater load.

Another important aspect is operational safety. The most important risk in heater operation is in feeding fuel gas to a heater in the absence of flame. This will create an explosive atmosphere in the firebox, which could explode when the burner ignition is attempted.

All modern fired heaters have safety systems based on PLC logic that handles the startup operations, performing the air purge before any ignition attempt and reading the signals coming from the flame detectors that close the fuel valves in case of flame failure. These systems are called *burner management systems* (BMSs).[7]

13-6 ELEMENTS FOR THE DESIGN AND VERIFICATION OF THE CAPACITY OF A FIRED HEATER

Fired heater design normally is performed by specialized companies. Each manufacturer has its own design techniques, and some commercial software with variable degrees of sophistication is available to perform this task. However, even though the design in all cases must be accomplished by a qualified vendor, sometimes users and purchasers must perform their own calculations to

- Estimate the size of a heater and fuel consumption in a preliminary process design stage
- Evaluate and compare the consistency of vendor proposals
- Predict the effect of changes in flow rates, physical properties, and other operating variables
- Anticipate the effect of modifications on existing heaters

In what follows, we shall present some tools that can be used to verify the heat transfer capacity of a process fired heater. These tools must be considered as primary elements for the preceding purposes. The final design of the heater, however, in all cases must be performed by specialized companies.

13-6-1 Radiant-Zone Model

The radiant heat exchange in the firebox can be represented by a model such as that shown in Fig. 13-16. The figure represents a gas at temperature T_G acting as energy source, a surface A_2 absorbing heat, and reradiant refractory walls. It is evident that owing to the complexity of the heater geometry, the principal difficulty is the calculation of the several view factors. We shall study some simplifications to calculate them.

Equivalent Gray-Plane Simplification.   Usually, the heat-absorbing surface (the tubes) is located before a refractory wall, as shown in Fig. 13-17. Let's consider the imaginary planar surface 1–1 tangent to the tubes. From the total radiation that reaches this surface, a portion (1) is directly incident on the tube surface and, if the tubes are considered a blackbody, will be absorbed completely. Another fraction (2) reaches the refractory wall behind the tubes and is reflected. Part of this reflected energy (4) strikes on the backside of the tube surface and is absorbed, whereas the rest of the reflected energy (3) comes back to the source.

If the bank of tubes is exchanging energy with a parallel plane, the graph in Fig. 13-18 (which was already introduced in Chap. 12 as Fig. 12-15) allows one to calculate each one of these fractions and to obtain a factor $\bar{F}_{12}$ representing the fraction of the energy absorbed with respect to the incident energy, including that absorbed directly and that absorbed after reflection in the refractory wall.

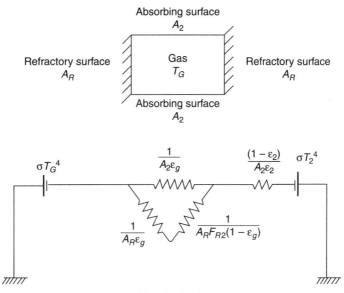

FIGURE 13-16   Model of a radiant heater.

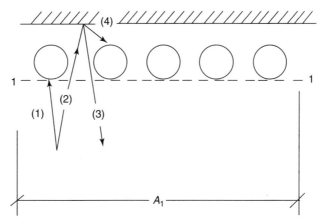

FIGURE 13-17   Equivalent plane simplification.

Let's consider a system consisting of a bank of tubes backed by a refractory wall, as shown in Fig. 13-17. This system is exchanging radiant energy with a planar surface at temperature T_E parallel to the tubes. If T_2 is the temperature of the tube surface, the exchanged energy per unit time is

$$Q = A_1 \bar{F}_{12} \sigma \left(T_E^4 - T_2^4 \right) \tag{13-6-1}$$

If, instead of the tube bank, the surface absorbing energy were another planar surface such as 1–1 at temperature T_2, the heat transfer rate would be

$$Q' = A_1 \sigma \left(T_E^4 - T_2^4 \right) \tag{13-6-2}$$

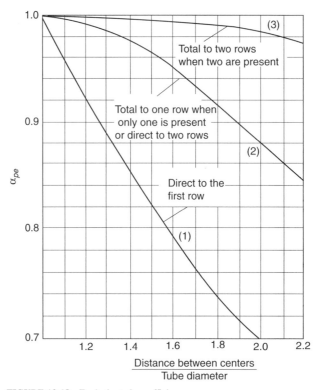

FIGURE 13-18   Equivalent plane efficiency.

because the view factor is unity. If we compare both equations, we see that it is possible to substitute the tube bank, including the refractory wall, by a planar surface with area A_1 at the same temperature as the tubes but affecting this area by a certain equivalent plane efficiency, which shall be called α_{pe}. Thus

$$Q = A_1 \alpha_{pe} \, \sigma \left(T_E^4 - T_2^4 \right) \tag{13-6-3}$$

It is evident, comparing Eqs. (13-6-1) and (13-6-3), that

$$\alpha_{pe} = \bar{F}_{12} \tag{13-6-4}$$

We thus could have continued designating this variable as $\bar{F}_{12}$, avoiding a new name. However, it is considered that the new meaning we are assigning to this parameter (as efficiency of the equivalent plane) is more clearly represented by the new nomenclature, and we prefer not to continue designating it as a view factor.

It then would be possible to substitute the heater surfaces where the tubes are located by an equivalent plane tangent to the tubes. This plane must be affected, in the radiant heat transfer equations, by an efficiency factor numerically equal to the total view factor obtained from Fig. 13-18 using the curve that corresponds to the geometric array of the tubes.

The temperature of this equivalent plane will be equal to the tube temperature. In the preceding explanation, we considered that the surface of the tubes was a blackbody, but if, additionally, the tubes have a certain emissivity ε_2, the same emissivity must be assigned to the equivalent plane.

The reasoning we have presented does not pretend to be a rigorous demonstration of the validity of the substitution performed. However, despite the weakness in the theoretical fundamentals, the procedure has given good results in practice.

The radiant heat exchange in the heater firebox thus will be represented by the analog electric circuit in Fig. 13-19. If the equivalent resistance enclosed within the dashed-line envelope is called

$$\frac{1}{A_1\alpha_{pe}\overline{F}_{1G}}$$

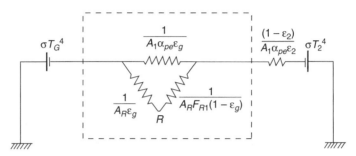

FIGURE 13-19 Equivalent electric circuit of the firebox.

and designating $(1/A_1\alpha_{pe}\mathcal{F}_{1G})$ as the resistance of the complete circuit, we have

$$\frac{1}{A_1\alpha_{pe}\mathcal{F}_{1G}} = \frac{1}{A_1\alpha_{pe}\overline{F}_{1G}} + \frac{1-\varepsilon_2}{A_1\alpha_{pe}\varepsilon_2} \tag{13-6-5}$$

To obtain the value of $1/A_1\alpha_{pe}\overline{F}_{1G}$, it is necessary to obtain the equivalent resistance as

$$\frac{1}{\dfrac{1}{A_1\alpha_{pe}\overline{F}_{1G}}} = \frac{1}{\dfrac{1}{A_1\alpha_{pe}\varepsilon_g}} + \frac{1}{\dfrac{1}{A_R\varepsilon_g}+\dfrac{1}{A_RF_{R1}(1-\varepsilon_g)}} \tag{13-6-6}$$

This means that

$$A_1\alpha_{pe}\overline{F}_{1G} = A_1\alpha_{pe}\varepsilon_g + \frac{A_R\varepsilon_g}{1+\dfrac{\varepsilon_g}{F_{R1}(1-\varepsilon_g)}} \tag{13-6-7}$$

or

$$\overline{F}_{1G} = \varepsilon_g\left[1+\left(\frac{A_R}{A_1\alpha_{pe}}\right)\frac{1}{1+\dfrac{\varepsilon_g}{F_{R1}(1-\varepsilon_g)}}\right] \tag{13-6-8}$$

The total heat flux transferred by radiation from the gas to the tubes then can be calculated as

$$Q = A_1\alpha_{pe}\mathcal{F}_{1G}\sigma\left(T_G^4 - T_2^4\right) \tag{13-6-9}$$

The subject still unsolved is calculation of the view factor F_{R1}, which does not look like a simple task owing to the complexity of the geometries involved. Later on in the text we shall present a simplified method, originally proposed by Lobo and Evans, that allows an easy estimation of this factor.

Radiation to the Shield Tubes.   The shield tubes that also receive radiation from the firebox require a special treatment. These tubes do not have a back-reradiant refractory wall. This means that all the energy not absorbed directly by the shield tubes passes to the convection section to be finally absorbed there. Thus, from the model point of view, we can consider that α_{pe} for the equivalent plane that replaces the shield tubes is unity because all the energy is absorbed. Thus α_{pe} must be applied only to the tubes having a back-refractory surface. The area of the equivalent plane for the shield tubes is taken with its total value.

Convective Effects.   Even though most of the heat in the firebox is transferred by radiation, the combustion gases also transfer some convective heat to the surface of the tubes. It is thus necessary to add the convective contribution to Eq. (13-6-9). The final expression is

$$Q = A_1\alpha_{pe}\mathcal{F}_{1G}\sigma\left(T_G^4 - T_2^4\right) + A_2 h_G(T_G - T_2) \tag{13-6-10}$$

where h_G is the convective heat transfer coefficient between the gases and the tube wall.

13-6-2 Heat-Flux Density in the Radiant Section

The first step in heater design is to select the average heat-flux density. This is the amount of energy per unit area and per unit time received by the tubes. This can be defined as the quotient between the total heat delivered to the process fluid in the radiant section and the exterior area of the tubes (Q/A_2).

As mentioned earlier, the higher this radiant heat density, the smaller will be the required heat transfer area and the cost of the heater. However, to achieve a high heat-flux density, it is necessary to operate with high flue-gas temperature, and this will result in high maintenance costs. Since refractories and tube supports will be exposed to elevated temperatures, their service life will be shorter. Additionally, high tube-wall temperatures also reduce the service life of the tubes and increase the probability of coke formation on the interior surface of the tubes.

This is why the selection of heat-flux density to be used in the design must be agreed between vendor and purchaser. Typical values of the average heat-flux density for several services are shown in Table 13-2.

It also must be considered that the heat flux density is not uniform in the entire tube circumference. The variation depends on the ratio between tube diameter and spacing and whether the tubes receive heat from only one side or both. API Standard 530 ("Calculation of Heater Tube Thicknesses in Petroleum Refineries") includes a graph to calculate the ratio between maximum and average radiant flux depending on tube arrangement. For single-tube-row heaters, the maximum heat flux ranges between 1.5 and 3 times the average value.

13-6-3 Design and Verification of Fired-Heater Capacity: The Method of Lobo and Evans[2]

Generalities, Hypotheses, and Limitations.   The Lobo-Evans method, since its publication in 1939, has been a simplified design approach that is very popular despite of its limitations, which are important to clearly point out. They are

• It considers that the average temperature of the combustion gases into the firebox is equal to the temperature at which they exit it. This hypothesis gives good results in cabin- or box-type heaters having an approximately square section. However, in vertical cylindrical heaters or even in box-type heaters in which the flue-gas path is longer than the transverse dimensions, this is not true. This gave rise to a series of investigations from the 1950s to the 1970s (when these types of heaters became more popular) to find calculation methods suitable to these geometries. A simple suggestion was made by Wimpress[3] and consists of using the Lobo-Evans method but assuming a temperature difference between the flue gases leaving the firebox and the average temperature of the flue gases inside the firebox. In other words, when writing the energy balance it is assumed that the flue gases leave the firebox at a temperature which is somewhat lower than the average flue gas temperature used for heat transfer calculations. This difference can be as high as 100°C, but the adoption of a design value for this temperature difference is a very important subject and has to be based on a sound experience in similar types of heaters.

• The method only allows to calculate a mean value for the heat transfer rate in the radiant section assuming a uniform tube-wall temperature. The maximum temperature on the tube surface is an important parameter for heater design because it must be used for selection of the tube material. This method does not predict the temperature distribution over the tube surface and this maximum value.

• The original publication of the method only included information to estimate the equivalent mean hemispherical beam length for six geometries characterized by the ratios between the main dimensions of the firebox (Table 13-3).

TABLE 13-3   Equivalent Mean Hemispherical Beam Length

Dimensional Ratio (Length, Width, Height in Any Order)	L
Rectangular furnaces	
1-1-1 to 1-1-3	⅔(heater volume)$^{1/3}$
1-1-4 to 1-1-∞	1 × smaller dimension
1-2-5 to 1-2-8	1.3 × smaller dimension
1-3-3 to 1-∞-∞	1.8 × smaller dimension
Cylindrical furnaces	
$d \times d$	⅔ × diameter
$d \times 2d$ to $d \times \infty$	1 × diameter

Calculation of the View Factor F_{R1}.   The most important contribution of the method consists of the introduction of simplified equations to calculate the view factor F_{R1}. Lobo and Evans correlated this variable as a function of the parameter $A_R/A_1\alpha_{pe}$ as follows:

If

$$0 \leq \frac{A_R}{A_1\alpha_{pe}} \leq 1$$

then

$$F_{R1} = \frac{A_1\alpha_{pe}}{A_T} \tag{13-6-11}$$

where

$$A_T = A_1\alpha_{pe} + A_R$$

If

$$3 \leq \frac{A_R}{A_1\alpha_{pe}} \leq 6.5$$

then

$$F_{R1} = \frac{A_1\alpha_{pe}}{A_R} \tag{13-6-12}$$

If

$$1 \leq \frac{A_R}{A_1 \alpha_{pe}} \leq 3$$

then

$$\frac{A_1 \alpha_{pe}}{A_R} > F_{R1} > \frac{A_1 \alpha_{pe}}{A_T} \tag{13-6-13}$$

Gas Emissivity ε_g.   The gas emissivity included in the equations of the model can be evaluated with the McAdams graphs introduced in Chap. 12. However, it is also possible to use simpler methods. When the flue gases are combustion products of usual refinery fuels, the ratio of vapor pressures of water and CO_2 is quite constant.

This allows use of a single graph such as that shown in Fig. 13-20, in which the gas emissivity is obtained as a function of the radiant chamber temperature. Even though, as explained in Chap. 12, the gas emissivity also depends on the temperature of the cold surface, for the construction of the graph in Fig. 13-20, it was considered that the tube wall temperature in a refinery heater is in most cases between 300 and 650°C. Within this range, the influence of the tube wall temperature on calculation of the gas emissivity is less than 1 percent. Thus a constant value was assumed, and the variable is not included in the graph.

The partial pressures of water vapor and CO_2 can be calculated from the stoichiometry of the combustion, as explained earlier. However, to speed up manual calculations, the graph in Fig. 13-21 is offered. This graph can be used for usual refinery fuels, and with only the excess air known, the sum of the partial pressures of water vapor and CO_2 can be obtained.

Tube Wall Temperature.   Equation (13-6-10) allows calculation of the heat flux from the combustion gases at temperature Tg to the tube wall surface at temperature T_2. As in any heat transfer device, it is necessary to couple this equation with that representing the heat transfer between the tube wall and the process fluid. If t_F is the mean temperature of the fluid inside the tubes, the temperature drop between the external

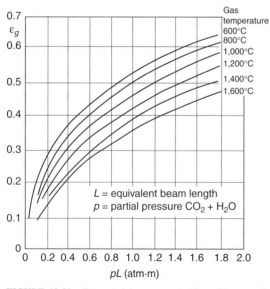

FIGURE 13-20 Gas emissivity versus *pL*. (*From Wimpress N: "Generalized Method Predicts Fired Heaters Performance,"* Chem Eng *May 22:95–102 1978. Reproduced with permission.*)

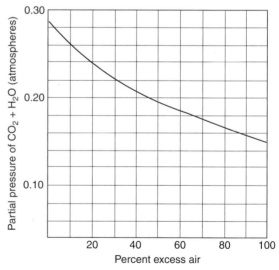

FIGURE 13-21   Partial pressure of $CO_2 + H_2O$ versus excess air for usual refinery fuels. (*From Wimpress N: "Generalized Method Predicts Fired Heaters Performance,"* Chem Eng *May 22:95–102 1978. Reproduced with permission.*)

tube surface and t_F takes place through the metallic wall, the fouling or coke deposit at the internal surface, and the boundary layer corresponding to the internal fluid. Thus

$$Q = \frac{(T_2 - t_f)}{\dfrac{e_m}{k_m} + \dfrac{1}{h_{io}} + R_{fi}} A_2 \qquad (13\text{-}6\text{-}14)$$

This equation allows calculation of the wall temperature T_2.

Heaters with Tubes in a Central Row.   For the foregoing discussion, the furnace is assumed to be of the type indicated in Figs. 13-7 or 13-8 with refractory-backed tubes around the periphery of the firebox. Certain adjustments are necessary in the case of a heater in which the tubes are arranged as in Figs. 13-9(*b*) and 13-9(*c*) with a single or double row of centrally mounted tubes. In such cases, advantage is taken of the fact that a plane of symmetry in the firebox can be replaced, for computational purposes, by a refractory wall (indicated by the dashed line in Fig. 13-22). Therefore, the equivalent plane area A_1 for the tube bank is twice the projected equivalent plane area because each side is figured separately. On this basis, α_{pe} for a single row is obtained from curve 3 and for a double row from curve 2 in Fig. 13-18. Similarly, the mean beam length L must be calculated on the basis of half a furnace.

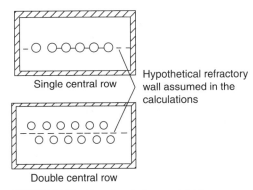

Single central row

Hypothetical refractory wall assumed in the calculations

Double central row

FIGURE 13-22   Case of centrally mounted tubes.

13-6-4 Calculation Procedure for All-Radiant Heaters

We shall now see how the previously explained concepts can be arranged in a calculation procedure to size an all-radiant heater (no convection zone). The heater will be designed to deliver a certain heat flux Q to a process stream whose flow rate is W_c and whose inlet and outlet enthalpies will be i_i and i_o corresponding to temperatures t_i and t_o, respectively. This means that

$$Q = W_c(i_o - i_i)$$

The design steps are as follows:

1. As in any other heat transfer equipment, a trial-and-error design procedure must be followed. After proposing a certain geometric configuration, the heat transfer capacity must be verified and checked against the required capacity. In the case of a fired heater, we have already explained that the principal design parameter is the average heat-flux density. Thus the design procedure starts by adopting a desired Q/A_2, and this allows calculation of A_2.

2. With the value of A_2, it is possible to define a first heater configuration, adopting the number, length, and diameter of the tubes. The separation between tubes normally depends on the type of accessories used for the return bends. Usually the separation:diameter ratio is in between 1.8 and 2.5. It is then possible to estimate the firebox dimensions and complete the geometric configuration.

3. The number of heater passes must be adopted. This is defined by process fluid velocity and pressure-drop considerations.

4. The convective heat transfer coefficient for the process fluid then can be calculated. This can be easy in the case of no phase change. In vaporization service, the methods of Chap. 11 can be employed.

5. With the mean temperature of the process fluid and the calculated h_{io}, it is possible to find the tube wall temperature T_2 with Eq. (13-6-14). This is the required tube wall temperature to transfer the desired heat flux to the process fluid in a heater with area A_2.

6. With Eq. (13-6-10), T_G is calculated. This is the temperature the combustion gases should have to transfer the required heat flux to the tubes with the proposed geometry.

To calculate T_G with Eq. (13-6-10), an iterative procedure must be employed because gas emissivity is, in turn, a function of T_G. It could be possible to use T_G as the iteration variable, but its value is somewhat difficult to estimate. Another option is to start the iteration procedure assuming a value for $\mathcal{F}_{1G}$. A good starting point is to assume 0.57 as initial guess.

Another variable included in Eq. (13-6-10) is the convection flue-gas film coefficient h_G. But owing to the smaller importance of the convective term in comparison with the radiation term, it is suggested to adopt $h_G = 11.5 \text{ W/(m}^2 \cdot \text{K)}$ without any further verification.

The complete procedure is then as follows:

a. With the tube arrangement (tube diameter and separation), calculate α_{pe} with Fig. 13-18.

b. With the assumed heater geometry, calculate the equivalent plane area A_1, refractories area A_R, total area A_T, and factor F_{R1} with Eqs. (13-6-11) through (13-6-13).

c. With an assumed $\mathcal{F}_{1G} = 0.57$ and $h_G = 11.5 \text{ W/(m}^2 \cdot \text{K)}$, calculate T_G from Eq. (13-6-10).

d. With this T_G, calculate the gas emissivity with Fig. 13-20.

e. By means of Eq. (13-6-8), calculate $\overline{F}_{1G}$.

f. By means of Eq. (13-6-5), calculate $\mathcal{F}_{1G}$ and compare with the assumed value. If they differ, go back to step 7c with the new value and recalculate up to convergence. The final T_G is the combustion gas temperature required to reach the desired heat flux with the proposed geometry.

As explained earlier, the higher the temperature T_G, the smaller is the combustion efficiency because more heat will be lost with the flue gas at the stack outlet. The following step is then necessary to calculate the amount of fuel that has to be burned (in other words, the heater efficiency).

7. In box-type heaters, it can be assumed that T_G is equal to the flue-gas exit temperature T_H. For cylindrical heaters with high L/D ratios, it is suggested to assume that the mean temperature of the gas in the radiation chamber is some degrees higher than the exit temperature. However, this must be supported by designer expertise in similar heaters. If this information is not available, it must be assumed conservatively that both temperatures are equal. Then the heater efficiency and the necessary amount of fuel can be calculated with Eq. (13-2-6).

8. If the calculated efficiency is not satisfactory, it is necessary to modify the proposed design either by adopting a different value for the radiant heat-flux density or by changing some of the geometric relationships that may improve $\mathcal{F}_{1G}$.

It is worth it to compare the heat transfer equations of a fired heater and a heat exchanger. The coefficient $\mathcal{F}_{1G}$ plays a similar role as the heat transfer coefficient U. Remembering the heat exchanger design equation, namely,

$$Q = UA\Delta T$$

we can see that to increase the heat flux, we can either increase the heat transfer area A or improve the heat transfer coefficient U or increase the temperature difference ΔT. In the case of a fired heater, it is possible to increase the heat flux by either increasing area A_2 or improving $\mathcal{F}_{1G}$ (e.g., increasing the tube pitch or the volume of the radiant chamber to improve gas emissivity) or increasing the temperature driving force by increasing T_G. The first two options will raise the cost of the heater, and the last one will increase fuel consumption.

9. Once the heat transfer design is done, the pressure drop for the process fluid has to be checked.

Example 13-3 Estimate the size of the radiant section of a cabin-type heater to heat up 45 kg/s of a 26°API crude oil from 250–357°C and the necessary fuel consumption. The mean specific heat of the process fluid is 2,978 J/kg.K. The fuel will be a refinery gas whose net heating value is 60.3 × 10^6 J/(Nm³) and whose density is 0.88 kg/(Nm³); 25 percent excess air will be used.

Four-inch nominal diameter tubes with exterior diameters of 0.1143 m and interior diameters of 0.103 m will be used. The tube length will be 11.5 m and the tube separation will be 203 mm (8 in).

In order to have a reasonable pressure drop in the process fluid, four parallel passes will be used. With this arrangement, the interior convection film coefficient results: 757 W/(m²·K). The design fouling resistance is 0.0003 (m²·K)/W. It is desired to have an average heat-flux density of 35,500 W/m².

Solution The heater will be a radiant convective type, but we shall only design the radiant zone. The oil enters the radiant zone at 250°C and exits at 357°C. Then

$$Q = W_c c_c (t_2 - t_1) = 45 \times 2,978 \times (357 - 250) = 14,339,000 \text{ J/s}$$

Estimation of the heat transfer area: With a heat-flux density $q = 35,500$ W/m², the necessary area will be

$$A_2 = Q/q = 14,339,000/35,500 = 403.9 \text{ m}^2$$

The number of tubes will be

$$N = A_2/(\pi D_o L) = 403.9/(3.14 \times 0.1143 \times 11.75) = 95.7 \text{ tubes (We adopt 96.)}$$
$$A_2 = 96 \times \pi \times D_o \times L = 96 \times 3.14 \times 0.1143 \times 11.75 = 404.8 \text{ m}^2$$

Determination of the geometric and thermal parameters: A cabin-type heater with horizontal tubes shall be used. We shall arrange the 96 tubes of the radiant zone along the perimeter of the heater cross section. Since there is a convective zone, we shall assume six shield tubes. Since we adopted the tubes separation, we now can calculate the necessary dimensions (Fig. 13-23).

The total area of the radiation chamber is

$$A_T = 2 \times [(6.96 \times 5.08) + {}^1\!/_2(6.96 + 1.22) \times 2.87)] + 2$$
$$\times [(4.06 + 5.08) \times 12.2)] + 6.96 \times 12.2 = 402.1 \text{ m}^2$$

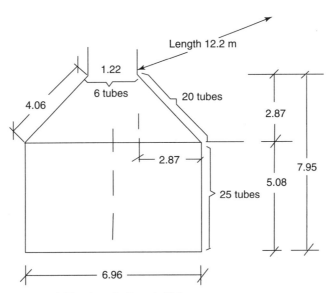

FIGURE 13-23 Figure for Example 13-3.

The area of the equivalent plane will be the sum of the areas of the radiant tube planes and the shield tube plane. The first one is

$$A_1 = \text{tube length} \times \text{tube number} \times \text{tube separation} = 90 \times 11.75 \times 0.203 = 214.8 \text{ m}^2$$

And the second one is

$$A_1 = 11.75 \times 6 \times 0.203 = 14.3 \text{ m}^2$$

For the tubes on the walls,

$$\alpha_{pe} = 0.92 \text{ (with } c/D = 0.203/0.1143)$$

For the shield tubes, we shall adopt $\alpha_{pe} = 1$, because it is assumed that all the radiation that is incident on the shield tube plane is absorbed either by the shield tubes or by the convective tubes behind. Then

$$A_1\alpha_{pe} = 214.8 \times 0.92 + 14.3 = 211.9$$
$$A_R = A_T - A_1\alpha_{pe} = 402.1 - 211.9 = 190.2$$
$$A_R/A_1\alpha_{pe} = 190.2/211.9 = 0.897$$
$$\therefore F_{R1} = A_1\alpha_{pe}/A_T = 211.9/402.1 = 0.526$$

Equivalent mean hemispherical beam length:

$$\text{Dimensions ratios } \frac{6.96}{6.96}:\frac{7.95}{6.96}:\frac{12.2}{6.96} = 1:1.15:1.76$$

$$L = \frac{2}{3}(\text{heater volume})^{1/3}$$

$$V = [(6.96 \times 5.08) + \frac{1}{2}(6.96 + 1.22) \times 2.87] \times 12.2 = 574.5 \text{ m}^3$$

$$L = \frac{2}{3}(574.5)^{1/3} = 5.54 \text{ m}$$

Calculation of the tube wall temperature: The heat flux through area A_2 can be expressed as

$$Q = \frac{T_1 - t}{\left(\dfrac{1}{h_{io}} + R_f\right)} A_2$$

where t is the average temperature of the process fluid $= \frac{1}{2}(250 + 357) = 303.5°C$. Then

$$\left(\frac{1}{h_{io}} + R_f\right)^{-1} = \left(\frac{0.114}{0.103 \times 757} + 0.0003\right)^{-1} = 567 \text{ W/(m}^2 \cdot \text{K)}$$

The tube wall temperature will be

$$T_1 = t + Q/(A_2 \times 567) = 303.5 + 14,339,000/(404.8 \times 567) = 365°C$$

Now, the flue-gas temperature must be calculated with an iterative procedure (we shall only show the last iteration steps). Thus

$$Q = A_1\alpha_{pe}\mathcal{F}_{1G}\sigma\left(T_G^4 - T_2^4\right) + A_2 h_G (T_G - T_2)$$

$$\therefore \mathcal{F}_{1G} = \frac{Q - A_2 h_G (T_G - T_2)}{A_1\alpha_{pe}\sigma(T_G^4 - T_2^4)}$$

Assuming that $T_G = 880°C$,

$$\mathcal{F}_{1G} = \frac{14,339,000 - 11.5 \times 404.8(880 - 366)}{211.9 \times 5.66 \times 10^{-8} \left[(880 + 273)^4 - (366 + 273)^4 \right]} = 0.62$$

Verification of $\mathcal{F}_{1G}$: From Fig. 13-20 we get $p_{CO_2} + p_{H_2O} = 0.23$ atm. Then

$$(p_{CO_2} + p_{H_2O})L = 0.23 \times 5.54 = 1.27 \text{ atm} \cdot \text{m}$$

With $T_G = 880°C$ and $pL = 1.27$, from Fig. 13-20 we obtain $\varepsilon_g = 0.53$. Then

$$\bar{F}_{1G} = \varepsilon_g \left[1 + \left(\frac{A_R}{A_1 \alpha_{pe}} \right) \frac{1}{1 + \dfrac{\varepsilon_g}{F_{R1}(1 - \varepsilon_g)}} \right] = 0.53 \left[1 + \frac{190.2}{211.9} \frac{1}{1 + \dfrac{0.53}{0.526(1 - 0.53)}} \right] = 0.681$$

$$\frac{1}{\mathcal{F}_{1G}} = \frac{1}{\bar{F}_{1G}} + \frac{1 - \varepsilon_2}{\varepsilon_2} = \frac{1}{0.681} + \frac{1 - 0.9}{0.9} = 1.58$$

$$\mathcal{F}_{1G} = 0.633$$

This value agrees reasonably with the 0.62 calculated previously.
Energy balance:

$$\frac{Q}{W_F} = \frac{W_A}{W_F} c_A (T_A - T_o) + c_F (T_F - T_o) + \text{NHV} \left(1 - \frac{\gamma}{100} \right) - \frac{W_H}{W_F} c_H (T_H - T_o)$$

Since $T_A = T_F = T_0$,

$$\frac{Q}{W_F} = \text{NHV} \left(1 - \frac{\gamma}{100} \right) - \frac{W_H}{W_F} c_H (T_H - T_o)$$

Net heating value (NHV) = 50.3×10^6 J/(Nm³)/[0.88 kg/(Nm³)] = 57.2×10^6 J/kg

In order to use the graph of Fig. 13-2, the energy balance can be written as

$$\frac{Q}{W_F} = \text{NHV} \left[1 - \frac{\gamma}{100} - \frac{W_H}{W_F \text{NHV}} c_H (T_H - T_0) \right]$$

The last term within the square brackets is what we get from Fig. 13-2. We enter the graph with 25 percent excess air and a gas temperature of 880°C, and we obtain an ordinate value of 0.45. Considering 2 percent conduction heat loss through the heater walls, we get

$$\frac{Q}{W_F} = \text{NHV}(1 - 0.02 - 0.45) = 57.3 \times 10^6 \times 0.53 = 30.3 \times 10^6 \text{ J/kg}$$

$$W_F = \frac{Q}{Q/W_F} = \frac{14,339,000}{30.3 \times 10^6} = 0.473 \text{ kg/s} = 1704 \text{ kg/h}$$

$$\eta = \frac{14,339,000}{57.2 \times 10^6 \times 0.473} = 0.53$$

This efficiency corresponds only to the radiant zone.

13-6-5 Radiant-Convection Heaters

At present, most process heaters have a convection section. The first two or three rows of the convection bank are the shield tubes. These tubes normally are arranged in a triangular pattern and protect the rest of the convection bank from direct radiation from the firebox.

The usual practice is to consider these tubes as part of the radiation section and to include them in the calculation of the radiation-absorbing surface. Afterwards, when calculating the convection section, the convective heat contribution from the flue gas is added.

The shield tubes are normally plain tubes (not finned) because they are exposed to a high heat-flux density. The rest of the convection bank tubes can be plain, finned, or studded.

Despite the fact that most of the heat transfer in the convection zone is convective heat, when plain tubes are employed, it is also necessary to consider the effect of radiation from the hot gases circulating through the bank and reradiation from the convection-zone walls. When finned tubes are employed, the radiation effects can be neglected against convection.

In previous chapters, correlations to calculate the heat transfer coefficient for tube banks, either plain or finned, have been presented. All these correlations were developed for heat exchangers, and predictions are not good when they are applied to greater-diameter tubes with higher fin spacing and high-temperature gases. For this reason, some correlations specifically developed for process fired heaters will be presented below.

Heat Transfer Correlations for a Finned-Tube Convection Bank. The following correlations are based on a paper from Weierman[1] and were adopted by the Escoa Company. They are valid for solid helicoidal fins. The publication also includes correlations for serrated fins.

Weierman recommends, for calculation of the heat transfer coefficient, a correlation of the type

$$h_f = jc_H G_H \operatorname{Pr}_H^{-0.67} \tag{13-6-15}$$

where G_H is the mass flow density. That is,

$$G_H = W_H / a_s \tag{13-6-16}$$

where a_s is the flow area in the central plane of a tube row. Calling

W = width of the tube bank
L = length of the tube bank
n_{tf} = number of tubes per row
n_f = number of tube rows
D_o = exterior diameter of tubes
D_b = fin diameter
b = fin thickness
H = fin height = $(D_b - D_o)/2$
N_m = number of fins per meter

then

$$a_S = WL - n_{tf} L \left[D_o + N_m (D_b - D_o) b \right] \tag{13-6-17}$$

The j factor included in Eq. (13-6-15) can be obtained as a function of a Reynolds number. When we studied correlations for air coolers, different Reynolds numbers for the calculation of film heat transfer coefficients and friction factors were defined. These Reynolds numbers were based on different equivalent diameters. Weierman's suggestion is to use the same Reynolds number for both heat transfer and friction factor, employing the tube diameter in its definition. Then

$$\operatorname{Re} = \frac{D_o G_H}{\mu_H} \tag{13-6-18}$$

The proposed correlation follows:

A_o = plain tube area per unit length = πD_o.
A_D = tube exposed surface per unit length = $\pi \times D_o \times (1 - N_m \times b)$.
A_f = fin surface per unit tube length = $\pi \times 2 \times N_m \times (D_b^2 - D_o^2)/4$.
a_p = projected finned tube area per unit tube length = $D_o + (D_b - D_o) \times b \times N_m$.
s = clearance between fins = $1/N_m - b$.
S_T = distance between tube axes measured in a direction perpendicular to the flow.
S_F = distance between tube axes measured in a direction parallel to the flow.

For Triangular Patterns (Staggered Arrays).

$$C_1 = 0.091 \mathrm{Re}^{-0.25} \tag{13-6-19}$$

$$C_3 = 0.35 + 0.65 \exp\left[-\frac{0.125(D_b - D_o)}{s}\right] \tag{13-6-20}$$

$$C_5 = 0.7 + \left[0.7 - 0.8\exp(-0.15n_f^2)\right]\exp\left(-\frac{S_F}{S_T}\right) \tag{13-6-21}$$

$$j = C_1 C_3 C_5 \left(\frac{D_b}{D_o}\right)^{0.5}\left(\frac{T}{T_f}\right)^{0.25} \tag{13-6-22}$$

In the last equation, T is the mean temperature of the gas around the tubes, and T_f is the average temperature of the fin. These temperatures must be in kelvins, and the term represents the correction for viscosity at the solid wall temperature.

For Square Patterns (Nonstaggered Arrays).

$$C_1 = 0.053 \mathrm{Re}^{-0.21}\left[1.45 - 2.9\left(\frac{S_F}{D_o}\right)^{-2.3}\right] \tag{13-6-23}$$

$$C_3 = 0.20 + 0.65\exp\left[-\frac{0.125(D_b - D_o)}{s}\right] \tag{13-6-24}$$

$$C_5 = 1.1 - \left[0.75 - 1.5\exp(-0.7n_f)\right]\exp\left(-2\frac{S_F}{S_T}\right) \tag{13-6-25}$$

$$j = C_1 C_3 C_5 \left(\frac{D_b}{D_o}\right)^{0.5}\left(\frac{T}{T_f}\right)^{0.5} \tag{13-6-26}$$

Correction for Fouling and Fin Efficiency.   The effect of the fouling resistance must be added to the value of h_f calculated with Eq. (13-6-15), and then it must be corrected by the fin efficiency. This means

$$\frac{1}{h'_f} = \frac{1}{h_f} + R_{fo} \tag{13-6-27}$$

$$h'_{fo} = h'_f (A_D + \Omega A_f)/A_o \tag{13-6-28}$$

The fin efficiency can be calculated with the expressions developed in Chap. 9:

$$m = \sqrt{\frac{2h'_f}{k_f b}} \tag{13-6-29}$$

$$Y = \left(H + \frac{b}{2} \right)\left(1 + 0.35 \ln \frac{D_b}{D_o} \right) \tag{13-6-30}$$

$$\Omega = \frac{\tanh(mY)}{mY} \tag{13-6-31}$$

Calculation of the Flue-Gas Pressure Drop in the Convection Bank (Finned Tubes). To define the stack height, it is necessary to calculate the flue-gas pressure drop through the convection bank. This pressure drop is the sum of a friction term (calculated with a friction factor f) and the pressure drop owing to acceleration losses, which for the present purposes can be neglected. The following calculation algorithm has been proposed by Weierman[1]:

$$\Delta p = (f + a) n_f G_H^2 / \rho_m \tag{13-6-32}$$

$$a = \frac{(1+\beta^2)}{4 n_f} \rho_m \left(\frac{1}{\rho_2} - \frac{1}{\rho_1} \right) \tag{13-6-33}$$

$$\beta = \frac{a_S}{WL} \tag{13-6-34}$$

where ρ_m is the mean density of the flue gas in the convection bank, ρ_1 is the inlet density, and ρ_2 is the outlet density.

The friction factor f can be correlated as a function of the Reynolds number. The proposed correlation for staggered triangular arrays is

$$C_2 = 0.075 + 1.85 \text{Re}^{-0.3} \tag{13-6-35}$$

$$C_4 = 0.11 \left(0.05 \frac{S_T}{D_o} \right)^m \tag{13-6-36}$$

$$m = -0.7 \left(\frac{H}{s} \right)^{0.2} \tag{13-6-37}$$

$$C_6 = 1.11 + \left[1.8 - 2.1 \exp(-0.15 n_f^2) \right] \exp\left(-2 \frac{S_F}{S_T} \right) - \left[0.7 - 0.8 \exp(-0.15 n_f^2) \right] \exp\left(-0.6 \frac{S_F}{S_T} \right) \tag{13-6-38}$$

$$f = C_2 C_4 C_6 \frac{D_b}{D_o} \left(\frac{T}{T_f} \right)^{-0.25} \tag{13-6-39}$$

Heat Transfer Correlations for a Convection Bank with Plain Tubes. The first rows of the convection bank are the shield tubes, and they normally do not have fins because they are exposed to high heat-flux density. Since additionally the flue-gas temperature in this zone is high, it is necessary to consider the radiation from the hot gas circulating through the tube bank.

A correlation usually used in this zone is due to Monrad.[5] This author proposes to calculate the heat transfer coefficient between the gas and the tubes as the sum of a radiant and a convective contribution. The convective film coefficient is calculated with the following dimensional correlation:[4]

$$h_c = 1.268 \frac{G_H^{0.6} T_H^{0.28}}{D_o^{0.4}} \tag{13-6-40}$$

where G_H = mass flow density of the flue gas evaluated at the central plane of a tube row, expressed in kg/(s·m²)

T_H = average flue-gas temperature in K

D_o = tube diameter in meters

h_c = results in W/(m²·K)

The radiant coefficient can be obtained as

$$h_R = 0.0254 t_H - 2.37 \tag{13-6-41}$$

where t_H is the average flue-gas temperature, but this time in °C.

There is also a radiation effect coming from the walls of the convection section. This usually increases the heat flux by about 10 percent. Then the total coefficient may be calculated[4] as

$$h = 1.1(h_c + h_R) \tag{13-6-42}$$

Rating of a Radiant Convective Heater.   The design of a radiant convective heater is a more complex procedure because it is necessary to select a larger number of parameters. It is necessary to define the percentage of the total heat that will be delivered in the convection and radiant zones and to adapt the convection-bank dimensions to have the necessary heat transfer coefficients with a pressure drop that does not require excessive stack height.

It would be cumbersome to present a methodology and recommendations to select all these variables. In addition, since the process heater design is normally performed by the vendor company, it is not a usual task for the process engineer. It is more frequently necessary to perform a rating of an existing heater to check if the design can be adapted to new process conditions. Thus we shall analyze how to perform a thermal rating of an existing heater whose geometry is already defined for a certain process service.

The following steps shall be followed:

1. The heat duty Q is a process datum. This duty shall be partially delivered in the radiation zone (Q_R) and partially in the convection zone (Q_C), where $Q = Q_C + Q_R$. A value for Q_R will be assumed, and Q_C will be calculated, as well as the process-stream temperatures in each zone. Then the combustion-gas temperature necessary to transfer the radiation heat duty will be calculated with the Lobo-Evans model. This requires us to know the gas emissivity, so it is necessary to calculate the partial pressure of H_2O and CO_2 in the flue gas with the excess air and the stoichiometry of the combustion or using the graph in Fig. 13-20.

2. With the heat balance for the radiation zone and the temperature of the combustion gas, it is possible to calculate the fuel consumption and the flue-gas flow rate. With the fuel consumption and a heat balance for the complete heater, the flue-gas temperature at the outlet of the convection zone is obtained.

3. With the flow rate and average temperature of the flue gas in the convection zone, it is possible to calculate the convection film coefficient. Since all the temperatures are known, it is possible to check if the heat transfer area in the convection zone allows the convection heat duty Q_C.

4. If the heat transfer area in the convection zone is not enough for the convection heat duty, it will be necessary to increase the heat duty in the radiant zone Q_R, which will require a higher combustion gas temperature. It is necessary then to come back to step 1 and repeat the calculations.

5. It is also necessary to check that the heat-flux density in the radiation zone is below the allowable maximum. If it is necessary to operate with a combustion gas temperature that exceeds this limit, the heater is not suitable for the required service.

The other parameter to be checked is the thermal efficiency. This value should be compatible with an economic operation. These steps will be understood more easily in the following example.

Example 13-4   It is desired to rate a process heater for the heating of a hot oil stream. The process conditions are the following:

	Inlet	Outlet
Flow rate (kg/h)	407,580	407,580
Temperature (°C)	172.9	287.8
Pressure [kg/(cm² (a)]	5.1	3.4
Heat capacity [J/(kg · K)]	2,450	2,850
Thermal conductivity [W/(m · K)]	0.1105	0.0970
Viscosity (cP)	0.975	0.366
Density (kg/m³)	769	683
Pr	21.6	10.7

The unit is a vertical cylindrical heater with four passes in both the radiant and the convection sections. The radiant chamber is 7.445 m in diameter and 18.08 m high. The exterior diameter of the tubes is 168.1 mm, and the internal diameter is 153.9 mm. The radiant chamber has 72 straight tubes, 17 m long, arranged in a 6.99-m-diameter circumference. The convection bank presents a frontal area 6.09 m long and 3.657 m wide. It has two plain tube rows acting as shield tubes. Each row has 12 tubes similar to those in the radiant section. The rest of the convection bank consists of 7 finned tube rows. The tubes are also similar to the others but are provided with solid fins 25.4 mm high and 2.76 mm thick with 157 fins/m. Fouling resistances of 0.0002 shall be assumed for the process fluid and for the external side of the convection bank. The fuel is a refinery gas with the following molar composition:

$$CH_4 = 0.9286$$
$$C_2H_6 = 0.03391$$
$$C_3H_8 = 0.00822$$
$$C_4H_{10} = 0.00352$$
$$C_5H_{12} = 0.00114$$
$$C_6H_{14} = 0.00185$$
$$N_2 = 0.01011$$
$$CO_2 = 0.01264$$

Fifteen percent excess air will be used.
Note: A downloadable spreadsheet with the calculations of this example is available at http://www.mhprofessional.com/product.php?isbn=0071624082

Solution   *Combustion calculations:* The heat duty is

$$Q = W_c c_c (t_2 - t_1) = 407,580/3,600 \times 2,650 \times (287.8 - 172.9) = 34.47 \times 10^6 \text{ W}$$

Following the procedure of Example 13-1, it is possible to solve the combustion stoichiometry, and it can be seen that with 15 percent excess air, for each kilomole of fuel, the following combustion products are obtained:

$$N_2 = 8.95 \text{ kmol}$$
$$H_2O = 2.029 \text{ kmol}$$
$$CO_2 = 1.0646 \text{ kmol}$$
$$O_2 = 0.31 \text{ kmol}$$

The amount of flue gas is 12.354 kmol for each kilomole of fuel. The net heating value of the fuel is 826,159 kJ/kmol, and the molecular weight is 27.83. The mole fraction of $CO_2 + H_2O$ is $(2.029 + 1.0646)/12.354 = 0.25$. Thus the partial pressure of these gases is about 0.25 atm.

Rating of the radiation zone: The heat transfer area of the radiant zone is

$$A_2 = 72 \times (\pi D_o L_{ef})$$

where L_{ef} is the effective tube length. For the 17-m straight tube length, we shall assume 17.45-m effective tube length to account for the return bends. Then

$$A_2 = 72 \times (\pi D_o L_{ef}) = 72 \times (\pi \times 0.168 \times 17.45) = 663.1 \text{ m}^2$$

We know that for a hot-oil heater the allowable radiant heat-flux density is 34,000 W/m² according to Table 13-2. We shall start the calculations assuming this value. Then the radiant heat duty will be

$$Q_R = 34,000 \times A_2 = 34,000 \times 663.1 = 22.55 \times 10^6 \text{ W}$$

Since in A_2 we did not include the shield tubes, it is possible to assume a little higher value. If we consider an additional 5 percent, we get

$$Q_R = 23.6 \times 10^6 \text{ W}$$

Then, for the rest of the heater, it must be

$$Q_C = 34.47 \times 10^6 - 23.6 \times 10^6 = 10.87 \times 10^6 \text{ W}$$

Considering a mean heat capacity of 2,650 J/(kg·K), it is possible to calculate the process fluid temperature at the outlet of the convection zone as

$$t_1 + Q_C/(W_c \cdot c_c) = 172.9 + 10.87 \times 10^6/(113.1 \times 2,650) = 209.1°C$$

We shall now calculate the gas temperature that is necessary to allow this duty in the radiant zone. The equivalent plane area corresponding to the vertical tubes is

$$\pi \times 6.99 \times 17 = 373.12$$

The separation between tubes can be calculated from the diameter of the tube allocation circumference:

$$\pi \times 6.99/72 = 0.3048 \text{ m}$$

The separation:diameter ratio is $0.3048/0.1681 = 1.81$. With this ratio we enter the graph in Fig. 13-18 and obtain $\alpha_{pe} = 0.911$. The equivalent plane area corresponding to the shield tubes is equal to the front area of the convection zone, which is

$$6.09 \times 3.657 = 22.27 \text{ m}^2$$

For this section, $\alpha_e = 1$. Then the total corrected equivalent plane area will be

$$A_1 \alpha_{pe} = 0.911 \times 373.12 + 22.27 = 362.18 \text{ m}^2$$

The equivalent length for this cylindrical geometry is equal to the radiant chamber diameter: 7.445 m, Then

$$P_{(CO_2 + H_2O)} \times L = 0.25 \times 7.445 = 1.86 \text{ atm} \cdot \text{m}$$

To calculate the gas temperature, a trial-and-error procedure must be followed. A temperature is assumed, and with Eq. (13-6-10) we can check if the desired heat duty is obtained. After some trials, we assume 840°C temperature. Gas emissivity (from Fig. 13-21) is $\varepsilon_g = 0.65$. The area of the heater wall is $\pi \times 7.445 \times 18.08 = 422.66$. The floor area is $\pi \times 7.445^2/4 = 43.51$. The area of the arch is $\pi \times 7.445^2/4 - 22.27 = 21.24$. The total area is

$$A_T = 422.66 + 43.51 + 21.24 = 487.4 \text{ m}^2$$

Then

$$A_R = A_T - A_1\alpha_{pe} = 487.4 - 362.18 = 125.22$$

$$A_R/A_1\alpha_{pe} = 125.22/362.18 = 0.345 < 1$$

Then

$$F_{R1} = A_1\alpha_{pe}/A_T = 0.74$$

$$\overline{F}_{1G} = \varepsilon_g \left[1 + \left(\frac{A_R}{A_1\alpha_{pe}} \right) \frac{1}{1 + \dfrac{\varepsilon_g}{F_{R1}(1-\varepsilon_g)}} \right] = 0.65 \left[1 + 0.345 \frac{1}{1 + \dfrac{0.65}{0.74(1-0.65)}} \right] = 0.712$$

$$\frac{1}{\mathcal{F}_{1G}} = \frac{1}{\overline{F}_{1G}} + \frac{1-\varepsilon_2}{\varepsilon_2} = \frac{1}{0.712} + \frac{1-0.9}{0.9} = 1.5$$

$$\therefore \mathcal{F}_{1G} = 0.66$$

Now we shall calculate the tube wall temperature. The convection film coefficient for the process fluid inside the tubes is obtained as

$$h_i = 0.023 \frac{k}{D_i} \text{Re}^{0.8} \text{Pr}^{0.33}$$

The internal tube diameter is 153.9 mm. Then the flow area will be

$$a_t = \pi \times 0.1539^2/4 = 0.0185 \text{ m}^2$$

Since the heater has four flow passes, only one-quarter of the total flow circulates through each tube. Thus

$$\text{Re} = \frac{D_i W}{\mu a_t} = \frac{0.1539 \times 407,580/(4 \times 3,600)}{0.67 \times 10^{-3} \times 0.0185} = 351,400 \quad \text{(an average viscosity was considered)}$$

If we consider an average Prandtl number = 16.1, then

$$h_i = 0.023 \frac{0.10}{0.1539} \times 351,400^{0.8} \times 16.1^{0.33} = 1,021 \text{ W/(m}^2 \cdot \text{K)}$$

And the sum of all heat transfer resistances referred to the external area of the tubes is

$$\frac{D_o}{h_i D_i} + \frac{e}{k_m} + R_{fi} \frac{D_o}{D_i} = \frac{168.1}{153.9 \times 1,021} + \frac{0.007}{45} + 0.0002 \frac{168.1}{153.9} = 0.0014$$

The mean process fluid temperature in the radiant zone is

$$(209.1 + 287.8)/2 = 248°C$$

With an average heat-flux density of 34,000 W/m² and a mean process fluid temperature of 248°C, the mean wall temperature of the external surface of the tubes is

$$248 + 34,000 \times 0.0014 = 296°C$$

The heat transfer rate will be

$$Q = A_1 \alpha_{pe} \mathcal{F}_{1G} \sigma \left(T_G^4 - T_2^4\right) + A_2 h_G (T_G - T_2) = 362.18 \times 0.66 \times 5.66 \times 10^{-8} \times [(840 + 273)^4 - (296 + 273)^4]$$
$$+ 663.1 \times 11.2 \times (840 - 296) = 19.34 \times 10^6 + 4.04 \times 10^6 = 23.38 \times 10^6 \text{ W}$$

which agrees with the assumed value.

Heat balances in the convection zone: We have already explained that the flue-gas outlet temperature from the radiation zone is somewhat lower than the mean temperature of the gases into that zone. The adoption of a value for this difference is based on designer experience and in many cases is neglected. We shall assume a 15°C difference; then the flue-gas outlet temperature would be 825°C. By knowing the radiant heat duty (23.38×10^6 W) and this flue-gas outlet temperature, we can calculate the fuel consumption and the flue-gas flow rate. Equation (13-2-6) shall be used, following an iterative procedure. We assume a fuel-gas flow W_F and calculate the outlet temperature with a heat balance:

$$\frac{Q_R}{W_F(\text{NHV})} = \frac{W_A c_A (T_A - T_o)}{W_F(\text{NHV})} + \frac{W_V c_V (T_V - T_o)}{W_F(\text{NHV})} + \frac{c_F(T_F - T_o)}{\text{NHV}} + \left(1 - \frac{\gamma}{100}\right) - \frac{W_H c_H (T_H - T_o)}{W_F(\text{NHV})}$$

The value of W_F is changed until T_H coincides with the assumed 825°C. The procedure can be easily followed with the help of a spreadsheet.

The final results are

Fuel consumption = 174.5 kmol/h
Heat fired = $W_F \times$ NHV = 174.5 × 826,159/3,600 = 40,045 kW (34,520 kcal/h)
Flue gas molar flow = 12.354 × 174.5 = 2,155 kmol/h

The molecular weight of the flue gas is 27.83; then the flue-gas mass flow rate is

$$27.83 \times 2,155 = 59,973 \text{ kg/h} = 16.65 \text{ kg/s}$$

Flue gas enthalpy at 825°C = 1.412×10^7 kcal/h = 16,379 kW. Conduction heat loss is 1.5 percent of 40,045 = 606 kW. We assume that combustion air is at 15°C; then its enthalpy equals 0.

Heat balance:

Heat fired:	40,045	
Heat absorbed:	23,380	
Flue-gas enthalpy:	16,379	
Heat loss:	606	
Total	40,366	OK

Knowing the flue-gas mass flow, we can check the heat transfer capacity of the convection zone. The front area of the convective bank is 6.09 × 3.65. To calculate the flow area, we must subtract the projected area of the tubes.

Shield tube zone: In the shield tube zone, each row has 12 tubes with 168 mm diameter. The flow area thus is

$$6.09 \times 3.65 - 12 \times 0.168 \times 6.09 = 9.95 \text{ m}^2$$

The mass velocity is $G_H = 59,973/(3,600 \times 9.95) = 1.67$ kg/s m^2. The flue-gas inlet temperature to the convection zone is 825°C. The gases cool down when delivering heat to the tubes. We shall assume a mean flue-gas temperature in the shield tube zone of 750°C (= 1,023 K) (to be verified). Thus

$$h_c = 1.268 \frac{G_H^{0.6} T_H^{0.28}}{D_o^{0.4}} = 1.268 \, \frac{1.67^{06} \times 1,023^{0.28}}{0.168^{0.4}} = 24.51 \text{ W/(m}^2 \cdot \text{K)}$$

$$h_R = 0.0254 t_H - 2.37 = 0.0254 \times 750 - 2.37 = 16.68 \text{ W/(m}^2 \cdot \text{K)}$$

$$h_o = 1.1(h_c + h_R) = 1.1(24.51 + 16.58) = 45.29 \text{ W/(m}^2 \cdot \text{K)}$$

Considering that the combined heat transfer resistance of the internal boundary layer, metal wall, and fouling is similar to that of the radiant section [0.0014 (m$^2 \cdot$ K)/W], and with the addition of an external fouling resistance of 0.0002 (m$^2 \cdot$ K)/W, the overall heat transfer coefficient will be

$$U = \left(\frac{1}{45.29} + 0.0014 + 0.0002 \right)^{-1} = 42.14 \text{ W/(m}^2 \cdot \text{K)}$$

The heat transfer area of the 24 shield tubes is

$$A = \pi \times D_o \times L \times 24 = \pi \times 0.168 \times 6.09 \times 24 = 77 \text{ m}^2$$

We have calculated that the process fluid temperature at the outlet of the convection zone is 209.1°C. As a first approximation, we can consider that the mean temperature difference between the flue gas and the process fluid in this zone is $750 - 209.1 = 541$°C (to be later checked). Then the heat duty transferred in this zone will be

$$Q = UA\Delta T = 42.14 \times 77 \times 541 = 1.755 \times 10^6 \text{ W}$$

With this first approximation, we can correct the hot-oil and flue-gas temperatures. The hot-oil inlet temperature to the shield tubes would be

$$209.1 - 1.755 \times 10^6/(W_c c_c) = 209.1 - 1.755 \times 10^6/(113.2 \times 2,650) = 203.2°C$$

And the flue-gas temperature in that zone would be

$$825 - 1.755 \times 10^6/(W_H c_H) = 825 - 1.755 \times 10^6/(16.65 \times 1,300) = 744$$

[For the present purposes, an approximate value for the flue-gas heat capacity is good enough. We assumed 1,300 J/(kg · K).] With these temperatures, the LMTD for the shield tubes would be

$$\text{LMTD} = \frac{(825 - 209.1) - (744 - 203.2)}{\ln \dfrac{825 - 209.1}{744 - 203.2}} = 577°\text{C}$$

Recalculating with this value, we get

$$Q = UA(\text{LMTD}) = 42.14 \times 77 \times 577 = 1.872 \times 10^6 \text{ W}$$

And correcting again, the hot-oil inlet temperature to the shield tubes would be

$$209.1 - 1.872 \times 10^6/(W_c c_c) = 209.1 - 1.872 \times 10^6/(113.2 \times 2,650) = 202.8°C$$

And the flue-gas outlet temperature from the shield tube zone would be

$$825 - 1.872 \times 10^6/(W_H c_H) = 825 - 1.872 \times 10^6/(16.65 \times 1,300) = 738.5°C$$

Finned tube bank: The flue gases enter this zone at 738.5°C. We shall use the following physical properties for the flue gas in this zone:

Thermal conductivity = 0.0535 W/(m·K)
Heat capacity = 1,221 J/(kg·K)
Viscosity = 0.032 cP
Pr = 0.74

The convective bank geometry is

W = tube-bank width = 3.657 m
L = tube-bank length = 6.09 m
n_{tf} = number of tubes per row = 12
n_f = number of tube rows = 7
D_o = tube external diameter = 0.168 m
D_b = fin diameter = 0.219 m
b = fin thickness = 0.00276 m
H = fin height = $(D_b - D_o)/2$ = 0.0254 m
N_m = number of fins per meter = 157

Then

$$a_S = WL - n_{tf}L\left[D_o + N_m(D_b - D_o)b\right] = 3.657 \times 6.09 - 12 \times 6.09$$
$$\times (0.168 + 157 \times 0.051 \times 0.00276) = 8.38 \text{ m}^2$$

$$G_H = W_H/a_s = 16.65/8.38 = 1.986 \text{ kg/(s·m}^2)$$

$$\text{Re} = \frac{D_o G_H}{\mu_H} = \frac{0.168 \times 1.986}{0.032 \times 10^{-3}} = 10,297$$

and

A_o = plain tube surface per unit length = $\pi D_o = \pi \times 0.168 = 0.527 \text{ m}^2/\text{m}$
A_D = exposed tube surface per unit length = $\pi \times D_o \times (1 - N_m \times b) = 0.527 \times (1 - 157 \times 0.00276) =$
0.298 m²/m
A_f = fin surface per unit tube length = $\pi \times 2N_m(D_b^2 - D_o^2)/4 = 2\pi \times 157 \times (0.219^2 - 0.168^2)/4 =$
4.865 m²/m
s = fin clearance = $1/N_m - b = 1/157 - 0.00276 = 0.0036$ m
S_T = distance between tube axes in direction perpendicular to flow = 0.304 m
S_F = distance between tube axes in direction parallel to flow = 0.263 m

Then

$$C_1 = 0.091 \text{Re}^{-0.25} = 0.091 \times 10,297^{-0.25} = 0.009$$

$$C_3 = 0.35 + 0.65\exp\left[-\frac{0.125(D_b - D_o)}{s}\right] = 0.35 + 0.65\exp\left(-\frac{0.125 \times 0.051}{0.0036}\right) = 0.461$$

$$C_5 = 0.7 + \left[0.7 - 0.8\exp(-0.15n_f^2)\right]\exp\left(-\frac{S_F}{S_T}\right) =$$
$$= 0.7 + \left[0.7 - 0.8\exp(-0.15 \times 7^2)\right]\exp(-0.263/0.304) = 0.994$$

$$j = C_1 C_3 C_5 \left(\frac{D_b}{D_o}\right)^{0.5}\left(\frac{T}{T_f}\right)^{0.25} = 0.009 \times 0.461 \times 0.994 \times (0.219/0.168)^{0.5} = 4.73 \times 10^{-3}$$

(The wall tube temperature correction was neglected.)

$$h_f = jc_H G_H \, Pr_H^{-0.67} = 4.73 \times 10^{-3} \times 1{,}221 \times 1.986 \times 0.74^{-0.67} = 14.05 \text{ W/(m}^2 \cdot \text{K)}$$

$$h'_f = \left(\frac{1}{h_f} + R_{fo}\right)^{-1} = \left(\frac{1}{14.05} + 0.0002\right)^{-1} = 14.01 \text{ W/(m}^2 \cdot \text{K)}$$

Calculation of the fin efficiency:

$$Y = \left(H + \frac{b}{2}\right)\left(1 + 0.35 \, \ln\frac{D_b}{D_o}\right) = \left(0.0254 + \frac{0.00276}{2}\right)\left(1 + 0.35 \, \ln\frac{0.219}{0.168}\right) = 0.0293$$

$$m = \sqrt{\frac{2h'_f}{k_f b}} = \sqrt{\frac{2 \times 14.01}{45.5 \times 0.00276}} = 14.94$$

$$\Omega = \frac{\tanh(mY)}{mY} = \frac{\tanh(14.94 \times 0.0293)}{14.94 \times 0.0293} = 0.94$$

$$h'_{fo} = h'_f (A_D + \Omega A_f)/A_o = 14.01(0.298 + 0.94 \times 4.865)/0.527 = 129.2$$

Assuming the same internal film coefficient and fouling factor as for the rest of the heater, the overall heat transfer coefficient will be

$$U = \left(\frac{1}{129.2} + 0.0014 + 0.0002\right)^{-1} = 107 \text{ W/(m}^2 \cdot \text{K)}$$

The heat transfer area is

$$A = A_o n_f n_{tf} L = 0.527 \times 7 \times 12 \times 6.09 = 269.6 \text{ m}^2$$

The heat capacities of both streams are

Flue gas $= W_H c_H = 16.65$ kg/s $\times 1{,}221$ J/(kg·K) $= 20{,}329$ W/K

Hot oil $= W_c c_c = 113.2 \times 2{,}650 = 300{,}005$ W/K
In order to check the heat transfer capacity, we can use the efficiency method developed in Chap. 7:

$$NTU = UA/(Wc)_{min} = 269.6 \times 107/20{,}329 = 1.41$$

$$R' = (Wc)_{min}/(Wc)_{max} = 20{,}329/300{,}005 = 0.0677$$

For a countercurrent arrangement, we get an efficiency $\varepsilon = 0.74$. The flue gas and hot-oil inlet temperatures for this section are, respectively, 738.5°C (previously calculated as the shield tube section outlet temperature) and 172.9°C. Then

$$Q = (Wc)_{min} \varepsilon \, (738.5 - 172.9) = 8.508 \times 10^6 \text{ W}$$

Adding this heat duty to the shield tube heat duty, we get for both convection zones

$$Q_C = 8.508 \times 10^6 + 1.872 \times 10^6 = 10.38 \times 10^6 \text{ W}$$

The value we calculated previously as the necessary heat duty for the convection zone was 10.87×10^6 W. This indicates that the heater is slightly undersized for this service. The amount of heat that could be delivered would be

$$Q = 10.38 \times 10^6 + 23.03 \times 10^6 = 33.41 \times 10^6 \text{ W}$$

To increase the heat duty, it would be necessary to increase the fire. This means a higher fuel consumption and a higher heat-flux density in the radiant zone. However, since the radiant heat-flux density is at the allowable maximum, this would result in a reduction in the service life.
The heater efficiency is

$$\eta = \frac{Q}{WF(\text{NHV})} = \frac{33.41 \times 10^6}{40.045 \times 10^6} = 83.4\%$$

Calculation of the stack height: Usually in cylindrical vertical heaters the height of the radiant chamber provides enough draft to overcome the pressure drop of the burners. The stack height is calculated to maintain a subatmospheric pressure of about 1.2–1.7 mmH$_2$O at the arch level. Then the draft must equal the pressure drop of the convective bank and the stack conduct, including the regulation damper. The flue gas density at any temperature is calculated as

$$\rho = \frac{Mp}{RT} = \frac{27.83}{0.082T}$$

and then we have

Convection bank inlet $T = 825°C$ (1,098 K) $\times \rho = 0.309$ kg/m^3
Inlet to the finned tube section $T = 738°C$ (1,011K) $\times \rho = 0.335$ kg/m^3
Stack inlet $T = 330°C$ $\rho = 0.562$ kg/m^3
Stack outlet (estimated) $= 330 - 40 = 290°C \times \rho = 0.602$
Mean density in the stack $= 0.582$ kg/m^3

The density of the atmospheric air ($M = 29$), assuming 30°C, is 1.16 kg/m^3. The frictional pressure drops are

Plain tube $G_H = 1.67$ kg/(s · m^2)
Number of tube rows $= 2$

Considering 0.2 velocity heads per row,

$$\Delta p = 0.2 \times 2 \times \frac{G_H^2}{2\rho} = 0.2 \times 2 \times \frac{1.67^2}{2 \times 0.309^2} = 5.8 \text{ Pa}$$

Finned tubes: Following the algorithm of Eqs. (13-6-32) through (13-6-37), we get $C_2 = 0.19$, $C_4 = 1.32$, $C_6 = 1.00$, $a = -0.022$, $\beta = 0.376$, and $f = 0.288$. Then, for the seven tube rows, we obtain

$$\Delta p = 16.38 \text{ Pa}$$

Stack conduct: For a mass velocity of 5 kg/(s · m^2), the required cross section is $16.65/5 = 3.32$ m^2 ($D = 2.05$ m). Considering three velocity heads for the inlet, outlet, and regulation damper, and assuming 30 m height, we get

$$\Delta p_{\text{stack}} \text{ (Pa)} = 2.76 \times 10^{-5} \left(\frac{G_H^2 T}{D} \right) L = 2.76 \times 10^{-5} \times \frac{5^2 \times (310 + 273)}{2.05} \times 30 = 5.88 \text{ Pa}$$

Additionals:

$$3\frac{G_H^2}{2\rho} = 3 \times \frac{5^2}{2 \times 0.582} = 64 \text{ Pa}$$

If it is desired to have a subatmospheric pressure at the arch level of 1.5 mmH$_2$O (15 Pa), the total draft is = 15 + 5.8 + 16.38 + 5.88 + 64 = 107 Pa. The draft produced by the stack is:

$$\frac{\Delta p}{L} = 3,465\left(\frac{1}{T_{air}} - \frac{1}{T_{flue\ gas}}\right) = 3,465\left(\frac{1}{303} - \frac{1}{583}\right) = 5.49 \text{ Pa/m}$$

Thus, to provide the 107 Pa, the stack height has to be 20 m above the shield tubes. It is important to mention that frequently the stack height is defined by environmental regulations dealing with the dispersion of the flue-gas contaminants such as SO$_2$ and NO$_x$.

GLOSSARY

A = area (m^2)

A_R = refractory area (m^2)

A_1 = equivalent plane area (m^2)

e = thickness (m)

$F, \mathcal{F}$ = view factors defined in text

f = friction factor

G = mass velocity [kg/(m$^2 \cdot$s)]

h = convection film coefficient [W/(m$^2 \cdot$K)]

h_{io} = internal film coefficient referred to external surface [W/(m$^2 \cdot$K)]

h_c = convective contribution to the external h [W/(m$^2 \cdot$K)]

h_f = film coefficient over the finned surface [W/(m$^2 \cdot$K)]

h'_f = same as above but corrected for fouling [W/(m$^2 \cdot$K)]

h_R = radiation contribution to the external h [W/(m$^2 \cdot$K)]

i = specific enthalpy (J/kg)

k = thermal conductivity [W/(m$\cdot$K)]

Convective tube bank geometry:

A_o = plain tube surface per unit length = πD_o

A_D = exposed tube surface per unit length = $\pi D_o(1 - N_m b)$

A_f = fin area per straight-tube unit length = $\pi 2 N_m (D_b^2 - D_o^2)/4$

a_p = projected finned-tube area per unit length = $D_o + (D_b - D_o)b N_m$

b = fin thickness

D_o = tube external diameter

D_b = fin diameter

H = fin height = $(D_b - D_o)/2$

L = tube length

n_{tf} = number of tubes per row

n_f = number of tube rows

N_m = number of fins per meter

s = clearance between fins = $1/N_m - b$

S_T = distance between tube axes in direction perpendicular to flow

S_F = distance between tube axes in direction parallel to flow

Subscripts

A = air

E = equivalent plane

F = fuel

g, G = gas

H = flue gas

m = metal

R = refractory

V = vapor

REFERENCES

1. Weierman C.: "Correlations ease the selection of finned tubes," -*Oil and Gas Journal* Sept 6:97–100, 1976.

2. Lobo W., Evans J.: "Heat transfer in the radiant section of petroleum heaters," *Trans AIChE* 35:743–778, 1939.

3. Wimpress N.: "Generalized method predicts fired heater performance," *Chem Eng* May 22:95–102, 1978.

4. Berman H.: in *Process Heat Exchange*: Mc Graw Hill, New York 1978, Reprinted from *Chem Eng* June 19, July 31, Aug 14, an Sept 11, 1978.

5. Monrad C. C.: "Heat transmission in convection section of pipe stills," *Ind Eng Chem* 24:505, 1932.

6. API Standard 560: "Fired Heaters for General Refinery Service," API, Washington, 1996.

7. API Standard RP 566: "Instrumentation and Control Systems for Fired Heaters," API, Washington, 1997.

8. Zink J.: in Baukal C (ed): *Combustión Handbook*. Boca Raton: CRC Press, Fla., 2001.

9. API Standard 530: "Calculation of Heater Tube Thicknesses in Petroleum Refineries," API, Washington, 2004.

10. Wimpress R. N.: "Rating fired heaters," *Hydr. Proc & Petr Ref* 42 (10)115–126, 1963.

APPENDIX A
DISTILLATION

In Chaps. 10 and 11 we make reference to the distillation process, and it is necessary for a better comprehension of these chapters to have at least a basic knowledge of this subject. Since in the curricula of most chemical engineering colleges, mass-transfer operation are studied after heat transfer, it was considered convenient to include this appendix for readers not yet familiar with the distillation operation.

Distillation operations are performed in fractionation columns. A usual type is the perforated-trays column. In these units a liquid stream and a gas stream are put in contact in a series of countercurrent stages, which are the trays (Fig. A-1).

A perforated-trays column is a cylindrical vertical vessel in which perforated plates or trays are installed. The trays are connected by conducts called *downcomers,* which in two consecutive trays are located diametrically opposed.

Thus the liquid introduced in the upper stage circulates in crossflow through the trays and downward from one tray to the next tray below through the downcomers. The liquid comes into the downcomers overflowing above a weir thus a liquid level is maintained in each tray.

The trays are perforated, and the gas stream bubbles into the liquid passing through these perforations. This creates a foam layer in each tray with great interface area. Both streams exchange their chemical components through this interface.

Let's consider the equilibrium-curve temperature-composition for a binary mixture such as that shown in Fig. A-2. The abscissa $x = y = 1$ corresponds to the lighter component (the component with the lower boiling or condensation temperature).

It can be appreciated in the graph that at low temperatures, both phases are richer in the light component, whereas at high temperatures, both phases are richer in the heavy component. This means that in a binary system, if we can keep a high-temperature zone and a low-temperature zone, the heavy component will concentrate in the former, whereas the light component will predominate in the latter.

In a multicomponent mixture, the same principle is valid. Lighter components will accumulate in the cold zone, and heavy components will concentrate in the warmer zone. This is the basis of the distillation operation.

Let's assume that we have a multicomponent liquid mixture that we want to fractionate in two streams, one rich in heavy components and the other rich in light components. The mixture is loaded in the bottom of a fractionation column, and heat is delivered to the mixture to produce a partial vaporization by means of a piece of equipment called a *reboiler.* The resulting vapors will ascend through the column up to the top.

At the top of the column, these vapors pass into another piece of equipment, the *condenser,* that removes heat from the vapor stream, resulting in its condensation. The condensed liquid is returned to the column. This stream is called the *reflux* (Fig. A-3).

This liquid stream circulates downward through the column, coming in contact with the ascending vapors, and reaches the column bottoms, where it is revaporized in the reboiler. In this way, a warm lower end and a cold upper end are maintained in the column, and an internal circulation of ascending vapor and descending liquid is established.

As it was explained earlier, the light components will concentrate in the cold upper end, and the heavy components will concentrate in the warm bottoms. At the top of the column, both streams, vapor and liquid, are rich in light compounds, whereas at the bottom, they will be rich in heavies. Both streams exchange

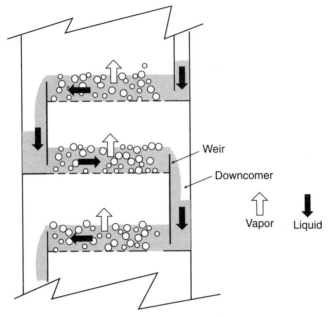

FIGURE A-1   Perforated trays.

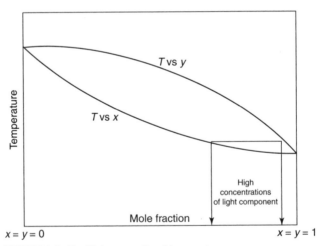

FIGURE A-2   Equilibrium curve for a binary system.

components when they circulate through the column. The vapor stream extracts preferably light compounds from the liquid while it travels upward, and the liquid extracts heavy components from the vapor while it moves downward.

A composition gradient is established along the column, and at a certain intermediate tray, the compositions are similar to that of the original mixture. Once the column reaches a steady state, it is possible to start feeding the column continuously into that tray and at the same time withdraw an equivalent amount from the top and bottom.

If the flow rates of the inlet and outlet streams are small in comparison with those of the internal column streams, this will not substantially alter the concentration profiles, and the column will reach a steady state in which the stream entering the column is fractionated into light and heavy streams.

Since the internal streams are the vehicle that allows the mass transfer, the efficiency of the fractionation will depend on the ratio between the flow rates of internal streams and the withdrawn streams. The higher this ratio, the more efficient will be the separation. This is measured by a parameter called the *reflux ratio,* which is the quotient between the reflux molar flow and the molar flow of the stream extracted from the top of the column.

The other parameter that affects the separation is the number of trays. The larger the stream path within the column (larger number of trays), the better will be the separation. This means that column design requires adopting a combination of number of trays and reflux ratio to obtain the desired degree of fractionation. It is possible to reduce the reflux ratio and increase the number of trays, and vice versa. If a high reflux ratio is adopted, we shall have higher internal traffic into the column per unit mass of processed feed. This is achieved at the expense of higher energy consumption in the reboiler and condenser. Increasing the number of stages implies higher capital cost.

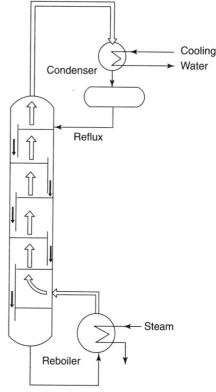

FIGURE A-3   Column in total reflux.

CONTROL SCHEME

Figure A-4 represents a complete fractionation column with its control system. To maintain a steady state, it is necessary to guarantee that the amount of vapor and liquid contained into the column do not change during operation. This is achieved by keeping the liquid levels and pressure into the column constant. The elements indicated as LC are level controllers, which maintain constant liquid levels in the reflux accumulator and in the column bottom by operating the product outlet valves. The liquid levels in the trays are held constant by the presence of the tray weirs.

The element indicated as PC is a pressure controller, which keeps a constant pressure in the column by removing some vapor from the top. (We shall later see that some columns operate without top vapor extraction.)

The three remaining valves are those which the column operator can handle, thus changing the top-product extraction rate and the amount of heat removed or supplied in the condenser and reboiler. A change in the opening position of any of these valves will have an effect on the performance of the column, modifying the efficiency of the separation.

Thus we can say that in a column such as that represented in Fig. A-4, with a definite number of trays and operating with a certain feed at a fixed pressure in steady state, there are three degrees of freedom that are the variables the operator can manipulate to modify the result of the operation. For each set of values of these variables, we shall have a different operating state with different flow rates and compositions of the outlet streams.

These degrees of freedom obviously should be adopted such that the compositions of the products result within the required specifications. (Please note that the level-control loop of the reflux accumulator can be configured either with the reflux stream or with the liquid top product. The sum of these two streams equals the amount of liquid the condenser is capable of condensing. Once any of these streams is fixed, the other one is also defined.)

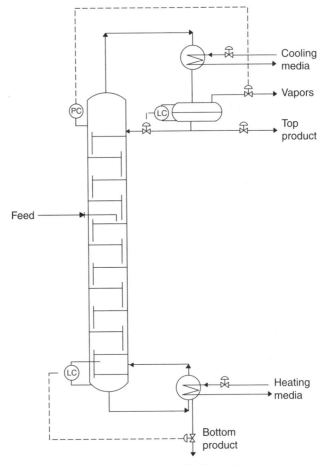

FIGURE A-4   Basic control scheme of a distillation column.

TOTAL AND PARTIAL CONDENSER

The column represented in Fig. A-4 allows extraction of top product in vapor and liquid phases. However, the most common situation is that the top product is extracted only as a liquid or a vapor.

For example, let's consider the case in Fig. A-5. This column has a total condenser with no vapor extraction.

In this case, the pressure control loop is performed with the heat removed in the condenser. (It is always necessary to have a pressure control loop to maintain the steady state.) This control loop adjusts the amount of heat removed in the condenser such that it condenses what is strictly necessary to maintain the pressure in the column. In this case, only two valves are left to be manipulated by the operator, so the system has two degrees of freedom.

SIMULATION OF DISTILLATION COLUMNS

Calculation of the composition of the outlet streams of a distillation column is usually performed with simulation programs. These programs solve the mass and energy balances of each column tray, calculating the composition, temperatures, and flow rates of the gas and liquid streams.

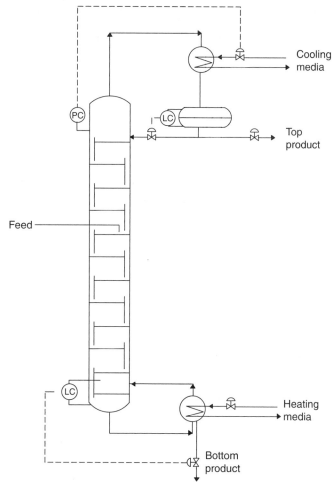

FIGURE A-5   Column with total condenser.

The concept of "theoretical tray" is commonly used, which implies an assumption that each tray has a perfect efficiency and that the liquid and vapor leaving the tray are in thermodynamic equilibrium. To calculate the actual number of trays, this result is affected by tray efficiency.

PHASE EQUILIBRIUM IN A MULTICOMPONENT SYSTEM

The equilibrium of phases in a multicomponent system is usually expressed by the equilibrium constants. For any component in a two-phase system in equilibrium, this constant is the ratio of the mole fractions of that component in vapor and liquid phases. This is

$$k_i = \frac{y_i}{x_i}$$

In highly ideal systems that follow Raoult's law, k_i is an exclusive function of the temperature. In real systems, the equilibrium constant (which is very far from being a constant) is a function not only of the temperature but also of the complete composition of both phases.

There is a thermodynamic property called *chemical potential* that can be defined for any component in each phase as

$$\mu_i = \left(\frac{\partial G}{\partial n_i} \right)_{p,T,nj}$$

where G is the Gibbs free energy and n_i and n_j are the number of moles of component i or j in the considered phase. It can be demonstrated by thermodynamic considerations that in equilibrium, the chemical potential of every component in both phases must be the same.

As was explained in Chap. 2, if we know empirically the function that expresses the dependence of any thermodynamic property (e.g., volume) with the state variables (i.e., pressure, temperature, and composition), it is possible to mathematically calculate any other property (e.g., free energy).

It is then possible to calculate chemical potentials and find the set of compositions that makes the chemical potential of every component equal in both phases. Equilibrium constants thus can be calculated. The mathematical treatment is very complex, and it was made possible in the last years with the development of simulation programs.

MASS AND ENERGY BALANCES

Let's consider a tray of a fractionation column such as that shown in Fig. A-6. A schematic representation of this system is shown in the figure on the right. This stage receives a liquid coming from the tray above, with molar flow rate L_{j-1} and a vapor coming from the stage below with molar flow rate V_{j+1}.

Let's call x_{ij} the mole fraction of component i in the liquid leaving tray j, whereas y_{ij} refers to the vapor composition. To make the analysis even more general, we have assumed that there exists the possibility to perform a liquid extraction in each tray and to deliver or remove heat, even though these streams usually do not exist.

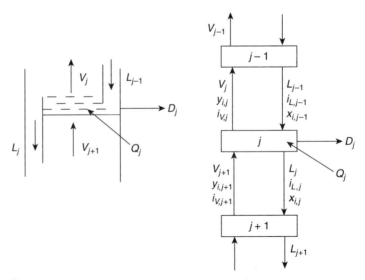

FIGURE A-6   Tray schematic.

The vapor and liquid phases leaving the stage are in thermodynamic equilibrium. If the inlet streams are known (which means we know composition, flow rates, and temperatures of streams L_{j-1} and V_{j+1}), as well as the amount of heat Q_j and liquid D_j extracted from the tray, the outlet streams could be calculated with the following equations system:

1. Equations of the mass balance (one equation for any of the i components):

$$V_{j+1} \cdot y_{i,j+1} + L_{j-1} x_{i,j-1} = L_j \cdot x_{i,j} + V_j y_{i,j} + D_j x_{i,j} \qquad \text{(A-1)}$$

where V_j, L_j, and D_j are the molar flows and $x_{i,j}$ and $y_{i,j}$ are the mole fractions. Subscript j denotes the tray number.

2. Equilibrium relations (one equation for each component):

$$k_{i,j} = \frac{y_{i,j}}{x_{i,j}} \qquad \text{(A-2)}$$

3. Enthalpy balance:

$$V_{j+1} \cdot i_{v,j+1} + L_{j-1} i_{L,j-1} = L_j \cdot i_{L,j} + V_j \cdot i_{V,j} + D_j \cdot i_{L,j} + Q_j \qquad \text{(A-3)}$$

4. Summation of the liquid mole fractions:

$$\sum_{i=1}^{N} x_{i,j} = 1 \qquad \text{(A-4)}$$

where N is the number of components.

5. Summation of the vapor mole fractions:

$$\sum_{i=1}^{N} y_{i,j} = 1 \qquad \text{(A-5)}$$

This means that for any tray we have $2N + 3$ equations. The $2N + 3$ unknowns are the compositions x_{ij} and y_{ij}, the molar flows L_j and V_j, and the temperature of the tray. (Even though temperature does not appear explicitly in the equations system, it must be known to calculate enthalpies and equilibrium constants.)

The column condenser and reflux accumulator can be treated with the same equations system as if they were an additional stage. If we consider the condenser as stage 1, identification of the streams for this stage is shown in Fig. A-7, where it can be noted that Q_1 and D_1 are the independent variables for this stage.

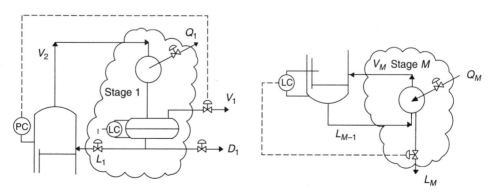

FIGURE A-7   Condenser and reboiler as additional stages.

The reboiler can be treated as an additional stage as well. If this is identified as M, the stream nomenclature for this stage is also shown in the figure, where it can be seen that the heat delivered to the reboiler is the only independent variable that the operator can manipulate.

Thus, for the complete modeling of the column, it is necessary to solve M sets of equations like Eqs. (A-1) through (A-5), thus resulting a system with $2MN + 3M$ equations with the same number of unknowns. The independent variables are the heat loads on each stage and the liquid extractions, which in the present case are only the heat loads of the condenser and reboiler and the top-product molar flow, as explained earlier.

Process simulators solve these systems and calculate flow rates, compositions, and temperatures of all trays in the column. The principal difficulty is that this set of equations must be solved simultaneously with the thermodynamic models that allow calculation of the equilibrium constants. This results in a highly nonlinear equation system.

Of course, the operation must be conducted in such a way that the products are within specifications. Thus the designer usually prefers to fix as independent variable the concentration of one component, either at the top or at the bottom. Most process simulators have the capability of accepting specifications. This means that in the equation system it is possible to choose which variables will act as independent and which will act as dependent.

For example, it is possible to specify the desired mole fraction of a certain component in stage 1 (the condenser) and let the program calculate the heat load of the condenser, which in this case acts as dependent variable. The program output will include all the internal streams that are necessary for design of the condenser and reboiler.

CONDENSATION AND BOILING CURVES

To perform the condenser and reboiler designs, it is not only necessary to know their inlet and outlet streams, but it is also necessary to know how the temperature and vapor fraction change along the unit while the process stream receives or delivers heat. This information is usually supplied as boiling and condensation curves or tables.

These curves are also part of the information that the simulation program delivers to the heat exchanger designer. To calculate a condensation curve, the condenser is divided in intervals in which a fixed amount of heat is removed. (For example, if five intervals are assumed, in each interval, one-fifth of the total heat is removed.)

Each interval receives the vapor and liquid coming from the preceding one (Fig. A-8). For each interval, the following set of equations can be written:

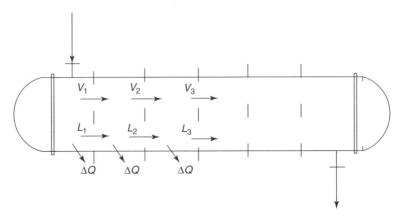

FIGURE A-8   Condenser model.

$$V_{j-1}y_{i,j-1} + L_{j-1}x_{i,j-1} = V_j y_{i,j} + L_j x_{i,j} \quad (N \text{ equations for each interval}) \tag{A-6}$$

$$y_{i,j} = K_{i,j} x_{i,j} \quad (N \text{ equations for each interval}) \tag{A-7}$$

$$V_{j-1}i_{V,j-1} + L_{j-1}i_{L,j-1} = V_j i_{V,j} + L_j i_{L,j} + \Delta Q \tag{A-8}$$

$$\sum_i y_{i,j} = 1 \tag{A-9}$$

$$\sum_i x_{i,j} = 1 \tag{A-10}$$

The resolution of this $2N + 3$ equation system allows calculating for each interval the vapor and liquid flow rates, the mole fraction of the N components in the vapor and liquid, and the temperature, knowing the values of the preceding interval.

In this model, it is assumed that the vapor and condensate travel together along the unit. This is called Integral Condensation Model. It is also possible to define the model assuming that the condensate is withdrawn from the system as it is formed. In this case we have a Differential Condensation Model. We shall come to the same equations system but assuming $L_i - 1 = 0$. The adoption of one model or the other depends on the type of condenser that will be employed.

LMTD CORRECTION FACTORS FOR E-SHELL HEAT EXCHANGERS

This appendix contains graphs for calculation of the logarithmic mean temperature difference correction factor for heat exchangers with TEMA type E shells. For other configurations, consult the *Standards of the Tubular Exchangers Manufacturers Association*.

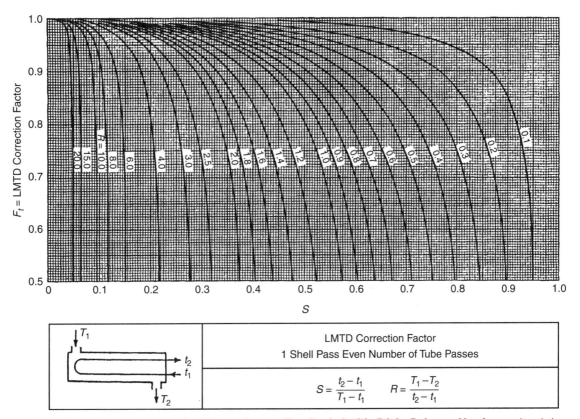

FIGURE B-1 LMTD correction factors for 1-2 heat exchangers. (*From Standards of the Tubular Exchangers Manufacturers Association with permission from TEMA.*)

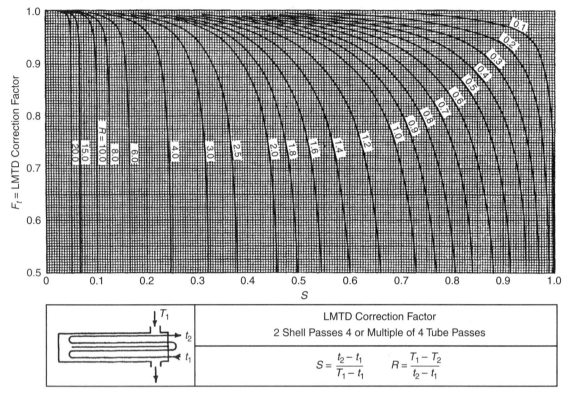

FIGURE B-2   LMTD correction factors for 2-4 heat exchangers. (*From Standards of the Tubular Exchangers Manufacturers Association with permission from TEMA.*)

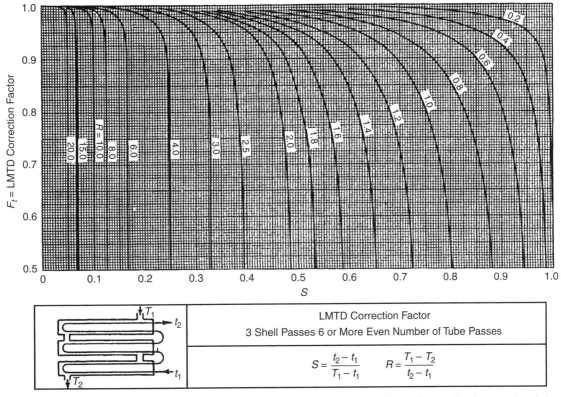

FIGURE B-3   LMTD correction factors for 3-6 heat exchangers. (*From Standards of the Tubular Exchangers Manufacturers Association with permission from TEMA.*)

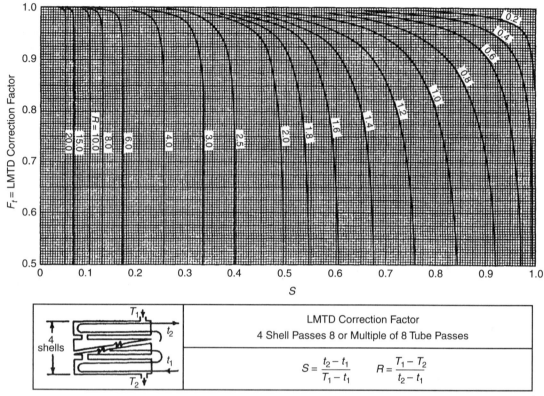

FIGURE B-4 LMTD correction factors for 4-8 heat exchangers. (*From Standards of the Tubular Exchangers Manufacturers Association with permission from TEMA.*)

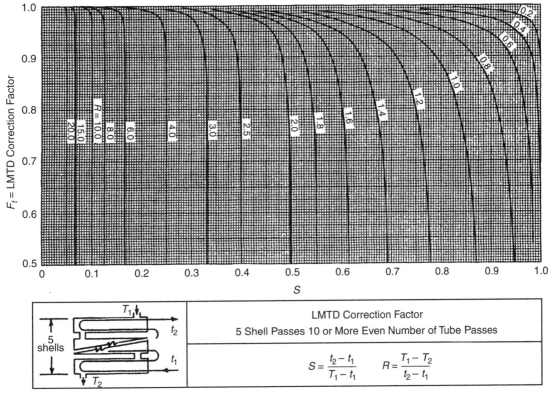

FIGURE B-5 LMTD correction factors for 5-10 heat exchangers. (*From Standards of the Tubular Exchangers Manufacturers Association with permission from TEMA.*)

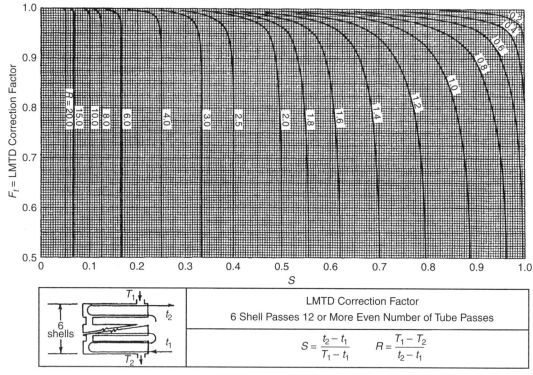

FIGURE B-6 LMTD correction factors for 6-12 heat exchangers. (*From Standards of the Tubular Exchangers Manufacturers Association with permission from TEMA.*)

APPENDIX C
LMTD CORRECTION FACTORS FOR AIR COOLERS

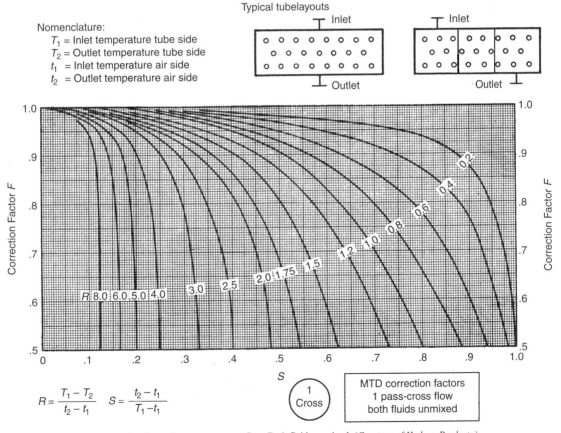

Nomenclature:
T_1 = Inlet temperature tube side
T_2 = Outlet temperature tube side
t_1 = Inlet temperature air side
t_2 = Outlet temperature air side

Typical tubelayouts

$$R = \frac{T_1 - T_2}{t_2 - t_1} \quad S = \frac{t_2 - t_1}{T_1 - t_1}$$

1 Cross

MTD correction factors
1 pass-cross flow
both fluids unmixed

FIGURE C-1   LMTD correction factor for one-pass cross-flow. Both fluids unmixed. (*Courtesy of Hudson Products.*)

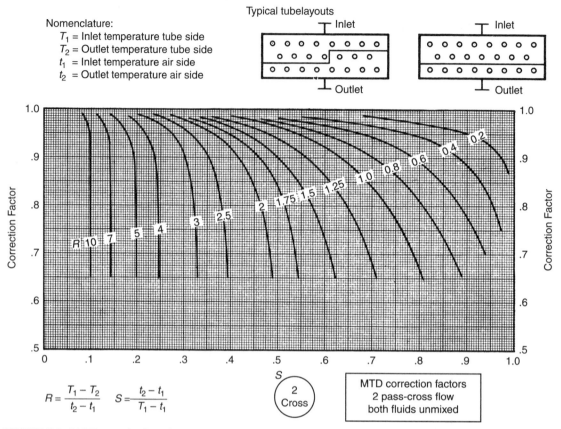

Nomenclature:
T_1 = Inlet temperature tube side
T_2 = Outlet temperature tube side
t_1 = Inlet temperature air side
t_2 = Outlet temperature air side

Typical tubelayouts

$$R = \frac{T_1 - T_2}{t_2 - t_1} \qquad S = \frac{t_2 - t_1}{T_1 - t_1}$$

MTD correction factors
2 pass-cross flow
both fluids unmixed

FIGURE C-2 LMTD correction factor for two-pass cross-flow. Both fluids unmixed. (*Courtesy of Hudson Products.*)

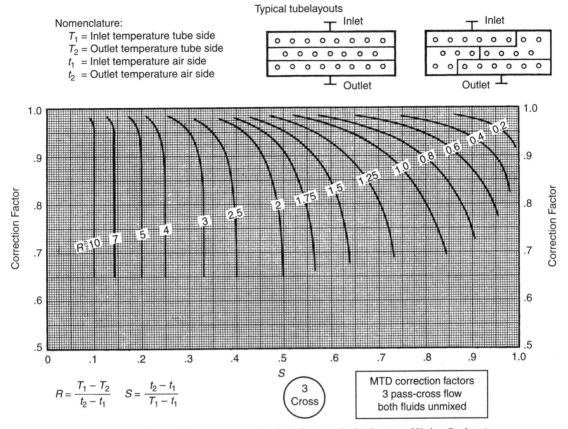

Nomenclature:

T_1 = Inlet temperature tube side
T_2 = Outlet temperature tube side
t_1 = Inlet temperature air side
t_2 = Outlet temperature air side

Typical tubelayouts

$$R = \frac{T_1 - T_2}{t_2 - t_1} \qquad S = \frac{t_2 - t_1}{T_1 - t_1}$$

3 Cross

MTD correction factors
3 pass-cross flow
both fluids unmixed

FIGURE C-3   LMTD correction factor for three-pass cross-flow. Both fluids unmixed. (*Courtesy of Hudson Products.*)

APPENDIX D
TUBE COUNT TABLES

This appendix is intended as a help in preliminary heat exchanger design to find the number of tubes that can be allocated in a certain shell diameter. It must be noted that the tables included in this appendix indicate only the maximum number of tubes for each shell diameter because no allowance was made to install impingement plates or to satisfy TEMA requirements in relation to the maximum bundle entrance and shell entrance velocities (see Sec. 6-4 and Fig. 6-7).

Since most removable bundles are constructed with square tube patterns to allow mechanical cleaning of the tube exteriors, tables for triangular arrays are only included in fixed tube-sheet designs (TEMA type L).

The pass-partition arrangements on which the tables are based are shown in Fig. D-1. Some of the tube distributions resulting from these tables are drawn in App. E.

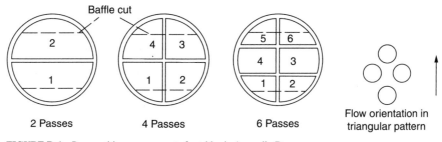

FIGURE D-1 Pass partition arrangements for tables in Appendix D.

TABLE D-1 Tube Count $\frac{3}{4}$-in (19.05-mm) Tubes on Triangular (30-Degree) Array, Pitch 1 in (25.4 mm) TEMA Type L

Internal Shell Diameter		Number of Tube Passes			
in	mm	1	2	4	6
8	203.2	38	36	32	24
10	254	69	62	56	48
12	304.8	105	94	88	76
13¼	336.5	129	120	108	104
15¼	387.3	181	166	154	148
17¼	438.15	235	218	206	198
19¼	488.9	295	280	262	252
21¼	539.7	356	344	330	314
23¼	590.5	431	418	398	388
25	635	504	492	462	446
27	685.8	597	578	550	538
29	736.6	694	674	646	634
31	787.4	799	778	750	732
33	838.2	919	888	854	834
35	889	1031	1004	968	946
37	939.8	1149	1128	1084	1052
39	990.6	1284	1258	1216	1202
42	1066.8	1499	1452	1416	1382
45	1143	1727	1686	1640	1616

TABLE D-2 Tube Count $\frac{3}{4}$-in (19.05-mm) Tubes on Square (90-Degree) Array, Pitch 1 in (25.4 mm) TEMA Type S

Internal Shell Diameter		Number of Passes		
in	mm	2	4	6
8	203.2	22	20	18
10	254	40	40	36
12	304.8	70	64	70
13¼	336.5	90	84	84
15¼	387.3	126	114	114
17¼	438.15	164	158	156
19¼	488.9	218	210	198
21¼	539.7	268	262	260
23¼	590.5	334	326	314
25	635	392	378	364
27	685.8	462	450	426
29	736.6	544	534	510
31	787.4	634	610	594
33	838.2	716	702	674
35	889	816	796	780
37	939.8	914	896	882
39	990.6	1028	1012	988
42	1066.8	1194	1168	1156
45	1143	1390	1364	1336

TABLE D-3 Tube Count ¾-in (19.05-mm) Tubes on Square (90-Degree) Array, Pitch 1 in (25.4 mm) TEMA Type U

Internal Shell Diameter		Number of Passes		
in	mm	2	4	6
8	203.2	26	20	20
10	254	48	44	40
12	304.8	74	68	68
13¼	336.5	94	88	84
15¼	387.3	126	122	114
17¼	438.15	172	170	160
19¼	488.9	218	218	218
21¼	539.7	280	278	268
23¼	590.5	346	338	330
25	635	408	390	386
27	685.8	478	470	454
29	736.6	560	546	534
31	787.4	646	634	618
33	838.2	744	730	716
35	889	840	820	816
37	939.8	946	932	918
39	990.6	1060	1048	1032
42	1066.8	1240	1220	1192
45	1143	1430	1404	1388

TABLE D-4 Tube Count 1-in (25.4-mm) Tubes on Triangular (30-Degree) Array, Pitch 1.25 in (31.75 mm) TEMA Type L

Internal Shell Diameter		Number of Passes			
in	mm	1	2	4	6
8	203.2	22	22	12	12
10	254	40	36	32	28
12	304.8	61	58	48	48
13¼	336.5	81	72	68	58
15¼	387.3	104	100	94	82
17¼	438.15	145	132	122	114
19¼	488.9	181	174	162	152
21¼	539.7	224	218	202	198
23¼	590.5	270	264	246	242
25	635	317	306	286	280
27	685.8	376	362	338	328
29	736.6	431	414	398	388
31	787.4	500	492	462	458
33	838.2	571	556	530	510
35	889	641	630	604	582
37	939.8	737	708	680	664
39	990.6	813	796	764	748
42	1066.8	941	918	888	868
45	1143	1091	1068	1028	1008

TABLE D-5 Tube Count 1-in (25.4-mm) Tubes on Square (90-Degree) Array, Pitch 1.25 in (31.75 mm) TEMA Type U

Internal Shell Diameter		Number of Passes		
in	mm	2	4	6
8	203.2	12	8	
10	254	26	20	20
12	304.8	40	36	28
13¼	336.5	56	48	48
15¼	387.3	76	70	66
17¼	438.15	102	98	94
19¼	488.9	126	130	118
21¼	539.7	164	166	156
23¼	590.5	206	202	198
25	635	248	242	232
27	685.8	294	282	274
29	736.6	346	334	322
31	787.4	400	390	382
33	838.2	458	446	438
35	889	522	508	496
37	939.8	584	576	548
39	990.6	648	648	628
42	1066.8	758	744	742
45	1143	898	880	854

TABLE D-6 Tube Count 1-in (19.05-mm) Tubes on Square (90-Degree) Array, Pitch 1 in (31.75 mm) TEMA Type S

Internal Shell Diameter		Number of Passes		
in	mm	2	4	6
8	203.2	12	8	
10	254	26	24	22
12	304.8	40	40	36
13¼	336.5	56	48	48
15¼	387.3	76	70	66
17¼	438.15	102	98	94
19¼	488.9	126	130	118
21¼	539.7	164	166	156
23¼	590.5	206	202	198
25	635	248	242	232
27	685.8	294	282	274
29	736.6	346	334	322
31	787.4	400	390	382
33	838.2	458	446	438
35	889	522	508	496
37	939.8	584	576	548
39	990.6	648	648	628
42	1066.8	758	744	742
45	1143	898	880	854

APPENDIX E
TUBE LAYOUT

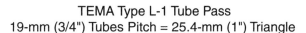

TEMA Type L-1 Tube Pass
19-mm (3/4") Tubes Pitch = 25.4-mm (1") Triangle

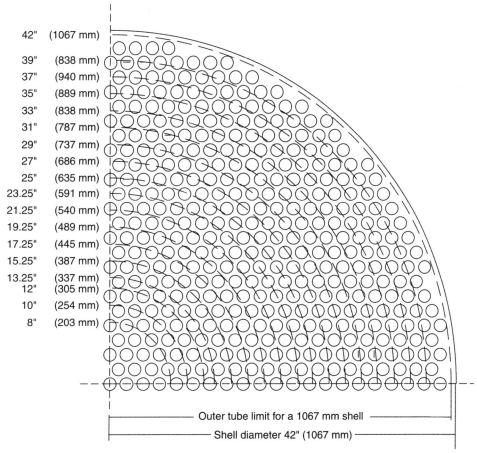

42"	(1067 mm)
39"	(838 mm)
37"	(940 mm)
35"	(889 mm)
33"	(838 mm)
31"	(787 mm)
29"	(737 mm)
27"	(686 mm)
25"	(635 mm)
23.25"	(591 mm)
21.25"	(540 mm)
19.25"	(489 mm)
17.25"	(445 mm)
15.25"	(387 mm)
13.25"	(337 mm)
12"	(305 mm)
10"	(254 mm)
8"	(203 mm)

Outer tube limit for a 1067 mm shell

Shell diameter 42" (1067 mm)

Dashed lines represent the outer tube limits for the indicated shell diameters.
Tubes must be removed for the installation of tie rods, impingement plates,
and to satisfy entrance velocity requirements

FIGURE E-1   3/4-in tube distribution in 1-in triangular pattern, one-pass configuration. Rear head type L.

TEMA Type L-2 Tube Passes
19-mm (3/4") Tubes Pitch = 25.4-mm (1") Triangle

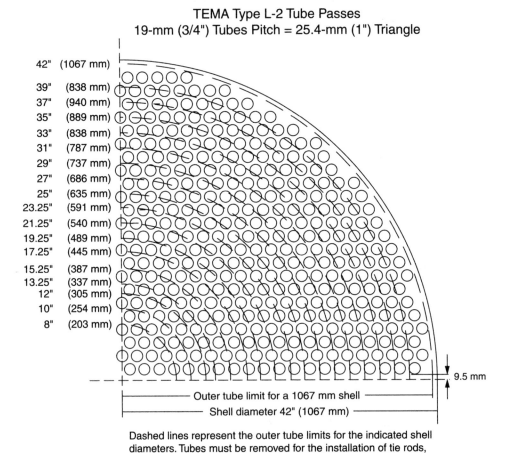

42"	(1067 mm)
39"	(838 mm)
37"	(940 mm)
35"	(889 mm)
33"	(838 mm)
31"	(787 mm)
29"	(737 mm)
27"	(686 mm)
25"	(635 mm)
23.25"	(591 mm)
21.25"	(540 mm)
19.25"	(489 mm)
17.25"	(445 mm)
15.25"	(387 mm)
13.25"	(337 mm)
12"	(305 mm)
10"	(254 mm)
8"	(203 mm)

9.5 mm

Outer tube limit for a 1067 mm shell
Shell diameter 42" (1067 mm)

Dashed lines represent the outer tube limits for the indicated shell
diameters. Tubes must be removed for the installation of tie rods,
impingement plates, and to satisfy entrance velocity requirements

FIGURE E-2 ³/₄-in tube distribution in 1-in triangular pattern, two-pass configuration. Rear head type L.

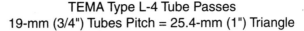

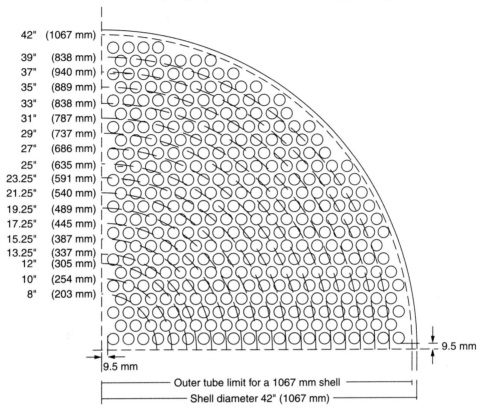

Dashed lines represent the outer tube limits for the indicated shell
diameters. Tubes must be removed for the installation of tie rods,
impingement plates, and to satisfy entrance velocity requirements

FIGURE E-3 $3/4$-in tube distribution in 1-in triangular pattern, four-pass configuration. Rear head type L.

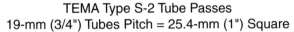

TEMA Type S-2 Tube Passes
19-mm (3/4") Tubes Pitch = 25.4-mm (1") Square

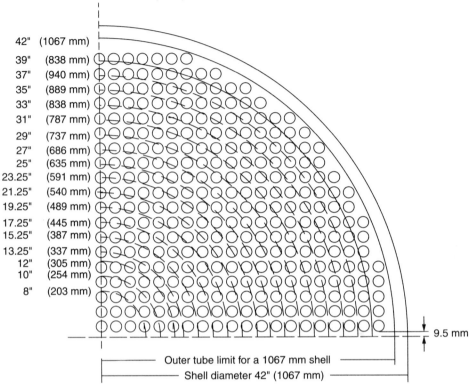

42"	(1067 mm)
39"	(838 mm)
37"	(940 mm)
35"	(889 mm)
33"	(838 mm)
31"	(787 mm)
29"	(737 mm)
27"	(686 mm)
25"	(635 mm)
23.25"	(591 mm)
21.25"	(540 mm)
19.25"	(489 mm)
17.25"	(445 mm)
15.25"	(387 mm)
13.25"	(337 mm)
12"	(305 mm)
10"	(254 mm)
8"	(203 mm)

9.5 mm

Outer tube limit for a 1067 mm shell

Shell diameter 42" (1067 mm)

Dashed lines represent the outer tube limits for the indicated shell
diameters. Tubes must be removed for the installation of tie rods,
impingement plates, and to satisfy entrance velocity requirements

FIGURE E-4   3/4-in tube distribution in 1-in square pattern, two-pass configuration. Rear head type S.

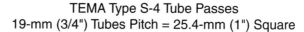

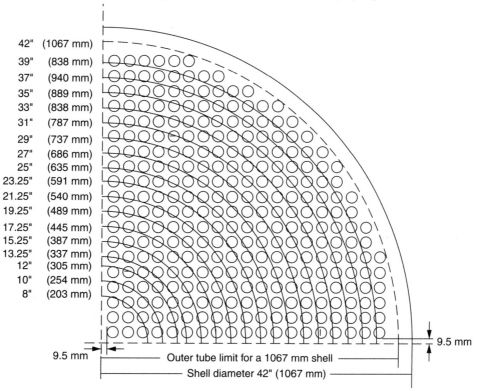

Dashed lines represent the outer tube limits for the indicated shell diameters. Tubes must be removed for the installation of tie rods, impingement plates, and to satisfy entrance velocity requirements

FIGURE E-5   3/4-in tube distribution in 1-in square pattern, four-pass configuration. Rear head type S.

TEMA Type U-2 Tube Passes
19-mm (3/4") Tubes Pitch = 25.4-mm (1") Square

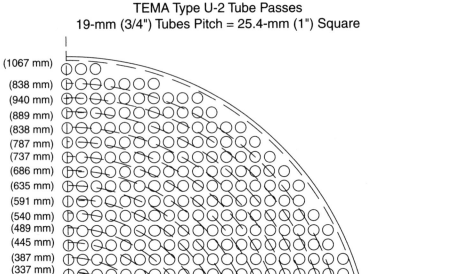

42"	(1067 mm)
39"	(838 mm)
37"	(940 mm)
35"	(889 mm)
33"	(838 mm)
31"	(787 mm)
29"	(737 mm)
27"	(686 mm)
25"	(635 mm)
23.25"	(591 mm)
21.25"	(540 mm)
19.25"	(489 mm)
17.25"	(445 mm)
15.25"	(387 mm)
13.25"	(337 mm)
12"	(305 mm)
10"	(254 mm)
8"	(203 mm)

19 mm

Outer tube limit for a 1067 mm shell
Shell diameter 42" (1067 mm)

Dashed lines represent the outer tube limits for the indicated shell
diameters. Tubes must be removed for the installation of tie rods,
impingement plates, and to satisfy entrance velocity requirements

FIGURE E-6 3/4-in tube distribution in 1-in square pattern, two-pass configuration. Rear head type U.

TEMA Type U-4 Tube Passes
19-mm (3/4") Tubes Pitch = 25.4-mm (1") Square

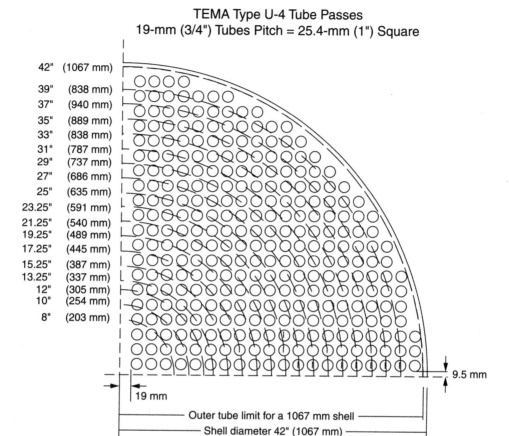

42"	(1067 mm)
39"	(838 mm)
37"	(940 mm)
35"	(889 mm)
33"	(838 mm)
31"	(787 mm)
29"	(737 mm)
27"	(686 mm)
25"	(635 mm)
23.25"	(591 mm)
21.25"	(540 mm)
19.25"	(489 mm)
17.25"	(445 mm)
15.25"	(387 mm)
13.25"	(337 mm)
12"	(305 mm)
10"	(254 mm)
8"	(203 mm)

9.5 mm

19 mm

Outer tube limit for a 1067 mm shell

Shell diameter 42" (1067 mm)

Dashed lines represent the outer tube limits for the indicated shell
diameters. Tubes must be removed for the installation of tie rods,
impingement plates, and to satisfy entrance velocity requirements

FIGURE E-7 3/4-in tube distribution in 1-in square pattern, four-pass configuration. Rear head type U.

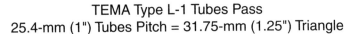

TEMA Type L-1 Tubes Pass
25.4-mm (1") Tubes Pitch = 31.75-mm (1.25") Triangle

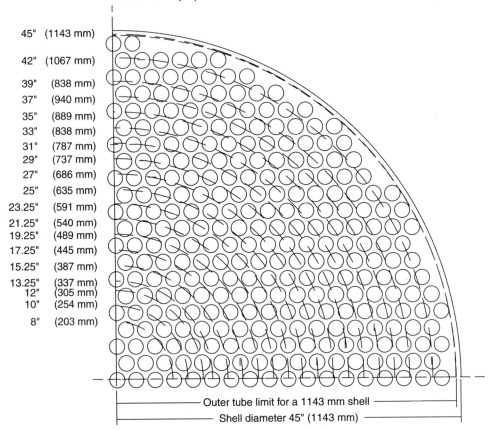

45"	(1143 mm)
42"	(1067 mm)
39"	(838 mm)
37"	(940 mm)
35"	(889 mm)
33"	(838 mm)
31"	(787 mm)
29"	(737 mm)
27"	(686 mm)
25"	(635 mm)
23.25"	(591 mm)
21.25"	(540 mm)
19.25"	(489 mm)
17.25"	(445 mm)
15.25"	(387 mm)
13.25"	(337 mm)
12"	(305 mm)
10"	(254 mm)
8"	(203 mm)

Outer tube limit for a 1143 mm shell

Shell diameter 45" (1143 mm)

Dashed lines represent the outer tube limits for the indicated shell diameters.
Tubes must be removed for the installation of tie rods, impingement plates,
and to satisfy entrance velocity requirements

FIGURE E-8 1-in tube distribution in 1.25-in triangular pattern, one-pass configuration. Rear head type L.

TEMA Type L-2 Tube Passes
25.4-mm (1") Tubes Pitch = 31.75-mm (1.25") Triangle

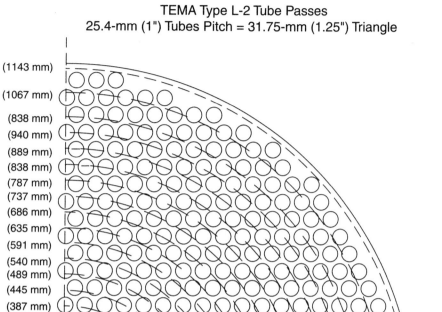

45"	(1143 mm)
42"	(1067 mm)
39"	(838 mm)
37"	(940 mm)
35"	(889 mm)
33"	(838 mm)
31"	(787 mm)
29"	(737 mm)
27"	(686 mm)
25"	(635 mm)
23.25"	(591 mm)
21.25"	(540 mm)
19.25"	(489 mm)
17.25"	(445 mm)
15.25"	(387 mm)
13.25"	(337 mm)
12"	(305 mm)
10"	(254 mm)

Outer tube limit for a 1143 mm shell

Shell diameter 45" (1143 mm)

Dashed lines represent the outer tube limits for the indicated shell diameters.
Tubes must be removed for the installation of tie rods, impingement plates,
and to satisfy entrance velocity requirements

FIGURE E-9 1-in tube distribution in 1.25-in triangular pattern, two-pass configuration. Rear head type L.

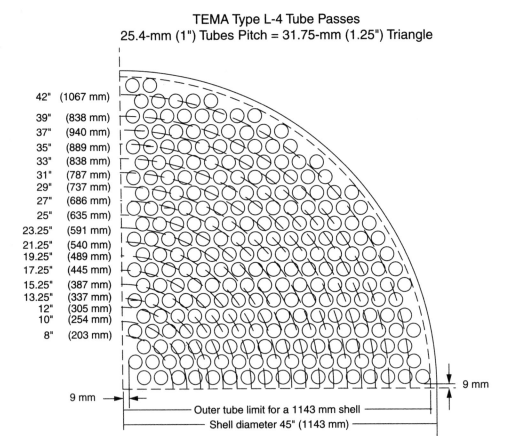

TEMA Type L-4 Tube Passes
25.4-mm (1") Tubes Pitch = 31.75-mm (1.25") Triangle

42" (1067 mm)
39" (838 mm)
37" (940 mm)
35" (889 mm)
33" (838 mm)
31" (787 mm)
29" (737 mm)
27" (686 mm)
25" (635 mm)
23.25" (591 mm)
21.25" (540 mm)
19.25" (489 mm)
17.25" (445 mm)
15.25" (387 mm)
13.25" (337 mm)
12" (305 mm)
10" (254 mm)
8" (203 mm)

9 mm

9 mm

Outer tube limit for a 1143 mm shell
Shell diameter 45" (1143 mm)

Dashed lines represent the outer tube limits for the indicated shell
diameters. Tubes must be removed for the installation of tie rods,
impingement plates, and to satisfy entrance velocity requirements

FIGURE E-10   1-in tube distribution in 1.25-in triangular pattern, four-pass configuration. Rear head type L.

TEMA Type S-2 Tube Passes
25.4-mm (1") Tubes Pitch = 31.75-mm (1.25") Square

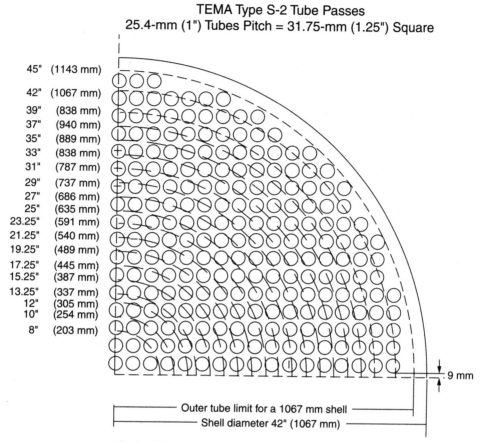

45"	(1143 mm)
42"	(1067 mm)
39"	(838 mm)
37"	(940 mm)
35"	(889 mm)
33"	(838 mm)
31"	(787 mm)
29"	(737 mm)
27"	(686 mm)
25"	(635 mm)
23.25"	(591 mm)
21.25"	(540 mm)
19.25"	(489 mm)
17.25"	(445 mm)
15.25"	(387 mm)
13.25"	(337 mm)
12"	(305 mm)
10"	(254 mm)
8"	(203 mm)

9 mm

Outer tube limit for a 1067 mm shell
Shell diameter 42" (1067 mm)

Dashed lines represent the outer tube limits for the indicated shell
diameters. Tubes must be removed for the installation of tie rods,
impingement plates, and to satisfy entrance velocity requirements

FIGURE E-11   1-in tube distribution in 1.25-in square pattern, two-pass configuration. Rear head type S.

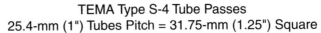

TEMA Type S-4 Tube Passes
25.4-mm (1") Tubes Pitch = 31.75-mm (1.25") Square

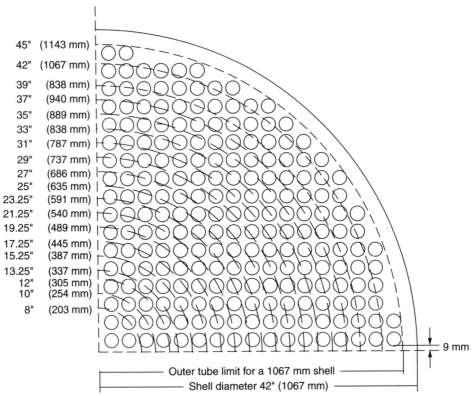

45"	(1143 mm)
42"	(1067 mm)
39"	(838 mm)
37"	(940 mm)
35"	(889 mm)
33"	(838 mm)
31"	(787 mm)
29"	(737 mm)
27"	(686 mm)
25"	(635 mm)
23.25"	(591 mm)
21.25"	(540 mm)
19.25"	(489 mm)
17.25"	(445 mm)
15.25"	(387 mm)
13.25"	(337 mm)
12"	(305 mm)
10"	(254 mm)
8"	(203 mm)

9 mm

Outer tube limit for a 1067 mm shell
Shell diameter 42" (1067 mm)

Dashed lines represent the outer tube limits for the indicated shell
diameters. Tubes must be removed for the installation of tie rods,
impingement plates, and to satisfy entrance velocity requirements

FIGURE E-12   1-in tube distribution in 1.25-in square pattern, four-pass configuration. Rear head type S.

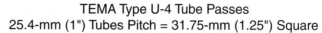

TEMA Type U-4 Tube Passes
25.4-mm (1") Tubes Pitch = 31.75-mm (1.25") Square

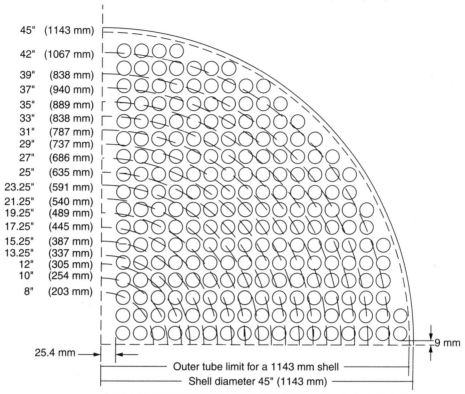

45"	(1143 mm)
42"	(1067 mm)
39"	(838 mm)
37"	(940 mm)
35"	(889 mm)
33"	(838 mm)
31"	(787 mm)
29"	(737 mm)
27"	(686 mm)
25"	(635 mm)
23.25"	(591 mm)
21.25"	(540 mm)
19.25"	(489 mm)
17.25"	(445 mm)
15.25"	(387 mm)
13.25"	(337 mm)
12"	(305 mm)
10"	(254 mm)
8"	(203 mm)

9 mm

25.4 mm

Outer tube limit for a 1143 mm shell

Shell diameter 45" (1143 mm)

Dashed lines represent the outer tube limits for the indicated shell diameters. Tubes must be removed for the installation of tie rods, impingement plates, and to satisfy entrance velocity requirements

FIGURE E-13 1-in tube distribution in 1.25-in square pattern, two-pass configuration. Rear head type U.

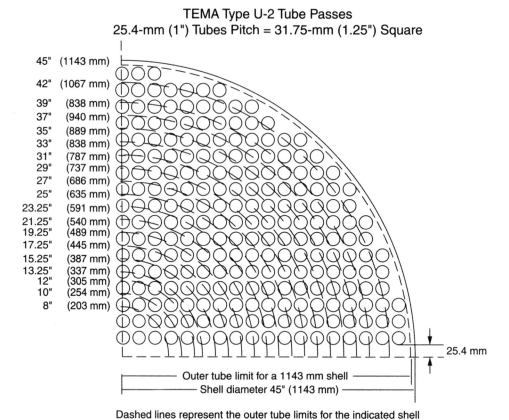

TEMA Type U-2 Tube Passes
25.4-mm (1") Tubes Pitch = 31.75-mm (1.25") Square

45"	(1143 mm)
42"	(1067 mm)
39"	(838 mm)
37"	(940 mm)
35"	(889 mm)
33"	(838 mm)
31"	(787 mm)
29"	(737 mm)
27"	(686 mm)
25"	(635 mm)
23.25"	(591 mm)
21.25"	(540 mm)
19.25"	(489 mm)
17.25"	(445 mm)
15.25"	(387 mm)
13.25"	(337 mm)
12"	(305 mm)
10"	(254 mm)
8"	(203 mm)

25.4 mm

Outer tube limit for a 1143 mm shell
Shell diameter 45" (1143 mm)

Dashed lines represent the outer tube limits for the indicated shell
diameters. Tubes must be removed for the installation of tie rods,
impingement plates, and to satisfy entrance velocity requirements

FIGURE E-14 1-in tube distribution in 1.25-in square pattern, four-pass configuration. Rear head type U.

APPENDIX F
FOULING FACTORS

Fouling Resistances
[Values in $(K \cdot m^2)/W$]
From *Standards of the Tubular Exchangers Manufacturers Association*.
Reproduced with permission.

Fouling Resistances for Water

Temperature of the Heating Medium	Up to 115°C		115–205°C	
	52°C or less		More than 52°C	
	Velocity, m/s		Velocity, m/s	
Temperature of Water	1 or less	>1	1 or less	>1
Sea water	0.00009	0.00009	0.0002	0.0002
Brackish water	0.0004	0.0002	0.0005	0.0004
Cooling tower and artificial spray pond				
Treated makeup	0.0002	0.0002	0.0004	0.0004
Untreated	0.0005	0.0005	0.0009	0.0007
City or well water	0.0002	0.0002	0.0004	0.0004
River water				
Mínimum	0.0004	0.0002	0.0005	0.0004
Average	0.0005	0.0004	0.0007	0.0005
Muddy or silty	0.0005	0.0004	0.0007	0.0005
Hard (over 15 grains/gal)	0.0005	0.0005	0.0009	0.0009
Engine jacket	0.0002	0.0002	0.0002	0.0002
Distilled or closed cycle				
Condensate	0.00009	0.00009	0.00009	0.00009
Treated boiler feedwater	0.0002	0.00009	0.0002	0.0002
Boilers purge	0.0004	0.0004	0.0004	0.0004

Note: If the heating medium temperature is over 400°F and the cooling medium is known to scale, these ratings should be modified accordingly.

Fouling Resistances for Industrial Fluids

Oils		Gases and vapors	
Fuel oil no. 2	0.0004	Manufactured gas	0.002
Fuel oil no. 6	0.0009	Diesel engine exhaust gas	0.002
Transformer oil	0.0002	Steam (nonoil bearing)	0.00009
Engine lube oil	0.0002	Exhaust team (oil bearing)	0.0004
Quench oil	0.0007	Compressed air	0.0002
		Refrigerant vapors in refrigerant cycle condensers	0.0004
Liquids		Ammonia vapor	0.0002
Refrigerant liquids	0.0002	CO_2 vapor	0.0002
Hydraulic fluids	0.0002	Chlorine vapor	0.0004
Industrial organic heat transfer media	0.0002	Coal fuel gas	0.002
Molten salts	0.00009	Natural gas flue gas	0.0009
Ammonia, liquid	0.0002		
Ammonia, liquid (oil bearing)	0.0005		
Calcium chloride solutions	0.0005		

Fouling Resistances for Chemical Processing Streams

Liquids		Gases and vapors	
MEA and DEA solutions	0.0004	Acid gases	0.0004
DEG and TEG solutions	0.0004	Solvent vapors	0.0002
Stable side-draw and bottom products	0.0002	Stable overhead products	0.0002
Caustic solutions	0.0004		
Vegetable oils	0.0005		

Fouling Resistances for Natural Gas and Gasoline Processing Streams

Liquids		Gases and vapors	
Rich oil	0.0002	Natural gas	0.0002–0.0004
Lean oil	0.0004	Column overhead products	0.0002
Natural gasoline and LPG	0.0002		

Fouling Resistances for Oil Refinery Streams

	Crude Oil											
Temp.	0–120°C			120–180°C			180–230°C			>230°C		
	Velocity (m/s)			Velocity (m/s)			Velocity (m/s)			Velocity (m/s)		
	<0.60	0.60–1.20	>1.20	<0.60	0.60–1.20	>1.20	<0.60	0.60–1.20	>1.20	<0.60	0.60–1.20	>1.20
Dry	0.0005	0.0004	0.0004	0.0005	0.0004	0.0004	0.0007	0.0005	0.0004	0.0009	0.0007	0.0005
Salt	0.0005	0.0004	0.0004	0.0009	0.0007	0.0007	0.001	0.0009	0.0007	0.0012	0.001	

Fouling Resistances for Oil Refinery Streams (cont.)

Crude and vapor unit gases and vapors

Atmospheric tower overhead vapors	0.0002
Light naphthas	0.0002
Vacuum tower overhead vapors	0.0004

Crude and vapor unit liquids

Gasoline	0.0004
Naphtha and light distillates	0.0005
Kerosene	0.0005
Light gas oil	0.0005
Heavy gas oil	0.0005–0.0008
Heavy fuel oil	0.0008–0.0012

Asphalt and residuum

Vacuum tower bottoms	0.002
Atmospheric tower bottoms	0.0014

Cracking and coking unit streams

Overhead vapors	0.0004
Light-cycle oil	0.0004
Heavy-cycle oil	0.0005–0.0007
Light coker gas oil	0.0005–0.0007
Heavy coker gas oil	0.0007–0.0009
Bottom slurry oil (minimum 4.5 ft/s)	0.0005
Light liquid products	0.0005

Naphtha hydrotreater

Feed	0.0005
Effluent	0.0004
Naphthas	0.0004
Overhead vapors	0.0003

Visbreaker

Overhead vapor	0.0005
Visbreaker bottoms	0.0020

Catalitic re-forming, hydrocracking, and hydrodesulfurization streams

Re-former charge	0.0003
Re-former effluent	0.0003
Hydrocracker charge and effluent*	0.0004
Recycle gas	0.0002
Hydrodesulfurization charge and effluent	0.0004
Overhead vapors	0.0002
Liquid products more than 50° API	0.0002
Liquid products from 30–50° API	0.0004

Light end processing streams

Overhead vapors and gases	0.0002
Liquid products	0.0002
Absortion oils	0.0004
Reboiler streams	0.0005
Alkylation trace acid streams	0.0004

Lube oil processing streams

Feed stock	0.0004
Solvent feed mix	0.0004
Solvent	0.0002
Extract	0.0005
Raffinate	0.0002
Asphalt	0.0009
Refined lube oil	0.0002
Wax slurries	0.0005

APPENDIX G
TYPICAL HEAT TRANSFER COEFFICIENTS

Approximate Values of the Overall Heat Transfer Coefficients (U) in Shell-and-Tube Heat Exchangers[*]

Application	Hot Fluid	Cold Fluid	U [W/(m$^2 \cdot$ K)]
Single phase	Gas	Gas	10–120
	Gas	Light hydrocarbon	15–200
	Light hydrocarbon	Water	250–750
	Heavy oil	Water	30–420
	Gas	Water	15–250
	Water	Water	800–1,600
	Light hydrocarbon	Brine	150–500
	Light hydrocarbon	Light hydrocarbon	120–350
	Heavy oil	Heavy oil	45–250
Boiling	Steam	Light hydrocarbon	450–1,000
	Steam	Water	2,000–4,000
	Steam	Heavy hydrocarbon	150–300
	Water	Refrigerant fluid	400–800
	Hydrocarbon	Refrigerant fluid	150–600
Steam heaters	Steam	Light hydrocarbon	250–800
	Steam	Water	1,500–4,000
	Steam	Gas	20–200
	Steam	Heavy oil	50–450
	Steam	Organic solvents	600–1,200
Water-cooled condensers	Steam	Water	2,000–4,000
	Organic solvents at atmospheric pressure	Water	550–1,100
	Turbine exhaust steam (subatmospheric)	Water	1,500–3,000
	Organic solvents (subatmospheric with some inerts)	Water	250–700
	Organic compounds diluted in a noncondensable stream (subatmospheric)	Water	50–300
	Light hydrocarbons (above atmospheric pressure)	Water	100–450
	Heavy hydrocarbons (subatmospheric)	Water	50–150

[*]The values indicated in this table should not be interpreted as limits but are only guide values to be used as a first approximation to design. They include total fouling factors of about 0.0005 (m$^2 \cdot$ K)/W.

Approximate Values of the Heat Transfer Coefficients h_{io} or h_o [W/(m^2 · K)]

Single Phase	Boiling	Condensation
Water 1,500–11,000	Water 4,500–11,000	Steam 5,500–17,000
Gas 15–250	Organic solvents 500–1,500	Organic solvent 800–2,800
Organic solvents 350–2,000	Light hydrocarbon 850–1,700	Heavy organics, subatmospheric 110–250
Oils 60–700	Heavy oil 60–250	Ammonia 2,500–5,000
	Ammonia 1,000–2,000	

APPENDIX H
DIMENSIONS OF TUBES ACCORDING TO BWG STANDARD

D_o (in)	BWG	D_o (m)	D_i (m)
1/2	12	0.0127	0.00716
"	14	"	0.00848
"	16	"	0.00940
"	18	"	0.01020
"	20	"	0.0109
3/4	10	0.0190	0.0127
"	11	"	0.0129
"	12	"	0.0135
"	13	"	0.0142
"	14	"	0.0148
"	15	"	0.0153
"	16	"	0.0157
"	17	"	0.0161
"	18	"	0.0165
1	8	0.0254	0.0170
"	9	"	0.0178
"	10	"	0.0185
"	11	"	0.0193
"	12	"	0.0198
"	13	"	0.0205
"	14	"	0.0211
"	15	"	0.0217
"	16	"	0.0221
"	17	"	0.0224
"	18	"	0.0229

APPENDIX I
PHYSICAL PROPERTIES OF PURE SUBSTANCES

This appendix includes selected physical properties of the most common chemical compounds. At present, process engineers usually work with process simulation programs, and the acquisition of physical properties of pure compounds is done using these tools. Simulation programs also select the most suitable mixing rules to calculate the physical properties of multicomponent mixtures and perform the necessary calculations.

In the particular case of the petroleum industry, where mixtures are characterized by their distillation curves, the use of process simulation software is almost the only conceivable option because it avoids the tedious calculations necessary in the past.

This appendix is only offered as support material for comprehension and analysis of the different topics covered in the book and to be used in the example problems included in the text.

The physical properties included in this appendix were calculated using a UniSim process simulator from Honeywell. Since they are not the result of direct measurement, some discrepancies with tabulated data from other sources are possible. However, these data are considered within the accuracy limits usually necessary for the design of heat transfer equipment in the process industry.

TABLE I-1   Physical Properties of Liquids Calculated with Honeywell UniSim Simulation Software. With authorization

Hydrocarbons

Ethane

Temperature	°C	0	20
Vapor pressure	bar	24.01	37.87
Density	kg/m³	402	341
Viscosity	cP	6.25E-02	4.78E-02
Thermal conductivity	W/m.K	8.8E-02	7.2E-02
Specific heat	KJ/kg.K	4.13	7.40
Heat of vaporization	KJ/kg.K	299	192

Propane

Temperature	°C	0	20	40	60	80
Vapor pressure	bar	4.73	8.36	13.73	21.29	31.53
Density	kg/m³	529	500	468	428	375
Viscosity	cP	0.13	0.10	8.6E-02	7.0E-02	5.5E-02
Thermal conductivity	W/m.K	0.11	9.8E-02	8.6E-02	7.4E-02	6.3E-02
Specific heat	KJ/kg.K	2.53	2.77	3.12	3.75	5.62
Heat of vaporization	KJ/kg.K	377	347	308	257	179

Isobutane

Temperature	°C	0	20	40	60	80	100	120
Vapor pressure	bar	1.56	3.00	5.28	8.66	13.44	19.93	28.49
Density	kg/m³	581	557	531	502	470	430	374
Viscosity	cP	0.21	0.18	0.14	0.12	9.9E-02	8.1E-02	6.4E-02
Thermal conductivity	W/m.K	9.4E-02	8.7E-02	8.0E-02	7.3E-02	6.4E-02	5.3E-02	3.8E-02
Specific heat	KJ/kg.K	2.23	2.38	2.56	2.80	3.13	3.72	5.60
Heat of vaporization	KJ/kg.K	355	336	315	288	254	209	141

n-Butane

Temperature	°C	0	20	40	60	80	100	120	140
Vapor pressure	bar	1.03	2.07	3.77	6.37	10.12	15.32	22.27	31.33
Density	kg/m³	601	579	555	529	500	467	426	366
Viscosity	cP	0.21	0.17	0.14	0.12	0.10	8.7E-02	7.2E-02	5.8E-02
Thermal conductivity	W/m.K	0.10	9.5E-02	8.9E-02	8.1E-02	7.3E-02	6.4E-02	5.3E-02	3.6E-02
Specific heat	KJ/kg.K	2.23	2.38	2.54	2.73	2.97	3.34	4.04	6.63
Heat of vaporization	KJ/kg.K	386	369	349	325	296	259	210	133

i-Pentane

Temperature	°C	0	20	40	60	80	100	120	140	160	180
Vapor pressure	bar	0.35	0.77	1.51	2.71	4.53	7.16	10.79	15.63	21.93	29.95
Density	kg/m³	642	621	600	578	553	527	497	463	419	349
Viscosity	cP	0.28	0.22	0.19	0.15	0.13	0.11	9.1E-02	7.5E-02	6.1E-02	4.6E-02
Thermal conductivity	W/m.K	0.11	0.10	9.6E-02	9.0E-02	8.4E-02	7.7E-02	6.9E-02	6.0E-02	4.8E-02	2.9E-02
Specific heat	KJ/kg.K	2.06	2.18	2.32	2.46	2.62	2.80	3.05	3.43	4.20	8.52
Heat of vaporization	KJ/kg.K	359	347	333	317	298	276	249	215	168	89

Temperature	°C	0	20	40	60	80	100	120	140	160	180
Vapor pressure	bar	0.24	0.56	1.15	2.13	3.67	5.92	9.09	13.39	19.05	26.34
Density	kg/m³	645	626	605	583	560	535	508	476	438	385

n-Pentane

Property	Units	0	20	40	60	80	100	120	140	160	180	200
Viscosity	cP		0.28	0.23	0.19	0.17	0.14	0.12	0.11	9.1E-02	7.8E-02	6.4E-02
Thermal conductivity	W/m.K		0.11	0.11	0.10	9.5E-02	8.9E-02	8.2E-02	7.4E-02	6.6E-02	5.5E-02	4.0E-02
Specific heat	KJ/kg.K		2.12	2.24	2.36	2.49	2.64	2.81	3.03	3.33	3.88	5.47
Heat of vaporization	KJ/kg.K		383	370	356	340	322	301	275	242	200	138

n-Hexane

Property	Units	0	20	40	60	80	100	120	140	160	180	200
Temperature	°C	0	20	40	60	80	100	120	140	160	180	200
Vapor pressure	bar	0.06	0.16	0.37	0.76	1.41	2.44	3.97	6.13	9.09	13.02	18.10
Density	kg/m³	680	662	643	623	603	582	559	534	506	475	436
Viscosity	cP	0.38	0.31	0.26	0.22	0.19	0.16	0.14	0.12	0.10	8.8E-02	7.5E-02
Thermal conductivity	W/m.K	0.12	0.12	0.11	0.10	9.9E-02	9.3E-02	8.7E-02	8.0E-02	7.3E-02	6.4E-02	5.3E-02
Specific heat	KJ/kg.K	2.04	2.15	2.26	2.37	2.49	2.62	2.76	2.93	3.14	3.44	3.98
Heat of vaporization	KJ/kg.K	377	366	355	342	329	314	296	275	251	220	180

n-Heptane

Property	Units	0	20	40	60	80	100	120	140	160	180	200
Temperature	°C	0	20	40	60	80	100	120	140	160	180	200
Vapor pressure	bar	0.016	0.049	0.13	0.28	0.57	1.06	1.83	2.98	4.61	6.86	9.84
Density	kg/m³	699	682	665	647	628	609	589	567	544	519	490
Viscosity	cP	0.52	0.41	0.33	0.28	0.23	0.20	0.17	0.14	0.12	0.11	9.1E-02
Thermal conductivity	W/m.K	0.13	0.12	0.12	0.11	0.11	0.10	9.5E-02	9.0E-02	8.3E-02	7.6E-02	6.9E-02
Specific heat	KJ/kg.K	2.00	2.10	2.21	2.31	2.42	2.53	2.65	2.78	2.92	3.09	3.31
Heat of vaporization	KJ/kg.K	371	362	352	341	330	317	304	288	270	249	224

n-Octane

Property	Units	0	20	40	60	80	100	120	140	160	180	200
Temperature	°C	0	20	40	60	80	100	120	140	160	180	200
Vapor pressure	bar	4.22E-03	1.47E-02	4.26E-02	0.11	0.23	0.47	0.86	1.48	2.41	3.73	5.55
Density	kg/m³	720	704	688	671	653	635	616	596	575	553	529
Viscosity	cP	0.71	0.55	0.43	0.35	0.29	0.24	0.21	0.18	0.15	0.13	0.11
Thermal conductivity	W/m.K	0.13	0.13	0.12	0.12	0.11	0.11	0.10	9.7E-02	9.1E-02	8.5E-02	7.8E-02
Specific heat	KJ/kg.K	1.98	2.09	2.19	2.29	2.40	2.50	2.61	2.72	2.84	2.98	3.13
Heat of vaporization	KJ/kg.K	368	359	350	340	330	319	308	295	281	265	246

n-Nonane

Property	Units	0	20	40	60	80	100	120	140	160	180	200
Temperature	°C	0	20	40	60	80	100	120	140	160	180	200
Vapor pressure	bar	1.16E-03	4.62E-03	1.50E-02	4.14E-02	9.94E-02	0.21	0.42	0.76	1.29	2.08	3.21
Density	kg/m³	736	721	705	689	672	655	637	618	599	579	557
Viscosity	cP	0.94	0.71	0.56	0.45	0.37	0.31	0.26	0.22	0.19	0.17	0.14
Thermal conductivity	W/m.K	0.14	0.13	0.13	0.12	0.12	0.11	0.10	0.10	9.7E-02	9.2E-02	8.6E-02
Specific heat	KJ/kg.K	1.97	2.07	2.16	2.26	2.36	2.45	2.55	2.65	2.75	2.87	2.99
Heat of vaporization	KJ/kg.K	364	355	346	337	328	318	308	297	285	272	257

n-Decane

Property	Units	0	20	40	60	80	100	120	140	160	180	200
Temperature	°C	0	20	40	60	80	100	120	140	160	180	200
Vapor pressure	bar	3.25E-04	1.48E-03	5.39E-03	1.64E-02	4.28E-02	9.89E-02	0.21	0.40	0.71	1.18	1.89
Density	kg/m³	746	731	716	700	684	668	651	633	615	596	576
Viscosity	cP	1.23	0.91	0.70	0.55	0.45	0.37	0.31	0.26	0.23	0.20	0.17
Thermal conductivity	W/m.K	0.14	0.13	0.13	0.12	0.11	0.11	0.10	0.10	9.9E-02	9.4E-02	8.8E-02

(*Continued*)

TABLE I-1 Physical Properties of Liquids Calculated with Honeywell UniSim simulation software. With authorization (Continued)

Substance	Property	Units											
	Specific heat	KJ/kg.K	1.96	2.05	2.15	2.24	2.33	2.42	2.52	2.61	2.71	2.81	2.91
	Heat of vaporization	KJ/kg.K	359	351	342	334	325	316	307	297	286	275	262
Benzene	Temperature	°C	0	20	40	60	80	100	120	140	160	180	200
	Vapor pressure	bar	3.87E-02	0.11	0.25	0.53	1.01	1.79	2.97	4.67	7.04	10.22	14.36
	Density	kg/m^3	899	877	856	834	811	787	763	737	710	681	650
	Viscosity	cP	0.88	0.65	0.50	0.39	0.31	0.25	0.21	0.17	0.14	0.12	0.09
	Thermal conductivity	W/m.K	0.14	0.13	0.13	0.12	0.12	0.11	0.11	1.0E-01	9.4E-02	8.7E-02	8.0E-02
	Specific heat	KJ/kg.K	1.41	1.50	1.59	1.68	1.77	1.87	1.97	2.07	2.19	2.33	2.50
	Heat of vaporization	KJ/kg.K	432	422	412	402	391	378	365	349	332	312	288
Toluene	Temperature	°C	0	20	40	60	80	100	120	140	160	180	200
	Vapor pressure	bar	1.02E-02	3.16E-02	8.32E-02	0.19	0.40	0.75	1.32	2.19	3.44	5.18	7.52
	Density	kg/m^3	887	869	850	831	812	792	771	749	726	702	676
	Viscosity	cP	0.75	0.58	0.46	0.38	0.31	0.26	0.22	0.19	0.17	0.14	0.12
	Thermal conductivity	W/m.K	0.14	0.14	0.13	0.13	0.12	0.12	0.11	0.11	0.10	9.5E-02	8.9E-02
	Specific heat	KJ/kg.K	1.45	1.54	1.63	1.71	1.80	1.89	1.98	2.07	2.17	2.28	2.40
	Heat of vaporization	KJ/kg.K	412	404	395	386	376	366	355	342	329	313	296
o-Xylene	Temperature	°C	0	20	40	60	80	100	120	140	160	180	200
	Vapor pressure	bar	2.07E-03	7.50E-03	2.26E-02	5.84E-02	0.13	0.28	0.52	0.92	1.53	2.42	3.67
	Density	kg/m^3	898	882	864	847	829	811	792	772	752	731	709
	Viscosity	cP	1.08	0.81	0.63	0.50	0.41	0.34	0.29	0.24	0.21	0.18	0.15
	Thermal conductivity	W/m.K	0.15	0.14	0.14	0.14	0.13	0.13	0.12	0.12	0.11	0.11	0.10
	Specific heat	KJ/kg.K	1.57	1.65	1.72	1.80	1.88	1.95	2.03	2.11	2.20	2.28	2.37
	Heat of vaporization	KJ/kg.K	408	399	391	383	374	365	356	346	335	323	310
m-Xylene	Temperature	°C	0	20	40	60	80	100	120	140	160	180	200
	Vapor pressure	bar	2.37E-03	8.59E-03	2.58E-02	6.65E-02	0.15	0.31	0.59	1.03	1.72	2.71	4.09
	Density	kg/m^3	883	866	848	831	812	793	774	754	733	711	688
	Viscosity	cP	0.79	0.62	0.50	0.41	0.34	0.29	0.24	0.21	0.18	0.16	0.14
	Thermal conductivity	W/m.K	0.15	0.14	0.14	0.13	0.13	0.12	0.12	0.11	0.11	0.10	9.6E-02
	Specific heat	KJ/kg.K	1.53	1.61	1.69	1.77	1.86	1.94	2.02	2.11	2.20	2.29	2.39
	Heat of vaporization	KJ/kg.K	407	399	390	381	372	363	353	342	330	317	303
p-Xylene	Temperature	°C	0	20	40	60	80	100	120	140	160	180	200
	Vapor pressure	bar	2.41E-03	8.72E-03	2.61E-02	6.72E-02	0.15	0.31	0.59	1.04	1.72	2.72	4.10
	Density	kg/m^3	879	862	845	827	809	790	771	751	730	708	685
	Viscosity	cP	0.83	0.64	0.51	0.42	0.34	0.29	0.25	0.21	0.18	0.16	0.14
	Thermal conductivity	W/m.K	0.15	0.14	0.14	0.13	0.13	0.12	0.12	0.11	0.10	0.10	9.6E-02
	Specific heat	KJ/kg.K	1.53	1.61	1.69	1.77	1.84	1.92	2.01	2.09	2.18	2.27	2.37
	Heat of vaporization	KJ/kg.K	407	398	389	380	371	362	352	341	329	316	302
	Temperature	°C	0	20	40	60	80	100	120	140	160	180	200
	Vapor pressure	bar	1.89E-03	6.89E-03	2.09E-02	5.42E-02	0.12	0.26	0.49	0.87	1.45	2.31	3.50

Compound	Property	Units											
Styrene	Density	kg/m³	922	905	888	870	852	834	815	795	775	753	731
	Viscosity	cP	1.04	0.77	0.59	0.47	0.38	0.31	0.26	0.22	0.18	0.16	0.13
	Thermal conductivity	W/m.K	0.15	0.15	0.14	0.14	0.13	0.13	0.12	0.12	0.11	0.11	0.10
	Specific heat	KJ/kg.K	1.57	1.62	1.67	1.73	1.79	1.86	1.93	2.00	2.07	2.16	2.24
	Heat of vaporization	KJ/kg.K	418	410	401	393	384	375	366	356	345	333	320

Miscellaneous Organic Compounds

Compound	Property	Units											
Acetaldehyde	Temperature	°C	0	20	40	60	80	100	120	140			
	Vapor pressure	bar	0.44	1.00	2.04	3.80	6.58	10.72	16.62	24.70			
	Density	kg/m³	798	772	746	718	688	655	618	576			
	Viscosity	cP	0.28	0.22	0.18	0.14	0.12	0.09	0.07	0.05			
	Thermal conductivity	W/m.K	0.18	0.17	0.16	0.15	0.13	0.12	0.11	0.094			
	Specific heat	KJ/kg.K	2.20	2.27	2.35	2.45	2.57	2.73	2.94	3.28			
	Heat of vaporization	KJ/kg.K	606	585	564	542	518	494	466	434			
Formic acid	Temperature	°C	0	20	40	60	80	100	120	140	160	180	200
	Vapor pressure	bar	0.01	0.04	0.11	0.26	0.53	1.00	1.75	2.91	4.61	7.00	10.28
	Density	kg/m³	1241	1212	1183	1154	1123	1092	1059	1025	989	950	909
	Viscosity	cP	2.46	1.60	1.11	0.81	0.62	0.48	0.38	0.31	0.25	0.21	0.17
	Thermal conductivity	W/m.K	0.20	0.20	0.19	0.18	0.17	0.16	0.15	0.15	0.14	0.13	0.12
	Specific heat	KJ/kg.K	1.45	1.48	1.50	1.53	1.55	1.58	1.62	1.66	1.71	1.78	1.87
	Heat of vaporization	KJ/kg.K	528	517	507	498	488	479	470	461	451	441	430
Acetone	Temperature	°C	0	20	40	60	80	100	120	140	160	180	200
	Vapor pressure	bar	0.09	0.25	0.57	1.16	2.15	3.72	6.05	9.36	13.88	19.90	27.72
	Density	kg/m³	807	785	763	740	716	690	663	633	601	563	518
	Viscosity	cP	0.40	0.32	0.26	0.22	0.18	0.16	0.13	0.11	0.10	0.08	0.07
	Thermal conductivity	W/m.K	0.17	0.16	0.15	0.14	0.13	0.13	0.12	0.11	0.10	0.08	0.07
	Specific heat	KJ/kg.K	2.04	2.10	2.18	2.25	2.34	2.43	2.54	2.68	2.86	3.11	3.54
	Heat of vaporization	KJ/kg.K	550	533	516	498	481	463	443	423	401	375	344
Acetic acid	Temperature	°C	0	20	40	60	80	100	120	140	160	180	200
	Vapor pressure	bar	0.00	0.02	0.05	0.12	0.27	0.57	1.08	1.93	3.25	5.21	8.02
	Density	kg/m³	1089	1066	1042	1018	993	968	942	914	886	855	823
	Viscosity	cP	1.72	1.22	0.90	0.70	0.55	0.44	0.37	0.30	0.26	0.22	0.18
	Thermal conductivity	W/m.K	0.18	0.18	0.17	0.16	0.16	0.15	0.14	0.13	0.13	0.12	0.11
	Specific heat	KJ/kg.K	1.46	1.50	1.55	1.59	1.63	1.68	1.73	1.77	1.83	1.89	1.96
	Heat of vaporization	KJ/kg.K	435	426	418	410	403	395	388	380	372	364	356
Ethyl acetate	Temperature	°C	0	20	40	60	80	100	120	140	160	180	200
	Vapor pressure	bar	0.03	0.10	0.25	0.56	1.12	2.04	3.47	5.57	8.51	12.48	17.72
	Density	kg/m³	920	896	872	847	821	793	764	734	700	662	619
	Viscosity	cP	0.56	0.44	0.36	0.30	0.25	0.21	0.18	0.15	0.13	0.11	0.09
	Thermal conductivity	W/m.K	0.16	0.15	0.14	0.14	0.13	0.12	0.12	0.11	0.10	0.09	0.08
	Specific heat	KJ/kg.K	1.85	1.92	1.98	2.06	2.13	2.21	2.30	2.40	2.53	2.68	2.91

(Continued)

TABLE I-1 Physical Properties of Liquids Calculated with Honeywell UniSim simulation software. With authorization (*Continued*)

	Property	Units											
	Heat of vaporization	KJ/kg.K	418	405	391	378	364	350	335	320	303	285	264
Carbon tetrachloride	Temperature	°C	0	20	40	60	80	100	120	140	160	180	200
	Vapor pressure	bar	0.04	0.12	0.29	0.59	1.12	1.96	3.20	4.96	7.36	10.52	14.60
	Density	kg/m³	1624	1588	1551	1513	1474	1433	1390	1345	1296	1244	1186
	Viscosity	cP	1.13	0.89	0.72	0.59	0.50	0.42	0.36	0.31	0.27	0.24	0.21
	Thermal conductivity	W/m.K	0.11	0.10	0.10	0.09	0.09	0.08	0.08	0.07	0.07	0.06	0.06
	Specific heat	KJ/kg.K	0.81	0.82	0.83	0.84	0.86	0.87	0.89	0.91	0.93	0.97	1.01
	Heat of vaporization	KJ/kg.K	216	211	205	200	194	189	183	177	171	164	157
Ethyl chloride	Temperature	°C	0	20	40	60	80	100	120				
	Vapor pressure	bar	0.62	1.34	2.61	4.63	7.66	11.97	17.84				
	Density	kg/m³	944	915	884	851	816	777	734				
	Viscosity	cP	0.31	0.26	0.22	0.19	0.16	0.14	0.12				
	Thermal conductivity	W/m.K	0.14	0.13	0.13	0.12	0.11	0.10	0.09				
	Specific heat	KJ/kg.K	1.55	1.60	1.67	1.74	1.83	1.95	2.10				
	Heat of vaporization	KJ/kg.K	391	378	365	351	337	322	305				
Phenol	Temperature	°C	20	20	40	60	80	100	120	140	160	180	200
	Vapor pressure	bar	1.02E-04	5.07E-04	2.05E-03	7.02E-03	2.09E-02	5.53E-02	0.13	0.28	0.54	0.97	1.64
	Density	kg/m³	1101	1083	1064	1045	1025	1006	986	965	944	922	900
	Viscosity	cP	6.91	4.24	2.70	1.85	1.33	1.00	0.78	0.62	0.50	0.41	0.34
	Thermal conductivity	W/m.K	0.15	0.15	0.14	0.14	0.14	0.13	0.13	0.12	0.12	0.11	0.11
	Specific heat	KJ/kg.K	1.73	1.80	1.87	1.93	1.99	2.06	2.12	2.18	2.25	2.31	2.37
	Heat of vaporization	KJ/kg.K	613	598	584	570	556	542	528	514	500	486	472
Chlorobenzene	Temperature	°C	0	20	40	60	80	100	120	140	160	180	200
	Vapor pressure	bar	0.00	0.01	0.04	0.09	0.20	0.40	0.73	1.26	2.05	3.18	4.72
	Density	kg/m³	1122	1101	1080	1059	1037	1015	991	967	943	917	890
	Viscosity	cP	1.29	0.98	0.77	0.62	0.51	0.43	0.36	0.31	0.27	0.23	0.20
	Thermal conductivity	W/m.K	0.13	0.13	0.12	0.12	0.11	0.11	0.11	0.10	0.10	0.09	0.09
	Specific heat	KJ/kg.K	1.25	1.31	1.36	1.41	1.46	1.52	1.57	1.62	1.68	1.73	1.79
	Heat of vaporization	KJ/kg.K	384	375	366	357	348	339	330	321	312	303	293
Aniline	Temperature	°C	0	20	40	60	80	100	120	140	160	180	200
	Vapor pressure	bar	0.00	0.00	0.00	0.01	0.03	0.06	0.14	0.28	0.52	0.91	1.52
	Density	kg/m³	1042	1025	1008	990	972	954	935	916	897	877	856
	Viscosity	cP	7.74	4.50	2.69	1.76	1.23	0.89	0.68	0.52	0.41	0.33	0.27
	Thermal conductivity	W/m.K	0.15	0.15	0.15	0.14	0.14	0.13	0.13	0.13	0.12	0.12	0.11
	Specific heat	KJ/kg.K	1.68	1.76	1.83	1.90	1.97	2.04	2.10	2.17	2.24	2.30	2.37
	Heat of vaporization	KJ/kg.K	560	547	535	523	511	500	488	476	464	452	440

TABLE I-2 Physical Properties of Selected Gases

Methane												
Temperature	°C	0	25	50	75	100	125	150	175	200	225	250
Viscosity	cP	0.010	0.011	0.012	0.013	0.014	0.014	0.015	0.016	0.016	0.017	0.018
Thermal conductivity	W/m.K	0.03	0.03	0.04	0.04	0.05	0.05	0.05	0.06	0.06	0.07	0.07
Specific heat	J/kg.K	2.19	2.25	2.31	2.38	2.45	2.53	2.61	2.70	2.79	2.88	2.97

Ethane												
Temperature	°C	0	25	50	75	100	125	150	175	200	225	250
Viscosity	cP	0.009	0.009	0.010	0.011	0.012	0.013	0.013	0.014	0.015	0.016	0.016
Thermal conductivity	W/m.K	0.019	0.021	0.024	0.027	0.030	0.034	0.037	0.041	0.044	0.048	0.052
Specific heat	J/kg.K	1.68	1.77	1.86	1.96	2.06	2.16	2.27	2.37	2.47	2.58	2.68

Propane												
Temperature	°C	0	25	50	75	100	125	150	175	200	225	250
Viscosity	cP	0.007	0.008	0.009	0.010	0.010	0.011	0.012	0.012	0.013	0.014	0.015
Thermal conductivity	W/m.K	0.015	0.018	0.020	0.023	0.026	0.029	0.032	0.035	0.038	0.042	0.045
Specific heat	J/kg.K	1.60	1.71	1.82	1.92	2.02	2.13	2.23	2.33	2.43	2.52	2.62

i-Butane												
Temperature	°C	0	25	50	75	100	125	150	175	200	225	250
Viscosity	cP	0.007	0.007	0.008	0.009	0.009	0.010	0.011	0.011	0.012	0.013	0.014
Thermal conductivity	W/m.K	0.013	0.016	0.018	0.021	0.023	0.026	0.029	0.032	0.035	0.038	0.041
Specific heat	J/kg.K	1.57	1.69	1.81	1.92	2.03	2.14	2.25	2.35	2.45	2.55	2.65

n-Butane												
Temperature	°C	0	25	50	75	100	125	150	175	200	225	250
Viscosity	cP	0.007	0.007	0.008	0.009	0.009	0.001	0.010	0.011	0.012	0.013	0.013
Thermal conductivity	W/m.K	0.013	0.016	0.018	0.020	0.023	0.026	0.029	0.032	0.035	0.038	0.041
Specific heat	J/kg.K	1.58	1.70	1.82	1.93	2.04	2.14	2.25	2.35	2.44	2.54	2.63

Acetylene												
Temperature	°C	0	25	50	75	100	125	150	175	200	225	250
Viscosity	cP	0.009	0.010	0.011	0.012	0.013	0.014	0.014	0.015	0.016	0.017	0.018
Thermal conductivity	W/m.K	0.021	0.023	0.026	0.028	0.031	0.033	0.036	0.038	0.041	0.043	0.046
Specific heat	J/kg.K	1.70	1.75	1.80	1.84	1.88	1.92	1.96	2.00	2.04	2.07	2.11

Hydrogen												
Temperature	°C	0	25	50	75	100	125	150	175	200	225	250
Viscosity	cP	0.008	0.009	0.009	0.010	0.011	0.011	0.012	0.013	0.013	0.014	0.014
Thermal conductivity	W/m.K	0.163	0.175	0.186	0.197	0.208	0.218	0.229	0.238	0.248	0.257	0.267
Specific heat	J/kg.K	14.08	14.10	14.13	14.16	14.19	14.22	14.25	14.28	14.32	14.35	14.38

Oxygen												
Temperature	°C	0	25	50	75	100	125	150	175	200	225	250
Viscosity	cP	0.019	0.021	0.022	0.024	0.025	0.026	0.027	0.028	0.030	0.031	0.032
Thermal conductivity	W/m.K	0.024	0.026	0.028	0.030	0.032	0.034	0.036	0.037	0.039	0.041	0.043
Specific heat	J/kg.K	0.91	0.91	0.92	0.93	0.94	0.94	0.95	0.96	0.97	0.97	0.98

Nitrogen												
Temperature	°C	0	25	50	75	100	125	150	175	200	225	250
Viscosity	cP	0.017	0.018	0.019	0.021	0.022	0.023	0.024	0.025	0.026	0.026	0.027
Thermal conductivity	W/m.K	0.024	0.026	0.028	0.029	0.031	0.032	0.034	0.036	0.037	0.039	0.040
Specific heat	J/kg.K	1.04	1.04	1.05	1.05	1.06	1.06	1.06	1.07	1.07	1.08	1.08

(Continued)

TABLE I-2   Physical Properties of Selected Gases (*Continued*)

Argon	Temperature	°C	0	25	50	75	100	125	150	175	200	225	250
	Viscosity	cP	0.022	0.023	0.025	0.026	0.027	0.029	0.030	0.031	0.033	0.034	0.035
	Thermal conductivity	W/m.K	0.017	0.018	0.019	0.020	0.021	0.022	0.023	0.025	0.025	0.026	0.027
	Specific heat	J/kg.K	0.52	0.52	0.52	0.52	0.52	0.52	0.52	0.52	0.52	0.52	0.52
Helium	Temperature	°C	0	25	50	75	100	125	150	175	200	225	250
	Viscosity	cP	0.019	0.021	0.023	0.027	0.032	0.037	0.045	0.054	0.065	0.078	0.093
	Thermal conductivity	W/m.K	0.113	0.120	0.126	0.132	0.138	0.144	0.150	0.156	0.161	0.167	0.172
	Specific heat	J/kg.K	5.20	5.20	5.20	5.20	5.20	5.20	5.20	5.20	5.20	5.20	5.20
NH_3	Temperature	°C	0.00	25.00	50.00	75.00	100.00	125.00	150.00	175.00	200.00	225.00	250.00
	Viscosity	cP	0.01	0.01	0.01	0.00	0.01	0.01	0.01	0.01	0.01	0.01	0.02
	Thermal conductivity	W/m.K	0.02	0.03	0.03	0.03	0.04	0.04	0.04	0.05	0.05	0.05	0.06
	Specific heat	J/kg.K	2.08	2.11	2.15	2.18	2.22	2.26	2.30	2.34	2.39	2.44	2.48
CO_2	Temperature	°C	0	25	50	75	100	125	150	175	200	225	250
	Viscosity	cP	0.013	0.015	0.016	0.017	0.019	0.020	0.022	0.023	0.024	0.026	0.027
	Thermal conductivity	W/m.K	0.015	0.017	0.019	0.021	0.023	0.025	0.027	0.029	0.031	0.033	0.035
	Specific heat	J/kg.K	0.86	0.87	0.89	0.91	0.93	0.94	0.96	0.97	0.99	1.00	1.02
CO	Temperature	°C	0	25	50	75	100	125	150	175	200	225	250
	Viscosity	cP	0.017	0.018	0.019	0.020	0.021	0.022	0.023	0.024	0.025	0.026	0.027
	Thermal conductivity	W/m.K	0.024	0.025	0.027	0.028	0.030	0.032	0.033	0.035	0.036	0.038	0.039
	Specific heat	J/kg.K	1.04	1.04	1.04	1.04	1.05	1.05	1.05	1.06	1.06	1.07	1.07
SO_2	Temperature	°C	0	25	50	75	100	125	150	175	200	225	250
	Viscosity	cP	0.011	0.012	0.013	0.014	0.015	0.016	0.017	0.019	0.020	0.021	0.022
	Thermal conductivity	W/m.K	0.009	0.010	0.012	0.013	0.014	0.015	0.016	0.017	0.019	0.020	0.021
	Specific heat	J/kg.K	0.61	0.63	0.64	0.66	0.67	0.68	0.70	0.71	0.72	0.73	0.74
SO_3	Temperature	°C	0	25	50	75	100	125	150	175	200	225	250
	Viscosity	cP	15.799	6.142	0.012	0.013	0.014	0.015	0.016	0.017	0.018	0.020	0.021
	Thermal conductivity	W/m.K	0.139	0.131	0.011	0.012	0.014	0.015	0.016	0.018	0.019	0.020	0.022
	Specific heat	J/kg.K	1.36	1.40	0.67	0.69	0.71	0.73	0.75	0.77	0.79	0.81	0.82
SH_2	Temperature	°C	0	25	50	75	100	125	150	175	200	225	250
	Viscosity	cP	0.011	0.012	0.013	0.014	0.015	0.016	0.017	0.018	0.019	0.020	0.021
	Thermal conductivity	W/m.K	0.015	0.016	0.018	0.020	0.021	0.023	0.024	0.026	0.027	0.029	0.030
	Specific heat	J/kg.K	1.01	1.01	1.02	1.03	1.04	1.05	1.06	1.07	1.08	1.09	1.11
HCN	Temperature	°C	0	25	50	75	100	125	150	175	200	225	250
	Viscosity	cP	0.232	0.207	0.006	0.006	0.007	0.007	0.008	0.008	0.009	0.010	0.010
	Thermal conductivity	W/m.K	0.230	0.213	0.014	0.016	0.017	0.018	0.020	0.021	0.023	0.024	0.026
	Specific heat	J/kg.K	3.26	3.37	1.39	1.42	1.45	1.47	1.50	1.52	1.54	1.56	1.59

TABLE I-2 Physical Properties of Selected Gases (*Continued*)

	Temperature	°C	0	25	50	75	100	125	150	175	200	225	250
ClH	Viscosity	cP	0.012	0.014	0.015	0.016	0.017	0.018	0.020	0.021	0.022	0.023	0.024
	Thermal conductivity	W/m.K	0.016	0.017	0.018	0.020	0.021	0.022	0.024	0.025	0.026	0.027	0.028
	Specific heat	J/kg.K	0.82	0.82	0.81	0.81	0.81	0.81	0.81	0.81	0.82	0.82	0.82
	Temperature	°C	0	25	50	75	100	125	150	175	200	225	250
NO₂	Viscosity	cP	0.506	0.010	0.011	0.012	0.013	0.014	0.016	0.017	0.019	0.020	0.022
	Thermal conductivity	W/m.K	0.175	0.013	0.014	0.016	0.017	0.019	0.020	0.022	0.023	0.024	0.026
	Specific heat	J/kg.K	2.91	0.81	0.83	0.85	0.86	0.88	0.90	0.91	0.93	0.94	0.96
	Temperature	°C	0	25	50	75	100	125	150	175	200	225	250
N₂O	Viscosity	cP	0.013	0.014	0.016	0.017	0.018	0.019	0.021	0.022	0.023	0.025	0.026
	Thermal conductivity	W/m.K	0.016	0.017	0.019	0.021	0.023	0.025	0.027	0.028	0.030	0.032	0.034
	Specific heat	J/kg.K	0.86	0.88	0.91	0.93	0.95	0.97	0.99	1.01	1.03	1.04	1.06
	Temperature	°C	0	25	50	75	100	125	150	175	200	225	250
COS	Viscosity	cP	0.011	0.012	0.013	0.014	0.015	0.016	0.017	0.018	0.019	0.020	0.021
	Thermal conductivity	W/m.K	0.010	0.011	0.012	0.014	0.015	0.016	0.017	0.019	0.020	0.021	0.022
	Specific heat	J/kg.K	0.67	0.69	0.71	0.73	0.75	0.76	0.78	0.79	0.80	0.81	0.82
	Temperature	°C	0	25	50	75	100	125	150	175	200	225	250
CS₂	Viscosity	cP	0.468	0.364	0.011	0.012	0.013	0.014	0.015	0.016	0.016	0.017	0.018
	Thermal conductivity	W/m.K	0.140	0.133	0.009	0.010	0.010	0.011	0.012	0.013	0.014	0.015	0.016
	Specific heat	J/kg.K	0.93	0.96	0.62	0.63	0.64	0.65	0.66	0.67	0.68	0.69	0.70
	Temperature	°C	0	25	50	75	100	125	150	175	200	225	250
Cl₂	Viscosity	cP	0.013	0.014	0.015	0.016	0.017	0.018	0.019	0.021	0.022	0.023	0.024
	Thermal conductivity	W/m.K	0.008	0.009	0.010	0.011	0.012	0.013	0.013	0.014	0.015	0.016	0.017
	Specific heat	J/kg.K	0.48	0.48	0.49	0.49	0.50	0.50	0.50	0.51	0.51	0.51	0.51
	Temperature	°C	0	25	50	75	100	125	150	175	200	225	250
Br₂	Viscosity	cP	0.240	0.194	0.155	0.019	0.021	0.022	0.023	0.025	0.026	0.028	0.029
	Thermal conductivity	W/m.K	0.098	0.093	0.089	0.006	0.007	0.007	0.008	0.008	0.009	0.009	0.009
	Specific heat	J/kg.K	0.40	0.40	0.41	0.23	0.23	0.23	0.23	0.23	0.23	0.23	0.23

TABLE I-3 Physical Properties of Amine and Glycol Solutions

Solution	Property	Unit	20	30	40	50	60	70	80	90	100	120	130
Mono Ethanol Amine 20% w/w	Temperature	°C	20	30	40	50	60	70	80	90	100	120	130
	Density	kg/m³	1004	1002	999	995	991	986	980	974	968	953	946
	Viscosity	cP	0.69	0.66	0.64	0.61	0.59	0.56	0.54	3.26	2.74	0.45	0.43
	Thermal conductivity	W/m.K	0.52	0.53	0.54	0.55	0.56	0.57	0.58	0.58	0.59	0.59	0.59
	Specific heat	KJ/kg.K	3.85	3.90	3.95	4.00	4.05	4.11	4.16	4.21	4.26	4.36	4.41
Mono Ethanol Amine 30% w/w	Temperature	°C	20	30	40	50	60	70	80	90	100	120	130
	Density	kg/m³	1007	1005	1002	999	995	990	985	979	973	959	951
	Viscosity	cP	0.72	0.69	0.66	0.64	0.61	0.59	4.02	3.26	0.51	0.46	0.44
	Thermal conductivity	W/m.K	0.48	0.49	0.50	0.51	0.52	0.52	0.53	0.54	0.54	0.55	0.55
	Specific heat	KJ/kg.K	3.72	3.77	3.82	3.87	3.92	3.97	4.02	4.06	4.11	4.21	4.26
Di Ethanol Amine 20% w/w	Temperature	°C	20	30	40	50	60	70	80	90	100	120	130
	Density	kg/m³	1022	1019	1015	1011	1005	1000	993	986	979	962	954
	Viscosity	cP	436.04	205.42	0.65	0.63	0.60	0.58	0.55	0.53	0.51	0.46	0.44
	Thermal conductivity	W/m.K	0.50	0.51	0.53	0.54	0.55	0.55	0.56	0.57	0.57	0.58	0.58
	Specific heat	KJ/kg.K	3.78	3.83	3.89	3.94	3.99	4.04	4.09	4.14	4.19	4.30	4.35
Di Ethanol Amine 30% w/w	Temperature	°C	20	30	40	50	60	70	80	90	100	120	130
	Density	kg/m³	1034	1031	1027	1022	1017	1010	1004	996	989	972	963
	Viscosity	cP	436.04	205.42	105.35	58.16	0.64	0.61	0.59	0.56	0.54	0.49	0.47
	Thermal conductivity	W/m.K	0.45	0.46	0.47	0.48	0.49	0.50	0.51	0.51	0.52	0.52	0.53
	Specific heat	KJ/kg.K	3.62	3.67	3.72	3.77	3.82	3.87	3.92	3.97	4.02	4.11	4.16
Di Ethanol Amine 40% w/w	Temperature	°C	20	30	40	50	60	70	80	90	100	120	130
	Density	kg/m³	1046	1043	1039	1034	1028	1021	1014	1007	999	982	972
	Viscosity	cP	436.04	205.42	0.71	0.71	0.69	0.66	0.63	0.60	0.57	0.52	2.89
	Thermal conductivity	W/m.K	0.40	0.41	0.42	0.43	0.44	0.45	0.46	0.47	0.47	0.48	0.48
	Specific heat	KJ/kg.K	3.46	3.51	3.56	3.61	3.65	3.70	3.75	3.79	3.84	3.93	3.97
Methyl di Ethanol Amine (MDEA) 30% w/w	Temperature	°C	20	30	40	50	60	70	80	90	100	120	130
	Density	kg/m³	1026	1022	1018	1012	1006	999	991	983	975	957	947
	Viscosity	cP	0.77	0.74	0.71	0.68	0.65	0.96	0.60	0.57	0.54	0.50	0.47
	Thermal conductivity	W/m.K	0.42	0.44	0.45	0.46	0.47	0.48	0.48	0.49	0.49	0.50	0.50
	Specific heat	KJ/kg.K	3.61	3.67	3.72	3.78	3.83	3.89	3.95	4.00	4.06	4.18	4.24

TABLE I-3 Physical Properties of Amine and Glycol Solutions (*Continued*)

	°C	20	30	40	50	60	70	80	90	100	120	130
Methyl di Ethanol	Temperature											
Amine (MDEA)	Density	kg/m³	1036	1031	1026	1020	1013	1006	998	989	980	961	951
40% w/w	Viscosity	cP	6.50	0.79	0.76	0.73	1.73	0.67	0.64	0.61	0.59	0.53	0.51
	Thermal conductivity	W/m.K	0.37	0.38	0.39	0.40	0.41	0.42	0.43	0.43	0.44	0.45	0.45
	Specific heat	KJ/kg.K	3.45	3.50	3.56	3.61	3.67	3.72	3.78	3.84	3.90	4.02	4.08

	°C	20	30	40	50	60	70	80	90	100	120	130
Ethylene glycol	Temperature											
50% w/w	Density	kg/m³	1075	1067	1059	1051	1042	1034	1026	1017	1008	990	981
	Viscosity	cP	3.63	2.79	2.20	1.78	1.45	1.21	1.02	0.87	0.74	0.57	0.50
	Thermal conductivity	W/m.K	0.52	0.53	0.53	0.54	0.54	0.55	0.55	0.55	0.55	0.55	0.55
	Specific heat	KJ/kg.K	3.40	3.43	3.46	3.48	3.50	3.52	3.55	3.57	3.60	3.67	3.72

	°C	20	30	40	50	60	70	80	90	100	120	130
Ethylene Glycol	Temperature											
80% w/w	Density	kg/m³	1105	1098	1090	1082	1074	1066	1058	1050	1042	1024	1016
	Viscosity	cP	10.89	7.96	5.74	4.29	3.30	2.60	2.09	1.71	1.42	1.02	0.88
	Thermal conductivity	W/m.K	0.41	0.41	0.41	0.41	0.41	0.41	0.41	0.40	0.40	0.39	0.39
	Specific heat	KJ/kg.K	2.91	2.96	3.00	3.04	3.08	3.12	3.16	3.20	3.25	3.34	3.39

TABLE I-4 Properties of Liquids at Low Temperatures

Fluid	Property	Unit											
Ethylene Glycol 80% w/w	Temperature	°C	-30	-25	-20	-15	-10	-5	0	5	10	15	20
	Density	kg/m³	1143	1139	1135	1132	1128	1124	1121	1117	1113	1109	1105
	Viscosity	cP	66.57	51.19	40.35	32.50	26.66	22.23	18.80	16.10	13.93	12.18	10.73
	Thermal conductivity	W/m.K	0.40	0.40	0.41	0.41	0.41	0.41	0.41	0.41	0.41	0.41	0.41
	Specific heat	KJ/kg.K	2.72	2.74	2.76	2.77	2.79	2.81	2.83	2.84	2.86	2.89	2.91
Ethylene glycol 50% w/w	Temperature	°C	-30	-25	-20	-15	-10	-5	0	5	10	15	20
	Density	kg/m³	1113	1109	1105	1101	1098	1094	1090	1086	1082	1078	1075
	Viscosity	cP	16.93	14.40	12.40	10.79	9.47	7.79	6.50	5.48	4.68	4.02	3.49
	Thermal conductivity	W/m.K	0.47	0.47	0.48	0.48	0.49	0.50	0.50	0.51	0.51	0.51	0.52
	Specific heat	KJ/kg.K	3.27	3.28	3.29	3.30	3.32	3.33	3.34	3.32	3.35	3.38	3.40
Tri Ethylene glycol 50% w/w	Temperature	°C	-30	-25	-20	-15	-10	-5	0	5	10	15	20
	Density	kg/m³	1138	1134	1130	1126	1122	1117	1113	1109	1105	1100	1096
	Viscosity	cP	21.18	18.03	15.53	13.53	11.89	10.54	8.89	7.50	6.39	5.50	4.78
	Thermal conductivity	W/m.K	0.48	0.49	0.50	0.51	0.52	0.52	0.53	0.54	0.54	0.55	0.55
	Specific heat	KJ/kg.K	3.16	3.17	3.18	3.19	3.21	3.22	3.22	3.21	3.23	3.26	3.28
Ammonia	Temperature	°C	-30	-25	-20	-15	-10	-5	0	5	10	15	20
	Vapor pressure	bar	1.17	1.49	1.87	2.33	2.87	3.50	4.24	5.10	6.08	7.21	8.49
	Density	kg/m³	670	663	657	651	645	639	632	626	619	612	605
	Viscosity	cP	0.24	0.22	0.21	0.20	0.19	0.18	0.17	0.16	0.15	0.14	0.13
	Thermal conductivity	W/m.K	0.61	0.59	0.58	0.57	0.56	0.55	0.54	0.53	0.51	0.50	0.49
	Specific heat	KJ/kg.K	4.47	4.50	4.54	4.57	4.61	4.66	4.70	4.76	4.81	4.88	4.94
Propane	Temperature	°C	-30	-25	-20	-15	-10	-5	0	5	10	15	20
	Vapor pressure	bar	1.68	2.03	2.44	2.91	3.44	4.05	4.73	5.50	6.35	7.30	8.36
	Density	kg/m³	567	561	555	549	542	536	529	522	515	508	500
	Viscosity	cP	0.17	0.16	0.16	0.15	0.14	0.13	0.13	0.12	0.12	0.11	0.10
	Thermal conductivity	W/m.K	0.13	0.12	0.12	0.12	0.12	0.11	0.11	0.11	0.10	0.10	0.10
	Specific heat	KJ/kg.K	2.27	2.31	2.35	2.39	2.44	2.48	2.53	2.59	2.64	2.70	2.77

TABLE I-5 Physical Properties and Vapor-Liquid Equilibrium of Water Calculated with Honeywell UniSim Simulation Software ASME Steam Property Package. With authorization

| Temp. | Pressure | Liquid Properties | | | | | | | Vapor Properties | | | | |
| | | Density | Thermal Conductivity | Viscosity | Heat Capacity | Surface Tension | Heat of Vaporization | Enthalpy | Enthalpy | Vapor Density | Thermal Conductivity | Viscosity | Heat Capacity |
°C	bar	kg/m³	W/m.K	cP	KJ/kg.K	dyn/cm	kJ/kg	KJ/kg	KJ/kg	kg/m³	W/mK	cP	KJ/kg.K
0	0.0061	1000	0.57		4.22	76.4	2502	−15910	−13411	4.85E-03	1.76E-02	8.04E-03	1.86
5	0.0087	1000	0.58	1.50	4.21	75.5	2490	−15889	−13401	6.79E-03	1.79E-02	8.24E-03	1.86
10	0.012	999	0.59	1.30	4.20	74.7	2479	−15869	−13392	9.39E-03	1.82E-02	8.45E-03	1.86
15	0.017	999	0.60	1.14	4.19	73.8	2467	−15848	−13383	1.28E-02	1.85E-02	8.65E-03	1.86
20	0.023	998	0.60	1.00	4.19	73.0	2455	−15827	−13374	1.73E-02	1.88E-02	8.85E-03	1.87
25	0.032	997	0.61	0.89	4.18	72.1	2443	−15806	−13365	2.30E-02	1.91E-02	9.06E-03	1.87
30	4.24E-02	995	0.62	0.80	4.18	71.2	2431	−15785	−13356	3.04E-02	1.95E-02	9.26E-03	1.88
35	5.62E-02	994	0.63	0.72	4.18	70.4	2419	−15764	−13347	3.96E-02	1.98E-02	9.46E-03	1.88
40	7.38E-02	992	0.63	0.65	4.18	69.5	2408	−15743	−13338	5.11E-02	2.01E-02	9.67E-03	1.89
45	9.58E-02	990	0.64	0.59	4.18	68.6	2396	−15722	−13329	6.54E-02	2.04E-02	9.87E-03	1.89
50	1.23E-01	988	0.64	0.54	4.19	67.7	2384	−15701	−13320	8.30E-02	2.08E-02	1.01E-02	1.90
55	1.57E-01	985	0.65	0.50	4.19	66.9	2371	−15680	−13311	1.04E-01	2.11E-02	1.03E-02	1.91
60	1.99E-01	983	0.65	0.46	4.19	66.0	2359	−15660	−13303	0.130198	2.15E-02	1.05E-02	1.92
65	2.50E-01	980	0.66	0.43	4.19	65.1	2347	−15639	−13294	0.161187	2.18E-02	1.07E-02	1.93
70	3.12E-01	977	0.66	0.40	4.19	64.2	2335	−15618	−13285	0.198112	2.22E-02	1.09E-02	1.94
75	3.85E-01	974	0.67	0.37	4.20	63.3	2322	−15597	−13277	0.241825	2.26E-02	1.11E-02	1.95
80	4.74E-01	971	0.67	0.35	4.20	62.3	2309	−15576	−13268	0.293254	2.29E-02	1.13E-02	1.96
85	5.78E-01	968	0.67	0.33	4.21	61.4	2297	−15555	−13260	0.353409	2.33E-02	1.15E-02	1.98
90	7.01E-01	965	0.68	0.31	4.21	60.5	2284	−15534	−13252	0.423381	2.37E-02	1.17E-02	1.99
95	8.45E-01	961	0.68	0.29	4.21	59.5	2271	−15513	−13244	0.504347	2.41E-02	1.19E-02	2.01
100	1.01E+0	958	0.68	0.28	4.22	58.6	2258	−15492	−13236	0.597567	2.45E-02	1.20E-02	2.03
105	1.21E+0	954	0.68	0.27	4.23	57.7	2244	−15471	−13229	0.704393	2.49E-02	1.22E-02	2.05
110	1.43E+0	950	0.68	0.25	4.23	56.7	2231	−15449	−13221	0.826265	2.53E-02	1.24E-02	2.07
115	1.69E+0	947	0.69	0.24	4.24	55.7	2217	−15428	−13214	0.964755	2.57E-02	1.26E-02	2.10
120	1.99E+0	943	0.69	0.23	4.25	54.7	2203	−15407	−13206	1.12137	2.61E-02	1.28E-02	2.12
125	2.32E+0	938	0.69	0.22	4.26	53.8	2189	−15386	−13199	1.297955	2.65E-02	1.30E-02	2.15
130	2.70E+0	934	0.69	0.21	4.27	52.8	2174	−15365	−13192	1.496294	2.69E-02	1.32E-02	2.18
135	3.13E+0	930	0.69	0.20	4.28	51.8	2160	−15343	−13186	1.718316	2.73E-02	1.34E-02	2.21
140	3.61E+0	926	0.69	0.19	4.29	50.7	2145	−15322	−13179	1.966059	2.77E-02	1.35E-02	2.24
145	4.16E+0	921	0.69	0.19	4.30	49.7	2129	−15300	−13173	2.241671	2.82E-02	1.37E-02	2.28
150	4.76E+0	917	0.69	0.18	4.31	48.7	2114	−15279	−13167	2.54742	2.86E-02	1.39E-02	2.32
155	5.43E+0	912	0.69	0.17	4.33	47.6	2098	−15257	−13161	2.8857	2.90E-02	1.41E-02	2.36
160	6.18E+0	907	0.68	0.17	4.34	46.6	2082	−15235	−13156	3.259035	2.95E-02	1.43E-02	2.40
165	7.01E+00	902	0.68	0.16	4.36	45.5	2065	−15214	−13150	3.670091	2.99E-02	1.44E-02	2.45

(Continued)

TABLE I-5 Physical Properties and Vapor-Liquid Equilibrium of Water Calculated with Honeywell UniSim Simulation Software ASME Steam Property Package. With authorization (*Continued*)

170	7.92E+00	897	0.68	0.16	4.38	44.4	2048	-15192	-13145	4.121682	3.04E-02	1.46E-02	2.49
175	8.92E+00	892	0.68	0.15	4.39	43.4	2031	-15170	-13141	4.616783	3.08E-02	1.48E-02	2.54
180	1.00E+01	887	0.68	0.15	4.41	42.3	2014	-15148	-13136	5.158543	3.13E-02	1.50E-02	2.60
185	1.12E+01	881	0.67	0.15	4.43	41.2	1996	-15126	-13132	5.750259	3.17E-02	1.51E-02	2.66
190	1.26E+01	876	0.67	0.14	4.45	40.1	1977	-15103	-13128	6.395574	3.22E-02	1.53E-02	2.72
195	1.40E+01	870	0.67	0.14	4.48	38.9	1958	-15081	-13125	7.098134	3.27E-02	1.55E-02	2.78
200	1.55E+01	864	0.66	0.13	4.50	37.8	1939	-15059	-13121	7.861967	3.31E-02	1.56E-02	2.85

TABLE I-6 Constants for the Ideal-Gas Heat Capacity in the Equation $cp = A + B.T + C^*T^{**}2 + D^*T^{**}3$ where cp is in Kcal/kmol.Kelvin and T is Absolute Temperature in kelvins

		A	B	C	D
Combustion gases					
H_2	HYDROGEN	6.483	2.22E-03	−3.30E-06	1.83E-09
Ar	ARGON	4.969	−7.67E-06	1.23E-08	0
O_2	OXYGEN	6.713	−8.79E-07	4.17E-06	−2.54E-09
N_2	NITROGEN	7.44	−3.24E-03	6.40E-06	−2.79E-09
H_2O	WATER	7.701	4.59E-04	2.53E-06	−8.59E-10
CO_2	CARBON DIOXIDE	4.728	1.75E-02	−1.34E-05	4.10E-09
SO_2	SULFUR DIOXIDE	5.697	1.60E-02	−1.19E-05	3.17E-09
CO	CARBON MONOXIDE	7.373	−3.07E-03	6.66E-06	−3.04E-09
N_2 and S compounds					
NO	NITRIC OXIDE	7.009	−2.24E-04	2.33E-06	−1.00E-09
NO_2	NITROGEN DIOXIDE	5.788	1.16E-02	−4.97E-06	7.00E-11
CS_2	CARBON DISULFIDE	6.555	1.94E-02	−1.83E-05	6.38E-09
COS	CARBONYL SULFIDE	5.629	1.91E-02	−1.68E-05	5.86E-09
SH_2	HYDROGEN SULFIDE	7.629	3.43E-04	5.81E-06	−2.81E-09
NH_3	AMMONIA	6.524	5.69E-03	4.08E-06	−2.83E-09
SO_3	SULFUR TRIOXIDE	5.697	1.60E-02	−1.19E-05	3.18E-09
Halogenated inorganic compounds					
Cl_2	CHLORINE	6.432	8.08E-03	−9.24E-06	3.70E-09
Br_2	BROMINE	8.087	2.69E-03	−2.85E-06	1.08E-09
I_2	IODINE	8.501	1.56E-03	−1.67E-06	6.77E-10
F_2	FLUORINE	5.545	8.73E-03	−8.27E-06	2.88E-09
ClH	HYDROGEN CHLORIDE	7.235	−1.72E-03	2.98E-06	−9.31E-10
BrH	HYDROGEN BROMIDE	7.32	−2.26E-03	4.11E-06	−1.49E-09
IH	HYDROGEN IODIDE	7.442	−3.41E-03	7.10E-06	−3.23E-09
FH	HYDROGEN FLUORIDE	6.941	1.58E-04	−4.85E-07	5.98E-10
Saturated hydrocarbons					
CH_4	METHANE	4.598	1.25E-02	2.86E-06	−2.70E-09
C_2H_6	ETHANE	1.292	4.25E-02	−1.66E-05	2.08E-09
C_3H_8	PROPANE	−1.009	7.32E-02	−3.79E-05	7.68E-09
nC_4H_{10}	n BUTANE	2.266	7.91E-02	−2.65E-05	−6.74E-10
iC_4H_{10}	iso BUTANE	−0.332	9.19E-02	−4.41E-05	6.92E-09
C_5H_{12}	n PENTANE	−0.886	1.16E-01	−6.16E-05	1.27E-08
Unsaturated hydrocarbons					
C_2H_2	ACETYLENE	6.406	1.81E-02	−1.20E-05	3.37E-09
C_2H_4	ETHYLENE	0.909	3.74E-02	−1.99E-05	4.19E-09
C_3H_6	PROPYLENE	0.886	5.60E-02	−2.77E-05	5.27E-09
C_4H_8	1-BUTENE	−0.715	8.44E-02	−4.75E-05	1.07E-08
C_4H_8	CIS-2-BUTENE	0.105	7.05E-02	−2.43E-05	−1.47E-09
C_4H_8	TRANS-2-BUTENE	4.375	6.12E-02	−1.68E-05	−2.15E-09
Halogenated organic compounds					
CH_3Cl	METHYL CHLORIDE	3.314	2.42E-02	9.29E-06	6.13E-10
CH_3F	METHYL FLUORIDE	3.302	2.06E-02	−4.95E-06	−4.79E-10
CH_3Br	METHYL BROMIDE	3.446	2.61E-02	−1.29E-05	2.39E-09
C_2H_3Cl	VINYL CHLORIDE	1.421	4.82E-02	−3.67E-05	1.14E-08
C_2H_5Br	ETHYL BROMIDE	1.59	5.61E-02	−3.52E-05	9.09E-09
C_2H_5Cl	ETHYL CHLORIDE	−0.132	6.23E-02	−4.39E-05	1.33E-08
Oxigenated organic compounds					
CH_2O	FORMALDEHYDE	5.607	7.54E-03	7.13E-06	−5.49E-09
CH_2O_2	FORMIC ACID	2.798	3.24E-02	−2.01E-05	4.82E-09

(*Continued*)

TABLE I-6 Constants for the Ideal-Gas Heat Capacity in the Equation $cp = A + B.T + C^*T^{**}2 + D^*T^{**}3$ where cp is in Kcal/kmol.Kelvin and T is Absolute Temperature in kelvins (*Continued*)

CH_4O	METHANOL	5.052	1.69E-02	6.18E-06	−6.81E-09
C_2H_4O	ACETALDEHYDE	1.843	4.35E-02	−2.40E-05	5.69E-09
C_2H_4O	ETHYLENE OXIDE	−1.796	5.31E-02	−3.00E-05	6.19E-09
$C_2H_4O_2$	ACETIC ACID	1.156	6.09E-02	−4.19E-05	1.18E-08
C_2H_6O	ETHANOL	2.153	5.11E-02	−2.00E-05	3.28E-10
$C_2H_6O_2$	ETHYLENE GLYCOL	8.526	5.93E-02	−3.58E-05	7.19E-09
C_3H_8O	1-PROPANOL	0.59	7.94E-02	−4.43E-05	1.03E-08
C_3H_6O	ACETONE	1.505	6.22E-02	−2.99E-05	4.87E-09
Others					
CH_4S	METHYL MERCAPTAN	3.169	3.48E-02	−2.04E-05	4.96E-09
C_2H_6S	ETHYL MERCAPTAN	3.564	5.62E-02	−3.24E-05	7.55E-09
CH_5N	METHYL AMINE	2.741	3.41E-02	−1.27E-05	1.14E-09
C_2H_7N	ETHYL AMINE	0.882	6.57E-02	−3.78E-05	9.10E-09
C_2H_7NO	MONOETHANOLAMINE	2.224	7.19E-02	−4.34E-05	1.11E-08

*Values Extracted from *The Properties of Gases an Liquids* R.Reid, J.Prausnitz, and T.Sherwood McGraw Hill Book Company—Reproduced with Permisssion

APPENDIX J
HEAT EXCHANGER DATA SHEET

This appendix includes the form to be used in the specification of a shell-and-tube heat exchanger according to the TEMA Standards. It contains a blank form and a form filled with data corresponding to Example 7-5 of the text.

		HEAT EXCHANGER SPECIFICATION SHEET		
1	Customer			Job No:
				Reference No:
2				Proposal No
3	Address			
4	Plant Location			Date
5	Service of Unit			Tag:
6	Size:		Type	Connected in:
7	Surface/Unit (Gross/Eff)		Shells/Unit	Surf/Shell
8	PERFORMANCE OF ONE UNIT			
9			Shell Side	Tubes Side
10	Fluid Name			
11	Fluid Quantity Total, kg/s			
12	Vapor (In/Out)			
13	Líquid(In/Out)			
14	Steam			
15	Water			
16	Noncondensables			
17	Temp (In/Out), °C			
18	Specific Gravity			
19	Viscosity Liquid, cP			
20	Molecular Weight Vapor			
21	Molecular Weight Noncondensable			
22	Specific Heat, $J/(kg \cdot °C)$			
23	Thermal Conductivity, $W/(m \cdot °C)$			
24	Latent Heat, J/kg			
25	Inlet Pressure, kPa (abs)			
26	Velocity, m/s			
27	Pressure Drop Allow/Calc, kPa			
28	Fouling resistance (min), $(m^2 \cdot °C)/W$			
29	Heat Exchanged, kW		MTD corrected	°C
30	Transfer Rate, Service		Clean	$W/(m^2 \cdot K)$
31	CONSTRUCTION OF ONE SHELL			
32		Shell Side	Tube Side	Sketch
33	Design/Test pressure, kPa(g)			
34	Design Temp (Max/Min), °C			
35	No. of Passes per Shell			
36	Corrosion Allowance, mm			
37	Connections	In		
38	Size &	Out		
39	Rating	Intermediate		
40	Tube No. OD, mm; Thk, mm; Length Pitch, mm,			30° 60° 90° 45°
41	Tube Type Plain		Material	
42	Shell ID OD, mm		Shell Cover (Integr) (Remov)	
43	Channel or Bonnet		Channel Cover	
44	Tubesheet Stationary		Tubesheet Floating	
45	Floating Head Cover		Impingement Protection	
46	Baffles Cross Segment Type % Cut(Diam/Area) Spacing:c/c inlet mm			
47	Baffles Long Seal Type			
48	Supports Tube U bend			Type
49	Bypass Seal Arrangement Tube to Tubesheet Joint			
50	Expansion Joint Type			
51	ρv^2 Inlet Nozzle Bundle entrance Bundle Exit			
52	Gaskets Shell Side Tube Side			
53	Floating Head			
54	Code Requirements TEMA Class			
55	Weight Shell Filled with Water Bundle			
56	Remarks			

	HEAT EXCHANGER SPECIFICATION SHEET			
1	Customer		Job No:	
			Reference No:	
2			Proposal No	
3	Address			
4	Plant Location		Date	
5	Service of Unit	*Secondary Heat Recovery*	Tag:	
6	Size: *609 × 4266*	Type *AES*	Connected in:	
7	Surface/Unit (Gross/Eff)	Shells/Unit *1*	Surf/Shell (Gross/Eff)	
	105.4/103.2		*105.4/103.2*	
8		PERFORMANCE OF ONE UNIT		
9		Shell Side	Tubes Side	
10	Fluid Name	*Process stream A*	*Process stream B*	
11	Fluid Quantity Total, kg/s	*33.33*	*38.88*	
12	Vapor (In/Out)			
13	Liquid(In/Out)	*33.33*	*38.88*	
14	Steam			
15	Water			
16	Noncondensables			
17	Temp (In/Out), °C	*104/85*	*40/61*	
18	Specific Gravity	*0.578*	*0.716*	
19	Viscosity Liquid, cP	*0.16*	*0.62*	
20	Molecular Weight Vapor			
21	Molecular Weight Noncondensable			
22	Specific Heat, J/(kg · °C)	*2640*	*2140*	
23	Thermal Conductivity, W/(m · °C)	*0.0917*	*0.129*	
24	Latent Heat, J/kg			
25	Inlet Pressure, kPa (abs)	*850*	*500*	
26	Velocity, m/s			
27	Pressure Drop Allow/Calc, kPa	*50/39.4*	*10/5.7*	
28	Fouling resistance (min), $(m^2 \cdot °C)/W$	*5E-4*	*4E-4*	
29	Heat Exchanged *1664* kW	MTD corrected *43* °C		
30	Transfer Rate, Service *374*	Dirty *417* Clean *668* W/(m² · K)		
31		CONSTRUCTION OF ONE SHELL		
32		Shell Side	Tube Side	Sketch
33	Design/Test Pressure, kPa(g)	*1000/code*	*600/Code*	
34	Design Temp (Max/Min), °C	*200*	*160*	
35	No. of Passes per Shell	*1*	*1*	
36	Corrosion Allowance, mm	*3*	*3*	
37	Connections In	*8"/#150*	*8"/#150*	
38	Size & Out	*8"/#150*	*8"/#150*	
39	Rating Intermediate			
40	Tube No *414* OD *19* mm; Thk *1.6* mm; Length *4267* mm; Pitch *25.4* mm 30° ~~60°~~ ~~90°~~ ~~45°~~			
41	Tube Type Plain Material *Carbon steel*			
42	Shell CS ID *609* OD mm	Shell Cover (Integr) (Remov)		
43	Channel or Bonnet *CS*	Channel Cover *CS*		
44	Tubesheet Stationary *CS*	Tubesheet Floating *CS*		
45	Floating Head Cover	Impingement Protection None		
46	Baffles Cross Segment Type *25* % Cut(Diam/~~Area~~) Spacing:c/c *234* Inlet *633* mm			
47	Baffles Long none Seal Type			
48	Supports Tube none U bend Type			
49	Bypass Seal Arrangement *seal strip* Tube to Tubesheet Joint *expanded*			
50	Expansion Joint none Type			
51	ρv² Inlet Nozzle *1760 kg/(m² · s)* Bundle entrance *593.88* Bundle Exit *992.25*			
52	Gaskets Shell Side Tube Side			
53	Floating Head			
54	Code Requirements TEMA Class *C*			
55	Weight Shell *3647* Filled with Water *5058* Bundle *1479*			
56	Remarks			
57	*Corresponds to data of Example 7-5*			

APPENDIX K
UNIT CONVERSIONS

Temperature		
From°C to °F	$T\ (°F) = T\ (°C) \times 1.8 + 32$	E.g.: 100°F is converted as $T\ (°F) = 100 \times 1.8 + 32 = 212$
From °F to °C	$T\ (°C) = [T\ (°F) - 32] \times 0.555$	Eg.: 212°F is converted as $T\ (°C) = (212 - 32) \times 0.555 = 100$

Length	m	in	ft
m	1	39.37	3.2808
in	0.0254	1	0.0833
ft	0.3048	12	1

Mass
1lb = 0.454 kg
1 kg = 2.202 lb

Area	m^2	in^2	ft^2
m^2	1	1550	10.764
in^2	6.451×10^{-4}	1	6.944×10^{-3}
ft^2	0.0929	144	1

Density
$1 kg/m^3 = 0.0625\ lb/ft^3$
$1\ lb/ft^3 = 16\ kg/m^3$
$1\ kg/dm^3 = 62.5 lb/ft^3$

Volume	m^3	in^3	ft^3	cm^3	liters
m^3	1	61,023.7	35.314	10^6	1,000
in^3	1.638×10^{-5}	1	5.787×10^{-4}	16.38	0.01638
ft^3	0.02831	1,728	1	28,317	28.32
cm^3	10^{-4}	0.061	3.53×10^{-5}	1	10^{-3}
liters	10^{-3}	61.023	0.0353	1,000	1

Energy					
	J	kWh	kcal	Btu	HP·h
J	1	2.77×10^{-7}	2.3901×10^{-4}	9.478×10^{-4}	3.725×10^{-7}
kWh	3.6×10^6	1	860.4	3.4122×10^{-3}	1.3410
kcal	4,184	1.1622×10^{-3}	1	3.9657	1.558×10^{-3}
Btu	1,055	2.93×10^{-4}	0.2522	1	3.9302×10^{-4}
HP·h	2.686×10^6	7.457×10^{-1}	641.62	2,544.5	1

Pressure								
	bar	N/m²	atm	psi	kg/cm²	mmHg	inHg	lb/ft²
bar	1	10^5	0.9869	14.504	1.0197	750.06	29.526	2,088
N/m²	10^{-5}	1	9.87×10^{-6}	1.454×10^{-4}	1.0197×10^{-6}	7.5×10^{-3}	2.952×10^{-4}	0.02088
atm	1.013	1.013×10^5	1	14.696	1.033	760	29.92	2,116
psi	0.06894	6,894.7	0.06804	1	0.07031	51.715	2.036	144
kg/cm²	0.9807	98,068	0.9678	14.22	1	735.6	28.96	2,048.16
mmHg	1.333×10^{-3}	133.32	1.315×10^{-3}	1.933×10^{-2}	1.359×10^{-3}	1	3.937×10^{-2}	2.7845
inHg	3.386×10^{-2}	3,386.4	3.34×10^{-2}	0.49116	0.03453	25,400	1	70.727
lb/ft²	4.788×10^{-4}	47.88	4.725×10^{-4}	6.944×10^{-2}	4.882×10^{-4}	0.35913	1.414×10^{-2}	1

Force			
	N	kgf	lbf
N	1	0.10198	0.2246
kgf	9.806	1	2.202
lbf	4.4519	0.454	1

Specific Heat			
	J/kg.K	kcal/kg°C	Btu/lb°F
J/kg.K	1	2.39×10^{-4}	2.39×10^{-4}
kcal/kg°C	4,184	1	1
Btu/lb°F	4,184	1	1

Power-Heat Flux			
	W	kcal/h	Btu/h	HP
W	1	0.8604	3.413	1.341×10^{-3}
kcal/h	1.163	1	3.9657	1.555×10^{-3}
Btu/h	0.2926	0.2522	1	3.929×10^{-4}
HP	745.7	641.62	2,545	1

Physical Constants

Ideal Gas Constant R
$= 0.082$ atm.m³/kmol.K
$= 4.968 \times 10^4$ lb.ft².s⁻².lbmol⁻¹.°R
$= 1.987$ kcal/kmol.K
$= 1.987$ Btu/lbmol°·R
Gravity Acceleration
$= 9.806$ m/s²
$= 32.174$ ft/s²
$= 4.18 \times 108$ ft/h²

Thermal Conductivity					
	W/(m²·°C)/m	Btu/(h·ft²·°F)/ft	cal/(s·cm²·°C)/cm	W/(cm²·°C)/cm	kcal/(h·m²·°C)/m
W/(m²·°C)/m	1	0.5779	0.002388	0.01	0.86
Btu/(h·ft²·°F)/ft	1.731	1	0.004134	0.01731	1.488
cal/(s·cm²·°C)/cm	418.7	241.9	1	4.187	360
W/(cm²·°C)/cm	100	57.79	0.2388	1	86
kcal/(h·m²·°C)/m	1.163	0.672	0.002778	0.01163	1

Heat Transfer Coefficients					
	W/(m²·°C)	Btu/(h·ft²·°F)	cal/(s·cm²·°C)	W/(cm²·°C)	kcal/(h·m²·C)
W/(m²·°C)	1	0.1761	0.00002388	0.0001	0.86
Btu/(h·ft²·°F)	5.678	1	0.0001355	0.0005678	4.882
cal/(s·cm²·°C)	41,870	7,373	1	4.187	36,000
W/(cm²·°C)	10,000	1.761	0.2388	1	8,600
kcal/(h·m²·°C)	1.163	0.2048	0.00002778	0.0001163	1

Heat Flux Density					
	W/m²	Btu/(h·ft²)	cal/(s·cm²)	W/cm²	
W/m²	1	0.3170	0.00002388	0.0001	0.86
Btu/(h·ft²)	3.154	1	0.00007535	0.0003154	2.712
cal/(s·cm²)	41,870	13,272	1	4.187	36,000
W/cm²	100,000	3,170	0.2388	1	8,600
kcal/(h·m²)	1.163	0.3687	0.00002778	0.0001163	1

Viscosity						
	(N·s)/m²	cP	lb/(s·ft)	(lbf·s)/ft²	lb/(h·ft)	kg/(h·m)
(N·s)/m²	1	1,000	0.672	0.0209	2,420	3,600
cP	0.001	1	0.000672	0.0000209	2.42	3.60
lb/(s·ft)	1.49	1,490	1	0.0311	3,600	5,350
(lbf·s)/ft²	47.88	47,880	32.2	1	116,000	172,000
lb/(h·ft)	0.0004134	0.4134	0.000278	0.00000864	1	1.49
kg/(h·m)	0.000278	0.278	0.000187	0.0000581	0.672	1

INDEX